MXenes: Emerging 2D Materials

AF670285

Om Prakash Pandey · Piyush Sharma
Editors

MXenes: Emerging 2D Materials

Springer

Editors
Om Prakash Pandey
Department of Physics and Materials Science (DPMS)
Thapar Institute of Engineering and Technology
Patiala, Punjab, India

Piyush Sharma
Division of Research and Development
Lovely Professional University
Phagwara, Punjab, India

ISBN 978-981-97-4066-6 ISBN 978-981-97-4064-2 (eBook)
https://doi.org/10.1007/978-981-97-4064-2

© The Editor(s) (if applicable) and The Author(s), under exclusive license to Springer Nature Singapore Pte Ltd. 2024

This work is subject to copyright. All rights are solely and exclusively licensed by the Publisher, whether the whole or part of the material is concerned, specifically the rights of translation, reprinting, reuse of illustrations, recitation, broadcasting, reproduction on microfilms or in any other physical way, and transmission or information storage and retrieval, electronic adaptation, computer software, or by similar or dissimilar methodology now known or hereafter developed.
The use of general descriptive names, registered names, trademarks, service marks, etc. in this publication does not imply, even in the absence of a specific statement, that such names are exempt from the relevant protective laws and regulations and therefore free for general use.
The publisher, the authors and the editors are safe to assume that the advice and information in this book are believed to be true and accurate at the date of publication. Neither the publisher nor the authors or the editors give a warranty, expressed or implied, with respect to the material contained herein or for any errors or omissions that may have been made. The publisher remains neutral with regard to jurisdictional claims in published maps and institutional affiliations.

This Springer imprint is published by the registered company Springer Nature Singapore Pte Ltd.
The registered company address is: 152 Beach Road, #21-01/04 Gateway East, Singapore 189721, Singapore

If disposing of this product, please recycle the paper.

Preface

The development of technologies depends upon the availability of suitable materials. The innovation in materials has brought revolutionary change in modern civilization. The emergence of two-dimensional (2D) materials has been nothing short of revolutionary change of this era. These ultrathin sheets, often consisting of just a single layer of atoms, exhibit a plethora of extraordinary properties that defy the conventions of classical bulk materials. In recent years, the exploration of '(2D)' materials has become one of the most dynamic and transformative fields in materials science and nanotechnology. Among the diverse array of 2D materials, MXenes have rapidly emerged as one of the most promising and intriguing families.

Gogotsi and colleagues at Drexel University first discovered the MXenes family of 2D transition metal carbides, nitrides, and carbonitrides in 2011. Since then, MXenes have garnered immense attention due to their unique combination of properties, including high electrical conductivity, excellent mechanical strength, and exceptional chemical stability. Over the 14 years that have passed since the discovery of MXenes, there have been more than 1500 papers published by scientists all around the world on this topic. Moreover, MXenes exhibit tunable surface chemistry, making them highly versatile for a wide range of applications spanning from energy storage and conversion to electronics, catalysis, sensing, and beyond.

This book delves into the fascinating realm of 2D MXenes, exploring their synthesis, properties, and diverse applications. It brings together contributions from leading researchers in the field, providing a detailed and up-to-date overview of this rapidly evolving area of study. The book presents the various methods for synthesizing MXenes, ranging from selective etching of MAX phases to various exfoliation strategies and highlights the key factors influencing their structural and morphological properties. Furthermore, the book addresses the rich variety of properties exhibited by MXene materials, including their electronic, optical, mechanical, and electrochemical behaviors. Understanding these properties is critical for harnessing MXen'es full potential in practical applications and tailoring their properties to meet specific technological requirements.

One of the most exciting aspects of MXene research is the broad spectrum of applications of these materials that enable to open new applications every day.

Throughout this book, detailed insights into the cutting-edge applications of MXenes across different fields are provided, including energy storage devices such as batteries and supercapacitors, electrocatalysis for water splitting and pollutant degradation, electromagnetic interference shielding, flexible electronics, biosensing, and many others.

As researchers around the globe unravel the mysteries of MXenes and uncover new functionalities, it becomes increasingly evident that these materials hold the key to addressing some of the most pressing challenges of our time. Whether it is enabling next-generation batteries with enhanced energy density, designing ultra-efficient catalysts for sustainable chemical transformations, or revolutionizing flexible electronics for wearable devices, MXenes offer boundless opportunities for innovation and progress.

In this volume, leading experts in the field present a comprehensive overview of the latest developments in MXene research, offering insights into both fundamental science and practical applications. Each chapter provides a detailed examination of specific aspects of MXene materials, ranging from synthesis techniques, structural characterization, and advanced applications. By bringing together diverse perspectives and expertise, this book aims to serve as a valuable resource for researchers, students, and professionals interested in exploring the vast potential of 2D MXene materials.

We hope that this book will inspire readers to embark on their own journeys into the captivating world of MXenes, fostering collaboration and driving further advancements that will shape the future of materials science and technology.

Patiala, India

Dr. Om Prakash Pandey
Senior Professor
oppandey@thapar.edu

Phagwara, India

Dr. Piyush Sharma
piyushsharma0135@gmail.com

Contents

Chapter 1
Wave of 2D MXene

Gurwinder Kaur, Piyush Sharma, and Om Prakash Pandey

1.1 Introduction

In the twentieth century, the refractory binary transition metal carbides and nitrides have emerged as a crucial industrial material [1]. These binary compounds are commonly named as MX phases, in which M refers to transition metals and X refers to C or N. The MX phases possess high elastic moduli, good mechanical properties, high oxidation, and corrosion resistance. These phases have shown great potential in a variety of industrial applications such as cutting and grinding tools, bearings, textile-machinery components, semiconductors, oxidation-resistant gas burners, etc. [2]. The discovery of their catalytic behavior in the 1970s sparked initial exploration of their applications in energy conversion and storage [3]. Later, various nanostructure designs were proposed to improve the surface properties and performance of transition metal carbides or nitrides [4, 5]. In addition, MX phases have limited thermal and electrical conductivity, are difficult to process, have little damage tolerance, and are exceedingly brittle, despite their many uses. To get around these drawbacks, MX phases with metallic characteristics are being developed as an alternative. Consequently, MAX phases (where A stands for metals) consisting of nano-laminated carbides and nitrides are discovered [6, 7]. Ti_3SiC_2 MAX phase was prepared by Barsoum and El-Raghy [8, 9], who sparked renewed attention to MAX phases. They investigated the material's mechanical, electrical, and thermal properties. The MAX phase was identified by this research team as a class of thermodynamically stable layered nanolaminate compounds. Subsequently, MAX phases captured the

G. Kaur · O. P. Pandey (✉)
Department of Physics and Materials Science, Thapar Institute of Engineering & Technology, Patiala 147004, India
e-mail: oppandey@thapar.edu

P. Sharma
Division of Research and Development, Lovely Professional University, Phagwara 144411, India
e-mail: piyushsharma0135@gmail.com

© The Author(s), under exclusive license to Springer Nature Singapore Pte Ltd. 2024
O. P. Pandey and P. Sharma (eds.), *MXenes: Emerging 2D Materials*,
https://doi.org/10.1007/978-981-97-4064-2_1

attention of the scientific community because of the exceptional blending of metal and ceramic characteristics [10]. These phases have excellent machinability, strong thermal shock and oxidation resistance, low weight, and good stiffness. Due to their unique combination of properties, MAX phases have a wide variety of potential applications in fields as diverse as high-temperature structural applications, protective coatings, sensors, low-friction surfaces, electrical connections, and many others [11–14].

In 2011, MAX phases were used as precursors for the synthesis of 2D MXenes. The first 2D titanium carbide ($Ti_3C_2T_x$) was discovered, which paved the way for the discovery of a new class of 2D transition metal carbides or nitrides called MXenes [15, 16]. MXenes are derived through etching of A layer from MAX phases, where A stands for metal (Al, Si, Ga, Ge, In, Sn). The hydrophilic nature of hydroxyl or oxygen-terminated surfaces and the properties of transition metal carbides make MXenes a particularly interesting material. The MXenes have shown tremendous potential in the field of energy, sensing, wastewater treatment, EMI shielding and biomedical. Nowadays, MXenes is one of the rapidly expanding families of 2D materials. The general formula for MXenes is $M_{n+1}X_nT_x$, where M stands for early transition metal, X for carbon or nitrogen, n is an integer and T_x stands for surface termination group (= O, −F, and −OH) [8]. As of now, MAX phases have been synthesized by using approximately 14 M elements, 16 A elements, and 2 X elements [9]. In this way, different MAX phases can be synthesized by trying out different combinations. The scope also widened with the substitution or doping and with the discovery of quaternary MAX phases. The chemical diversity of MAX phases plays an important role in the development of novel MXenes. In the present chapter, the fundamental aspects related to its basic structure and synthesis of MXenes have been discussed. The chapter aims to provide insight into chemical diversity of MXenes. Finally, the properties and the potential applications of MXenes have been addressed.

1.2 Fundamentals of MXenes

MXenes are a class of two-dimensional inorganic compounds. These materials consist of few atoms of thick layers of transition metal carbides, nitrides, or carbonitrides [10, 11]. These materials possess unique combination of properties of transition metal carbides and hydrophilic nature hydroxyl or oxygen-terminated surfaces [12]. The general formula of MXene is "$M_{n+1}X_nT_x$", here "M" represents early transition metal, "X" is carbon and/or nitrogen and "T_x" is the functional group attached with "M" element. The elements considered as "M", "X", and "T_x" in MXenes are shown in Fig. 1.1.

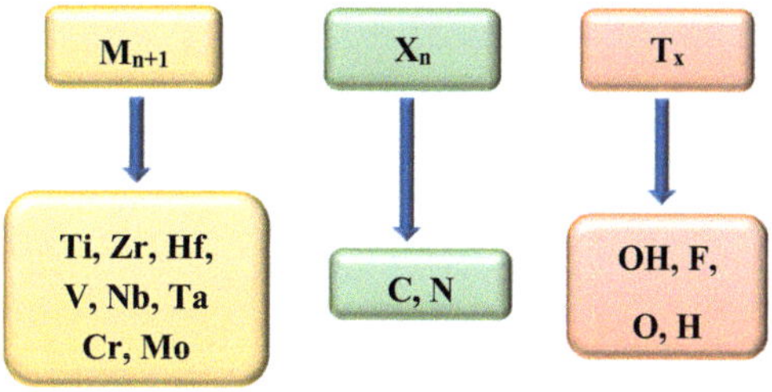

Fig. 1.1 The elements considered as "M", "X", and "T_x" in MXenes

1.2.1 Structure of MXenes

In general, MXenes surfaces are terminated with functional groups, which are obtained through etching of MAX phases. Figure 1.2 shows crystal structure of (M_3AX_2) MAX phase and their corresponding ($M_3X_2T_x$) MXene.

The general formula for MAX phases is "$M_{n+1}AX_n$", where "M" is the transition metal carbide, "A" is the metal, and "X" is carbon and/or nitrogen. MAX phases possess layered hexagonal crystal structures with two formula units per unit cell, for structures n equal to 1 to 3. The unit cell consist of M_6X octahedra with the X-atoms filling the octahedral sites between the M-atoms, which are identical to those found in the rock salt structure of the MX binaries. The octahedra alternates with layers of pure A-elements located at the centers of trigonal prisms that are slightly larger, and

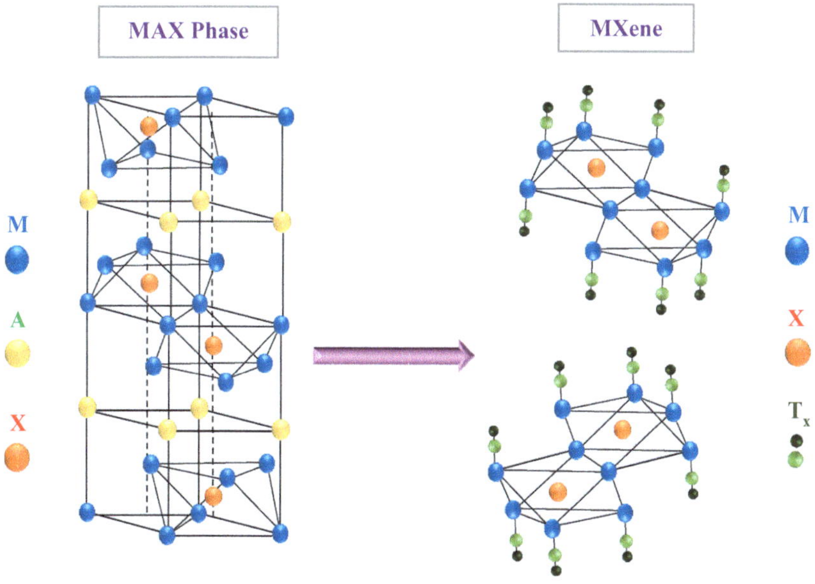

Fig. 1.2 Crystal structure of (M_3AX_2) MAX phase and their corresponding ($M_3X_2T_x$) MXene

thus more accommodating of the larger A-atoms. MXenes adopt three structures, as inherited from the parent MAX Phases: M_2C, M_3C_2, M_4C_3 [13].

1.2.2 Synthesis of MXenes

MXenes are synthesized by the following three different steps, i.e., (i) synthesis of MAX phase, (ii) exfoliation of "A" layer from MAX phase, and (iii) intercalation/delamination of MXenes layer, which will be discussed in Chap. 2. The schematic representation of the synthesis of MXenes is presented in Fig. 1.3. The brief description of the above steps for the synthesis of MXenes is as follows:

(i) *Synthesis of MAX Phases*

Several methods are reported for the synthesis of the MAX phases. Numerous researchers follow different sintering techniques such as conventional sintering, spark plasma sintering (SPS), hot isostatic pressing (HIP), and hot pressing (HP) to obtain highly pure nanolaminated MAX phases. In the conventional sintering process, no external mechanical pressure is employed on the material during heat treatment. In HIP and HP sintering techniques, the external mechanical pressure is induced [14]. Furthermore, electric field is applied to give heat treatment in SPS technique [17]. All the sintering processes are performed in the presence of argon environment to avoid the oxidation of the MAX phase. In SPS technique, a direct electric current is applied to heat up the electrically conductive devices in a graphite die [18].

(ii) *Exfoliation of "A" Layer from MAX Phase*

Exfoliation is the process in which "A" element can be etched from MAX phase, to separate the stacked layers into single flakes. The MAX is immersed and stirred or sonicated in acidic solution to obtain MXenes. After acidic treatment, the obtained

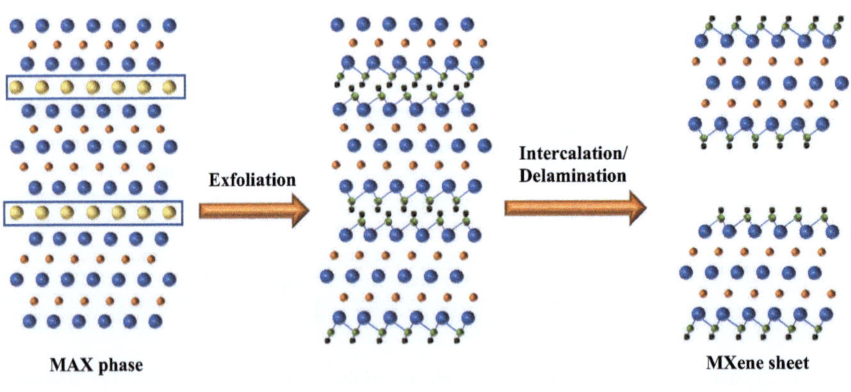

Fig. 1.3 The schematic representation of the synthesis of MXenes from 312 MAX phase

sample is washed several times in the centrifuge tube, to normalize the pH = 7. In this process, MXene layers are obtained but there is weak bonding between function group and transition metal. Hence, intercalation and delamination of exfoliated MAX phase should be done to obtain a single MXene layer or flake.

(iii) *Intercalation and Delamination of MXenes*

Exfoliated MXene layers are intercalated to increase the distance between two adjacent layers. In this process, MXene layers are intercalated with different polar organic molecules and metal ions (mono and multivalent cations such as Li^+, Na^+, K^+, NH^{4+}, and Mg^{2+}). After intercalation, a single layer of 2D MXene is obtained. Furthermore, MXene layers are treated with organic solvents for the delamination of MXene.

1.3 Chemical Diversity

The chemical diversity of MAX phases allows for a wide variety of possible MXenes configurations [9]. Substitution or doping in MAX phases further opens the door to novel MXenes. It was observed that the MAX phase becomes magnetic in nature after the doping of Mn or Fe at M site [19]. In such a way, the properties of resulting MXenes can also be tuned, which is discussed in Chap. 3. It is worth noting that the MAX phase solid solution provides several opportunities for novel MXenes. The potential for expansion of the family of 2D MXenes has also been sparked by the discovery of in-plane and out-of-plane ordered MAX phases. Figure 1.4 shows the periodic table highlighting the elements used for the synthesis of MXenes and MAX phases [20].

In MAX phases, M elements (highlighted in blue) are sandwiched between a single layer of A element (highlighted in red), with X (highlighted in grey) occupying the interlayer octahedral sites. In MXenes, the layer of M elements is interleaved with n layers of X elements having T_x (highlighted with orange) surface terminations. Over a hundred different stoichiometric MXene compositions and an infinite number of solid solutions provide not only novel property combinations but also offer possibility to tune properties by varying ratio of M or/and X elements. The cations that have so far been intercalated into MXenes are visible on the green backdrop in Fig. 1.4. The legend at the bottom of the figure, 1M and 1A are related to the formation possibility of a single M and A element MAX phase and/or MXene, respectively. SS stands for the solid solutions of MAX phases and MXenes. 2M indicates the presence of two transition metals to form an ordered MAX phase and/or MXene. The figure demonstrated how the elements can be used as building blocks to create a wide range of nanomaterials, with MXenes, their precursor MAX phases, and intercalated metal ions in MXenes as embodiments of the fundamental principles of chemistry.

In addition, MXene-based composites are also developed to overcome the issue of restacking of MXene layers resulting in oxidation and decrease in electrical conductivity. Several researchers are involved in developing MXene-based composite

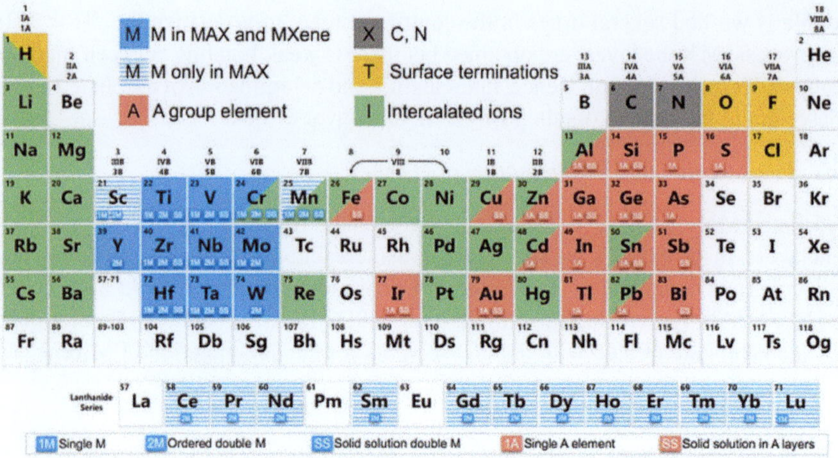

Fig. 1.4 The periodic table presents the elemental compositions of MXenes and MAX phases. Adapted with permission from [20]. Babak Anasori et al., Graphene and 2D Materials, MXenes: trends, growth, and future directions, 7, 75–79, 2022, Copyright (2022), Copyright the Authors, some rights reserved; exclusive licensee [Springer]

with a variety of materials including polymers, metals, and carbon-based materials. The resulting composites have shown improvement in their properties. A detailed discussion of MXene-based composites is presented in Chap. 4.

1.4 Properties of MXenes

MXenes have remarkable properties such as high Young's modulus, thermal conductivity, electrical conductivity, and the adjustable band gap [21]. The properties of MXenes are dependent on composition that includes solid solution formation and different transition metals "M" and "X" elements. In addition, MXene's properties also rely on thermal and chemical treatments, which are responsible for surface functionalization and variation in morphology [16]. The leading properties of MXene family exhibit very different properties as given here.

(i) *Electronic and Electrical Properties*

The two foremost interesting properties of MXenes are electronic and electric properties that can be modified by alteration of functional groups, stoichiometry, and solid solution formation. It has been observed that the experimental value of electrical conductivities of MXenes is similar to multi-layered graphene while higher than carbon nanotube materials. Moreover, it was witnessed that resistivity values increase with the rise in the number of layers and the presence of functional groups

[22], due to which the simulated conductivities have generally larger values than the one's observed experimental values [23].

The electrical conductivities of MXene are reported to vary due to changes in defect concentration, surface functional groups, delamination yield, d spacing between MXenes flakes and lateral sizes. The formation of MXenes with fewer defects and larger lateral sizes provides higher electrical conductivity because of lesser HF concentrations and etching time [24]. In case of electronic sensing materials [25], their conductivities are influenced from environmental humidity [24]. A useful way to enhance conductivity and electrical properties is surface modification via various treatments such as by altering the functional groups and intercalating the molecules.

(ii) *Mechanical Properties*

Mechanical properties of MXenes are highly interesting because of its strong bonding between atoms. An elastic constant (c11) of MXene is two times greater than MAX phases and other 2D materials along with high bending stiffness, which is good towards their use as reinforcements in composites [26, 27]. Apart from this, MXenes have functional groups due to which they interact better with polymeric matrixes than graphene for composite applications [23, 27]. With an increase in the number of layers of MXenes, Young Modulus of both MXene carbides and nitrides decreases but nitride-based compounds possess higher values than carbides [28]. The growth of critical deformations increases as the value of c11 decreases due to the occurrence of terminations. More importantly, these values of MXenes are higher as compared to graphene which possess the feature of flexibility in electronics. Although several methods for mechanical testing of bulk materials exist, the evaluation of mechanical properties of 2D is still challenging. Most of the time, mechanical properties of 2D nanomaterials were calculated by nanoindentation technique, in which an AFM tip applies force at the center of a 2D material film [29]. All in all, complete theoretical and experimental evaluations of mechanical properties with the different functionalization groups are still to be performed.

(iii) *Thermal Properties*

Thermal properties of MXenes are important particularly in electronic devices, which require faster heat dissipation. It plays an important role in electronics and energy-related heat dissipation devices due to their non-stop reduction in size [30]. Some studies suggested that MXenes have high thermal conductivities and low thermal expansion as compared to phosphorene and MoS_2 monolayer. Apart from this, it is observed that thermal conductivities of Ti_2CO_2, Zr_2CO_2, Hf_2CO_2, and Sc_2CF_2 vary from 22 to 472 $Wm^{-1}\ K^{-1}$ at room temperature. Interestingly, the thermal conductivity is dependent on length of MXenes as the thermal conductivity of Sc_2CF_2 increases from 298 to 722 $Wm^{-1}\ K^{-1}$ when the flake lengths vary from 1 to 50 μm [31]. Regarding the oxygen–terminated compounds, it was found that their thermal conductivities increase with the metal "M" atomic number. Overall, the synthesis procedure adopted and the resultant morphology are both substantial for thermal conductivity.

(iv) *Magnetic*

Several studies reported the variation in magnetic properties of MXenes using different theoretical calculations. In many compounds of MXenes, viz., Ti_4C_3, Ti_3CN, Fe_2C, Cr_2C, Ti_3N_2, Ti_2N, Zr_2C and Zr_3C_2 MXenes demonstrate retention in magnetic moments. Depending upon the surface terminations and functional groups, Ti_3CNT_x and $Ti_4C_3T_x$ turn into non-magnetic behavior, whereas Cr_2CT_x and Cr_2NT_x stay ferromagnetic at room temperature [32, 33]. In addition, Mn_2NT_x behaves as ferromagnetic irrespective of the surface terminations [34]. Due to limited synthesis and control on surface chemistry of MXene compounds, magnetic moments are calculated theoretically and have not observed experimentally yet.

(v) *Optical*

The absorption of visible and UV light is vital for transparent conductive electrode devices, photocatalytic, photovoltaic, and optoelectronic. In case of $Ti_3C_2T_x$ MXenes, it could absorb light in the UV region from 300 to 500 nm whereas 5 nm thickness film exhibited transmittance up to 91.2% [35]. Depending on the film thicknesses, it may present a strong and broad absorption band at around 700–800 nm which plays an important role in photothermal therapy (PTT) applications [36]. More importantly, with change in thickness and ion intercalation of MXene, the value of transmittance could be optimized. During the synthesis process, the value of transmittance decreases when hydrazine and DMSO are used as reducing agent whereas tetramethylammonium hydroxide (NMe_4OH) increased it from 74.9 to 92.0%. Apart from this, the presence of functional groups also influences the optical properties of these 2D compounds. For instance, –F and –OH terminations decrease absorption and reflectivity in the visible range whereas all terminations enhance the reflectivity in the UV region. MXenes are potential candidates for flexible transparent electrode applications as they have optical transparency in the visible region. Finally, it has the outstanding capability to convert light into heat which is further useful for biomedical and water evaporation applications [37].

1.5 Applications of MXenes

1.5.1 MXenes for Energy Conversion and Storage

Graphene-like materials such as TMCs, MXenes, and TMNs with better carrier mobility and an intrinsic layered structure have shown promise as a replacement for expensive noble metal electrocatalysts. TMCs and TMNs are intriguing because these materials share the same electronic structure as noble metals like Pt, making them a possible replacement for Pt and IrO_2/RuO_2 noble [3, 38]. The 2D MXenes have gained interest as a means of improving the hydrogen evolution reaction (HER) and oxygen evolution reaction (OER) activity as well as stability of TMCs and

TMNs electrocatalyst because of their remarkable inherent and extrinsic features. The enhanced water-splitting ability of MXene can be attributed to their similar band structure to Pt (caused by carbon inclusion in a metal lattice), their larger number of edge sites, and their improved active sites due to connected functional groups on the surface. Electrocatalyst for HER and OER has been developed using various TM-based MXene species (Ti, Mo, and V). MXene-based electrocatalysts are also used for other energy conversion reactions such as oxygen reduction reaction (ORR), methanol oxidation, nitrogen reduction reaction (NRR). It is worth noting that the electrochemical performance of MXenes can be tuned with the change in the functional group attached. Moreover, Bare MXene and most functionalized derivatives exhibit metallic or narrow band gap semiconducting characters, offering MXene inherent advantages in terms of electronic conductivity [39, 40]. This makes them suitable candidate for energy storage applications. Figure 1.5 shows the schematic representation of MXenes for the energy conversion and storage applications.

Among various MXenes, Ti_3C_2 is a promising candidate for lithium-ion battery (LIB) anodes with low operating voltage and diffusion barrier. Delaminating multilayer MXene into paper-like structure can effectively improve the electrochemical performance [41]. The pillared MXene with proper interlayer spacing can accommodate the ions of electrolytes, facilitate the adsorption and intercalation of electrolytes, and then improve the electrochemical performance. Due to the intrinsic

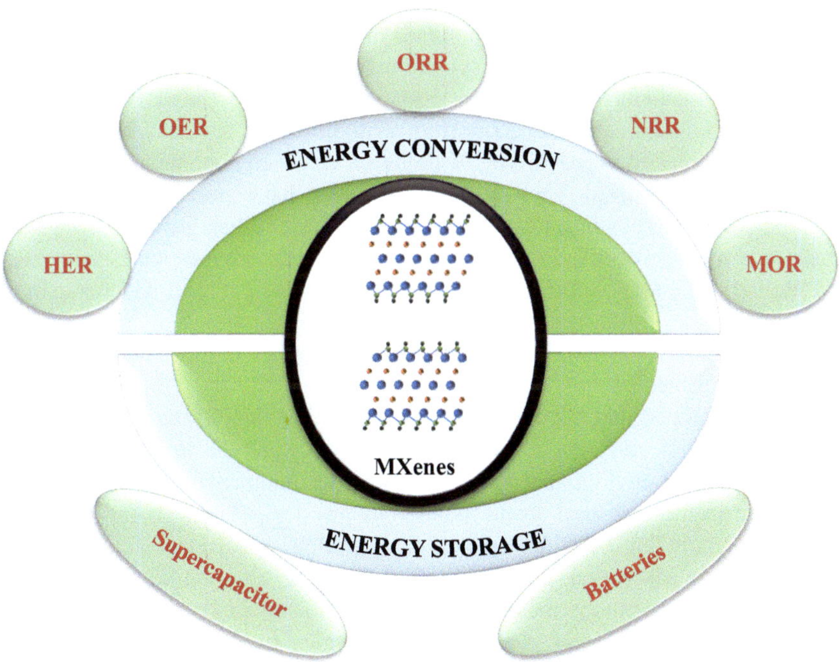

Fig. 1.5 Schematic representation of MXenes for the energy conversion and storage applications

metallic conductivity and the highly active 2D surface, MXene-based materials are expected to be promising hosts for Li–S batteries. Ti_2C MXene phase was effective as a cathode host material for Li–S batteries [42].

Due to the high power density, rapid charging/discharging rate, and long cycle life, MXene-based supercapacitors and pseudo-capacitors have attracted a lot of research interest [43]. Owing to their high power output, supercapacitors are always used in hybrid electric vehicles and fuel cell vehicles for start-stop control systems [43, 44]. Therefore, supercapacitors play a significant role in future energy storage systems. On the other side, pseudo-capacitors depend on fast and reversible Faradic processes. MXene-based pseudo-capacitors can store more energy at high charge/discharge rates and have attracted considerable attention [13]. The performance of pseudo-capacitive material mainly depends on the interaction between electrode materials and the electrolyte ions. However, the understanding of the interaction is still limited. A detailed discussion on energy conversion and storage is presented in Chap. 5.

1.5.2 MXenes for Wastewater Treatment

Water contamination has become a top problem for the industrialized world as a whole and a significant cause of worry for society and governmental agencies [45]. Water pollution happens when one or more pollutants are released into it that will negatively alter the water. These compounds may have negative effects on our environment and living organisms. Water contamination may be categorized in two ways depending on the main sources such as point and non-point. The first kind of pollution refers to contaminants that come from a single source, such as industrial water pollution, whereas the second type of pollution refers to contaminants that come from several sources [46]. There are several factors that contribute to water pollution, including energy consumption, radioactive waste, urban growth, sewage and wastewater management, industrial waste, mining operations, pesticides, and chemical fertilizers. Water pollution is inevitable simply because it is utilized for so many different purposes. Industrial, home, and agricultural operations all create wastewater that contains harmful contaminants. Water resources in this situation must be continuously protected.

In the developed nations, the regulation governing liquid industrial effluent is becoming harsher and now requires that all wastewater be treated before it is discharged into the environment [47]. As a result, the treatment of effluents has taken precedence in the industrial world. Numerous physical, chemical, and biological procedures, including flotation, precipitation, oxidation, solvent extraction, evaporation, carbon adsorption, ion exchange, membrane filtration, electrochemistry, biodegradation, and phytoremediation, have been documented during the last three decades. In general, physical, chemical, and biological methods are used to remove contaminants.

Recently, MXenes and MXene-based composites emerged as a potential catalyst for wastewater treatment. Various research groups are involved in the development of MXene-based catalyst for the removal of heavy metal ions, radionuclides, and organic pollutants (Fig. 1.6) [45]. Several factors (pH, dosage, oxidant concentration, temperature, and co-existing ions) are involved that govern the overall performance of MXenes for wastewater treatment. In addition, the chemical diversity of MXenes offers numerous possibilities to design and fabricate high-performance catalysts. A detailed discussion on the catalytic mechanisms of MXenes and MXene-based composites is presented in Chap. 6. Regarding their scalability for practical applications, the existing issues and future possibilities in wastewater treatment have also been presented in Chap. 6.

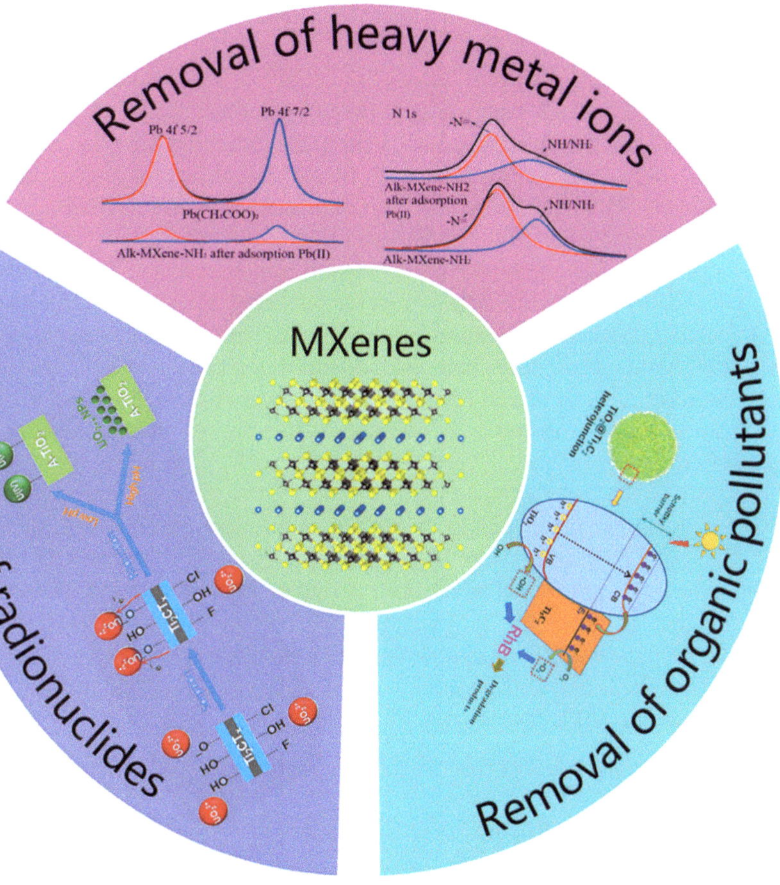

Fig. 1.6 MXenes for the removal of contaminants from the water. Adapted with permission from [45]. Copyright (2023), Copyright the Authors, some rights reserved; exclusive licensee [Elsevier]. Distributed under a Creative Commons Attribution License 4.0 (CC BY) https://creativecommons.org/licenses/by/4.0/

1.5.3 MXenes as Sensors

MXene materials have been proven as highly sensitive and selective detection platform for sensing applications in spite of their very short journey [48]. Morphology, extraordinary surface chemistry and excellent conducting properties, biocompatibility is among one of the most prominent features of MXenes which make them highly suitable matrix for fabrication of advance bio-sensing platforms [49]. MXene-based electrodes have been utilized as effective transducers for immobilization of biological receptors onto their surface and have shown considerable sensing characteristics. MXene biosensors enhance the catalytic performances of sensing matrix and help enzymes/protein to retain their bioactivity without altering the original native conformation.

Apart from electrochemical biosensors, MXenes have driven an impressive research in gas-sensing applications [48]. Determination of volatile organic compounds (VOCs) at ultra-trace level (ppb) for pollution analysis, detection of toxic gases, and therapeutic breath analysis for early identification of many diseases is highly crucial [44]. MXene materials have been considered as solid-state gas sensors with low electrical noise and well designed to detect these gaseous molecules with strong signal intensity, high manufacturing substrate flexibility and portability [50, 51]. Moreover, sensing approaches often include wearable nano-electronics, which have gained inexplicable attention for monitoring various health-related activities such as physical stimuli, weak pressure, and physiological signals [52]. MXenes are to be used as highly advanced platform for detection of phonations and substantial movements like walking, jumping, running or other human activities like coughing, joint bending [40].

In addition, MXenes are also used as a photoluminescence sensor. Photoluminescence is an effective technique that has often been employed in optical, biomedical, and cellular imaging areas [53]. However, recent advancements in materials sciences have made it possible to construct highly sensitive optical MXene-based quantum dots (MQDs) for bio-imaging applications [54]. In a recent study, an intracellular photoluminescent ratiometric pH sensor was developed using Ti_3C_2-MQDs for detection of cell metabolism [53]. A detailed description related to the MXene-based sensor is presented in Chap. 7.

1.5.4 MXenes in Biomedical Science

MXenes have many uses in the field of biomedical sciences. Optical and electrical signals of MXene-based sensor could be used for bioimaging, biosensing, therapy, drug delivery, etc. MXenes have become a known candidate for the development of extraordinary biosensors due to their high surface-to-volume ratio, unique electrical and optical properties, functional groups on the surface, and great hydrophilic behavior [55]. MXene-based sensor can detect hydrogen peroxide, nitride, glutathione, glucose, dopamine, epinephrine, organophosphate, and pesticides by sending electrical signals from a surface that has been changed by adding different analytes, which are made up of proteins and amino acids. MXenes are also used as optical sensors, in which signals can be picked up by two types: electrochemiluminescence (ECL) biosensors and surface plasmon resonance (SPR) biosensors. The ECL biosensors are used to detect proteins, enzymes, and DNA, while SPR biosensors are used to find and treat cancer cells by detecting carcinoembryonic antigen in human serum. Moreover, MXenes have superior electronic and mechanical properties that can change the way electrical signals behave in response to physical conditions like pressure and tension [56, 57]. Physical sensors, which are made with MXenes, are used to detect things like muscle strains and other physiological processes. A detailed insight related to the role of MXenes in biomedical science is presented in Chap. 8.

1.5.5 MXenes for Electromagnetic Shielding

The high electrical conductivity and distinctive multi-layer structure of MXene are primarily responsible for its remarkable EMI shielding capabilities. With a conductivity of roughly 0.3 S cm^{-1}, the Ti_2CT_x MXene has demonstrated the highest EMI SE over 70 dB when the thickness is 0.8 mm, exceeding most of the reported carbon- and metal-based materials [58]. The 2D structure and adequate functional groups on MXene's surface make them easily hybridized with other nanomaterials to accomplish the goal of tuning MXene's properties. It is difficult to get exceptional EMI shielding performance because of the homogeneous material. Therefore, the design of the material's structure and composition is crucial to address the need for MXene-based materials in EMI shielding. In Chap. 9, the use of MXenes for electromagnetic interference (EMI) shielding is discussed in depth.

1.5.6 MXenes in Biotechnology

The dielectric response of MXenes is greatly influenced by their metallic nature and interband transitions. It has been shown that the optical performance of MXenes is significantly influenced by surface functionalization (or functional groups) [59]. The MXene's absorption peaks are red-shifted due to the presence of functional group from the UV to the visible light spectrum by ~40% and depends on the location of the functional group. Further investigation based on thin film testing demonstrated the intrinsic electrochromic properties of titanium-based MXenes [60]. For the development of electroactive materials for biotechnological applications, controlling the band gap value is crucial. By inserting a metal atom, altering the material's surface terminals, or inducing hybridization, the band gap value of MXenes may be altered. The tunable band gap and exceptional electrical and transport properties are the promising prospects for their use as electroactive agents in biotechnology. The importance of MXenes to the biotechnology industry is discussed in Chap. 10.

1.5.7 MXenes for Separation Processes

Several materials with desirable properties have been studied to improve the separation efficiency. To fabricate high-performance membranes, which may be designed on a nanoscale or even sub-nanometer scale requires atomically thin 2D materials, which have been the subject of intense research. Hence, the trade-off constraint might potentially be overcome by lowering the transport resistance that will increase permeance and guarantee a clean separation process. According to the structure, 2D material-based membranes are often divided into two types: nanosheet membranes and lamellar membranes [61]. In the case of nanosheet membranes, solutes are carried through inherent or drilled holes in a membrane consisting of a single or few layers of nanosheets. It is predicted that the ultimate permeance and selectivity will be achieved by a nanosheet membrane with an atomic thickness. Nevertheless, it is still difficult to create nanopores with a consistent size and a high density. Graphene-based membranes have paved the way for this new class of membranes, which exhibit unprecedented permeance and selectivity. By using vacuum-assisted filtration (VAF), Li et al. [62] developed ultrathin graphene oxide (GO) membranes with thicknesses ranging from 1.8 to 20 nm. These membranes showed unprecedented separation performance toward hydrogen, with selectivities of H_2/CO_2 and H_2/N_2 pair gases reaching as high as 3400 and 900, respectively, and are quite stable across a wide range of temperatures. The separation performance surpasses that of reported polymeric membranes, hence dissolving the polymeric membranes' top bond. A wide range of 2D materials, beyond GO, have been explored for use in the fabrication of separation membranes. These include molybdenum disulfide, metal–organic frameworks (MOFs), covalent organic frameworks (COFs), and zeolites [61]. MXene, a relatively novel 2D material with unusual physical properties, has recently attracted considerable scientific interest. MXenes are well suited for separation application due to their hydrophilicity, remarkable flexibility, and abundance of surface groups. Separation applications such as gas separation, water treatment, organic solvent purification, nanofluidic ion transport, and osmotic energy conversion are discussed in detail in Chap. 11.

1.5.8 MXenes in Solution-Processed Optoelectronic Devices

New insights into the properties of 2D materials have shown an attractive platform for energy conversion and storage technologies of the future. Mechanical exfoliation of the bulk material revealed the unique characteristics of these materials as 2D systems. Recent developments in liquid phase exfoliation have provided a new avenue for research into these materials by allowing for the mass fabrication of solutions containing distributed nanosheets of 2D crystals [63]. Since these novel 2D materials may be applied in a liquid phase, it provides a new path to solution-process procedures, which can improve manufacturing efficiency and decrease production costs. Among various 2D materials, MXenes emerged as a suitable candidate for separation application. The optoelectronic properties of MXenes are tunable with the choice of intercalating agents. MXene-based films are the foundation of plasmonic application. MXenes (colloidal aggregate flakes) have an astonishing figure of merit (FOM) for their optoelectronic properties, which is twice as great as that of reduced graphene oxide. For high-performance solution-processed optoelectronic devices (organic/perovskite light-emitting diodes and solar cells), Chap. 12 highlights the many uses of MXenes in electrode, interface, and emitting/active layers.

1.5.9 Future Prospective and Research Avenues

There are several properties of MXene has been explored. However, considering its structural features which depends on the procedure adopted for preparation, many more applications needs to be explored. This has been highlighted in Chap. 13 of this book.

References

1. H. Kindlund, D.G. Sangiovanni, I. Petrov, J.E. Greene, L. Hultman, A review of the intrinsic ductility and toughness of hard transition-metal nitride alloy thin films. Thin Solid Films **688**, 137479 (2019). https://doi.org/10.1016/j.tsf.2019.137479
2. L. Toth, *Transition Metal Carbides and Nitrides* (1971)
3. R.B. Levy, M. Boudart, Platinum-like behavior of tungsten carbide in surface catalysis. Science **181**(80), 547–549 (1973)
4. Y. Zhong, X.H. Xia, F. Shi, J.Y. Zhan, J.P. Tu, H.J. Fan, Transition metal carbides and nitrides in energy storage and conversion. Adv. Sci. **3**, 1–28 (2015). https://doi.org/10.1002/advs.201500286
5. Y.G. Gogotsi, R.A. Andrievski, Materials science of carbides. Nitrides Borides (1999). https://doi.org/10.1007/978-94-011-4562-6
6. J.C. Schuster, H. Nowotny, C. Vaccaro, The ternary systems: Cr-Al-C, V-Al-C, and Ti-Al-C and the behavior of H-phases (M2AlC). J. Solid State Chem. **32**, 213–219 (1980). https://doi.org/10.1016/0022-4596(80)90569-1
7. J.C. Schuster, H. Nowotny, Investigations of the ternary systems (Zr, Hf, Nb, Ta)-Al-C and studies on complex carbides. Zeitschrift Fuer Met. Res. Adv. Tech. **71**, 341–346 (1980)
8. M. Ghidiu, M.R. Lukatskaya, M.Q. Zhao, Y. Gogotsi, M.W. Barsoum, Conductive two-dimensional titanium carbide 'clay' with high volumetric capacitance. Nat. **516**, 78–81 (2014). 2014 5167529. https://doi.org/10.1038/nature13970.
9. M. Sokol, V. Natu, S. Kota, M.W. Barsoum, On the chemical diversity of the MAX phases. Trends Chem. **1**, 210–223 (2019). https://doi.org/10.1016/j.trechm.2019.02.016

10. T. O'Mahony, P. Escardó-Serra, J. Dufour, Revisiting ISEW valuation approaches: the case of Spain including the costs of energy depletion and of climate change. Ecol. Econ. **144**, 292–303 (2018). https://doi.org/10.1016/j.ecolecon.2017.07.024
11. N. Kasmi, M. Majdoub, G.Z. Papageorgiou, D.N. Bikiaris, Synthesis and crystallization of new fully renewable resources-based copolyesters: poly(1,4-cyclohexanedimethanol-co-isosorbide 2,5-furandicarboxylate). Polym. Degrad. Stab. **152**, 177–190 (2018). https://doi.org/10.1016/j.polymdegradstab.2018.04.009
12. S. Niu, Z. Wang, M. Yu, M. Yu, MXene-based electrode with enhanced pseudocapacitance and volumetric capacity for power-type and ultra-long life lithium storage (2018). https://doi.org/10.1021/acsnano.8b01459
13. Y. Sun, D. Chen, Z. Liang, Two-dimensional MXenes for energy storage and conversion applications. Mater. Today Energy **5**, 22–36 (2017). https://doi.org/10.1016/j.mtener.2017.04.008
14. X. Wang, Y. Zhou, Solid–liquid reaction synthesis of layered machinable Ti_3AlC_2 ceramic. J. Mater. Chem. **12**, 455–460 (2002). https://doi.org/10.1039/b108685e
15. M. Naguib, M. Kurtoglu, V. Presser, J. Lu, J. Niu, M. Heon, L. Hultman, Y. Gogotsi, M.W. Barsoum, Two-dimensional nanocrystals produced by exfoliation of Ti_3ALC_2, in: *MXenes From Discovery to Applications of Two-Dimensional Metal Carbides and Nitrides* (2023), pp. 15–29. https://doi.org/10.1201/9781003306511-4
16. M. Naguib, O. Mashtalir, J. Carle, V. Presser, J. Lu, L. Hultman, Y. Gogotsi, M.W. Barsoum, Two-dimensional transition metal carbides. ACS Nano **6**, 1322–1331 (2012). https://doi.org/10.1021/nn204153h
17. N.C. Ghosh, S.P. Harimkar, Consolidation and synthesis of MAX phases by spark plasma sintering (SPS): a review. Woodhead Publishing Limited (2012). https://doi.org/10.1533/9780857096012
18. W.B. Zhou, B.C. Mei, J.Q. Zhu, X.L. Hong, Rapid synthesis of Ti_2AlC by spark plasma sintering technique. Mater. Lett. **59**, 131–134 (2005). https://doi.org/10.1016/j.matlet.2004.07.052
19. C.M. Hamm, J.D. Bocarsly, G. Seward, U.I. Kramm, C.S. Birkel, Non-conventional synthesis and magnetic properties of MAX phases $(Cr/Mn)_2AlC$ and $(Cr/Fe)_2AlC$. J. Mater. Chem. C. **5**, 5700–5708 (2017). https://doi.org/10.1039/c7tc00112f
20. B. Anasori, Y. Gogotsi, MXenes: trends, growth, and future directions. Graphene 2D Mater. **7**, 75–79 (2022). https://doi.org/10.1007/s41127-022-00053-z.
21. P. Eklund, J. Rosen, P.O.Å. Persson, Layered ternary $M_{n+1}AX_n$ phases and their 2D derivative MXene: an overview from a thin-film perspective. J. Phys. D Appl. Phys. **50**, 113001 (2017). https://doi.org/10.1088/1361-6463/aa57bc
22. M. Khazaei, M. Arai, T. Sasaki, M. Estili, Y. Sakka, Two-dimensional molybdenum carbides: potential thermoelectric materials of the MXene family. Phys. Chem. Chem. Phys. **16**, 7841–7849 (2014). https://doi.org/10.1039/c4cp00467a
23. X. Tang, X. Guo, W. Wu, G. Wang, 2D metal carbides and nitrides (MXenes) as high-performance electrode materials for lithium-based batteries. Adv. Energy Mater. **8** (2018). https://doi.org/10.1002/aenm.201801897
24. M. Ghidiu, S. Kota, V. Drozd, M.W. Barsoum, Pressure-induced shear and interlayer expansion in Ti_3C_2 MXene in the presence of water. Sci. Adv. **4** (2018). https://doi.org/10.1126/sciadv.aao6850
25. E.S. Muckley, M. Naguib, H.W. Wang, L. Vlcek, N.C. Osti, R.L. Sacci, X. Sang, R.R. Unocic, Y. Xie, M. Tyagi, E. Mamontov, K.L. Page, P.R.C. Kent, J. Nanda, I.N. Ivanov, Multimodality of structural, electrical, and gravimetric responses of intercalated MXenes to water. ACS Nano **11**, 11118–11126 (2017). https://doi.org/10.1021/acsnano.7b05264
26. M. Kurtoglu, M. Naguib, Y. Gogotsi, M.W. Barsoum, First principles study of two-dimensional early transition metal carbides. MRS Commun. **2**, 133–137 (2012). https://doi.org/10.1557/mrc.2012.25
27. M. Naguib, M. Kurtoglu, V. Presser, J. Lu, J. Niu, M. Heon, L. Hultman, Y. Gogotsi, M.W. Barsoum, Two-dimensional nanocrystals produced by exfoliation of Ti_3AlC_2. Adv. Mater. **23**, 4248–4253 (2011). https://doi.org/10.1002/adma.201102306

28. N. Zhang, Y. Hong, S. Yazdanparast, M.A. Zaeem, Superior structural, elastic and electronic properties of 2D titanium nitride MXenes over carbide MXenes: a comprehensive first principles study. 2D Mater. **5** (2018). https://doi.org/10.1088/2053-1583/aacfb3
29. G. Plummer, B. Anasori, Y. Gogotsi, G.J. Tucker, Nanoindentation of monolayer $Ti_{n+1}C_nT_x$ MXenes via atomistic simulations: the role of composition and defects on strength. Comput. Mater. Sci. **157**, 168–174 (2019). https://doi.org/10.1016/j.commatsci.2018.10.033
30. H. Wang, Y. Wu, X. Yuan, G. Zeng, J. Zhou, X. Wang, J.W. Chew, Clay-inspired MXene-based electrochemical devices and photo-electrocatalyst: state-of-the-art progresses and challenges. Adv. Mater. **30** (2018). https://doi.org/10.1002/adma.201704561
31. X.H. Zha, J. Zhou, Y. Zhou, Q. Huang, J. He, J.S. Francisco, K. Luo, S. Du, Promising electron mobility and high thermal conductivity in Sc_2CT_2 (T = F, OH) MXenes. Nanoscale **8**, 6110–6117 (2016). https://doi.org/10.1039/c5nr08639f
32. P. Urbankowski, B. Anasori, T. Makaryan, D. Er, S. Kota, P.L. Walsh, M. Zhao, V.B. Shenoy, M.W. Barsoum, Y. Gogotsi, Synthesis of two-dimensional titanium nitride Ti_4N_3 (MXene). Nanoscale **8**, 11385–11391 (2016). https://doi.org/10.1039/c6nr02253g
33. M. Khazaei, M. Arai, T. Sasaki, C.Y. Chung, N.S. Venkataramanan, M. Estili, Y. Sakka, Y. Kawazoe, Novel electronic and magnetic properties of two-dimensional transition metal carbides and nitrides. Adv. Funct. Mater. **23**, 2185–2192 (2013). https://doi.org/10.1002/adfm.201202502
34. H. Kumar, N.C. Frey, L. Dong, B. Anasori, Y. Gogotsi, V.B. Shenoy, Tunable magnetism and transport properties in nitride MXenes. ACS Nano **11**, 7648–7655 (2017). https://doi.org/10.1021/acsnano.7b02578
35. C.J. Zhang, B. Anasori, A. Seral-Ascaso, S.H. Park, N. McEvoy, A. Shmeliov, G.S. Duesberg, J.N. Coleman, Y. Gogotsi, V. Nicolosi, Transparent, flexible, and conductive 2D titanium carbide (MXene) films with high volumetric capacitance. Adv. Mater. **29** (2017). https://doi.org/10.1002/adma.201702678
36. C. Xing, S. Chen, X. Liang, Q. Liu, M. Qu, Q. Zou, J. Li, H. Tan, L. Liu, D. Fan, H. Zhang, Two-dimensional MXene (Ti_3C_2)-integrated cellulose hydrogels: toward smart three-dimensional network nanoplatforms exhibiting light-induced swelling and bimodal photothermal/chemotherapy anticancer activity. ACS Appl. Mater. Interfaces **10**, 27631–27643 (2018). https://doi.org/10.1021/acsami.8b08314
37. R. Li, L. Zhang, L. Shi, P. Wang, MXene Ti_3C_2: an effective 2D light-to-heat conversion material. ACS Nano **11**, 3752–3759 (2017). https://doi.org/10.1021/acsnano.6b08415
38. J.S. Lee, S.T. Oyama, M. Boudart, Molybdenum carbide catalysts. J. Catal. **106**, 125–133 (1987). https://doi.org/10.1016/0021-9517(87)90218-1
39. A. Sinha, H. Dhanjai, Y. Zhao, X. Huang, J. Lu, R. Chen, Jain, MXene: an emerging material for sensing and biosensing. TrAC - Trends Anal. Chem. **105**, 424–435 (2018). https://doi.org/10.1016/j.trac.2018.05.021
40. X. Zhang, Z. Zhang, Z. Zhou, MXene-based materials for electrochemical energy storage. J. Energy Chem. **27**, 73–85 (2018). https://doi.org/10.1016/j.jechem.2017.08.004
41. Y. Xie, M. Naguib, V.N. Mochalin, M.W. Barsoum, Y. Gogotsi, X. Yu, K. Nam, X. Yang, A.I. Kolesnikov, P.R.C. Kent, Role of surface structure on li-ion energy storage capacity of two-dimensional transition-metal carbides (2014)
42. D. Aurbach, Review of selected electrode–solution interactions which determine the performance of Li and Li ion batteries 206–218 (2000)
43. C. Du, J. Yeh, N. Pan, High power density supercapacitors using locally aligned carbon nanotube electrodes. Nanotechnology **16**, 350–353 (2005). https://doi.org/10.1088/0957-4484/16/4/003
44. J. Zhu, E. Ha, G. Zhao, Y. Zhou, D. Huang, G. Yue, L. Hu, N. Sun, Y. Wang, L.Y.S. Lee, C. Xu, K.Y. Wong, D. Astruc, P. Zhao, Recent advance in MXenes: a promising 2D material for catalysis, sensor and chemical adsorption. Coord. Chem. Rev. **352**, 306–327 (2017). https://doi.org/10.1016/j.ccr.2017.09.012
45. S. Yu, H. Tang, D. Zhang, S. Wang, M. Qiu, G. Song, D. Fu, B. Hu, X. Wang, MXenes as emerging nanomaterials in water purification and environmental remediation. Sci. Total Environ. **811** (2022). https://doi.org/10.1016/j.scitotenv.2021.152280

46. L. Ritter, K. Solomon, P. Sibley, K. Hall, P. Keen, G. Mattu, B. Linton, Sources, pathways, and relative risks of contaminants in surface water and groundwater: a perspective prepared for the Walkerton inquiry. J. Toxicol. Environ. Heal. - Part A. **65**, 1–142 (2002). https://doi.org/10.1080/152873902753338572
47. G. Crini, E. Lichtfouse, Advantages and disadvantages of techniques used for wastewater treatment. Environ. Chem. Lett. **17**, 145–155 (2019). https://doi.org/10.1007/s10311-018-0785-9
48. B. Xiao, Y. Li, X. Yu, J. Cheng, MXenes: reusable materials for NH_3 sensor or capturer by controlling the charge injection. Sens. Actuators B. Chem. (2016). https://doi.org/10.1016/j.snb.2016.05.062
49. L. Lorencova, T. Bertok, E. Dosekova, A. Holazova, D. Paprckova, A. Vikartovska, V. Sasinkova, J. Filip, P. Kasak, M. Jerigova, D. Velic, K.A. Mahmoud, J. Tkac, Electrochemical performance of $Ti_3C_2T_x$ MXene in aqueous media: towards ultrasensitive H_2O_2 sensing. Electrochim. Acta (2017). https://doi.org/10.1016/j.electacta.2017.03.073
50. J. Guo, Q. Peng, H. Fu, G. Zou, Q. Zhang, Heavy-metal adsorption behavior of two-dimensional alkalization-intercalated MXene by first-principles calculations (2015). https://doi.org/10.1021/acs.jpcc.5b05426
51. C. (John) Zhang, V. Nicolosi, Graphene and MXene-based transparent conductive electrodes and supercapacitors. Energy Storage Mater. **16**, 102–125 (2019). https://doi.org/10.1016/j.ensm.2018.05.003
52. R. Paradiso, G. Loriga, N. Taccini, A wearable health care system based on knitted integrated sensors **9**, 337–344 (2005)
53. X. He, X. Hu, T.D. James, Multiplexed photoluminescent sensors: towards improved disease diagnostics. Chem. Soc. Rev. (2017). https://doi.org/10.1039/c6cs00778c
54. C. Wolf, X. Mei, Synthesis of Conformationally Stable 1 , 8-Diarylnaphthalenes : Development of New Photoluminescent Sensors for Ion-Selective Recognition, (2003) 10651–10658.
55. N. Tyagi, G. Sharma, D. Kumar, P. Pratap Neelratan, D. Sharma, M. Khanuja, M.K. Singh, V. Singh, A. Kaushik, S.K. Sharma, 2D-MXenes to tackle wastewater: from purification to SERS-based sensing. Coord. Chem. Rev. **496**, 215394 (2023). https://doi.org/10.1016/j.ccr.2023.215394
56. Y. Wang, Y. Xu, M. Hu, H. Ling, X. Zhu, MXenes: focus on optical and electronic properties and corresponding applications. Nanophotonics. **9**, 1601–1620 (2020). https://doi.org/10.1515/nanoph-2019-0556
57. Y. Guo, M. Zhong, Z. Fang, P. Wan, G. Yu, A wearable transient pressure sensor made with MXene nanosheets for sensitive broad-range human-machine interfacing. Nano Lett. **19**, 1143–1150 (2019). https://doi.org/10.1021/acs.nanolett.8b04514
58. X. Li, X. Yin, S. Liang, M. Li, L. Cheng, L. Zhang, 2D carbide MXene Ti_2CT_x as a novel high-performance electromagnetic interference shielding material. Carbon N. Y. **146**, 210–217 (2019). https://doi.org/10.1016/J.CARBON.2019.02.003
59. G.R. Berdiyorov, M.E. Madjet, Structural, electronic transport and optical properties of functionalized quasi-2D TiC_2 from first-principles calculations. Appl. Surf. Sci. **390**, 1009–1014 (2016). https://doi.org/10.1016/j.apsusc.2016.08.179
60. G. Valurouthu, K. Maleski, N. Kurra, M. Han, K. Hantanasirisakul, A. Sarycheva, Y. Gogotsi, Tunable electrochromic behavior of titanium-based MXenes. Nanoscale **12**, 14204–14212 (2020). https://doi.org/10.1039/d0nr02673e
61. L. Huang, L. Ding, H. Wang, MXene-based membranes for separation applications. Small Sci. **1**, 2100013 (2021). https://doi.org/10.1002/smsc.202100013
62. H. Li, Z. Song, X. Zhang, Y. Huang, S. Li, Y. Mao, H.J. Ploehn, Y. Bao, M. Yu, Ultrathin, molecular-sieving graphene oxide membranes for selective hydrogen separation. Science **342**(80), 95–98 (2013). https://doi.org/10.1126/science.1236686
63. M. Mariano, O. Mashtalir, F.Q. Antonio, W.H. Ryu, B. Deng, F. Xia, Y. Gogotsi, A.D. Taylor, Solution-processed titanium carbide MXene films examined as highly transparent conductors. Nanoscale **8**, 16371–16378 (2016). https://doi.org/10.1039/c6nr03682a

Chapter 2
Strategies to Prepare 2D MXenes

Aydan Yeltik, Alp Yilmaz, Nihan Kosku Perkgoz, Feridun Ay, and Sina Rouhi

2.1 Introduction

Since the synthesis of the first MXene in 2011, its demand has been increasing. MXenes have been used in various applications such as batteries, supercapacitors, hydrogen storage, and biosensors due to their high electrical conductivity, high volumetric electrochemical capacitance, adjustable band gap, high thermal conductivity, and high strength resistance [1, 2]. MXenes are characterized by the chemical formula $M_{n+1}X_nT_x$ (M = Ti, V, Mo, Hf, Cr, etc.; X = C, N; T = –OH, –O, –F, etc.) and can be synthesized using a variety of top-down and bottom-up methods. Presently, top-down methods are favored due to their low cost, ease of use, and scalability for large-scale 2D MXene synthesis. However, the quality of products synthesized using top-down methods is often lower, leading to the preference for bottom-up methods in producing high-quality 2D MXenes, albeit at a higher complexity, cost, and smaller-scale production [3]. In this chapter, 2D MXene preparation strategies are described under two sections as top-down methods and bottom-up methods.

A. Yeltik (✉) · A. Yilmaz
Department of Material Science and Nanotechnology Engineering, TOBB University of Economics and Technology, Ankara 06560, Turkey
e-mail: ayeltik@etu.edu.tr

N. K. Perkgoz · F. Ay · S. Rouhi
Department of Electrical and Electronics Engineering, Eskisehir Technical University, Eskisehir 26555, Turkey

© The Author(s), under exclusive license to Springer Nature Singapore Pte Ltd. 2024

O. P. Pandey and P. Sharma (eds.), *MXenes: Emerging 2D Materials*,
https://doi.org/10.1007/978-981-97-4064-2_2

2.2 Top-Down Methods

Top-down methods stand out particularly in terms of being used in the industrial field, as they allow products to be synthesized on a larger scale and at lower cost. For the synthesis of 2D $M_{n+1}X_nT_x$ MXenes with these methods, the layers of A atoms are removed from the $M_{n+1}AX_n$ MAX phase (A = Al, Si, Zn, etc.) by selective etching. Top-down methods are presented here under the headings of wet-chemical synthesis of 2D MXene and other advanced top-down methods as well as fabrication of 2D MXene films, respectively. The wet-chemical synthesis part was examined under four subtitles, respectively, as fluorine-based acid etchants, fluorine-free base etchants, molten-salt etchants, and electrochemical etching.

2.2.1 Wet-Chemical Synthesis of 2D MXene

2.2.1.1 Fluorine-Based Acid Etchants

One of the frequently used methods for the synthesis of MXene from the MAX phase is the selective etching of A atoms using HF solvent as described in Fig. 2.1a. In 2011, Naguib et al. synthesized nanometer-thick Ti_3C_2 MXene crystal layers for the first time by removing Al from the Ti_3AlC_2 MAX phase [4]. The Al atoms in the MAX phase, which act as a bridge between the Ti_3C_2 layers and connect the layers with metallic bonds, were etched using HF solvent. Accordingly, it was observed that dangling bonds terminated with –OH or –F surface groups were formed in the synthesized 2D Ti_3C_2 layers. The chemical reactions occurring in the process are shown in Eqs. 2.1–2.3. After centrifugation, 2D layers were obtained by replacing stronger metallic bonds in the layered structure with weak van der Waals bonds.

$$Ti_3AlC_2 + 3HF \rightarrow AlF_3 + (3/2)H_2 + Ti_3C_2 \tag{2.1}$$

$$Ti_3C_2 + 2H_2O \rightarrow Ti_3C_2(OH)_2 + H_2 \tag{2.2}$$

$$Ti_3C_2 + 2HF \rightarrow Ti_3C_2F_2 + H_2 \tag{2.3}$$

As a further study to optimize the standard HF-based etching method and examine the underlying mechanism, Chang et al. observed the effect of HF treatment time on the etching process [5]. Accordingly, after 20 h of HF treatment, it was observed that the Ti_3AlC_2 MAX phase transformed into Ti_3C_2 MXene phase with a layered structure as a result of the selective etching of Al atoms, as shown in Fig. 2.1b–d. In addition, the synthesis of the MXene phase was proven with the mostly disappearance of the peaks of the Ti_3AlC_2 MAX phase after 20 h of treatment in the XRD pattern given in Fig. 2.1e. In another study, Liu et al. observed that the performance of the

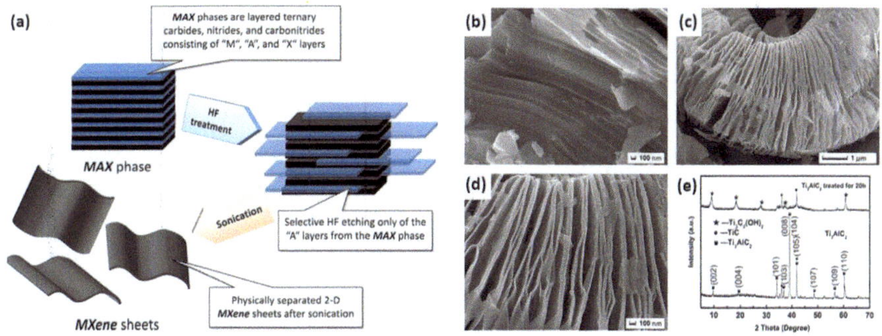

Fig. 2.1 **a** Schematic representation of MXene synthesis from the MAX phase using HF [7]. SEM images showing unetched and etched MAX phases; **b** before etching process, **c** after 20 h, and **d** high-resolution image of the etched sample, respectively. **e** XRD pattern of unetched and etched MAX phases. Adapted with permission from [5]. Copyright (2013), Copyright The Authors, some rights reserved; exclusive licensee [Elsevier]. Distributed under a Creative Commons Attribution License 4.0 (CC BY) https://creativecommons.org/licenses/by/4.0/

etching process increased due to the higher solubility of the HF solvent as a result of increasing the etching temperature [6]. In addition, it was observed that the number of bonds terminated with –OH and –F groups is higher in etching processes at higher temperatures, due to greater ion mobility.

As a variant of 2D MXene, $Hf_3C_2T_x$ (T_x; –OH, –O, –F groups), few-layered flakes with the 2–5 μm lateral dimensions were synthesized by Zhou et al. using HF from the $Hf_3(Al\text{-}Si)_4C_6$ MAX phase having the A layers doped with Si [8]. With this method different from the standard HF-based etching, the length of the bonds between the Hf-C atoms increased and weakened due to the doping of Si atoms, resulting in stronger bonds between the C-Al atoms with the adhesive energy decreasing at the etching interface which increased the efficiency of the etching process. As a result, $Hf_3C_2T_x$ MXene flakes were obtained by removing Al_4C_4 molecules from the structure.

In addition to using HF acid etchant directly, it can be formed as a product by using a mixture of HCl and LiF acid as shown in Eq. (2.4). In 2015, Anasori et al. examined the MXene structures of about 20 double transition metals carbides using density functional theory (DFT) calculations and synthesized Mo_2TiC_2Tx, $Mo_2Ti_2C_3T_x$, and $Cr_2TiC_xT_x$ double metal carbide 2D MXene flakes by etching with a mixture of HCl and LiF solvents followed by delamination including DMSO intercalation and sonication steps [9]. $Cr_2TiC_2T_x$ and $Mo_2Ti_2C_3T_x$ MXene flakes synthesized with this method were found to contain more contamination compared to $Mo_2TiC_2T_x$ MXene flakes.

$$LiF + HCl \rightarrow HF + LiCl \tag{2.4}$$

To investigate the electrochemical and heterogeneous electron transfer properties of MXenes, Nayak et al. synthesized few layer and multilayer 2D $Ti_3C_2T_x$ (T_x; –OH, –O, –F groups) MXene crystals using two methods named Clay and minimum intensive layer delamination (MILD), as depicted in Fig. 2.2a–b [10]. In the Clay method, the Ti_3AlC_2 MAX phase was etched using a mixture of LiF and HCl solvent to obtain the bulk $Ti_3C_2T_x$ phase, and then the ultrasonication method was used to break the weak bonds between the layers. The MILD method was carried out to synthesize few-layered MXene phases with simple shaking instead of sonification, using a mixture of LiF and HCl solvent with increased LiF ratio in the etching process. As a result, multilayer (10–13 layered) $Ti_3C_2T_x$ MXene flakes were obtained in the clay method, while few layer (1–2 layered) $Ti_3C_2T_x$ MXene flakes were obtained in the MILD method (Fig. 2.2c–f).

Furthermore, Shayesteh et al. developed the evaporated-nitrogen MILD (EN-MILD) process by adding dry nitrogen to the system to improve the MILD method and increase the stability, capacitance, and electrical conductivity of MXene flakes [11]. By using nitrogen environment in EN-MILD method, the acid concentration was increased by providing partial evaporation of the LiF and HCl solvent mixture used for etching, and hence, etching and delamination processes could be performed with higher efficiency. Moreover, by increasing the etching time, high-efficiency

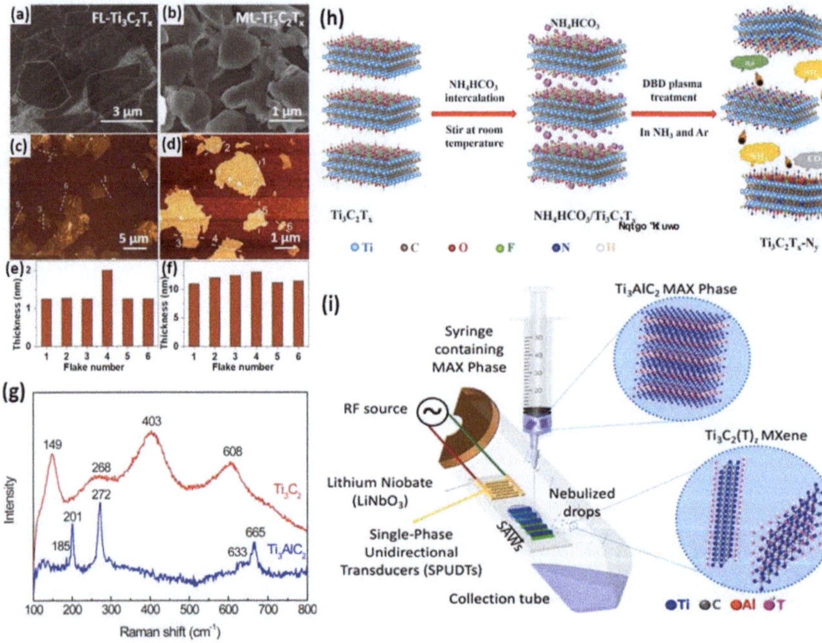

Fig. 2.2 SEM images of **a** few-layered and **b** multilayered MXene flakes of varying thickness, shown in **c, d** AFM images and **e, f** thickness distributions [10]. **g** Raman spectra of Ti_3AlC_2 and Ti_3C_2 after etching [13]. **h** Diagrams illustrating a plasma treatment-assisted DBD plasma system method used to obtain nitrogen-doped $Ti_3C_2T_x$, adapted with permission from [14]. Copyright (2021), Copyright: The Authors, some rights reserved; exclusive licensee [Elsevier]. Distributed under a Creative Commons Attribution License 4.0 (CC BY) https://creativecommons.org/licenses/by/4.0/, and **i** an etching process assisted by SAWs to synthesize $Ti_3C_2T_x$ MXene flakes [15]

etching was achieved even using a lower concentration solvent mixture. In addition, it was observed that the number of defects in the synthesized multilayer MXene flakes was lower while the sizes of the flakes were higher, and the number of TiO_2 intermediates formed during the synthesis was significantly reduced in the EN-MILD method compared to the MILD method.

Due to the low solubility of the LiF solvent in water and therefore the increased etching time, a high temperature is required to dissolve in the etching solution. Accordingly, when synthesizing 2D Mo_2CT_x (T_x; –OH, –O, –F groups) MXene flakes, Guo et al. decreased the etching time from several days to 24 h by using NaF, KF, and NH_4F solvents at lower temperatures than the solubility temperature of LiF [12]. Further, Wang et al. synthesized $Ti_3C_2T_x$ MXene flakes from Ti_3AlC_2 MAX phase via etching using NH_4F solvent in aqueous solution [13]. The water in the aqueous solution reacted with the NH_4F solvent and formed NH_3 and HF. It has been reported that etching is carried out when the HF compound reacts with Al atoms in the MAX phase, which is proved with Raman spectra as shown in Fig. 2.2g. Etching reactions were also observed to be more effective by increasing the NH_4F concentration or the reaction temperature or the reaction time for the synthesized few-layered $Ti_3C_2T_x$ flakes.

In another study, Chen et al. obtained nitrogen-doped $Ti_3C_2T_x$ MXene flakes using a plasma treatment-assisted dielectric barrier discharge (DBD) plasma system at low temperatures [14] (Fig. 2.2h). Multilayer $Ti_3C_2T_x$ MXene flakes were synthesized from Ti_3AlC_2 MAX phase by etching using a solvent mixture of LiF and HCl, and then $NH_4HCO_3/Ti_3C_2T_x$ MXene flakes were obtained by immersing the MXene flakes in the saturated NH_4HCO_3 solution and allowing NH_4HCO_3 to enter between $Ti_3C_2T_x$ layers. After that, nitrogen-doped $Ti_3C_2T_x$ MXene flakes were obtained by using $NH_4HCO_3/Ti_3C_2T_x$ MXene flakes in DBD plasma system with a mixture of NH_3 and Ar gas as plasma gas. Accordingly, the NH_4HCO_3 group, which is connected by weak bonds between the layers, decomposes into NH_3, CO_2, and H_2O with the effect of heat during the DBD plasma treatment. As a result, the layers are easily and quickly separated from each other and 2D MXene flakes can be obtained. At the same time, nitrogen atoms join the dangling bonds on the surfaces, forming bonds that terminate with nitrogen. Also, the number of active nucleation sites increases as a result of Ar ions bombarding the surface of the MXene flakes, and the nitrogen doping is controlled by adjusting the concentration of the NH_3/Ar gas mixture.

A reduced etching time from a few days to 15 min was obtained by a microwave-assisted method involving HCl/LiF etching while obtaining $Ti_3C_2T_x$ MXene flakes from the Ti_3AlC_2 MAX phase by Zhu et al. [2]. In the process, 15 min after etching with a LiF/HCl solvent mixture, the solution is rapidly heated using microwaves, resulting in a rapid and violent release of H_2 gas and H_2O vapor into the air environment. Thus, the weak bonds between the $Ti_3C_2T_x$ MXene flakes are easily broken and the layers are separated, eliminating the need for an additional delamination step and significantly shortening the synthesis time. In addition, Ghazaly et al. were able to synthesize monolayer $Ti_3C_2T_x$ MXene flakes in milliseconds using a very low concentration (0.05 M) LiF solution for etching with the assistance of surface acoustic waves (SAWs) [15] (Fig. 2.2i). Accordingly, the H_2O compound in the solution is decomposed into H^+ and OH^- ions with SAWs applied to the 0.05 M LiF aqueous solution containing the Ti_3AlC_2 MAX phase dispersed on the substrate surface. The H^+ protons lower the pH of the solution and allow the LiF solvent to more efficiently etch the Al element in the MAX structure. The resulting Li^+ and OH^- ions are weakly bound between the $Ti_3C_2T_x$ layers. Following the etching, the weak bonds between the layers are broken by local mechanical vibrations applied to the substrate surface, and the single-layer $Ti_3C_2T_x$ MXene flakes are suspended above the water and then collected.

2.2.1.2 Fluorine-Free Base Etchants

Solvents used in fluorine-based etching methods are very harmful to the environment and human health, as well as limiting their usage areas due to the presence of fluorine-terminated bonds on the surface of the material [13, 16, 17]. Furthermore, intermediates such as AlF_3 formed as a result of the reaction are quite difficult to be dissolved in any solvent [16]. As a result of such reasons, fluorine-free processes have been developed to synthesize 2D MXene structures. In 2014, Xie et al. conducted one of the early studies of fluorine-free processes using NaOH solvent to synthesize Pt nanoparticle-doped $Ti_3C_2T_x$ MXene layers [18]. Ti_3AlC_2 MAX phase was kept in NaOH solvent heated to 80 °C for 100 h and immersed in H_2SO_4 solution at 80 °C to remove the impurities formed by etching. Due to the inability of NaOH to be a good enough solvent, very limited etching of the Al atoms was observed. After that, Xuan et al. used tetramethylammonium hydroxide (TMAOH) solvent to synthesize hundred nanometer thick $Ti_3C_2T_x$ plate-like crystals [19] (Fig. 2.3a). Accordingly, while TMA^+ cations bond between $T_3C_2T_x$ layers, OH^- anions combine with Al atoms to form bonds terminated with $Al(OH)_4{}^-$ groups. Moreover, Li et al. synthesized high-quality $Ti_3C_2T_x$ nanosheets with large lateral size for the first time by fluorine-free process using a KOH solvent [16]. Water was used to easily remove the intermediate $KAlO_2$ formed in the reaction.

In another study, Li et al. utilized NaOH in water to obtain a high-quality multilayer $Ti_3C_2T_x$ (T_x; –OH, –O groups) MXene phase from Ti_3AlC_2 MAX phase [17]. In the etching process, while Al atoms in the MAX phase form aluminum hydroxide compounds with OH^- ions, Na atoms bond with weak bonds between layers, and bonds terminated with –O and –OH groups are formed after the delamination process. It was observed that the efficiency of the etching process decreased when the reaction temperature was reduced from 270 to 250 °C, and the impurities increased as the oxidation rate of Ti atoms increased due to the decrease in NaOH concentration at 270 °C (Fig. 2.3b). In addition, Li et al. used NaOH etching solvent to synthesize $Ti_3C_2T_x$ nanolayers with a high specific surface area in order to increase the electrocatalytic effect, and also used TMAOH intercalation reagent in the delamination process [20]. The effect of TMAOH intercalation reagent on the size of MXene flakes was compared using other intercalation reagents such as N, N-dimethylformamide (DMF), DMSO, and ethanol in the same process as shown in Fig. 2.3c–f, respectively. Accordingly, it was observed that, due to the presence of positively charged

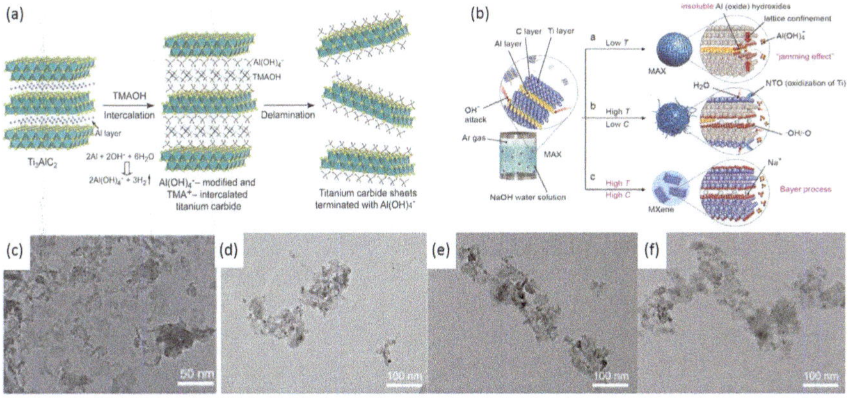

Fig. 2.3 **a** Schematic illustration showing the TMAOH intercalation and delamination processes to obtain MXene flakes from the Ti_3AlC_2 MAX phase. Adapted with permission from [19]. Copyright (2016), Copyright: The Authors, some rights reserved; exclusive licensee [John Wiley and Sons]. Distributed under a Creative Commons Attribution License 4.0 (CC BY) https://creativecommons.org/licenses/by/4.0/. **b** Schematic representation of etching processes in NaOH/water solution at different NaOH concentrations and temperatures. Adapted with permission from [17]. Copyright (2018), Copyright: The Authors, some rights reserved; exclusive licensee [John Wiley and Sons]. Distributed under a Creative Commons Attribution License 4.0 (CC BY) https://creativecommons.org/licenses/by/4.0/. MXene flakes were synthesized in the intercalation process using **c** TMAOH, **d** DMSO, **e** DMF, and **f** ethanol solvents, respectively [20]

TMA^+ ions in the TMAOH intercalation compound, it exfoliated more effectively than electroneutral DMSO, DMF, and ethanol, yet eventually smaller (50–100 nm) multilayer/few-layer MXene flakes yielded.

2.2.1.3 Molten-Salt Etchants

Particularly in the synthesis of 2D nitride MXenes, the use of acidic solution is not sufficient for an effective etching process due to the high formation energy of $Ti_{n+1}N_n$ and the strong binding of Al atoms to nitride atoms in $Ti_{n+1}AlN_n$ MAX phase [21, 22]. Therefore, another advanced method, the molten salt method, is used by applying a very high temperature for etching. In 2016, Urbankowski et al. synthesized the first $Ti_4N_3T_x$ nitride MXene flakes from the Ti_4AlN_3 MAX phase using the molten salt method [22]. They applied the method by heating a mixture of LiF, NaF, and KF fluorinated salts containing the MAX phase to 550 °C in an Ar atmosphere. As a result of selective etching, it was observed that bonds terminated with F groups were formed and Al atoms were removed from the structure. In order to obtain single-layered flakes from the multi-layered $Ti_4N_3T_x$ MXene structure and to eliminate the contamination caused by unetched MAX phases, the MXene structure was immersed in tetrabutylammonium hydroxide (TBAOH) solution and then 2D MXene flakes were obtained by breaking the weak bonds between Ti_4N_3

layers via sonication. Following that, washing process with deionized water was used to remove TBAOH residues and the synthesized 2D MXene flakes were collected by centrifugation. In addition, according to the DFT calculations, the probability of formation of bonds terminated with the –O group in $Ti_4N_3T_x$ structure was found to be considerably high compared to the –F and –OH groups. Further, it was observed that bare Ti_4N_3 layers that do not terminate with any group have the highest density of states.

Li et al. investigated in detail the mechanism of the molten salt method and succeeded in obtaining different multilayer MXenes using various Cl-based molten salts from many different MAX structures [23]. Accordingly, to synthesize $Ti_3C_2Cl_2$ multilayer MXene, the Ti_3SiC_2 MAX precursor was immersed in $CuCl_2$ molten salt at 750 °C (Fig. 2.4a–d). High temperature molten salt provides more active etching of the MAX phase. During the etching process, Si cations that are weakly bonded to Ti atoms combine with Cl^- anions to form $SiCl_4$ compounds, while Cu^{2+} cations and some Cl^- anions bind to Ti atoms in Ti_3C_2 layers to form the $Ti_3C_2Cl_2$/Cu multilayer MXene. The multilayer MXene is then dipped in ammonium persulfate (APS) solution to remove Cu atoms and other impurities. However, during the APS process, the –O functional group is attached to Ti atoms instead of the Cu atoms removed from the multilayer MXene. Also, multilayer structures of Ti_2CT_x, $Ti_3C_2T_x$, Ti_3CNT_x, Nb_2CT_x, Ta_2CT_x, Ti_2CT_x, $Ti_3C_2T_x$ MXene were synthesized using Ti_2AlC, Ti_3AlC_2, Ti_3AlCN, Nb_2AlC, Ta_2AlC, Ti_2ZnC, Ti_3ZnC_2 MAX phases and $CdCl_2$, $FeCl_2$, $CoCl_2$, $CuCl_2$, AgCl, $NiCl_2$ molten salts (Fig. 2.4e–j). In addition, the fact that it is very difficult to synthesize Ta_2CT_x and $Ti_3C_2T_x$ multilayer MXenes using Ta_2AlC and Ti_3SiC_2 MAX phases show how effective the molten salt etching method is. They also succeeded in synthesizing $Ti_3C_2T_x$ multilayer MXene using $CuBr_2$ and CuI molten salts and functionalized the surface of MXene structures as a result of different anion groups attached to the surface of the MXene layers. In another study, Li et al. synthesized 2D $Ti_3C_2Cl_2$ and Ti_2CCl_2 MXene flakes using $ZnCl_2$ molten salt solution and novel Ti_3ZnC_2, Ti_2ZnC, Ti_2ZnN, and V_2ZnC MAX phases [24]. The etching process was carried out by heating the prepared MAX/$ZnCl_2$ mixture to 550 °C in Ar gas environment. During etching, the weakly bound Zn atoms in the MAX phase dissolve in the $ZnCl_2$ molten salt solution, while the Cl atoms bond between the Ti_3C_2 layers. Then, $ZnCl_2$ residues were removed by washing with deionized water.

In the molten salt method, an extra cleaning process is required to remove the remaining contaminations from the structure after the reaction. To prevent this and control surface modification, Shen et al. developed the molten salt-assisted electrochemical etching method (MSE) [25] (Fig. 2.4k–l). Accordingly, there is a nickel cathode in the cuvette containing the LiCl/KCl molten salt mixture, and the Ti_3AlC_2 (or Ti_3SiC_2) MAX phase is used as the anode and immersed in the solution. In the process carried out in the Ar gas environment, the solution was heated to 450 °C and a voltage of 2 V was applied. Due to the thermochemical effect, Al atoms in the MAX phase were oxidized and formed the $AlCl_3$ compound, which evaporated at high temperature. At the same time, some –Cl anion groups were attached to the site of the etched A atoms, resulting in the formation of bonds terminating with the

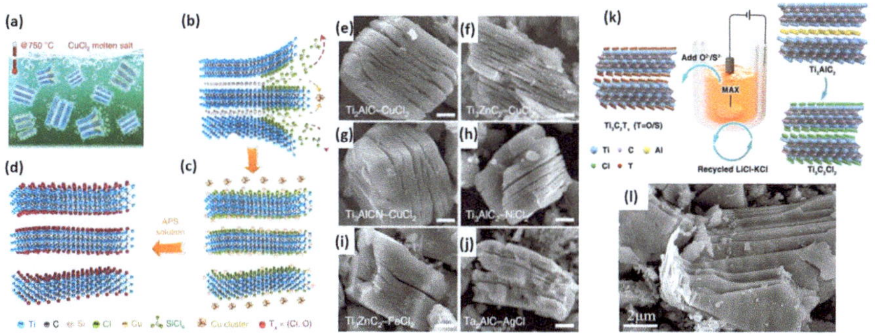

Fig. 2.4 **a–d** Schematic illustration of obtaining $Ti_3C_2T_x$ MXene flakes as a result of the reactions of Ti_3SiC_2 MAX phase immersed in $CuCl_2$ molten salt solution heated to 750 °C and washed with ammonium persulfate (APS) solution. **e–j** SEM images of various MXene flakes synthesized using different molten salts or MAX phases (Scale Bar: 2 μm) [23]. **k** Schematic representation of the effect of the MSE method on the MAX phase shows the surface terminations of the synthesized MXene flakes. **l** SEM image of MXene flakes synthesized using the MSE method. Adapted with permission from [25]. Copyright (2021), Copyright: The Authors, some rights reserved; exclusive licensee [John Wiley and Sons]. Distributed under a Creative Commons Attribution License 4.0 (CC BY) https://creativecommons.org/licenses/by/4.0/

–Cl group. In addition, Li/K cations are kept away from the MXene structure by being coated on the cathode due to the electrochemical effect. As a result, $Ti_3C_2Cl_2$ MXene flakes were synthesized without any contamination. To control the surface modification, the reaction was carried out under the same conditions by adding LiO_2 (or Li_2S) salt additive to the molten salt solution containing the synthesized $Ti_3C_2Cl_2$ MXene flakes.

Thus, depending on the type of additive salt used, bonds terminating with –O or –S group instead of –Cl group could be obtained, and it was observed that the surface modification property could be controlled. In addition, it has been observed that the molten salt solution used during the process can be used repeatedly, and it has been reported that the MSE process is an environmentally friendly and sustainable method.

2.2.1.4 Electrochemical Etching

Electrochemical etching is another method that does not use highly toxic chemicals and is used to synthesize 2D MXenes in a shorter time and quite simply [3]. In 2013, Sun et al. synthesized layered 2D Ti_2CT_x MXene structures from Ti_2AlC MAX phase for the first time by electrochemical etching using an HCl electrolyte solution [26] (Fig. 2.5a). Accordingly, while Ti_2AlC was used as the working (anode) electrode in the HCl electrolyte solution, Pt foil was used as the counter (cathode) electrode. Due to the electrochemical effect created by the applied electrical force, the bonds between Ti and Al atoms are broken and –Cl, –O, –OH groups are bonded instead

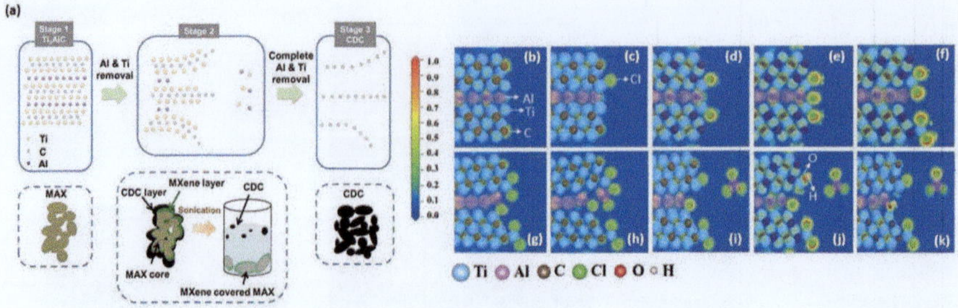

Fig. 2.5 **a** Schematic representation of the electrochemical etching process used to synthesize the 2D $Ti_2C_xT_x$ MXene flakes and the products formed according to the steps [26]. **b–k** Analysis of the electrochemical etching mechanism of the Ti_3AlC_2 MAX phase using DFT calculations. Adapted with permission from [27]. Copyright (2018), Copyright: The Authors, some rights reserved; exclusive licensee [John Wiley and Sons]. Distributed under a Creative Commons Attribution License 4.0 (CC BY) https://creativecommons.org/licenses/by/4.0/

of Al atoms. It has been observed that phase differences occur on the surfaces of the MAX structures during the reaction. Since the surface of the MAX structures was etched stronger than the inner regions, it was observed that not only the A atoms were eliminated but also the Ti atoms were etched and the carbon-derived carbide (CDC) phase was formed. It was also observed that increasing the etching time, applied voltage, or the concentration of the HCl solvent in the electrolyte solution increased the CDC ratio. A sonication bath with deionized water was used to remove the CDC layer after etching and layered 2D MXene structures were obtained.

Yang et al. succeeded in synthesizing single-layer TiC_2 MXene flakes by electrochemical etching method using a solvent mixture of NH_4Cl and TMAOH as an electrolyte solution [27]. The electrochemical etching mechanism was investigated in more detail by DFT calculations. When the Ti_3AlC_2 MAX phase used as the electrode is loaded with a positive charge, the Cl ions in the solution move towards the surface of the MAX phase (Fig. 2.5b–c). The number of Cl ions bound to Ti atoms on the surface of the MAX phase increases with time and the distance between the Ti_3C_2 layers increases (Fig. 2.5d–g). The resulting Al atoms form an $AlCl_3$ compound with Cl ions and move away from the MAX structure (Fig. 2.5h–i). Instead of Al atoms, OH anions with an ionic radius smaller than the distance between the layers are bonded between the Ti_3C_2 layers (Fig. 2.5j–k). At the same time, NH^{4+} and TMA^+ cations are collected on the other electrode. Furthermore, electrolyte solutions such as H_2SO_4, HNO_3, NaOH, $FeCl_3$ were used as well as NH_4Cl for comparison in the electrochemical etching process. Accordingly, it has been observed that electrolyte solutions containing Cl^- anion groups are more effective for electrochemical etching due to the effective bonding of Al atoms and Cl^- ions. Delamination with TMAOH solution is used to obtain single-layer TiC_2 MXene flakes from the multi-layered $Ti_3C_2T_x$ MXene structure obtained after etching. TMA^+ cations are weakly bonded between Ti_3C_2 layers, while the OH anion group is attached to dangling bonds on the surfaces of T_3C_2 layers. Single-layer Ti_3C_2 MXene flakes were obtained by mechanically breaking the weak bonds between the layers as a result of continuous stirring. The dimensions of the flakes collected by centrifugation go up to about 18 μm and their average size is 2.4 μm.

Additionally, Pang et al. were able to synthesize Ti_3CT_x, Cr_2CT_x, and V_2CT_x MXene plates using the thermal-assisted electrochemical etching method and develop a universal method for the synthesis of many different MXenes [28]. They used composite Ti_2AlC-CB/CFC electrode consisting of carbon black additive (CB) and carbon fiber cloth (CFC) to increase the efficiency of the electrochemical etching method. The composite electrode shows low Warburg resistance, causes increased ion diffusion to the electrode surface, and increases the efficiency of reactions in the electrochemical etching process. In addition, it was observed that the efficiency of the delamination process increased and Ti_3CT_x structures with rougher surfaces were obtained when the solution temperature was increased from 25 to 50 °C during the electrochemical etching process and with the applied voltage of 0.3 V. Furthermore, laminar Ti_3CT_x MXene structures were observed when the etching time was increased from 3 to 9 h. In the case of the more difficultly synthesized Cr_2CT_x and V_2CT_x MXenes, higher voltages of 1 V and 0.5 V were applied, respectively, for 9 h etching at 50 °C. The lateral size of the synthesized multilayer Cr_2CT_x plates is up to 25 μm, while the lateral sizes of the multilayer Ti_2CT_x and V_2CT_x MXene plates are greater than 1 μm.

2.2.2 *Other Advanced Top-Down Methods*

Besides the most commonly used methods described above, interesting methods such as halogen etching, thermal reduction, UV-induced selective etching, algae extraction, and ball milling are also used to synthesize high-quality 2D MXenes. Jawaid et al. succeeded in obtaining the MXene phase by using halogen (I_2, Br_2) and interhalogen compounds (IBr, ICl) to etch A atoms from the MAX phase at room temperature [29] (Fig. 2.6a). Accordingly, as a result of the significant increase in the activity of I_2, Br_2 halogens in high dielectric solvents (such as acetonitrile), $TiBr_x$ structures are formed as a result of the loss of selectivity due to the etching of the X layer as well as the A layer in the MAX phase. Therefore, nonpolar solvents such as cyclohexane (CH) are used to increase selectivity. In addition, the formation of more reactive by-products on the surface with the increase of Br_2 concentration in the solution reduces the efficiency of the etching process, while lowering the concentration value decreases the etching rate since it cannot be adsorbed on the MAX surface sufficiently. Tetrabutylammonium halides ($TBAX^-$; X; F^-, Cl^-, Br^-) are used to disperse the synthesized $Ti_3C_2T_x$ MXene flakes. In interhalogen compounds, the halogen atom with higher electronegativity participates in the etching process. Since interhalogens are stable in low dielectric organic solutions, MXene flakes can be synthesized using solvents such as tetrahydrofuran (THF), acetonitrile, chloroform. In addition, it has been stated that halogen etching method can be used to obtain

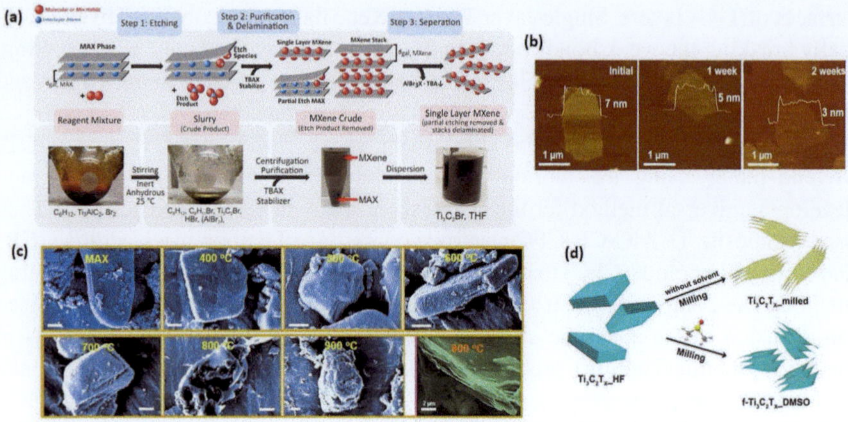

Fig. 2.6 **a** Schematic view and experimental images of the halogen etching process [29]. **b** AFM images show dissolution and reduction in the thickness of MXene flakes exposed to water over time; initial, 1 week, and 2 weeks, respectively. Adapted with permission from [30]. Copyright (2021), Copyright: The Authors, some rights reserved; exclusive licensee [John Wiley and Sons]. Distributed under a Creative Commons Attribution License 4.0 (CC BY) https://creativecommons.org/licenses/by/4.0/. **c** SEM images show the transformation in the structure of the MAX phase at different temperatures (Scale bar: 1 μm). Adapted with permission from [31]. Copyright (2020), Copyright: The Authors, some rights reserved; exclusive licensee [Elsevier]. Distributed under a Creative Commons Attribution License 4.0 (CC BY) https://creativecommons.org/licenses/by/4.0/. **d** Schematic illustration showing the synthesis of few-layered MXene flakes using DMSO solvent by the HEBM method [34]

MXene structures from Cr_2AlC and V_2AlC MAX phases as well as Ti_2AlC MAX phase by using halogen etching method.

Furthermore, Shi et al. synthesized $Ti_3C_2T_x$ ($Tx = O^-, OH^-$) MXene sheets from the MAX phase by etching with halogen (I_2) in anhydrous acetonitrile and delamination with HCl solvent [30]. Since some AlI_3 by-product remained as impurity between $Ti_3C_2T_x$ layers after etching, it was immersed into HCl solution to remove the impurity, and delamination was performed at the same time. In addition, it was observed that the performance increased proportionally with the increase in temperature, and the Al concentration decreased from 16.7 to 0.9 wt% when the temperature was increased from room temperature to 100 °C. The thickness of the oxygen-rich $Ti_3C_2T_x$ sheets synthesized by the I_2 halogen etching process is mostly a few layers and their lateral sizes are 1.8 μm. It has also been observed that MXene sheets have a high stability value for up to 2 weeks without decomposition in water due to the non-aqueous etching process (Fig. 2.6b).

In another study, Mei et al. synthesized multilayer Ti_2C MXene flakes from Ti_2SC MAX phase using a very simple, environmentally friendly, and easy-to-control thermal reduction method [31]. Accordingly, the Ti_2SC MAX phase is placed in a sealed tube furnace with Ar/H gas flow by a quartz boat, and the temperature is increased from 400 to 800 °C with a heating rate of 10 °C/min, and multi-layered

2D Ti_2C MXene flakes are synthesized in 2–4 h. The most important part of the thermal reduction method is to find the appropriate temperature value (Fig. 2.6c). It was observed that, when the temperature exceeded 500 °C, the S atoms in the MAX phase began to etch, and Ti_2C, MXene flakes and TiO_2 particles were formed at 800 °C. It has also been reported that MXene flakes return to the MAX phase when the temperature reaches 900 °C. However, it was noted that more S atoms were etched than MXene flakes left at 800 °C for 30 min and then at 900 °C for 30 min.

In another approach, Mei et al. succeeded in synthesizing mesoporous 2D Mo_2C plate-like structures using UV-induced selective etching from the ultraviolet (UV) wavelength-sensitive Mo_2Ga_2C MAX phase [32]. With this method, it has been shown that 2D MXene structures can be easily synthesized with UV-sensitive MAX phases in a very short time, within a few hours, without the use of any dangerous acidic chemicals. Accordingly, Ga atoms in the Mo_2Ga_2C MAX phase are etched under UV irradiation due to strong UV absorption. Etched Mo_2C MXene plates were delaminated using H_3PO_4 solvent. The synthesized few layered Mo_2C MXene plates have a graphene-like structure, and interestingly, the plates have mesoporous structures with a pore size in the 2–16 nm range. In addition, due to the solvent used in the delamination process, they have bonds that terminate with the –O group.

In addition to these studies, Zada et al. succeeded in synthesizing V_2C MXene nanoflakes for use as photothermal agents in photothermal therapy applications from the V_2AlC MAX phase using the algae extraction method [33]. Al atoms in the V_2AlC MAX phase are etched with an algae extraction mixture containing organic acids. Moreover, many biological compounds intercalate between the V_2C layers during the etching process and also accelerate the delamination process. The few-layer V_2C MXene nanoplates, synthesized after 48 h, have lateral sizes in the range of 50–100 nm. In addition, it was observed that MXene nanoplates with smaller lateral size and thinner thickness were synthesized with increasing etching and delamination times. Thus, a very low cost, completely environmentally friendly, simple and easy-to-control method has been developed.

In addition, Wu et al. used high energetic ball milling (HEBM) method in the exfloation process to obtain few-layer 2D $Ti_3C_2T_x$ nanoplates from multilayered $Ti_3C_2T_x$ MXene particles [34]. Accordingly, DMSO solvent was used in the HEBM method to prevent oxidation of MXene particles during the process and for an effective intercalation (Fig. 2.6d). MXene nanoplates were obtained by breaking the weak bonds between $Ti_3C_2T_x$ layers due to the mechanical force applied by ball milling method. In addition, when the ball milling method is applied without using DMSO solvent, it was observed that the MXene structure turned into TiO_2 and graphitic carbon structures. Further, it has been reported that multilayer MXene structures exist when different solvents such as DMF, ethanol, HF are used and the delamination process is not as efficient as when DMSO is used. In another study, He et al. used PDDA solution-assisted ball milling (P-BM) method in the delamination process to obtain functionalized few-layered MXene nanoplates from multilayer $Ti_3C_2T_x$ MXene particles [35]. The high viscosity of the PDDA solution prevents the steel balls from hitting the MXene particles violently and facilitates the exfoliation process by preventing the agglomeration of the newly exfoliated nanoplates. It was observed that several-layered $Ti_3C_2T_x$ MXene nanoplates thus synthesized had a surface functionalized with the PDDA group intercalated between Ti_3C_2 layers and had larger lateral size (hundreds of nm).

2.2.3 *Fabrication of 2D MXene Films*

Although a wide variety of advanced methods have been developed for the synthesis of 2D MXene flakes, the development of large-scale, low-cost, and low-toxicity production methods for high-quality 2D MXene thin films is of great importance. In 2014, Michael et al. showed that $Ti_3C_2T_x$ MXene 2D film can be produced via the rolling method using etched multilayer MXene particles called clay-like MXene paste [36] (Fig. 2.7a). In addition, it is stated that printing can be made on the desired surface by using clay-like MXene paste as ink. It has been also reported that clay-like MXene paste can be formed into the desired shape when wet and MXene structures in inch sizes can be obtained after drying. Thus, it has been shown that MXene structures can be produced easily, quickly, and at large scales (Fig. 2.7b–c). This opened a new page for producing 2D MXene films via many methods.

Vacuum filtration method was developed by Ling et al. to produce flexible, high electrical conductive 2D MXene films [37]. Accordingly, the suspended solution containing 2D MXene flakes is filtered with the help of vacuum pump, and a 2D film with a thickness of 13 μm is obtained, consisting of thousands of MXene flakes stacked on the filter. Moreover, the film thickness could be easily controlled by adjusting the volume of solution used, and 2D MXene films with thicknesses less than 1 μm could be produced. Thus, it has been shown that MXene 2D films with high electrical conductivity, high flexibility, and low density can be produced with a facile and cost-effective method. In another study, Kim et al. synthesized inch-size Mo-based MXene 2D films such as Mo_2CT_x, $Mo_2TiC_2T_x$, and $Mo_2TiC_3T_x$ by vacuum filtration method [38]. Moreover, the electrical properties of the 2D MXene film at different temperatures were investigated, and it was observed that the electrical properties of the thermally treated (800 K) films were improved. It has been stated that the reason for this is that the MXene flakes become very close to each other as a result of the evaporation of the intercalation compounds between the MXene layers and the functional groups on the surfaces at temperatures above 500 K. Thus, it was observed that Mo_2TiCT_x MXene 2D films thermally treated at 800 K have high electrical conductivity (1380 S/cm), large Seebeck coefficient (–47.3 μV/K) and high thermoelectric power (3.09×10^{-4} W m^{-1} K^{-2}). Furthermore, Shuck et al. carried out large-scale production using a specially designed vacuum filtration-assisted reactor to synthesize 2D $Ti_3C_2T_x$ MXene structures in different batch sizes (1 and 50 g), showing that large-scale production did not alter the properties of MXene [39].

In addition, Dillon et al. produced inch-sized MXene 2D films of different thicknesses on the substrate surface by spin-coating method using a colloidal solution containing single/few layer $Ti_3C_2T_x$ MXene plates [40] (Fig. 2.7d–f). The thickness of $Ti_3C_2T_x$ MXene 2D film was changed between 2 and 81 nm to examine

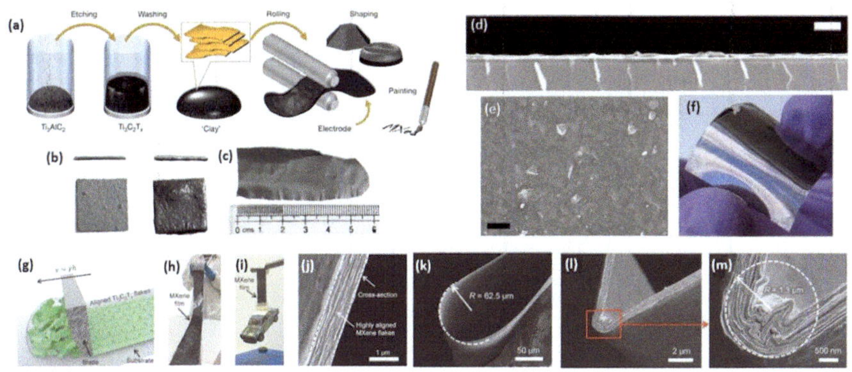

Fig. 2.7 **a** Schematic representation of clay-like $Ti_3C_2T_x$ MXene paste making and various uses. **b** Plate produced using clay-like MXene paste (left) swells and shrinks (left) during the drying process (top cross-section views). **c** The inch-size MXene film obtained after the rolling process [36]. SEM images of **d** lateral area (Scale bar: 500 nm) and **e** top surface area (Scale Bar: 1 μm) of MXene thin films produced by spin coating. **f** Image of MXene thin film coated on a flexible substrate (polyetherimide polymer). Adapted with permission from [40]. Copyright (2016), Copyright: The Authors, some rights reserved; exclusive licensee [John Wiley and Sons]. Distributed under a Creative Commons Attribution License 4.0 (CC BY) https://creativecommons.org/licenses/by/4.0/. **g** Schematic representation of the blade coating method. **h** Image of $Ti_3C_2T_x$ MXene thin film with a length of 1 m, a width of 10 cm, and a thickness of 940 nm, coated on Celgrad membrane by blade coating method. **i** Digital photograph showing thin film carrying 40 g in durability test. **j** Cross-sectional SEM image showing MXene flakes lined up in a highly coherent array. **k** SEM image showing that a 62.5 μm radius fold was formed by 180° bending of the thin film. **l** Low-magnification and **m** high-magnification SEM images show the formation of folds with extremely small radius down to 1.3 μm. Adapted with permission from [42]. Copyright (2020), Copyright: The Authors, some rights reserved; exclusive licensee [John Wiley and Sons]. Distributed under a Creative Commons Attribution License 4.0 (CC BY) https://creativecommons.org/licenses/by/4.0/

the relationship between thickness and electrical conductivity. It was observed that the electrical properties did not change with the increase in thickness, since few-layered film with a thickness of 2 nm showed bulk conductivity. It has also been reported that the inch size $Ti_3C_2T_x$ MXene 2D film has very high electrical conductivity (6500 ± 800 S/cm). Further, Hantanasirisakul et al. produced a 70 nm thick transparent $Ti_3C_2T_x$ MXene 2D film on a glass substrate by spray coating method using a suspended solution [41]. Also, Zhang et al. used blade coating method with few-layered $Ti_3C_2T_x$ MXene flakes to obtain high-quality 2D MXene film with high electrical conductivity and high mechanical strength [42]. Accordingly, since MXene flakes with a large area and thin thickness have a high aspect ratio, large flakes are dispersed in the solution. Large flakes dispersed in solution have been reported to self-assemble and form agglomerates. Also, the viscosity of the solution decreased when shear stress was applied to the gel-like solution. Thus, it was observed that the large flakes dispersed in the solution aligned very easily in a certain pattern in the direction of the shear stress applied by the blade (Fig. 2.7g). Therefore, several meters long and 940 nm thick 2D $Ti_3C_2T_x$ MXene film was synthesized with excellent electrical conductivity (15,100 S/cm), outstanding tensile strength (570 MPa), a record among thin films, and high Young modulus (20.6 GPa) (Fig. 2.7h–j). It has also been observed that the produced 2D MXene thin film has very high flexibility and high stability as a result of many repeated bending processes. The produced 2D MXene thin film was reported to retain its stability even after repeated 5000 times bending at 180° angle, a 62.5 μm radius curl is formed as a result of bending, and the large flakes are able to be compressed to form fold with an extremely small radius of 1.3 μm (Fig. 2.7k–m).

2.3 Bottom-Up Methods

The bottom-up synthesis method is applied to fabricate precise MXene structures by increasing the controllability of the growth parameters. Also, this approach provides a clean and acid reaction-free situation without using any functional group. One of the most important members of this approach is the well-known chemical vapor deposition (CVD) process.

2.3.1 Chemical Vapor Deposition Method

The CVD process has proven to be a reliable, inexpensive, and flexible method for growing two-dimensional structures such as nanotubes, nanofibers, graphene, and MXene. This method offers efficient uniformity due to more controllable parameters compared to other alternatives [43]. Compared to the top-down approaches, CVD-MXene offers fewer structural impurities and lower levels of defects and atomic disorder. Furthermore, the instability of MXenes synthesized via top-down techniques under atmospheric conditions limits their widespread application. The bottom-up approach described here involves exposing a reactive substrate to volatile precursors via the heated reaction chamber to affect chemical reactions and deposition on the surface. As a result, this approach has been used extensively in the preparation of carbide, nitride, and silicon-based compounds [44]. The synthesis strategies influence the controlled growth process of the MXene structure and its inherent features. With the emergence of novel MXene structures and specifications such as Mo_2C, W_2C, WC, TaC, or NbC, the emergence of a new approach towards bottom-up synthesis extends MXene capability to more sensitive applications. CVD is a well-established key process in the bottom-up approach. As a result, the significance of CVD in MXene synthesis is well understood [45] but still needs to be improved and optimized. A single-step CVD process with controllable parameters allows precise fine-tuning of each feature to achieve the required thickness, morphology, grain size, and 2D structure quality [46]. In addition, the CVD technique exhibits superior controllability over the thickness and size of MXene structures, facilitating the synthesis of extended lateral heterostructures of graphene and MXene [47].

The synthesis method of 2D MXene crystals influences grain boundaries, imperfections, crystal structure, and thickness. Furthermore, the shape formation regime is a major consequence of chemical stability, and the electrical conductivity of the MXene family is thickness-dependent [48]. Consequently, tight control of the growth process is essential to obtain the relevant properties that CVD offers for future MXene-based applications. Chemical reactions at high temperatures also generate vertical van der Waals band heterostructures, which are crucial for the specific properties of MXene. The CVD approach could also produce an MXene/graphene heterostructure in multiple steps using different gases at different thermal conditions. Cooling duration can also be adjusted as a critical parameter to control flake growth rate. Some CVD furnaces with an upstream heating zone provide some fixed chemical combinations for precursors [49]. Numerous complex factors need to be considered, including effective optimization of the synthesis process and consequent mass production of the MXene structure, which is still debated by researchers. In general, the MXene growth compound consists of stacked metal foils positioned in a quartz tube. Because of their higher potential for atomic surface mobility and reaction within the molten metal, the dominant transition metals used to control MXene production are preferably selected from the third through sixth periodic table element groups. The tube is closed on both sides and coupled with a gas control system and a mechanical vacuum pump as a pressure optimizer. In order to prevent the air pollution in the early stage, the vacuum pump is linked with the air exhaust system. The tube is preheated by the furnace after being filled with the appropriate carrier gas ratio and then the temperature rises to the set-point. In general, the temperature reported is the melting point of the catalyst or a temperature in the range of the needed values for MXene formation. The temperature stabilization performance also has an impact on the quality of the final product. Mo_2C is one of the most studied MXene structures, specifically by CVD where stacked Mo/Cu foils are used as a precursor and a catalyzer, respectively. Above 1085 °C, copper foil becomes liquid and covers the Mo foil completely. The temperature limits of reaction activation are typically near the melting point of the catalyst and the recrystallization temperature of the transition metal. The process ends with the deactivation of the furnace, which moves the sample out of the hot zone and cools it down in the accompanying gas flow. According to the latest study results, at least 800 °C is required for the development of carbon-based MXene structures [50]. Figure 2.8a–b explains the schemes of system configuration and growth mechanism of Mo_2C in different shapes and single or multilayer forms under different growth conditions. Figure 2.8c shows the possibility of a structural change with a multi-step CVD operation from Mo_2C to MoN, and Fig. 2.8d illustrates the Mo_2C growth inhibition by initial growth of a graphene layer. In conclusion, since the synthesis of 2D MXene structures with desired and high optoelectronic properties is of great importance, the important parameters that can be controlled in the CVD method will be discussed in this section.

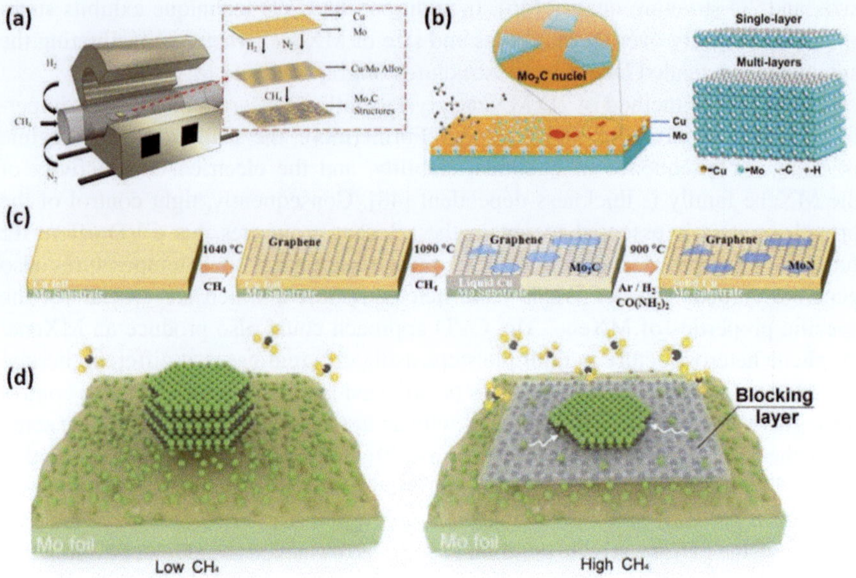

Fig. 2.8 The principles of bottom-up method CVD: **a** the system configuration [51], **b** the schematic of nucleation mechanism, adapted with permission from [52]. Copyright (2021), Copyright: The Authors, some rights reserved; exclusive licensee [Elsevier]. Distributed under a Creative Commons Attribution License 4.0 (CC BY) https://creativecommons.org/licenses/by/4.0/, **c** Multi-step MXene flakes growing by Mo_2C to MoN transformation, adapted with permission from [49]. Copyright (2022), Copyright: The Authors, some rights reserved; exclusive licensee [Elsevier]. Distributed under a Creative Commons Attribution License 4.0 (CC BY) https://creativecommons.org/licenses/by/4.0/, **d** is the MXene vertical growth rate blocking by initial growth graphene layer, adapted with permission from [53]. Copyright (2017), Copyright: The Authors, some rights reserved; exclusive licensee [John Wiley and Sons]. Distributed under a Creative Commons Attribution License 4.0 (CC BY) https://creativecommons.org/licenses/by/4.0/

2.3.1.1 Catalyzer Type and Thickness

Because transition metals such as Mo have high recrystallization temperatures, surface atoms require much activation energy for atomic transition [54]. The molten metal catalyst facilitates the diffusion of metal atoms to the gas/liquid interface. The atoms meet decomposed gas atoms at the liquid interface, resulting in a chemical reaction. The molten metal is a reactive environment that provides an atomic motion environment through the applied kinetic energy. Catalytic metals generally have little ability to perform chemical reactions by purging precursor gases. The kinetic energy of MXene formation is provided by the high temperature and atomic transition in molten metal. In addition, it is possible to dissolve gas atoms in molten metal to achieve deeper penetration and higher vertical growth rates. The early melting temperatures of the catalyst provide the earliest nucleation points and a general guideline for the controllable growth of large-scale, high-quality samples. Catalytic metals are typically copper, gold, indium, and their alloys. The lower melting point and hydrogen solubility increase the potential of MXene formation [55–57]. With a precisely engineered thickness, the catalyst thickness can be a reliable parameter to enable an affordable method of growing ultrathin 2D materials. So it can demonstrate the importance of the process by varying the thickness of the catalyst foil on the transition metal film, which readily diffuses to react with the hydrocarbon gas [58]. In addition, the applied catalyst thickness causes MXene flakes with a few nanometers of thickness to grow on the superlattice. The thicker the catalyst, the more crystallographic layers stack up and the more carbon vacancies there are. By modifying the thickness of the Cu foil, the Mo_2C vertical growth range can be varied from the superlattice to hundreds of nanometers thick. Due to their lower melting points, alloyed catalysts outperform pure catalytic metal catalysts. Temperature control is crucial in this process as there are few options for all-metal catalysts. Lowering the melting point of the catalyst allows for more diffusion of the supported transition metal towards the liquid interface. The alloyed catalysts are obtained to lower the growth temperature of MXene crystals. Several alloyed catalysts such as Ag-Cu and In-Cu alloys could be used for Mo_2C synthesis at temperatures below 1000 °C [55, 56]. By increasing the Sn content, Sn-Cu alloyed catalysts show higher carbon absorption than pure copper. The viscosity of the liquid Sn-Cu alloy is an important factor in reducing the resistance of the liquid to atomic movement and enabling faster diffusion of Mo. The density of MXene nucleation inversely increases with growth temperature, which has a non-linear effect on the viscosity of the alloyed catalyst. Accordingly, an optimal growth temperature is required to provide growth kinetics. As aforementioned, the growth temperatures can be reduced to 880 °C, with similar growth rates for pure Cu above 1085 °C [55].

2.3.1.2 Ratio of the Gases

The crystal morphology and flake thickness could be controlled by changing growth parameters such as the CH_4:H_2 gas combination ratio [56]. Injected gases are classified into two types: carriers and precursors. Carbides and nitrides are the two most important groups of transition metal compounds. According to research, the formation of carbide MXenes by catalytic metals is easier than that of the nitride group due to undesirable nitride compositions. Methane (CH_4) or ammonia (NH_3) flows into the quartz chamber as carbon and nitrogen sources during CVD operation [59]. As previously stated, the most researched CVD-grown MXenes are transition metal carbides such as Mo_2C, W_2C, WC, TaC, and NbC, but binary nitrides such as Co_3N, WN, Fe_3N, CrN, and VN have less relevance in CVD products. Because exothermic reactions of nitride composition demand less energy for catalytic activity, N_2 is a superior candidate for Ta_2C, Ta_2N, Ti_2C, and Ti_2N production than Mo_2C and W_2C [59]. $M_{n+1}C_n$ is the most prevalent chemical composition for carbides, while $M_{n+1}N_n$ is the chemical composition for nitrides. The functional groups (precursor gases) breakdown by interaction with the mass flow of hydrogen and/or nitrogen. The gas integration ratio is an important parameter in controlling lateral or vertical flake development as well as surface coverage. In the following, the carbide phase will be discussed more than the nitride phase.

2.3.1.3 CH_4 Flow Rate

The major process of crystal development on the crystal face is 2D nucleation due to carbon supersaturation. As a result, increasing the flow rate of carbon allows crystallographic faces with distinct crystal morphologies to vary, with the resultant flake forms shifting from triangle to polyhedron as the carbon precursor is increased [58]. Due to the extended rapid extension of graphene structure at low methane flow rates, graphene domains appear to dominate on the surface of molten metal. Graphene prevents the interfacial diffusion of Mo atoms and passivates the Mo interaction with the gaseous hydrocarbon. When the flow rate of CH_4 increases significantly to reach the surface saturation limits, Mo atoms diffuse along the liquid-metal interface and react with the hydrocarbon molecules, resulting in the formation of Mo_2C crystals based on the first nucleated graphene. Researchers demonstrate a wide range of methane flow rate utilization ranging from 0.3 to 25 sccm. Different flow rates support the formation of flakes ranging from triangular to polyhedral as the methane flow rate increases. In addition, methane decomposition increases with operating temperature, leading to carbon saturation [48, 53]. The methane flow rate is a critical factor in regulating the formation of the graphene/MXene heterostructure. The data supported the idea that crystal morphologies and crystallographic structures may be reformed by modifying the flow rate to the supersaturation of the precursors. Due to limited carbon diffusion, the initial Mo_2C nucleus gradually expands into a variety of different shaped flakes with low carbon content [58, 60, 61] Fig. 2.9a–c shows the diversity of MXene flakes formation at different CH_4 fluxes and carbon concentrations. Lower than 1 sccm methane ratio results in cubic-shaped flake, and with increasing flow rate the shapes begin to expand in star shape, and finally changing to hexagonal type above 5 sccm methane.

2.3.1.4 Carrier Gas Flow Rate

Hydrogen flow as a carrier gas causes lateral growth by decreasing nucleation density and increasing MXene flake domains. Hydrogen atoms interact preferentially with methane and dehydrogenate at high temperatures, which increases the proportion of carbon in the reaction tube. Due to the rapid deposition of graphene on the metal catalyst surface, the low fraction of H_2 flow rate inhibits growth of MXene single crystals. As a result, graphene structure predominates in the MXene/graphene heterostructure [48]. Meanwhile, in a hot atmosphere, hydrogen acts as a corrosive agent, resulting in negative consequences on the obtained compounds. Therefore, an optimal H_2 flow rate is required to achieve the desired structure [62]. Figure 2.9d–f shows the changes in flake growth and morphologies at a constant methane flow rate of 5 sccm by increasing the H_2 from 80 to 120 sccm at 1200 °C. The images show the direct

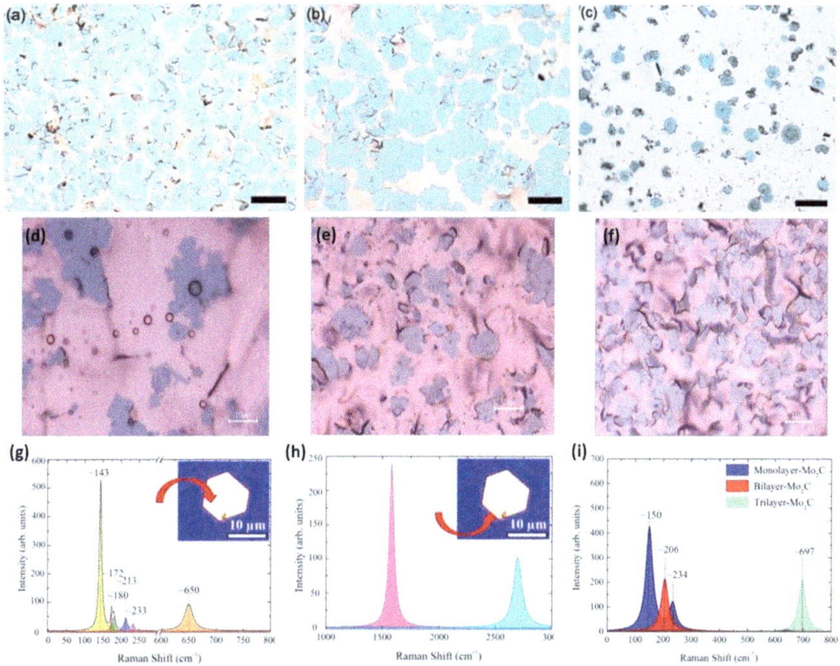

Fig. 2.9 MXene flake by changing the precursor flow rate; **a** under 1 sccm CH_4, **b** between 1 and 5 sccm CH_4, **c** above 5 sccm CH_4 (Scale Bar: 5 μm) [51], and **d** 80 sccm H_2, **e** 100 sccm H_2 **f** 120 sccm H_2 (Scale Bar: 10 μm). **g** Real state micro-Raman peaks of Mo_2C and **h** graphene micro-Raman peaks in context and **i** the calculated Mo_2C peaks [51]

effect of hydrogen flux on the lateral growth of flakes and eroded copper exposed to high hydrogen flows.

2.3.1.5 Operation Temperature

The dissolution of precursor atoms within the molten catalyst plays an important role in atomic diffusion and reaction with accumulated transition metal atoms. In addition, increasing the growth temperature promotes gas dissolution in the liquid metal. Melting point and viscosity in the molten state are the most important factors in varying metal catalyst type [56]. The increased wettability of the metal catalyst is beneficial for increasing flake coverage, but the high viscosity is disadvantageous for the MXene growth process. The lower melting point of some metallic catalysts, such as Au and In, affects their catalytic ability and determines the growth temperature of MXenes. It is evident that increasing the process temperature causes more metal diffusion due to increased atomic mobility. In addition, higher temperature (above the melting point of the catalyst metal) promotes the diffusion of transition metal atoms into the solid/liquid interface, increasing the probability of forming a carbide/nitride heterostructure [56, 57].

Additionally, the surface energy of the molten metal increases directly with temperature, which results in an increase in atomic dispersion and excessive precursor activity. By applying higher temperatures to the flakes, the vertical growth rate changed, resulting in large pyramidal growth regions with clearly defined terraces [61].

2.3.1.6 Growth Time

MXene structures, which are grown at different process durations, form in isolated flake structures and they are transformed into continuous films when the duration is increased. However, the prolonged deposition process allows more Mo atoms to enter the first-grown graphene domains and combine through interfacial diffusion to Mo_2C with hydrocarbon interaction. Typically, hexagonal Mo_2C flakes with a lateral range of about 50 μm appear after 30 min and expand to 100 μm after 60 min [53]. Figure 2.9g–i provides that the formation of graphene in context is almost inevitable due to the gradual temperature change.

CVD offers simplicity and diverse development opportunities as a bottom-up approach by tuning multiple parameters for the synthesis of 2D MXene structures with high potential for use in a wide variety of future electronic and optoelectronic applications. However, it is still challenging to obtain large-area and high-quality 2D MXene crystals with controlled thicknesses. The combination of multiple purified precursors and temperature ranges results in a scalable MXene synthesis with tunable thicknesses and excellent physical and chemical properties. Temperature and gas flow rate are among the most studied process parameters for CVD growth of 2D MXene crystals. But catalytic metals formed in shots (In, Ga) or as foils (Cu, Au) are placed on the surface of source transition metals, where such studies show the importance of the catalyzers for controllable MXene formation. In all situations, the thickness of the 2D MXene flakes decreases with increasing catalyst thickness. In addition, increasing the reaction time by choosing a lower melting point catalyst reduces the vertical growth rate.

2.3.2 Other Bottom-Up Methods to Produce 2D MXene

Bottom-up approaches are not limited to CVD where self-assembly, plasma-enhanced pulsed laser deposition (PEPLD), and template-assisted growth are also possible strategies to obtain 2D MXene structures [3]. Transition metal carbides (TMCs) and nitrides (TMNs) mainly consist of the crystalline structures of their bulk counterparts and the bottom-up ultrathin synthesis strategies are able to control their superior properties, in general avoiding functional groups. For example, as fabricated 2D MXene structures, the Mo_2C crystals show three crystalline structures including hexagonal (β), orthorhombic (α), and cubic phase allotropes. Occupying carbon atoms within the Mo interstitial octahedrons is a factor responsible for the phase variation. The PEPLD method is highly reliable to control the crystal structure and obtain multiphase allotropes [63].

A further advantage of interfacial self-assembly is that it allows high-performance composites to be organized into hierarchically ordered structures, for example, drug delivery, magnetic materials, or energy storage—without the risk of oxidation of MXene layers. The MXene sandwich composite structures such as Co-NiS/MXene nanosheets and nanorods constructed by self-assembly include hydrogen bonds within the structure and show good electrochemical performance for lithium and sodium ion storage [64].

The template-assisted growth approach essentially uses a metallic framework that could be removed by chemical etching to obtain a free-standing 3D MXene architecture. This method offers the possibility to generate macroscopic 3D MXene structures with high gravimetric capacity. This also allows the hydrophilicity of MXene to be tuned by exchanging H^+ with a cationic surfactant. In addition, the template-assisted growth MXene/binder polymerization leads to the formation of a porous monolithic cell structure [65]. Obviously, the main purpose of all bottom-up approaches is precise growth control of structure formation, and the resulting composition is a direct function of the synchronization of several growth parameters.

References

1. J. Zhang, N. Kong, S. Uzun et al., Scalable manufacturing of free-standing, strong $Ti_3C_2T_x$ MXene films with outstanding conductivity. Adv. Mater. **32**, 2001093 (2020). https://doi.org/10.1002/adma.202001093
2. J. Zhu, J. Zhang, R. Lin et al., Rapid one-step scalable microwave synthesis of Ti_3C_2T: XMXene. Chem. Commun. **57** (2021). https://doi.org/10.1039/d1cc04989e
3. L. Jiang, D. Zhou, J. Yang et al., 2D single- and few-layered MXenes: synthesis, applications and perspectives. J. Mater. Chem. A **10**, 13651–13672 (2022). https://doi.org/10.1039/D2TA01572B
4. M. Naguib, M. Kurtoglu, V. Presser et al., Two-dimensional nanocrystals produced by exfoliation of Ti_3AlC_2. Adv. Mater. **23** (2011). https://doi.org/10.1002/adma.201102306
5. F. Chang, C. Li, J. Yang et al., Synthesis of a new graphene-like transition metal carbide by de-intercalating Ti_3AlC_2. Mater. Lett. **109** (2013). https://doi.org/10.1016/j.matlet.2013.07.102
6. Y. Liu, X. Zhang, S. Dong et al., Synthesis and tribological property of $Ti_3C_2T_X$ nanosheets. J. Mater. Sci. **52** (2017). https://doi.org/10.1007/s10853-016-0509-0
7. M. Naguib, O. Mashtalir, J. Carle et al., Two-dimensional transition metal carbides. ACS Nano **6**, 1322–1331 (2012). https://doi.org/10.1021/nn204153h
8. J. Zhou, X. Zha, X. Zhou et al., Synthesis and electrochemical properties of two-dimensional Hafnium carbide. ACS Nano **11** (2017). https://doi.org/10.1021/acsnano.7b00030
9. B. Anasori, Y. Xie, M. Beidaghi et al., Two-dimensional, ordered, double transition metals carbides (MXenes). ACS Nano **9** (2015). https://doi.org/10.1021/acsnano.5b03591
10. P. Nayak, Q. Jiang, R. Mohanraman et al., Inherent electrochemistry and charge transfer properties of few-layered two-dimensional $Ti_3C_2T_x$ MXene. Nanoscale **10** (2018). https://doi.org/10.1039/c8nr01883a

11. A. Shayesteh Zeraati, S.A. Mirkhani, P. Sun et al., Improved synthesis of Ti_3C_2T: XMXenes resulting in exceptional electrical conductivity, high synthesis yield, and enhanced capacitance. Nanoscale **13** (2021). https://doi.org/10.1039/d0nr06671k
12. Y. Guo, S. Jin, L. Wang et al., Synthesis of two-dimensional carbide Mo_2CT_x MXene by hydrothermal etching with fluorides and its thermal stability. Ceram. Int. **46**, 19550–19556 (2020). https://doi.org/10.1016/j.ceramint.2020.05.008
13. L. Wang, H. Zhang, B. Wang et al., Synthesis and electrochemical performance of $Ti_3C_2T_x$ with hydrothermal process. Electron. Mater. Lett. **12**, 702–710 (2016). https://doi.org/10.1007/s13391-016-6088-z
14. X. Chen, X. Zhai, J. Hou et al., Tunable nitrogen-doped delaminated 2D MXene obtained by NH_3/Ar plasma treatment as highly efficient hydrogen and oxygen evolution reaction electrocatalyst. Chem. Eng. J. **420** (2021). https://doi.org/10.1016/j.cej.2021.129832
15. A. El Ghazaly, H. Ahmed, A.R. Rezk et al., Ultrafast, one-step, salt-solution-based acoustic synthesis of Ti_3C_2MXene. ACS Nano **15** (2021). https://doi.org/10.1021/acsnano.0c07242
16. L. Li, G. Li, L. Tan et al., Highly efficiently delaminated single-layered MXene nanosheets with large lateral size. Langmuir **33** (2017). https://doi.org/10.1021/acs.langmuir.7b01339
17. T. Li, L. Yao, Q. Liu et al., Fluorine-free synthesis of high-purity $Ti_3C_2T_x$ (T = OH, O) via alkali treatment. Angew. Chem. Int. Ed. **57** (2018). https://doi.org/10.1002/anie.201800887
18. X. Xie, Y. Xue, L. Li, et al., Surface Al leached Ti_3AlC_2 as a substitute for carbon for use as a catalyst support in a harsh corrosive electrochemical system. Nanoscale **6** (2014). https://doi.org/10.1039/c4nr02080d
19. J. Xuan, Z. Wang, Y. Chen, et al., Organic-base-driven intercalation and delamination for the production of functionalized titanium carbide nanosheets with superior photothermal therapeutic performance. Angew. Chem. Int. Ed. **55** (2016). https://doi.org/10.1002/anie.201606643
20. T. Li, X. Yan, L. Huang, et al., Fluorine-free Ti_3C_2T: X (T = O, OH) nanosheets (~50–100 nm) for nitrogen fixation under ambient conditions. J. Mater. Chem. A **7** (2019). https://doi.org/10.1039/c9ta03254a
21. Y. Wei, P. Zhang, R.A. Soomro et al., Advances in the synthesis of 2D MXenes. Adv. Mater. **33**, 2103148 (2021). https://doi.org/10.1002/adma.202103148
22. P. Urbankowski, B. Anasori, T. Makaryan et al., Synthesis of two-dimensional titanium nitride Ti_4N_3 (MXene). Nanoscale **8** (2016). https://doi.org/10.1039/c6nr02253g
23. Y. Li, H. Shao, Z. Lin et al., A general Lewis acidic etching route for preparing MXenes with enhanced electrochemical performance in non-aqueous electrolyte. Nat. Mater. **19**, 894–899 (2020). https://doi.org/10.1038/s41563-020-0657-0
24. M. Li, J. Lu, K. Luo et al., Element replacement approach by reaction with Lewis acidic molten salts to synthesize nanolaminated MAX phases and MXenes. J. Am. Chem. Soc. **141**, 4730–4737 (2019). https://doi.org/10.1021/jacs.9b00574
25. M. Shen, W. Jiang, K. Liang et al., One-pot green process to synthesize MXene with controllable surface terminations using Molten salts. Angew. Chem. Int. Ed. **60**, 27013–27018 (2021). https://doi.org/10.1002/anie.202110640
26. W. Sun, S.A. Shah, Y. Chen et al., Electrochemical etching of Ti_2AlC to Ti_2CT:X (MXene) in low-concentration hydrochloric acid solution. J. Mater. Chem. A **5** (2017). https://doi.org/10.1039/c7ta05574a
27. S. Yang, P. Zhang, F. Wang et al., Fluoride-free synthesis of two-dimensional titanium carbide (MXene) using a binary aqueous system. Angew. Chem. Int. Ed. **57** (2018). https://doi.org/10.1002/anie.201809662
28. S.Y. Pang, Y.T. Wong, S. Yuan et al., Universal strategy for HF-free facile and rapid synthesis of two-dimensional MXenes as multifunctional energy materials. J. Am. Chem. Soc. **141** (2019). https://doi.org/10.1021/jacs.9b02578
29. R.A. Vaia, A. Jawaid, A. Hassan et al., Halogen etch of Ti_3AlC_2 MAX phase for MXene fabrication. ACS Nano **15** (2021). https://doi.org/10.1021/acsnano.0c08630
30. H. Shi, P. Zhang, Z. Liu et al., Ambient-stable two-dimensional titanium carbide (MXene) enabled by iodine etching. Angew. Chem. Int. Ed. **60** (2021). https://doi.org/10.1002/anie.202015627

31. J. Mei, G.A. Ayoko, C. Hu, Z. Sun, Thermal reduction of sulfur-containing MAX phase for MXene production. Chem. Eng. J. **395**, 125111 (2020). https://doi.org/10.1016/j.cej.2020.125111
32. J. Mei, G.A. Ayoko, C. Hu et al., Two-dimensional fluorine-free mesoporous Mo_2C MXene via UV-induced selective etching of Mo_2Ga_2C for energy storage. Sustain. Mater. Technol. **25** (2020). https://doi.org/10.1016/j.susmat.2020.e00156
33. S. Zada, W. Dai, Z. Kai et al., Algae extraction controllable delamination of vanadium carbide nanosheets with enhanced near-infrared photothermal performance. Angew. Chem. Int. Ed. **59** (2020). https://doi.org/10.1002/anie.201916748
34. Y. Wu, P. Nie, J. Wang et al., Few-layer MXenes delaminated via high-energy mechanical milling for enhanced sodium-ion batteries performance. ACS Appl. Mater. Interfaces **9**, 39610–39617 (2017). https://doi.org/10.1021/acsami.7b12155
35. L. He, J. Wang, B. Wang et al., Large-scale production of simultaneously exfoliated and functionalized Mxenes as promising flame retardant for polyurethane. Compos. B Eng. **179** (2019). https://doi.org/10.1016/j.compositesb.2019.107486
36. M. Ghidiu, M.R. Lukatskaya, M.-Q. Zhao et al., Conductive two-dimensional titanium carbide 'clay' with high volumetric capacitance. Nature **516**, 78–81 (2014). https://doi.org/10.1038/nature13970
37. Z. Ling, C.E. Ren, M.Q. Zhao et al., Flexible and conductive MXene films and nanocomposites with high capacitance. Proc. Natl. Acad. Sci. U.S.A. **111** (2014). https://doi.org/10.1073/pnas.1414215111
38. H. Kim, B. Anasori, Y. Gogotsi, H.N. Alshareef, Thermoelectric properties of two-dimensional molybdenum-based MXenes. Chem. Mater. **29**, 6472–6479 (2017). https://doi.org/10.1021/acs.chemmater.7b02056
39. C.E. Shuck, A. Sarycheva, M. Anayee et al., Scalable synthesis of Ti_3C_2Tx MXene. Adv. Eng. Mater. **22** (2020). https://doi.org/10.1002/adem.201901241
40. A.D. Dillon, M.J. Ghidiu, A.L. Krick et al., Highly conductive optical quality solution-processed films of 2D titanium carbide. Adv. Func. Mater. **26** (2016). https://doi.org/10.1002/adfm.201600357
41. K. Hantanasirisakul, M.Q. Zhao, P. Urbankowski et al., Fabrication of Ti_3C_2Tx MXene transparent thin films with tunable optoelectronic properties. Adv. Electron. Mater. **2** (2016). https://doi.org/10.1002/aelm.201600050
42. J. Zhang, N. Kong, S. Uzun et al., Scalable manufacturing of free-standing, strong Ti_3C_2Tx MXene films with outstanding conductivity. Adv. Mater. **32** (2020). https://doi.org/10.1002/adma.202001093
43. Y. Fan, L. Li, Y. Zhang et al., Recent advances in growth of transition metal carbides and nitrides (MXenes) crystals. Adv. Func. Mater. **32**, 2111357 (2022). https://doi.org/10.1002/adfm.202111357
44. M. Hassan, G. Abbas, N. Li et al., Significance of flexible substrates for wearable and implantable devices: recent advances and perspectives. Adv. Mater. Technol. **7**, 1–45 (2022). https://doi.org/10.1002/admt.202100773
45. J. Chen, Y. Ding, D. Yan et al., Synthesis of MXene and its application for zinc-ion storage. SusMat **2**, 293–318 (2022). https://doi.org/10.1002/sus2.57
46. H. Şar, A. Özden, B. Yorulmaz et al., A comparative device performance assessment of CVD grown MoS_2 and WS_2 monolayers. J. Mater. Sci. Mater. Electron. **29**, 8785–8792 (2018). https://doi.org/10.1007/s10854-018-8895-5
47. R. Yang, X. Chen, W. Ke, X. Wu, Recent research progress in the structure, fabrication, and application of MXene-based heterostructures. Nanomaterials **12**, 1907 (2022). https://doi.org/10.3390/nano12111907
48. S. Chaitoglou, P. Tsipas, T. Speliotis et al., Insight and control of the chemical vapor deposition growth parameters and morphological characteristics of graphene/Mo_2C heterostructures over liquid catalyst. J. Cryst. Growth **495**, 46–53 (2018). https://doi.org/10.1016/j.jcrysgro.2018.05.015

49. Y. Ma, S. Lu, G. Han et al., Chemical vapor deposition of two-dimensional molybdenum nitride/graphene van der Waals heterostructure with enhanced electrocatalytic hydrogen evolution performance. Appl. Surf. Sci. **589**, 152934 (2022). https://doi.org/10.1016/j.apsusc.2022.152934
50. S. Biswas, P.S. Alegaonkar, MXene: evolutions in chemical synthesis and recent advances in applications. Surfaces **5**, 1–36 (2021). https://doi.org/10.3390/surfaces5010001
51. M. Öper, U. Yorulmaz, C. Sevik et al., Controlled CVD growth of ultrathin Mo_2C (MXene) flakes. J. Appl. Phys. **131** (2022). https://doi.org/10.1063/5.0067970
52. F. Jamil, H.M. Ali, M.M. Janjua, MXene based advanced materials for thermal energy storage: a recent review. J. Energy Storage **35**, 102322 (2021). https://doi.org/10.1016/j.est.2021.102322
53. D. Geng, X. Zhao, Z. Chen et al., Direct synthesis of large-area 2D Mo_2C on in situ grown graphene. Adv. Mater. **29**, 1700072 (2017). https://doi.org/10.1002/adma.201700072
54. S. Rouhi, J.E. Martinez-Medina, M. Ozdemir et al., A comprehensive study of molybdenum boats behavior during service life for continuous thermal evaporation technique, used in thin film technology. Vacuum **176**, 109167 (2020). https://doi.org/10.1016/j.vacuum.2020.109167
55. S. Chaitoglou, T. Giannakopoulou, T. Speliotis et al., Mo_2C/graphene heterostructures: low temperature chemical vapor deposition on liquid bimetallic Sn–Cu and hydrogen evolution reaction electrocatalytic properties. Nanotechnology **30**, 125401 (2019). https://doi.org/10.1088/1361-6528/aaf9e8
56. O.R. Caylan, G. Cambaz Buke, Low-temperature synthesis and growth model of thin Mo_2C crystals on indium. Sci. Rep. **11**, 1–7 (2021). https://doi.org/10.1038/s41598-021-87660-7
57. W. Sun, X. Wang, J. Feng et al., Controlled synthesis of 2D Mo_2C/graphene heterostructure on liquid Au substrates as enhanced electrocatalytic electrodes. Nanotechnology **30** (2019). https://doi.org/10.1088/1361-6528/ab2c0d
58. D. Geng, X. Zhao, L. Li et al., Controlled growth of ultrathin Mo_2C superconducting crystals on liquid Cu surface. 2D Mater. **4** (2017). https://doi.org/10.1088/2053-1583/aa51b7
59. M. Shao, Y. Shao, W. Chen et al., Efficient nitrogen fixation to ammonia on MXenes. Phys. Chem. Chem. Phys. **20**, 14504–14512 (2018). https://doi.org/10.1039/c8cp01396a
60. J. Zhang, F. Wang, V.B. Shenoy et al., Towards controlled synthesis of 2D crystals by chemical vapor deposition (CVD). Mater. Today **40**, 132–139 (2020). https://doi.org/10.1016/j.mattod.2020.06.012
61. Y. Fan, L. Huang, D. Geng, W. Hu, Controlled growth of Mo_2C pyramids on liquid Cu surface. J. Semicond. **41** (2020). https://doi.org/10.1088/1674-4926/41/8/082001
62. M. Oper, N. Kosku Perkgöz, Temperature dependent (83–483 K) Raman spectroscopy analysis of CVD Grown Ws_2 monolayers. Eskişehir Technical University J. Sci. Technol. A – Appl. Sci. Eng. **21**, 155–164 (2020). https://doi.org/10.18038/estubtda.675907
63. B. Anasori, Y. Gogotsi, *2D Metal Carbides and Nitrides (MXenes).* Springer International Publishing, Cham
64. Z. Zou, Q. Wang, J. Yan et al., Versatile interfacial self-assembly of $Ti_3C_2T_x$ MXene based composites with enhanced kinetics for superior lithium and sodium storage. ACS Nano **15** (2021). https://doi.org/10.1021/acsnano.1c03516
65. G. Tontini, M. Greaves, S. Ghosh et al., MXene-based 3D porous macrostructures for electrochemical energy storage. J. Phys. Mater. **3** (2020). https://doi.org/10.1088/2515-7639/ab78f1

Chapter 3
Properties of MXene

Shanli Nezami, Farzad Moazami, Maryam Helmi, Alireza Hemmati, and Ahad Ghaemi

3.1 Introduction

MXene is a group of two-dimensional materials of intermediate metal carbides and nitrides, which are the newest and most famous two-dimensional materials. In 2011, the first sample of MXene was produced and identified by selective extraction of layered crystal structures. MXene is derived from the selective removal of "A" from the MAX phase, which is added to the end of its name to emphasize its family with two-dimensional graphene materials [1]. MAX phase is a large family of ternary nitrides and carbides with the chemical formula $M_{n+1}AX_n$ (n = 1, 2, 3) and a hexagonal layered structure (M_2AX, M_3AX_2 or M_4AX_3, etc.). In this chemical formula, "A" represents an intermediary metal more than groups 13 and 14 of the periodic table, "X" is nitrogen, carbon, or both. So far, about 70 different MXene have been made experimentally. But theoretically, the structure and characteristics of dozens of cases have been predicted. Interestingly, recent research has shown that the surface of MXene is randomly covered by hydroxyl, oxygen, or fluorine, and the ratio of each of these compounds or elements is mainly determined by the synthesis method. Also, the comprehensive chemistry of MXene allows their properties to be suitable for specific applications such as energy storage, protection against electromagnetic interference, water, gas, biosensors, lubrication, and catalysis. The high electrical conductivity of MXene, diverse stacked structures, and controllable surface coatings also provide an attractive path for modifying material structure, properties, and sensing performance [1–4]. Figure 3.1 presents the initial reports of the structure of non-polar MXene.

The structure of MXene is a closed structure like the main MAX phase, where X atoms are placed in the octahedral interstitial positions. MXene has a hexagonal

---

S. Nezami · F. Moazami · M. Helmi · A. Hemmati (✉) · A. Ghaemi
School of Chemical, Petroleum and Gas Engineering, Iran University of Science and Technology, Tehran, Iran
e-mail: alireza_hemmati@iust.ac.ir

© The Author(s), under exclusive license to Springer Nature Singapore Pte Ltd. 2024
O. P. Pandey and P. Sharma (eds.), *MXenes: Emerging 2D Materials*,
https://doi.org/10.1007/978-981-97-4064-2_3

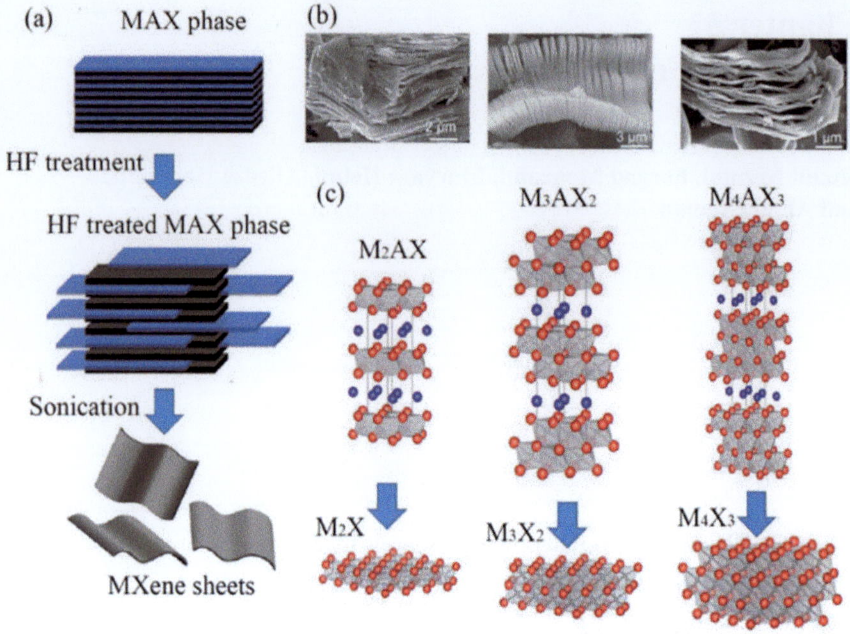

Fig. 3.1 **a** Report of three different MXene structures (inconclusive): M_2X, M_3X_2, and M_4X_3. **b** SEM images of Ti_2AlC, Ti_3AlC_2, and Ta_4AlC_3 after washing with HF (from right to left). **c** Schematic picture of MXene synthesis from MAX stages, The structure of MXenes. Adapted with permission from [1]. Copyright (2019), American Chemical Society

structure (HCP) in which M atoms are arranged in the ABABAB sequence in M_2X and ABCABC in M_3C_2 and M_4C_3(Similar to the cubic structure). The surface of the MXene layers is covered with acidic solutions containing fluoride with O^-, OH^- and F^- functional groups due to the synthesis and exfoliation of the A layer from the main MAX phase (Fig. 3.2a). The functional group on the surface of the MXene layers is determined by "T_x" with the formula $M_{n+1}X_nT_x$ MXene. Computational methods such as density functional theory (DFT) are used to understand the relationship between the properties and structure of MXene. These methods provide thermodynamic information to predict materials' chemical and physical properties at the atomic scale. In the computational study of MXene properties, the assumption of complete coverage of the MXene surface with a specific surface terminal such as ^-O, OH^- or ^-F is usually used. The theoretical calculation study proved that the structural stability of MXene is related to its surface performance. Thermodynamically, when the surface is formed, the formation energy decreases. For MXene Ti_2CT_x, a Ti_2CO_2 structure fully covered with oxygen is predicted to be more stable than partially oxygenated structures (Fig. 3.2a, c). A DFT study identifies other favorable Tx energy positions in $Ti_3C_2T_x$. Figure 3.2 shows the three types of structures of ^-OH groups in $Ti_3C_2T_x$. ^-OH groups are located above the hollow sites between the C atoms of structure I. In structure II, this functional group is placed higher than the C atom on both sides of the layers. While structure III is a mixed structure of both structures I and II, in terms of the ^{-}OH arrangement, the structural stability of these structures is estimated by comparing their DFT relative total energy. The results show that structure I was the most stable structure. While the II structure has the least stability due to the steric repulsion between the T_x groups and the underlying C layers [5–9].

In addition, DFT calculations show that F^{-} functional groups prefer to be in the hollow places between three C atoms and next to each other, and both OH^{H} and F^{-} end groups probably follow the I structure. The ratio of these functional groups on the surfaces of MXene will depend on the synthesis method and mainly the type of exfoliating solution. Hope and his colleagues investigated the performance of the $Ti_3C_2T_x$ surface using nuclear magnetic resonance (NMR) spectroscopy. In this investigation, two methods have been used for layering the samples. In the first method, 50% HF aqueous solution was used, and in the second method, LiF + HCl solution was used for layer removal. NMR analysis has quantitatively determined the random H and F surface terminals of $Ti_3C_2T_x$ MXene. Figure 3.2d shows that

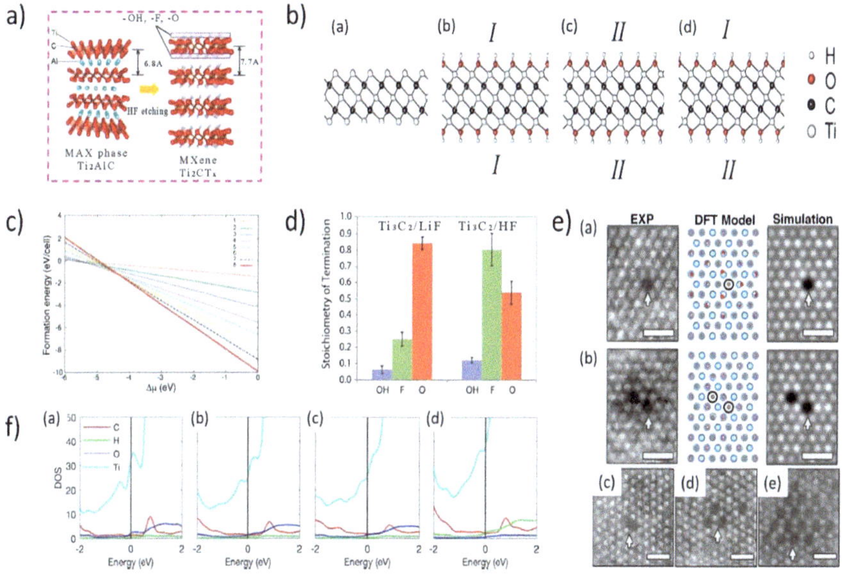

Fig. 3.2 **a** Schematic of Ti_2CT_x MXenes formation from Ti_2AlC [10]. **b** different structure layouts of MXenes surface atoms of: (a) Ti_3C_2, (b) I-Ti_3C_2 $(OH)_2$, (c) II-$Ti_2C(OH)_2$, and (d) III-$Ti_3C_2(OH)_2$ [11]. **c** The formation energy of the Ti_2C model containing 1–8 oxygen atoms as a function of the chemical potential of oxygen [3]. **d** The composition of functional groups on the $Ti_3C_2T_x$ surface resulting from Ti_3AlC_2 etching in HF and LiF + HCl solutions [5]. **f** HAADF STEM images of deficiency in $Ti_3C_2T_x$ monolayer. Comparison of experimental images of HAADF-STEM, crystal structure determined from DFT and simulated HAADF-STEM of VTi [12]. **e** Two adjacent VTi in the same substrate, Calculated DOS of OH monolayer Ti_3C_2: (a) uncoated, and coated with (b) V_2Ti, (c) V_4Ti, (d) V_6Ti [13]

after drying the samples in a vacuum, the proportion of the OH- functional group is relatively lower than other functional groups O^- and F^-. Also, washing the surface of the samples creates an obvious difference in the stoichiometric ratio of the surface functional groups of MXenes. In the synthesis with HF, the functional group F^- and in the synthesis with LiF + HCl, the ratio of the functional group O^- is very high and a small amount of OH^- and F^- is created on the surface of MXene. In addition, the type of washing solutions and their concentration also affect the number of injuries. It is noteworthy that laboratory MXene obtained with surface damage is not structurally complete. Such defects usually occur during washing with acid are created, affecting local surface chemistry [9, 14–16].

3.2 MXene Synthesis

Two-dimensional MXenes are synthesized as MAX phase by extracting and selectively exfoliating a large group of intermediate metal carbides and nitrides of the third group with a layered hexagonal structure. The MAX phase consists of layers of carbides or nitrides ($M_{n+1}X_n$) of the intermediate metal, which is mixed with layers of atoms of element A with the general formula $M_{n+1}AX_n$, where M is a metal.

The synthesis of MXene is carried out by selective exfoliation of A layers from the MAX phase using hydrofluoric acid or a solution containing it, which is the main method of MXene synthesis. Exfoliation of the A layer occurs due to the weak M-A bond compared to the bond between the M-X layers and the higher chemical reactivity of these layers.

$$M_{(n+1)}\,AlX_n + 3HF = M_{(n+1)}\,X_n + AlF_3 + 1.5H_2 \tag{3.1}$$

$$M_{n+1}\,X_n + 2H_2O = M_{n+1}X_n(OH)_2 + H_2 \tag{3.2}$$

$$M_{n+1}X_n + 2HF = M_{n+1}X_nF_2 + H_2 \tag{3.3}$$

In reactions 3.1–3.3 above, the typical reaction of the exfoliation process based on HF solution for the synthesis of MXene is presented. When the HF solution reacts with the MAX phase powders, hydrogen gas (H_2) is produced, which is usually observed during the experiments. After reaction 3.1, reactions 3.2 and 3.3 occur simultaneously. After the completion of these reactions, the resulting powders are multi-layer MXene particles. The exfoliation conditions with HF change with the atomic bonds of the materials [2, 17–22]. Usually, Ti-containing MXenes need weaker exfoliation than Cr-, Mo-, Nb-, Ta-, V-, and Zr-containing MXene. For example, Ti_2CT_x can be produced with 10 wt% HF in 10 h exfoliation. However, V_2CT_x with 50% by weight of HF requires 90 h. After exfoliation, these multi-layered powders are transformed into multi-layered nanosheets through the ultrasonic process disruption of large organic molecules. The prepared MXene is suitable for many applications. For example, active reaction sites are suitable and useful for sensor applications. Liquid exfoliation methods have been widely used to make layered MXene. Organic molecules, such as dimethyl sulfoxide (DMSO), isopropyl amine, chloroform, and tetrabutylammonium hydroxide (TBAOH), are converted into multiple layers of MXene, which increases the distance between the layers and weakens the interactions between the layers. It becomes M-X. Ultrasonic or mechanical vibration of multi-layer MXene is often used to produce single-layer or low-layer MXene. It should be noted that for the formation of layered MXene by wet chemical method, their surface is replaced with functional groups such as O-, OH- and -F groups. Acidic agents containing fluoride such as NH_4HF_2 and LiF + HCl can also be used instead of etching with HF. Through reaction 4, HF is formed. Then, reactions 1–3 follow [20, 21].

$$LiF + HCl = HF + LiCl \tag{3.4}$$

Acids containing fluoride eliminate the need for dangerous HF and the intermediate process. Because water molecules are transformed simultaneously in the structure of layers and as a result, layered MXene is easily formed. In this process, both Li^+ and NH^{+4} ions can be placed in the MXene layers and the distance between the layers can be expanded. Figure 3.3 shows the schematic of the synthesis and morphological characteristics of MXene. In short, MAX powders in acid selectively remove the A layer from their 3D structure [23].

The produced powders have a typical accordion-like morphology with a multi-layered structure in which organic molecules are converted into a multi-layered

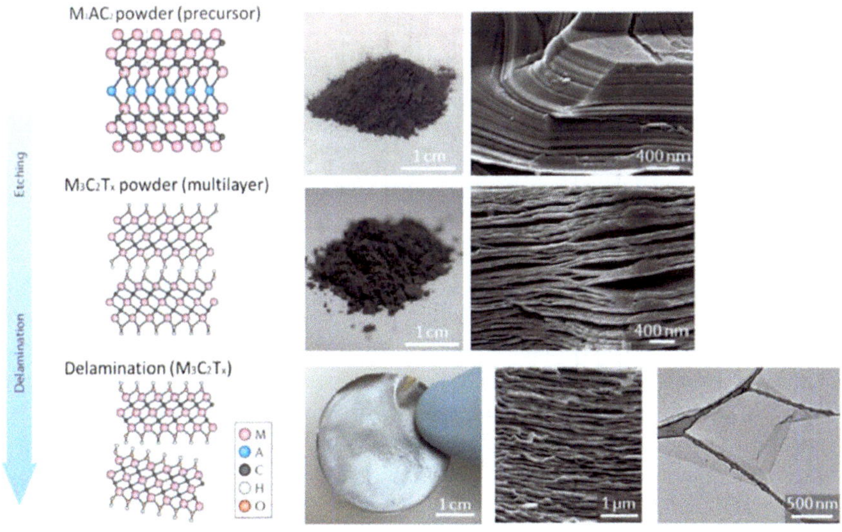

Fig. 3.3 Synthesis of MXenes. Image and images of MAX structures, multilayer MXenes, and layered MXenes [23]

structure to reduce interactions between MXene monolayers. Then, layered MXenes are dispersed in water by ultrasonication, and single or multilayered MXenes are separated by filtration. In contrast to the successful synthesis of MXenes from the wet exfoliation process, MXene layers are often chemically unstable due to prompt oxidation in the solution. Exfoliation methods must be carefully selected for a specific MXene compound and its applications. The intrinsic structure and chemical composition of MXene are derived from transition metals with d orbitals, which show attractive physical and chemical properties. Two-dimensional MXene has high electrical conductivity like the MAX phase due to the presence of intermediate metal, carbon and nitrogen in its structure. However, some MXene exhibits remarkable semiconducting behavior. These materials can be widely used in chemical gas sensors. Chemical synthesis significantly affects the type and number of surface terminals and the electrical properties of MXene. The surface performance is one of the attractive and unique features of MXenes, which can be adjusted by the synthesis method.

3.3 Properties of MXenes

The theoretical calculations show that the formation energy of M_2X, where X is carbon or nitrogen, is unchanged in the d orbitals of the intermediate metal (M), which shows metallic behavior. Most MXenes have p orbitals of carbon or nitrogen (X) below the d orbitals, which is a small bond separating them. By functionalizing the surface, new energy bands are created lower than the Fermi energy in which the p-orbitals of the F^- or O^- terminal are combined with the d-orbitals of the intermediate metal. Using simulation, it can also be predicted that the two terminals ^-F and ^-OH had a similar effect on the electronic structure at virgin and untouched MXenes because one electron can be transferred to each of them, while two electrons can be received with the O^- terminal. Figure .3.4 shows the prediction of the band structure and the density of states of Ti_2C, Ti_2CF_2, and Ti_2CO_2 in their conversion to semiconductors because functionalization transfers Fermi energy to the center of the gap in between It transfers d orbitals from M and p orbitals from X orbit. For example, the estimated energy gaps for Sc_2CT_x with F^-, OH^-, and ^-O terminals change to 1.03, 0.45, and 1.8 eV, respectively.

Apart from surface termination, the structure of electronic and transport properties of MXenes can be controlled as a function of the number of layers. The electronic structure and transport properties of M_2CO_2 multilayers (M = Ti, Zr, Hf) were evaluated using fundamental-principal methods. Considering the vertical electric field, due to the weak bond between the layers, the distance of different bands of M_2CO_2 decreases to some extent. In the I–V curve of p-n multilayers, Ti_2CO_2 illustrates higher conductivity with a lower starting voltage than others and Hf_2CO_2 shows stronger rectifying behavior.

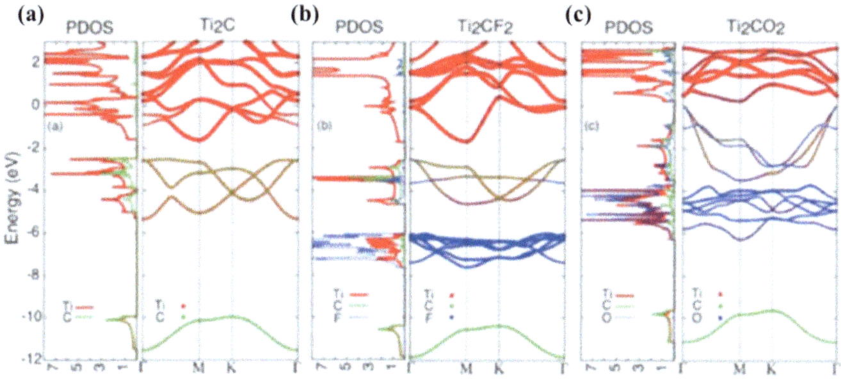

Fig. 3.4 Condensed states' prediction (different voltage states in each cell) and band structures' prediction for **a** Ti_2C, **b** Ti_2CF_2, and **c** Ti_2CO_2 where the Fermi energy is zero [24]

3.4 Mechanical Properties

The presence of free electrons attended with transition metal atoms indicates a rich and tunable surface chemistry for MXenes. Parameters such as MXene precursor, etchant solution, exfoliation method, and applied strategy to increase the distance between layers are very effective on structural properties, surface chemistry, magnetic properties, as well as chemical properties such as oxidation rate and environmental stability.

One of the most important parameters affecting the mechanical stability of MXenes is the strong bond between M-X and surface terminals (O^-, F^-, and OH^-). In other words, their mechanical stability can be increased by increasing the surface functional groups by functionalizing MXenes layers. Following this increase in mechanical stability, the breakdown of the atomic layer decreases [25, 26]. Based on theoretical DFT calculations, MXenes have a higher elastic modulus than their MAX precursor. Many researchers have investigated this issue based on experimental works [27]. Gogotsi and co-workers [28] experimentally showed that the composite prepared from $Ti_3C_2T_x$ with neutral polyvinyl alcohol could withstand a weight of about 15,000 times (Fig. 3.5). Cai et al. [29] also designed a $Ti_3C_2T_x$ carbon nanotube composite with more than 5000 cycles of tensile strength.

The existence of surface functional groups (O^-, F^-, and OH^-) has a decisive role in the magnetic nature of MXenes. There is a strong covalent bond between M and X, and the presence of surface functional groups (O^-, F^-, and OH^-) on the surface of MXenes makes them non-magnetic. By increasing the density of electrons near the d energy-related level with the M transition metal orbital or breaking a strong covalent bond such as the M-M and M-A bonds due to external plane strain, some MXenes can behave magnetically [30]. It has been reported that 2D Ti_3N_2 and Ti_3C_2 are antiferromagnetic Cr_2C and Cr_2N are ferromagnetic [31]. Due to their functionalization, the presence of functional groups on the layers of MXenes leads

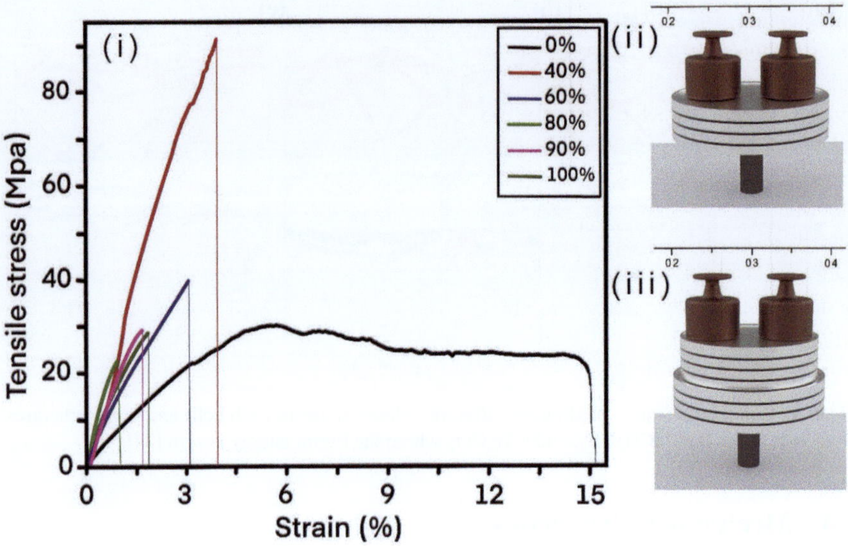

Fig. 3.5 Mechanical properties of flexible free-standing $Ti_3C_2T_x$, $Ti_3C_2T_x$/PVA, and cast PVA films, (i) stress–strain curves for $Ti_3C_2T_x$ /PVA films with different $Ti_3C_2T_x$ content. (ii) The 6-mm-diameter $Ti_3C_2T_x$ can support ~4000 times its own weight. (iii) The 6-mm-diameter 90% $Ti_3C_2T_x$ can support ~15,000 times its own weight

to a decrease in the electron density near the Fermi energy level and, as a result, the disappearance of the magnetic properties of MXenes.

3.5 Hydrogen Storage Properties

One of the big challenges for scientists and chemists is to obtain materials with excellent hydrogen adsorption/desorption behavior and hydrogen storage materials with high storage capacity, which can be used in environmental conditions. Reachable hydrogen storage materials with weak physical attraction for H_2 due to van der Waals forces, such as MOFs [30], covalent organic frameworks, carbon-based nanotubes [31], fullerenes [32], and graphene [33], can store hydrogen at low temperatures near the stored liquid nitrogen. Meanwhile, metal hydrides [32] need very high temperatures to dispose of the stored hydrogen because H_2 adsorption is done chemically. Recently, the MXene family has been shown to have a high potential for hydrogen storage [33–36]. H atoms in MXenes are confined in three ways: (i) physical adsorption, (ii) chemisorption, and (iii) Kubas-type interactions [33, 37]. In MXenes with hydrogen storage capacity, it is reversible under environmental conditions due to the Kubas-type interaction, whose binding energy is higher than physical adsorption binding energy and lower than chemical binding energy [15, 38]. The most trustworthy method to predict the physicochemical properties of MXenes are preliminary calculations based on DFT because some properties have either not been performed or confirmed by experimental tests. By performing systematic density functional calculations, Hu et al. [35] showed that the maximum storage capacity of hydrogen by Ti_2C_2 and Ti_2C is 8.6% adsorption at different locations and on both sides. Considering that 3.4% by weight of hydrogen molecules are through Kubas-type interaction, the storage is reversible. It can be released at ambient temperature (Fig. 3.5). Osti and Gogotsi et al. [39] investigated the potential of hydrogen storage in ambient conditions for MXene, synthesized $Ti_3C_2T_x$ with HF 48% and showed that about 0.3% of hydrogen weight, adsorption occurs during the synthesis. Also, the mechanisms of hydrogen storage are reported as follows: (i) Due to the two-dimensional crystalic and with a thin layer structure, H atoms can remain stable on the surface and move easily inside to become trapped between MXene layers; (ii) hydrogen inside probs (hollow sites or gaps or defects such as empty spaces and void clusters) due to morphological changes caused by the defect (such as changes caused by the using strong etchant during etching of MAX phase) to be trapped; (iii) H_2 is placed in large gaps or pores, or partially isolated H_2 molecules are located in small structural openings. For the first time, Wu et al. [40] made a 2D composite of Ti_3C_2 MXenes with NaAlH4 and showed that this composite's reversible hydrogen storage capacity is 4.9% by weight of hydrogen below 140 °C. The remarkable catalytic activity of $Nb_4C_3T_x$ with a layered structure leads to a special hydrogen storage capacity. Therefore, MXenes are known as promising candidates for use as hydrogen storage materials. They can be explored for developing high-capacity hydrogen storage materials with their great potential for modification and composite fabrication.

3.6 Electrical and Electronic Properties

Electrical and electronic properties are the two important characteristics of MXenes that have attracted much attention from researchers. These two characteristics can be adjusted by changing the active groups, the formation of solid solution, and the material balance. The electrical conductivity of MXenes with compact layers is approximately the same as that of multilayer graphene [41]. Meanwhile, the electrical conductivity of MXenes is higher than carbon nanotubes and materials extracted from graphene oxide. The electrical conductivity of MXenes functionalized with molecular groups increases [42].

Considering the increasing effect of MXenes synthesis on their electrical conductivity, it is possible to increase the electrical conductivity range of MXenes by minimally consuming HF solution and increasing the exfoliation time with minimal violations in the structure of MXenes and increasing the volumetric expansion [43]. The electrical properties can be modified by modifying the surface functional groups of MXenes through alkaline and thermal methods [44]. A two-fold increase in the conductivity magnitude of MXenes is achieved by adding or subtracting surface features and entangled ions/molecules [45].

3.7 Optical Properties

It is essential for photovoltaic, photocatalytic, transparent, and optically conducting electrode machines' visible/UV light spectrum adsorption. Even with a thickness of 10 nm, $Ti_3C_2T_x$ layers can adsorb Up to 91% energy of light in the visible region of UV from 300 to 500 nm wavelength [46]. Also, regardless of the thickness of the layer, these can adsorb light energy in a wide and strong area between 700 and 800 nm. This adsorption is indicated by the formation of green layers [47]. This amount of adsorption can be optimized by ion convergence and changing the thickness of MXene layers. For example, tetramethylammonium hydroxide increases the transmissive adsorption of $Ti_3C_2T_x$ layers, while DMSO, urea, and hydrazine decrease it [48].

3.8 Oxidative/Thermal Stability

The presence of water and oxygen on the MXene layers during the synthesis is very effective on the stability of MXene. Based on the experimental results, it has been shown that virgin MXenes are decomposed either at higher temperatures in the aqueous environment, by exposure to radiation in the air, or by melting for a long time in the oxygen environment. For example, in the conditions where $Ti_3C_2T_x$ MXene aqueous solution is placed under an Air atmosphere, the layered structure will be decomposed under oxidation and TiO_2 formation. The color of $Ti_3C_2T_x$ MXene changes from black/green to white TiO_2 upon nucleation and formation of anatase or rutile TiO_2. According to reports, the expansion of anatase TiO_2 takes place at temperatures of 500 °C, which transforms into rutile TiO_2 as the temperature increases under the air atmosphere. However, $Ti_3C_2T_x$ MXene is stable under an argon atmosphere up to 1300 °C. It is limited to using techniques such as milling with additional energy in the DMSO component or carbon nano-wetting to increase stability and maintain the original structure and morphology of MXenes. Using techniques such as dry air, degassed oxygen-less water, dark and cold refrigerator environments, or a full vacuum to increase the stability capacity of MXene and reduce oxidation can probably be effective [15, 37, 49].

References

1. Y. Gogotsi, B. Anasori, ACS Nano **13**, 8491 (2019); M. Naguib, V.N. Mochalin, M.W. Barsoum, Y. Gogotsi. Adv. Mater. **26**, 992 (2014)
2. M. Naguib et al., Two-dimensional transition metal carbides. ACS Nano **6**(2), 1322–1331 (2012)
3. M. Khazaei et al., Novel electronic and magnetic properties of two-dimensional transition metal carbides and nitrides. Adv. Func. Mater. **23**(17), 2185–2192 (2013)

4. M.-Q. Zhao et al., 2D titanium carbide and transition metal oxides hybrid electrodes for Li-ion storage. Nano Energy **30**, 603–613 (2016)
5. M.A. Hope et al., NMR reveals the surface functionalisation of Ti_3C_2 MXene. Phys. Chem. Chem. Phys. **18**(7), 5099–5102 (2016)
6. H.-W. Wang et al., Resolving the structure of $Ti_3C_2T_x$ MXenes through multilevel structural modeling of the atomic pair distribution function. Chem. Mater. **28**(1), 349–359 (2016)
7. Y. Xie et al., Role of surface structure on Li-ion energy storage capacity of two-dimensional transition-metal carbides. J. Am. Chem. Soc. **136**(17), 6385–6394 (2014)
8. C. Xu et al., Large-area high-quality 2D ultrathin Mo_2C superconducting crystals. Nat. Mater. **14**(11), 1135–1141 (2015)
9. X.-H. Zha et al., Intrinsic structural, electrical, thermal, and mechanical properties of the promising conductor Mo_2C MXene. J. Phys. Chem. C **120**(28), 15082–15088 (2016)
10. J. Chen et al., Recent progress and advances in the environmental applications of MXene related materials. Nanoscale **12**(6), 3574–3592 (2020)
11. M. Ikram, A. Raza, S. Ali, *2D-Materials for Energy Harvesting and Storage Applications* (Springer, 2022)
12. Z. Wu et al., MXene contact engineering for printed electronics. Adv. Sci. 2207174 (2023)
13. W. Cui et al., Atomic defects, functional groups and properties in MXenes. Chin. Chem. Lett. **32**(1), 339–344 (2021)
14. Y. Zhang et al., Adsorptive environmental applications of MXene nanomaterials: a review. RSC Adv. **8**(36), 19895–19905 (2018)
15. J.A. Kumar et al., Methods of synthesis, characteristics, and environmental applications of MXene: a comprehensive review. Chemosphere **286**, 131607 (2022)
16. Y. Lee, Y. Hwang, Y.-C. Chung, Achieving type I, II, and III heterojunctions using functionalized MXene. ACS Appl. Mater. Interfaces **7**(13), 7163–7169 (2015)
17. M. Naguib et al., Two-dimensional materials: 25th anniversary article: MXenes: a new family of two-dimensional materials (Adv. Mater. 7/2014). Adv. Mater. **26**(7), 982–982 (2014)
18. Y. Gogotsi, B. Anasori, *The Rise of MXenes* (ACS Publications, 2019), pp. 8491–8494
19. M. Naguib et al., Two-dimensional nanocrystals produced by exfoliation of Ti_3AlC_2. Adv. Mater. **23**(37), 4248–4253 (2011)
20. J. Halim et al., Q., E.J. Moon, J. Pitock, J. Nanda, S.J. May, Y. Gogotsi, M.W. Barsoum, Adv. Funct. Mater. **26**, 3118 (2016)
21. M. Alhabeb et al., Guidelines for synthesis and processing of two-dimensional titanium carbide ($Ti_3C_2T_x$ MXene). Chem. Mater. **29**(18), 7633–7644 (2017)
22. M. Naguib et al., New two-dimensional niobium and vanadium carbides as promising materials for Li-ion batteries. J. Am. Chem. Soc. **135**(43), 15966–15969 (2013)
23. B. Anasori, M.R. Lukatskaya, Y. Gogotsi, 2D metal carbides and nitrides (MXenes) for energy storage. Nat. Rev. Mater. **2**(2), 1–17 (2017)
24. Y. Yang, *Electronic and Magnetic Properties of Double Metal MXenes* (Drexel University, 2022)
25. X. Li et al., Functional MXene materials: progress of their applications. Chem. Asian J. **13**(19), 2742–2757 (2018)
26. J. Yang et al., MXene-based composites: synthesis and applications in rechargeable batteries and supercapacitors. Adv. Mater. Interfaces **6**(8), 1802004 (2019)
27. J. Zhu, Recent advance inMXenes: a promising 2Dmaterial for catalysis, sensor and chemical adsorptionCoord. Chem. Rev. **352**, 306–327 (2017)
28. Z. Ling et al., Flexible and conductive MXene films and nanocomposites with high capacitance. Proc. Natl. Acad. Sci. **111**(47), 16676–16681 (2014)
29. Y. Cai et al., Stretchable $Ti_3C_2T_x$ MXene/carbon nanotube composite based strain sensor with ultrahigh sensitivity and tunable sensing range. ACS Nano **12**(1), 56–62 (2018). https://doi.org/10.1021/acsnano.7b06251
30. M. Khazaei et al., Recent advances in MXenes: from fundamentals to applications. Curr. Opin. Solid State Mater. Sci. **23**(3), 164–178 (2019)

31. Y. Cai et al., Stretchable $Ti_3C_2T_x$ MXene/carbon nanotube composite based strain sensor with ultrahigh sensitivity and tunable sensing range. ACS Nano **12**(1), 56–62 (2018)
32. A. Schneemann et al., Nanostructured metal hydrides for hydrogen storage. Chem. Rev. **118**(22), 10775–10839 (2018)
33. Q. Hu et al., MXene: a new family of promising hydrogen storage medium. J. Phys. Chem. A **117**(51), 14253–14260 (2013)
34. Y. Liu et al., Superior catalytic activity derived from a two-dimensional Ti_3C_2 precursor towards the hydrogen storage reaction of magnesium hydride. Chem. Commun. **52**(4), 705–708 (2016)
35. G. Chen et al., Effects of two-dimension MXene Ti_3C_2 on hydrogen storage performances of MgH_2-$LiAlH_4$ composite. Chem. Phys. **522**, 178–187 (2019)
36. K. Xian et al., Superior kinetic and cyclic performance of a 2D titanium carbide incorporated $2LiH^+$ MgB_2 composite toward highly reversible hydrogen storage. ACS Appl. Energy Mater. **2**(7), 4853–4864 (2019)
37. S.K. Hwang et al., MXene: an emerging two-dimensional layered material for removal of radioactive pollutants. Chem. Eng. J. **397**, 125428 (2020)
38. A. Yadav et al., Study of 2D MXene Cr_2C material for hydrogen storage using density functional theory. Appl. Surf. Sci. **389**, 88–95 (2016)
39. N.C. Osti et al., Evidence of molecular hydrogen trapped in two-dimensional layered titanium carbide-based MXene. Phys. Rev. Mater. **1**(2), 024004 (2017)
40. R. Wu et al., Remarkably improved hydrogen storage properties of $NaAlH_4$ doped with 2D titanium carbide. J. Power. Sources **327**, 519–525 (2016)
41. X. Zou et al., A simple approach to synthesis Cr_2CT_x MXene for efficient hydrogen evolution reaction. Mater. Today Energy **20**, 100668 (2021)
42. J. Guo et al., Theoretical interpretation on lead adsorption behavior of new two-dimensional transition metal carbides and nitrides. J. Alloy. Compd. **684**, 504–509 (2016)
43. M. Boota et al., Pseudocapacitive electrodes produced by oxidant-free polymerization of pyrrole between the layers of 2D titanium carbide (MXene). Adv. Mater. **28**(7), 1517–1522 (2016)
44. P.W. Seo et al., Adsorptive removal of pharmaceuticals and personal care products from water with functionalized metal-organic frameworks: remarkable adsorbents with hydrogen-bonding abilities. Sci. Rep. **6**(1), 1–11 (2016)
45. G. Centi, S. Perathoner, Problems and perspectives in nanostructured carbon-based electrodes for clean and sustainable energy. Catal. Today **150**(1–2), 151–162 (2010)
46. X. Gao et al., Synthesis of NiCo-LDH/MXene hybrids with abundant heterojunction surfaces as a lightweight electromagnetic wave absorber. Chem. Eng. J. **419**, 130019 (2021)
47. K. Rasool et al., Water treatment and environmental remediation applications of two-dimensional metal carbides (MXenes). Mater. Today **30**, 80–102 (2019)
48. L. Verger et al., MXenes: an introduction of their synthesis, select properties, and applications. Trends Chem. **1**(7), 656–669 (2019)
49. J.-H. Yang et al., Adsorption activities of O, OH, F and Au on two-dimensional Ti_2C and Ti_3C_2 surfaces. Acta Phys. Chim. Sin. **31**(2), 369–376 (2015)

Chapter 4
MXene-Derived Composites and Their Application in Energy Storage and Catalysis

Rayees Ahmad Rather and Rameez Ahmad Mir

4.1 Introduction

Recent advances in materials science have demonstrated the development of composites between different materials as an effective approach to stimulate their activity. Composites are preferably developed by blending two or more material entities with variable thermodynamic, oxidation, and reduction properties to promote their efficiency for some specific applications. The composite formation is a striking strategy for enhancing stability, mechanical, electrical, and thermal properties of any material. In light-driven reactions, this strategy effectively promotes the lifetime of excited charge carriers. In recent years, composites of two-dimensional (2D) materials especially graphene [1, 2] had revealed their potential for several critical applications credited to their layered, flexible, and tunable structural features. However, the issues pertaining to production cost, fragility, and environmental concerns encouraged researchers to explore analogous metal-based layered structures. In this regard, the research progress in transition metal dichalcogenides (TMDCs) and transition metal carbides/nitrides (TMCs/TMNs) exploded as a research boom in 2D materials research specifically in the field of energy conversion and storage [3]. Among the latest and most advanced materials, MXenes have emerged [4] as an unconventional 2D material with the possibility of wider applicability in composite formation due to their layered structure, distinctive surface properties, and tunable morphology. MXenes, a special class of 2D layered TMCs/TMNs discovered by Gogtsi et al. in 2011 [5] are flexible structures that are ideal for forming multifunctional materials. MXenes are primarily composed of carbides, nitrides, and carbonitrides obtained

R. A. Rather (✉)
Department of Civil, Construction, and Environmental Engineering, San Diego State University (SDSU), San Diego, CA 92182, USA
e-mail: rrather@sdsu.edu; rzr0074@auburn.edu

R. A. Mir
School of Engineering, University of British Columbia (UBC), Okanagan, Canada

© The Author(s), under exclusive license to Springer Nature Singapore Pte Ltd. 2024

O. P. Pandey and P. Sharma (eds.), *MXenes: Emerging 2D Materials*,
https://doi.org/10.1007/978-981-97-4064-2_4

by selective etching of the MAX phase, where M is transition metal, A represents group IIIA or IVA elements and C is carbon or nitrogen. MXenes have a general formula, $M_{n+1}X_nT_n$ (n = 1–3) where M and X are the same as in the MAX phase and T is the respective functional group (–O–, –OH, and –F). MXenes are versatile materials that exhibit a combination of metallic and ceramic properties, making them suitable for use in a variety of applications such as batteries, energy storage, catalysis, sensors, capacitors, etc. [6, 7]. The electrical, optoelectronic, and induction properties of MXenes are optimized based on the ratio of M, X atomic species, which amenably improves their energy conversion and storage performance [8, 9]. In addition, MXene-based supercapacitors and batteries exhibit high energy storage performance even when the electrolyte contains larger cations because the wider spacing between the MXene layers allows easy intercalation and deintercalation of these larger ions. As a result, the larger ions can effectively store and release energy within the MXene layers [10, 11].

However, several issues in MXenes such as restacking of layers, oxidation of specific species, and reduced electrical conductivity of certain functional groups in out-of-plane directions constrain their performance not only in energy storage systems but also in other catalytic applications. The poor Vander Waals interactions between MXene layers tends to effect their mechanical properties thus limiting their use in electronic applications. To overcome these limitations, researchers have developed composites of MXenes with other potential materials, such as polymers, metals, and carbon-based materials. The blend of these materials possess improved mechanical strength, electrical conductivity, electrochemical and catalytic performance over their individual counterparts. However, the choice of the composite materials determines its potential application, for example, MXene/polymer composites can be used as flexible electrodes for energy storage devices, while MXene/metal composites can be used as catalysts for several reactions.

In this regard, this chapter presents a descriptive outlook of the MXene-based composites, their synthesis, and role in different energy storage and catalysis applications.

4.2 Synthesis of MXene Composites

The synergism between MXene and hetero material in a composite is decided based on certain factors such as (1) type of interfacial contact, (2) characteristics of space charge region, (3) interfacial electron transfer, (4) accessible surface area, and (5) path length, from electron transition level to its interaction with the adsorbed chemical entity. Therefore, these factors are usually considered while fabricating a functional MXene composite. Numerous procedures and techniques have been adopted for the synthesis of novel MXene composites.

Figure 4.1a, shows several reported composites of MXenes including with polymers [12], metal oxides [13], mesoporous carbon (mC) [14], graphene (atomic layer graphene, graphene oxide (GO), reduced GO (rGO) [15], carbon nitride (C_xN_y) [16],

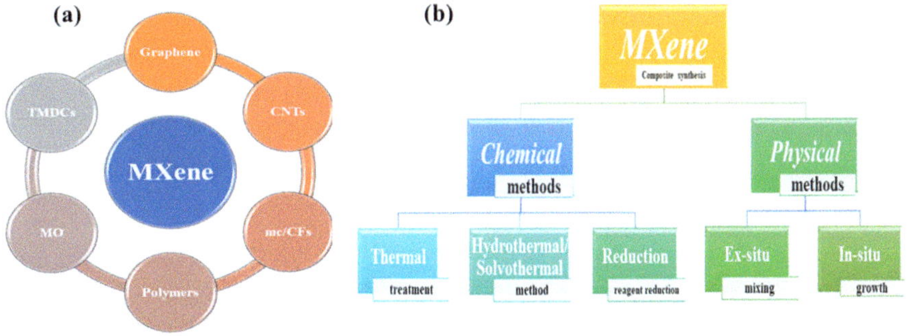

Fig. 4.1 **a** Schematic representation of different MXene composite species and **b** synthesis methods adopted

carbon nanotubes (CNTs) [17], and metal chalcogenides (TMDCs) [18]. Based on the type of additive material (0D, 1D, 2D, and 3D materials) and the intended end applications of the MXene composite, various synthesis protocols have been employed for composite formation (Fig. 4.1b).

Both chemical and physical methods are preferred for the synthesis MXene composites. However, the application requirement specifically determines the route of synthesis. Physical methods such as in-situ and ex-situ techniques are used for the development of a variety of MXene composites, as they avoid the formation of any impure phases or defects during the synthesis process.

In-situ composite synthesis involves the simultaneous formation of MXene and another material when precursors are subjected to reaction in a defined molar ratio. This method allows an integration of different functionalities of MXene and corresponding composite material enabling a significant change in the electrical and photoelectrical properties of the MXene composite. The in-situ process is a combination of several individual steps such as (1) precursor selection, (2) etching, (3) delamination, (4) addition of another composite component, (5) mixing and stabilization, (6) washing and drying. Typically etching and stabilization are two primarily important steps to regulate the formation of MXene structure and aggregation of particles. Characteristically, electrochemical and non-covalent interactions have been determined as the responsible forces for the formation of in-situ development of MXene composites [19].

In comparison to the in-situ method, the ex-situ route is a simpler procedure that involves the separate synthesis of MXene material and its subsequent mixing or dispersion into a matrix material in specific molar ratios to enable the interlayers of MXene to create a hybrid phase and enhance the overall activity of the composite. In ex-situ method, the etching is done on MAX phase only while as the dispersion and stability procedure remains like in-situ method. The ex-situ route involves the mixing or dispersion of MXene into different matrix materials such as polymers, metal oxides, carbonaceous materials, etc. In both cases, the developed hybrid composites exhibit higher interlayer spacing as compared to pure MXene [20, 21]. Additionally, multilayer MXenes, single layer is more efficient due to better surface hydrophilicity and compatibility with other materials.

Apart from the metal oxide, the MXene composites with other materials such as graphene and carbon nanotubes were also reported for several charge transfer reactions. In a study by Xu et al. [22], an electrochemical approach was used to develop flexible all-solid-state supercapacitors (SCs) made of a graphene oxide (rGO)/$Ti_3C_2T_x$ film, utilizing rGO as a binder to link the active conducting particles. Thick electrodes with high electrolyte accessibility were developed without the need for any special procedures to delaminate the MXene layers. The resulting composite material exhibited a higher specific capacitance than the individual ingredients (GO, rGO, or MXene).

In contrast to physical methods, chemical methods crucially improve the physicochemical and thermomechanical properties of MXene composites. The strength of the composite is determined by the hydrogen bonding between the negative charge of the functional group (–T) on the surface of MXene and the compounding species. The bonding and tailoring of the –T group with the composite species determine the utilization of the MXene composite for respective applications. Among the various additive materials used for MXene composite formation, carbon materials such as mesoporous carbon (mC), carbon fibers (CFs), carbon nanotubes (CNTs), and graphene have received special attention due to their easy availability, high surface area, and distinctive surface characteristics that work in larger voltage windows and possess better adsorption and charge transfer ability.

4.3 Applications of MXene-Derived Composites

MXene-based composites are being extensively explored to catalyze wide range of electronic and energy applications due to their unique electrokinetic, optoelectronic and storage properties. MXene derived composites offer higher effective surface area, high density of active sites, effective charge and mass transport. Majorly, the MXene-based composites are being tested for the following applications.

1. *Energy storage*: The MXene composites have high effective surface redox sites and conductivity, which makes them ideal for use in energy storage devices such as batteries and supercapacitors.
2. *Catalysis*: MXene based composites can be used as catalysts to drive several electrochemcial, thermal and photocatalytic reactions due to their high surface area and unique surface chemistry. MXene composites not only acts as the co-catalysts for certain reactions, but they also act as support for adsorption and promote the charge and mass transfer process. Therefore, MXene composites have the potential to promote the efficiency and yield in several environmental remediation and renewable energy processes.

3. *Water purification*: MXene composites are suitable adsorptive photocatalytic materials for sustainable treatment of heavy metals, organic pollutants and other recalcitrant pollutants.
4. *Biomedical applications*: MXene composites have potential applications in biomedical fields such as sensors, drug delivery, tissue engineering, and biosensing due to their biocompatibility and unique properties.

The MXene composites have a tremendous potential to scale up the efficiency of different reactions across different domains. However, in this chapter, a descriptive outlook of these composites for energy storage and catalysis is presented.

4.3.1 MXene Composites for Energy Storage

MXene composites are appropriate materials for energy storage and conversion applications due to their exceptional structural, morphological, and electrical properties. Research efforts are focused on optimizing the synthesis techniques to specify the molar ratios and enhance the chemical and electrostatic compatibility of individual ingredients in MXene composites, which in turn will improve the scalability of production which is a prerogative for practical implementation of these composites in energy storage and conversion applications.

Among various competitive options, the graphene-based MXene composites have gained significant attention due to their layered structures and high conductivity. These composites are attractive because they not only enhance performance by leveraging the superior properties of graphene but also maintain the unique two-dimensional structure of the composite, which further improves its overall activity and stability. The synergism and combination between MXene and graphene is a noteworthy, particularly with regards to the alignment of neighboring crystals, interface quality, and the abundance of edge sites. Zhou et al. [23] investigated the $Ti_3C_2T_x$ MXene phase to functionalize reduced graphene oxide (rGO) sheets through Ti–O–C covalent bonding, which was crosslinked by 1-aminopyrene-disuccinimidyl (AD) substrate to enhance the compactness and toughness of the resulting MXene-rGO (MrGO) sheets (Fig. 4.2a). A flexible symmetric supercapacitor was fabricated utilizing MrGO-AD sheets as electrodes (5 μm thick, 2 cm long, and 1 cm wide) and PVA/H_3PO_4 as the gel electrolyte (Fig. 4.2b). The developed cell demonstrated high cyclic stability even at a scan rate of 1000 mV s^{-1} (Fig. 4.2c). The supercapacitor performance estimated through GCD curves (Fig. 4.2d) exhibited volumetric capacitance of 645 F cm^{-3} at an applied current density of 1.0 A cm^{-3}. The cell also revealed a high stability with a capacitance retention of 75.3% at 8.6 A cm^{-3} and a high coulombic efficiency (CE ~100%) as shown in Fig. 4.2e, respectively. The capacitor also exhibited a very high cyclic stability with negligible loss of capacitance even after 20,000 CV cycles (Fig. 4.2f). The Ragone plot (Fig. 4.2g) revealed that the developed electrodes perform better than the commercial activated carbon (AC) and lithium (Li) batteries. The supercapacitor based on MXene and rGO sheets provides a combination of high toughness and strength electrodes with energy/power density essential for developing the highly efficient and durable flexible supercapacitors. The results revealed that the improved performance of the capacitor can be attributed to the use of MXene and π-π bridging, which increases the alignment of the MXene-rGO sheets. The capacitor's performance was evaluated at different bending angles (0°, 90°, and 180°), and it was found that bending and twisting did not affect its performance. Additionally, combining the three electrodes in series improved the voltage range threefold compared to using individual electrodes, providing a path to develop multilayer cells for practical applications. However, to further enhance the performance of the supercapacitors, it is crucial to consider ion accessibility on the electrode sites. The ion exchange transfer kinetics are notably low in organic electrolytes, especially at low temperatures (below freezing). Addressing this issue can lead to significant improvements in the performance of MXene-rGO-type supercapacitors.

The accessibility of the electrolyte ions in the electrodes plays a vital role for enhancing the charge storage performance especially at low temperatures. In this

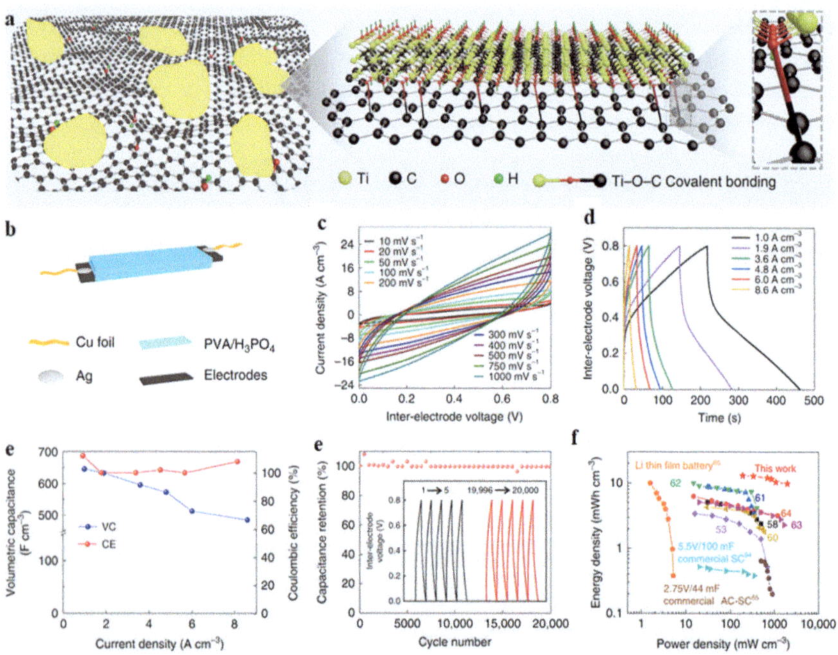

Fig. 4.2 **a** Schematic model of MXene-GO platelets showing the formation of Ti–O–C covalent bonding, **b** illustration of the components of a flexible supercapacitor based on MrGO-AD sheets. **c** Cyclic voltammetry (CV) curves at scan rates of 10–1000 mV s^{-1} for the MrGO-AD supercapacitor. **d** GCD curves for the MrGO-AD sheet supercapacitor. **e** Volumetric capacitance of the MrGO-AD sheet supercapacitor for current densities from 1.0 to 8.6 A cm^{-3}. **f** Capacitance retention during the cycling of a MrGO-AD supercapacitor. Figure adopted with permission from [23]

regard, Gao et al. [24] developed MXene knotted CNT composite as a high-rate performance electrodes (Fig. 4.3a–c). The breakage of the 2D alignment of the Ti_3C_2 MXene sheets by the special CNT knots prevents the restacking of the sheets and creates fast ion transport pathways. The capacitor performance of the MXene composite was tested through three and two electrodes set up in an organic electrolyte. The synthesized Ti_3C_2/CNT composite exhibited a capacitance performance of 128 F g^{-1} at 1 A g^{-1}. To check the performance of the composite at low operating temperatures, the asymmetric cell was fabricated with Ti_3C_2/CNT with 17% CNT composite as negative electrode and aligned CNT as positive electrode (Fig. 4.3d), respectively. The CV at multiple scan rates in a wide voltage window (0–3 V) shows a negligible loss of capacitance even at higher scan rates as shown in Fig. 4.3e. The quasi-rectangular shape of CV curve and symmetric shape of GCD curves (Fig. 4.3f) without a plateau reveals the pseudocapacitive behavior of the composite material. The asymmetric capacitor exhibited a capacitance of 130 F g^{-1} higher than aligned CNTs (94 F g^{-1}) and showed capacitance retention of more than 90% even after 8000 cycles. The cell capacitance at lower operating temperature (Fig. 4.3g) shows that capacitance decreased by 50% as compared that performed at 20 °C. The absence of the redox peaks at lower temperatures (Fig. 4.3h) shows that the double-layer mechanism is responsible for charge storage. The operational voltage window up to 4.2 V is higher than that achieved in activated carbon materials (3 V). The highest capacitance retention was observed in composite material with 9% CNT as compared to 17% (Fig. 4.3i) and bare MXene, respectively. The results suggested that conductivity plays a vital role on capacitive performance of the cell at lower operating temperatures. An impressive capacitance, energy density (59 Wh kg^{-1}), and a power density (9.6 kW kg^{-1}) were obtained at −30 °C. The performance is better than the values reported for supercapacitors with 2D electrode materials operating at low temperatures. The study has demonstrated that MXene/carbon composites made from pseudocapacitive materials can be utilized for high-rate energy storage in organic electrolytes by optimizing the electrode architecture. This is due to the exceptional hydrophobicity and conductivity of the MXene/graphene composites, which has led to extensive efforts to enhance their supercapacitor performance. Despite their potential, the use of MXene/graphene composites in practical applications is still limited by several factors, such as the complicated synthesis process and the high cost associated with the use of graphene [15].

To deliver on such bottlenecks, conductive polymers have garnered significant interest as alternative additive composite materials for creating flexible, lightweight, and conductive MXene/polymer composites that serve as electrodes for energy storage devices [25–27]. Adekoya et al. [28] conducted a study on the effects of polypyrrole (PPy) polymer on the properties and performance of MXene/PPy nanocomposites. However, to fabricate the MXene/PPy composites, in-situ polymerization is considered as a promising approach. Various types of flexible MXene/PPy composite electrodes were designed and fabricated. Furthermore, an in-depth investigation and comparison studies of MXene/PPy composites revealed that the incorporation of PPy enhances both the stability and the energy storage performance. Similarly, Li et al. [29] fabricated a textile-based electrode made of polyester fabric

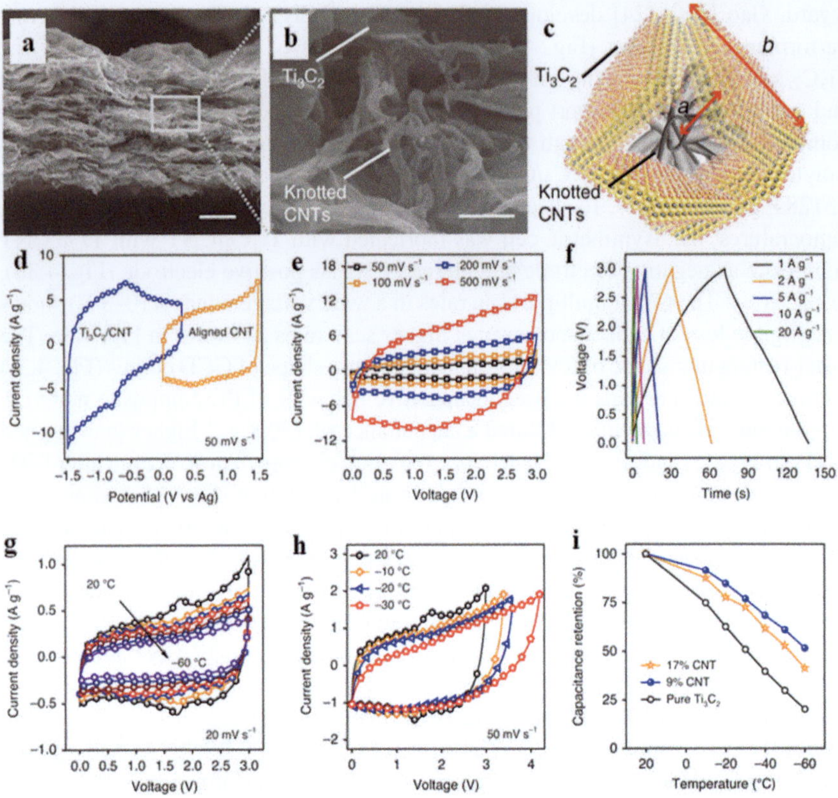

Fig. 4.3 **a** Cross-sectional SEM images of an MXene-knotted CNT composite electrode with 17% CNTs, **b** SEM image of composite, **c** schematic illustration of the Ti_3C_2 flakes was broken by the knotted CNTs to form 3D networked structure. "a" and "b" (~1:2 ratio), **d** full cell fabrication: three-electrode CVs of the MXene-knotted CNT (Ti_3C_2/CNT) composite electrode and aligned CNT electrode at 50 mV s^{-1}, **e** cyclic voltammograms for the full cell at different scan rates, **f** galvanostatic charge–discharge curves for the full cell at different current densities, **g** cyclic voltammograms at different temperatures at 20 mV s^{-1}, **h** cyclic voltammograms with larger voltage windows at low temperatures for the full cell using MXene-knotted CNT composite electrode with a CNT content of 17% as the negative electrode, and **i** capacitance retention of full cells with different Ti_3C_2 MXene-based electrodes as a function of temperature. Figure adopted with permission from Ref. [24]

modified with polyethyleneimine (PEI) Mxene/PPy composite (PPy/MXene/PMFF). The capacitor performance of the developed composite was estimated by a symmetric cell procedure with PPy/Ti_3C_2/PMFF as electrodes and advanced PVA/Na_2SO_4 gel as electrolyte. The polymer-encapsulated MXene provides more abundant sites and modified channels for electron transport to improve the capacitive performance. Based on the mass loading of the active material (PPy/MXene), the PPy/MXene/PMFF composite revealed a real capacitance of 1295 mF cm^{-2} and a gravimetric capacitance of 439 and 360 F g^{-1} at 1 and 50 mA cm^{-2}, respectively. Moreover, the polymer/MXene composite exhibited excellent stability with no significant drop in capacitance even at higher scan rates. The designed electrode shows only a drop of 6.3% in overall capacitance after 30,000 cycles. The higher rate performance, high energy/power density, high stability even different deformation angles, and excellent cyclic stability contribute to the significant advances in textile-based electrodes for wearable electronic devices. The comparison studies of different MXene polymer composites and their respective performance are given in Table 4.1.

Moreover, to improve the number of redox sites and enhance the performance of MXene electrodes, the synthesis of MXene/TMO composites is considered a promising approach. Typically, TMOs made of metals such as Manganese (Mn), Vanadium (V), Iron (Fe), Nickel (Ni), and Cobalt (Co) have been explored due to their high theoretical capacity, wide availability, low cost, multiple oxidation states, and higher number of redox sites. Various transition metal oxide-based MXene composites (MXene/TMOs) have been developed as electrode materials for supercapacitors, as presented in Table 4.1 [30].

However, despite the higher capacitance and energy density observed in the developed supercapacitors (SCs), still fall short of the energy density values obtained in batteries. To further improve the pseudocapacitive performance of the MXene electrodes, Wang et al. [31] produced a tungstate/MXene composite electrode for use in lithium and sodium-ion batteries. The electrode was free-standing and aligned to minimize ion-path tortuosity, enhancing ion transport, and ultimately leading to improved efficient pseudocapacitive storage performance. The developed flexible electrodes demonstrated high capacity and long-term performance, with a capacity of 267.0 mAh g^{-1} observed in the case of Li-ion batteries and a CE greater than 78%. Even at various deformation angles, negligible loss of capacity was observed. These results suggest the potential application of the developed composite material in wearable electronic devices.

In addition, energy storage properties of transition metal oxide/MXene composites were enhanced by increasing the number of redox-active sites due to the incorporation of metal oxide species, especially those of Ni, Co, Sn, and Fe-oxides. The charge storage mechanism in batteries is diffusion-controlled and depends on the redox reactions occurring deep within the electrode material. However, the charge storage mechanism is surface controlled process. A study presented by Wang et al. [32] showed a composite electrode consisting of carbon-coated iron oxide (Fe_3O_4) nanoparticles (C@Fe_3O_4 NPs) decorated on Ti_3C_2/MXene for Li-ion battery applications. Fe_3O_4 was chosen as an additive due to its high theoretical capacitance of 3625 F g^{-1}. The uniform distribution of C@Fe_3O_4 NPs improved the specific surface area from 7.8 to 30 m^2 g^{-1}. The C@Fe_3O_4 NPs/Ti_3C_2 composite exhibited a higher capacity of 231.5 mAh g^{-1} as an anode in Li-ion batteries compared to pristine Ti_3C_2 MXene (107.8 mAh g^{-1}). Similarly, Liu et al. [33] used the hydrothermal reaction route to develop VO_2-enriched V_2CT_x MXene as an electrode for Li-ion batteries. The partially etched V_2CT_x MXene phase was oxidized in the presence of H_2O_2 to enhance the layer spacing of the resultant oxidized phase as shown in Fig. 4.4a. The presence of the VO_2 phase was confirmed by the XRD pattern (Fig. 4.4b) and the enhancement of the layer spacing before and after hydrothermal treatment was

Table 4.1 Electrochemical performance of MXene/composite-based SCs. Adopted with permission from Refs. [15, 30]

Electrode(s)	Electrolyte	Capacitance	Current density	Retention (%)	Number of cycles
MXene/rGO composite	2M KOH	154.3 F g^{-1}		85	6000
MXene/rGO films	6M KOH	405 F g^{-1}		No change	10,000
MXene/rGO hybrid fiber	1M H_2SO_4	370 F cm^{-3}		No change	3000
MXene/rGO films	3M H_2SO_4	586.4 F cm^{-3}		No change	20,000
MXene/rGO composite aerogel	1M H_2SO_4	1040 F cm^{-3}		91	15,000
MXene/holey rGO composite	3M H_2SO_4	438 F g^{-1}		93	10,000
Co-MXene/rGO hybrid aerogel	6M KOH	345 F g^{-1}		85	10,000
MXene/rGO hydrogels	Saturated K_2SO_4	370 F g^{-1}		–	–
Mxene/MnO_2	Na_2SO_4	312 F cm^{-3}	5 mV s^{-1}	130.8	5000
MnO_2/Mxene	KOH	611.5 F g^{-1}	1 A g^{-1}	96	1000
Mxene/MnO_2 NR	KOH	130.5 F g^{-1}	0.2 A g^{-1}	100	1000
Mxene/MnO_2 NS	Na_2SO_4	340 F g^{-1}	1 A g^{-1}	87.6	2000
Mxene/Fe_3O_4	Li_2SO_4	46.4 mF cm^{-2}	0.5 mA cm^{-2}	96.3	85,000
MnO_2/Mxene NW	KOH	212 F g^{-1}	0.1 A g^{-1}	88	10,000
Co-Fe oxide/$Ti_3C_2T_x$	LiCl	2467.6 F cm^{-3}	0.2 mA cm^{-2}	88.2	10,000
$Ti_3C_2T_x$/Fe_2O_3	LiCl	405.4 F g^{-1}	2 A g^{-1}	97	2000
TiO_2/Mxene	KOH	143 F g^{-1}	5 mV s^{-1}	96	3000
V_2O_5/MXene	$LiNO_3$	217 F g^{-1}	0.1 mA cm^{-2}	104	5000

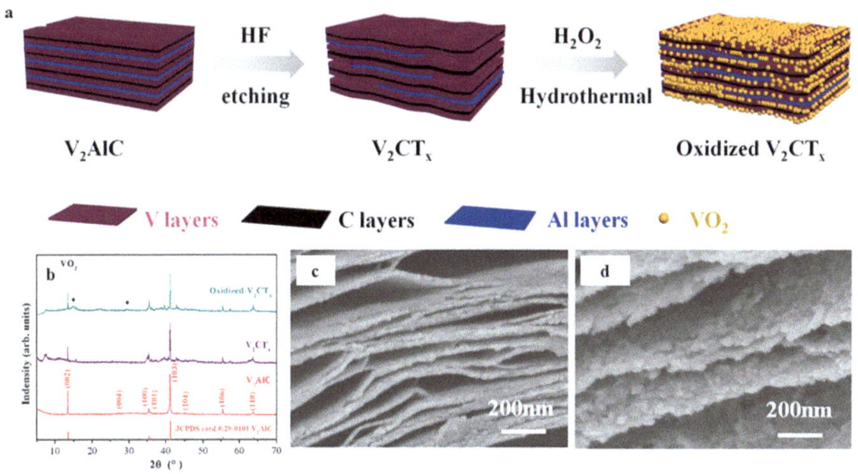

Fig. 4.4 **a** Schematic illustration of the synthesis of oxidized V_2CT_x, **b** XRD pattern, and **c–d** SEM images of the developed product. Adopted with permission from Ref. [33]

observed in the micrographs shown in Fig. 4.4c–d, respectively. The electrochemical tests were performed by fabricating a coin cell with oxidized V_2CT_x (80% by weight) as anode and Li-foil as cathode material. 1M $LiPF_6$ dissolved in (ethylene carbonate (EC), dimethyl-carbonate (DMC), and ethyl methyl carbonate (EMC) at a volume ratio of (1:1:1) as electrolyte and Celgard 2400 micro-porous membrane as separator.

The CV at a very low scan rate of 0.1 mV s^{-1} as shown in Fig. 4.5a (V_2CT_x), and Fig. 4.5b (oxidized V_2CT_x) shows the presence of redox peaks in the pristine MXene phase. However, the peaks disappeared in the oxidized phase which attributes to fact that the presence of O-terminated groups and VO_2 phase improved the Li^+ capture and improved the reversibility of the reactions during CV cycling. The specific capacity estimated at varying current densities in the oxidized V_2CT_x phase is higher than the pristine MXene and MAX phase (Fig. 4.5a–c). The rate performance (Fig. 4.5d) of the oxidized MXene reaches up to 318 mAh g^{-1}, which is more. The results revealed that the increased layer spacing in the oxidized phase plays a key role in the Li-intercalation/deintercalation process to increase the charge storage performance. However, for practical applications, cycling stability and long-term performance retention are necessarily required. The capacity of the MXene phase decreased for the first 100 cycles as shown in Fig. 4.5e. This decrease corresponds to the improved Li^+ accessibility with the number of cycles. The presence of the oxide terminated groups improves the reversible capacities than the pristine MXene. The VO_2 phase formed in the oxidized MXene (V_2CT_x) acts as the bridge between the V_2CT_x and Li^+, which provides more diffusion channels to improve the reversibility of the oxidized V_2CT_x phase. The oxidized V_2CT_x phase revealed an excellent stability up to 1000 cycles with negligible loss of capacitance as shown in Fig. 4.5f. The CE curve remained linear with no loss even after 1000 cycles.

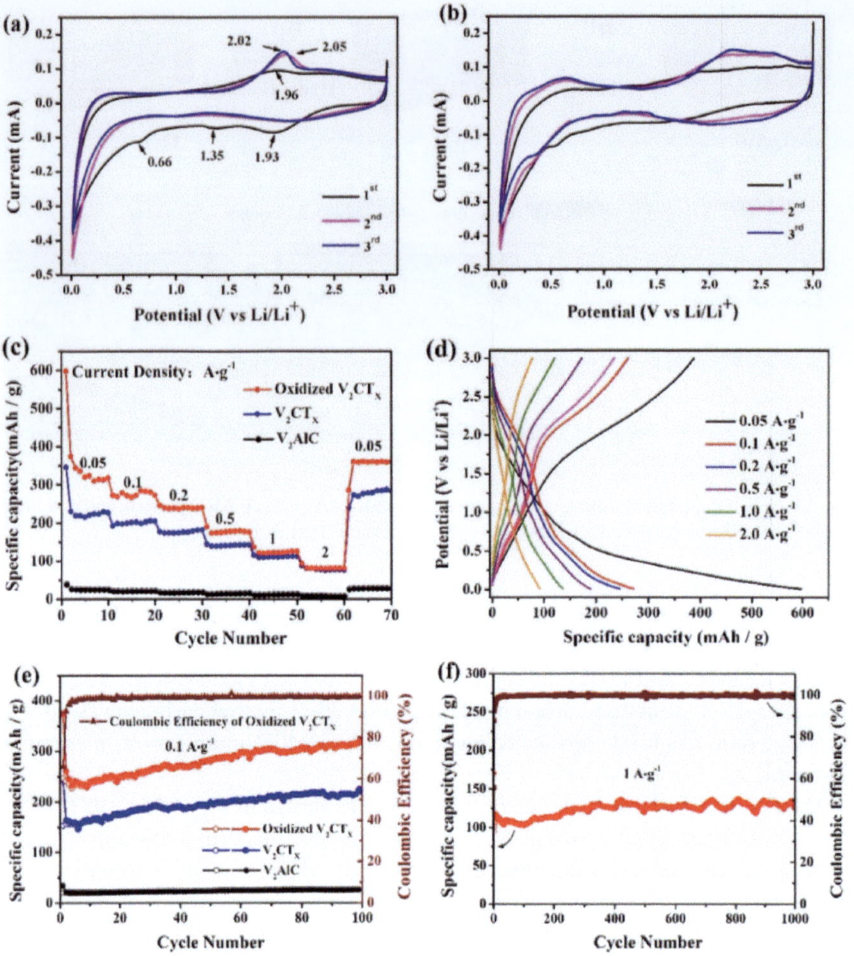

Fig. 4.5 The CV curves of V_2CT_x (**a**) and oxidized V_2CT_x **b** at 0.1 mV s^{-1} for the initial three cycles; **c** and **d** the rate performance at different current densities; **e** the cycle performance of V_2CT_x and oxidized V_2CT_x at a current density of 0.1 A g^{-1}. Adopted with permission from Ref. [33]

The oxidized V_2CT_x phase represented good performance stability and reversibility, justifying potential use as anode materials for lithium storage. Different other oxide materials have been tested along with MXenes as electrodes for Li, Na, and Zn batteries. The presence of the oxide materials as the composite additives improves both the activity and stability. The battery performance of the different MXene/TMO composite (Table 4.2) is higher than observed in pure V_2CT_x phase.

Table 4.2 MXene/TMO composite for battery applications. Adopted from Ref. [30]

Serial Number	MXene/TMOS material	First cycle capacity (mAh g^{-1})/ current density	First cycle capacity (mAh g^{-1})/current density	Last cycle capacity (mAh g^{-1})/ current density	Last cycle capacity (mAh g^{-1})/ current density
	$CoO/Co_2Mo_3O_8@Ti_3C_2T_x$	1008.8/0.1 A g^{-1}	936.2@100/0.1 A g^{-1}	545/2 A g^{-1}	1200
2	γ-$Fe_2O_3@Ti_3C_2T_x$	805.4/0.1 A g^{-1}	1060@400/0.5 A g^{-1}	466/2 A g^{-1}	800
3	Oxidized-V_2CT_x	318/0.05 A g^{-1}	318@100/0.1 A g^{-1}	125/1 A g^{-1}	1000
4	$Fe_3O_4@Ti_3C_2T_x$	517/0.2 C	–	236.7/5 C	300
5	$SnO_2@Ti_3C_2T_x$	829.4/0.2 A g^{-1}	829.4@100/0.2 A g^{-1}	533.9/1 A g^{-1}	500
6	$Fe_3O_4@Ti_3C_2T_x$	809.1/0.1 A g^{-1}	782.7@100/0.1 A g^{-1}	667.9/1 A g^{-1}	600
7	$Ti_3C_2T_x/TiO_2$	300/0.1 A g^{-1}	130@10/0.5 A g^{-1}	143/0.5 A g^{-1}	200
8	$Fe_3O_4@Ti_3C_2T_x$	954.7/0.5 A g^{-1}	701.8@200/0.2 A g^{-1}	609.9/0.5 A g^{-1}	600
9	$Co_3O_4/Ti_3C_2T_x$	1326/C/5	1005@300/1C	307	1000
10	TiO_{2-x}/Ti_3C_2	206/0.1 A g^{-1}	–	131/5 A g^{-1}	500
11	MXene-bonded Si@C	1660.6/ 0.42 A g^{-1}	–	1040/0.42 A g^{-1}	150
12	MXene derived TiO_2/Fe_2O_3	469.8/0.1 A g^{-1}	329.5@590/0.1 A g^{-1}	100/0.1 A g^{-1}	2000
13	$Co_3O_4@s$-$Ti_3C_2T_x$	747/0.1 A g^{-1}	550/@700/1 A g^{-1}	160/5 A g^{-1}	100
14	Cu_2O/Ti_2CT_x	790/0.01 A g^{-1}	–	143/1 A g^{-1}	250
15	SnO_2/Ti_2CT_x	736/0.5 A g^{-1}	–	258/0.5 A g^{-1}	50
16	$Li_4Ti_5O_{12}/Ti_2CT_x$	136/10 C	–	118/10 C	1000
17	N-doped TiO_2-MXene	305/0.03 A g^{-1}	–	369/0.03 A g^{-1}	100
18	$Ni_{0.8}Mn_{0.2}$-PBA/MXene	442/0.1 A g^{-1}	443@50/0.1 A g^{-1}	259.9	650
Zinc ion batteries (ZIBs)					
1	$Mn_xV_{10}O_{24}{\cdot}nH_2O@V_2CT_x$	424.4/0.5 A g^{-1}	358.2@300/1 A g^{-1}	289.6/10 A g^{-1}	25,000
2	$CC@MnO_2@Ti_3C_2T_x$	517/0.1 A g^{-1}	505.2@45/0.1 A g^{-1}	80.6/1 A g^{-1}	800

(continued)

Table 4.2 (continued)

Serial Number	MXene/TMOS material	First cycle capacity (mAh g^{-1})/ current density	First cycle capacity (mAh g^{-1})/current density	Last cycle capacity (mAh g^{-1})/ current density	Last cycle capacity (mAh g^{-1})/ current density
3	ZMO@$Ti_3C_2T_x$	~175/0.1 A g^{-1}	~175@100/0.1 A g^{-1}	172.6/1 A g^{-1}	5000
4	$H_2V_3O_8$/$Ti_3C_2T_x$	317.4/0.2 A g^{-1}	290.1@150/0.2 A g^{-1}	86.1/10 A g^{-1}	6600
5	3D $Ti_3C_2T_x$@MnO_2	301.2/0.1 A g^{-1}	–	202.2/0.1 A g^{-1}	2000
6	V_2O_5@V_2CT_x	397/0.5 A g^{-1}	~335@100/0.5 A g^{-1}	345/4 A g^{-1}	2000
7	V_2O_x@V_2CT_x	300/0.05 A g^{-1}	–	246/1 A g^{-1}	200
8	VO_x@V_2CT_x	423.5/1 A g^{-1}	398.1@80/5 A g^{-1}	283.7/30 A g^{-1}	2000
9	V_2CT_x@MnO_2	408.1/0.3 A g^{-1}	–	119.2/10 A g^{-1}	10,000
10	$(NH_4)_2V_{10}O_{25}\cdot 8H_2O$@ $Ti_3C_2T_x$	514.7/0.1 A g^{-1}	331.8@500/0.5 A g^{-1}	105.6/5 A g^{-1}	6000
11	VO_2/$Ti_3C_2T_x$	445/0.1 A g^{-1}	244.5@2600/ 20 A g^{-1}	225/30 A g^{-1}	4000
12	VO_2/$Ti_3C_2T_x$	228.5/0.2 A g^{-1}	–	126.6@2 A g^{-1}	700

4.3.2 MXene Composites for Catalysis

MXene composites had been reported as catalysts due to their unique physical and chemical properties. These composites have a high surface area, which provides many active and adsorption sites for catalytic reactions. Additionally, they exhibit excellent electronic conductivity and can act as conductive supports for catalyst nanoparticles. Several studies have reported the use of MXene composites as catalysts for various reactions. For example, $Ti_3C_2T_x$ composites have shown good performance for the oxygen evolution reaction (OER) and hydrogen evolution reaction (HER) [34]. These composites have shown excellent catalytic activity and stability, making them promising candidates for water electrolysis and other types of catalytically uphill reactions. Additionally, MXene composites were also used as catalysts for organic reactions like the reduction of nitroarenes or organic transformations [35]. The composites exhibited high catalytic activity and selectivity towards the desired reduction product, making them potential candidates for the synthesis of pharmaceuticals and agrochemicals. Three major domains where Mxene composites are explored as potential catalysts include, (1) photocatalysis, (2) catalytic sensors, and (3) pollutant remediation.

4.3.2.1 MXene Composites as Photocatalysts

Photocatalysis is a process in which light is used to drive a chemical reaction [36], among the widely studied reactions, the infamous water splitting, hydrogen production, CO_2 reduction, and nitrogen fixation are the best specimen reactions with uphill thermodynamic requirements that researchers try to undo via photocatalytic approach [37, 38]. To achieve higher quantum conversion efficiency, the photocatalysts need to possess some exceptional characteristics such as (1) higher photoactivation response, (2) perennial stability and greater charge transfer ability, (3) lower recombination rate of excited electron–hole pairs, (4) efficient separation of charge carriers and their interaction with reactants adsorbed on the surface, and above all (5) suitable band profile to execute specific oxidation and reduction reactions. Single-component MXenes can't meet all the criteria to be an effective photocatalyst. However, the surface hydrophilic properties of MXene due to the presence of –OH and –O groups enable them to form suitable photocatalytic composite systems with other photoactive co-catalysts to drive some reactions. MXene composites have been used to drive numerous uphill reactions such as water splitting, CO_2 reduction and N_2 reduction.

An et al. [39] used a combination of Pt and Ti_3C_2Tx co-catalyzed C_3N_4 MXene composite for photocatalytic hydrogen production. The composite showed improved conductivity and exhibited an excellent photocatalytic response for hydrogen production (5.5 mmol h^{-1} g^{-1}). The higher performance is attributed to the MXene-facilitated effective charge transfer and separation of e^-/h^+ pair. Similarly, a combination of Ti_3C_2 and cadmium sulfide (CdS) MXene composite was reported for the photocatalytic hydrogen production [40]. The Ti_3C_2–CdS composite was prepared by hydrothermal synthesis route, the morphology and structural interpretation of the composite revealed a heterojunction between Ti_3C_2 and CdS (Fig. 4.6a–b). The composite produced 14.3 mmol h^{-1} g^{-1} of hydrogen equaling and apparent quantum efficiency of 40.1% (Fig. 4.6c). The performance of the composite is credited to suitably aligned band profile of the Ti_3C_2–CdS photocatalytic system leading to an interfacial electron transfer from CdS to Mxene along the formation of space charge region (Fig. 4.6d). A Schottky junction is formed (Fig. 4.6d) after the fermi level equilibration allowing electrons to shuttle towards Ti_3C_2 surface leaving behind the positive holes, the excellent conductivity or the MXene allows the water reduction to generate H_2 gas. Furthermore, as shown in Table 4.4, several MXene-based composites have been reported for photocatalytic hydrogen production under half reaction conditions. The performance in most cases is ascribed to the co-catalytic role of MXene in the composite leading to the effective separation and prolonged lifetime of excited electron hole pairs.

Moreover, MXenes composites have been reported for the photocatalytic CO_2 reduction to produce C1 fine chemicals as the products in a sustainable way. For example, Ti_3C_2 was utilized as a co-catalyst to enhance the photocatalytic response

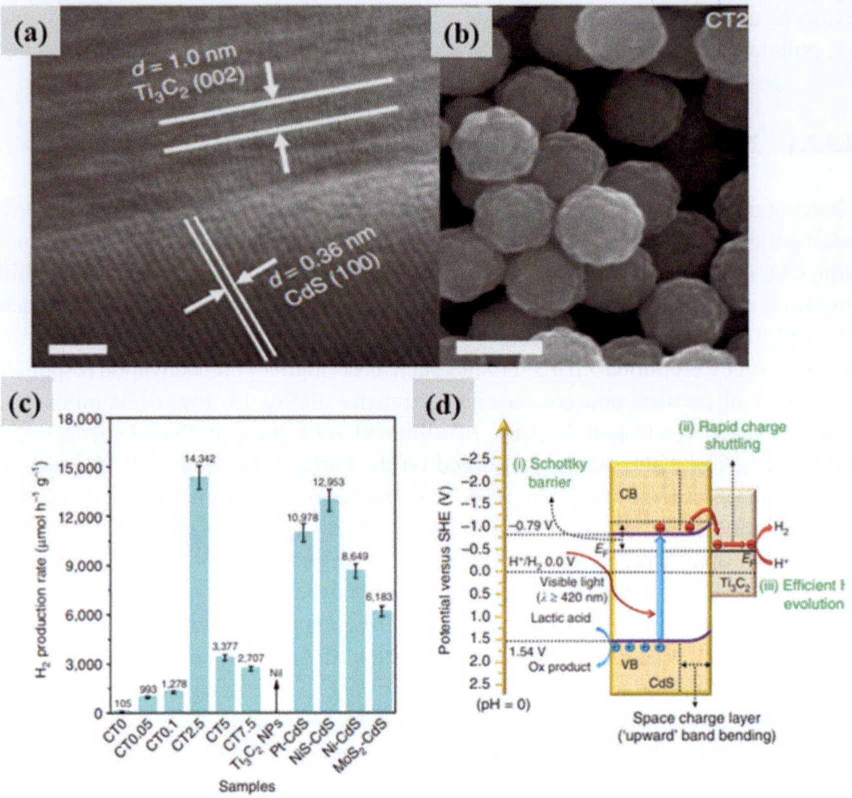

Fig. 4.6 **a** A high-resolution TEM image and **b** SEM image of CdS/Ti_3C_2 composite (CT2.5). **c** Photocatalytic activity of the CdS/Ti_3C_2 composite, **d** plausible mechanism of charge transfer in CdS/Ti_3C_2 under visible light irradiation. Adopted with permission from Ref. [40]

Table 4.4 Photocatalytic hydrogen production by different MXene composites in different reaction conditions and under visible light irradiation

MXene composite	Reaction medium	H_2 production (μmol g^{-1} h^{-1})	Ref.
TiO_2 nanofibers/$Ti_3C_2T_x$	Water/Methanol	6979	[42]
$Ti_3C_2T_x$ nanoparticles/CdS	Water/Lactic acid	14,342	[40]
TiO_2 /$Ti_3C_2T_x$/CoS	Water/Methanol	950	[43]
C3N4 /$Ti_3C_2T_x$/Pt	Water/TEOA	5100	[39]
g–C_3N_4@$Ti_3C_2T_x$	Water/TEOA	5111	[44]
ZnO nanorods /$Ti_3C_2T_x$	Water/Ethanol	456	[45]
$CdS/MoS_2/Ti_3C_2T_x$	Water/Sodium sulfate	9679	[46]
$CdLa_2S_4/Ti_3C_2T_x$	Water/Sodium sulfate	11,182	[47]
Ti_3C_2Tx/TiO_2/g–C_3N_4	Water/TEOA	1620	[48]
Mo_xS@TiO_2@$Ti_3C_2T_x$	Water/TEO/Acetone	10,505	[49]
Cu/TiO_2@Ti_3C_2T	Water/Methanol	764	[50]

of TiO_2 during a photocatalytic CO_2 reduction reaction [41]. As a result, the Ti_3C_2–TiO_2 composite produced CO (11.74 μmol g^{-1} h^{-1}) and CH_4 (16.61 μmol g^{-1} h^{-1}) as the reduction products. Again, the credit for the efficiency was attributed to the effective charge separation leveraged by the presence of MXenes. Several similar composites have been reported recently for the photocatalytic capture and conversion of CO_2 to fine chemicals.

4.3.2.2 MXene Composite Sensors

Studies have revealed that 2D materials are most popular for sensor and wearable applications due to their novel physical, morphological, and optoelectronic properties [51]. MXenes being a 2D class of layered materials have the potential tendency in sensor applications as they possess huge surface area, excellent stability, biocompatibility, and good thermal conductivity [52]. Moreover, the hydrophilic surface allows the surface modification and functionalization, which is a prerogative in any sensor application. Among the major sensing application, MXene composites have been reported as gas sensors, biosensors, wearables, etc. For example, Wang et al. [53] reported a TiO_2 modified Ti_3C_2 composite for the detection of H_2O_2. The composite showed an organ-like structure of Ti_3C_2 possessing a layered morphology resembling to that of exfoliated graphite, a suitable electrode of TiO_2–Ti_3C_2 was developed with the help of nafion binder which was beneficial for the immobilization of Hb. The specific surface area of the composite was also modified and due to these reasons, the prepared sensor showed a detection limit of 14 nM for H_2O_2. Similarly, a similar composite constituted of Ti_3C_5–Au was used for glucose detection. MXene-Au displayed the detection ability in the range of 0.1–18 mM with lower LOD at 5.9 μM. The sensing ability is attributed to the electrical conductivity of MXene, which was further enhanced by the uniform distribution of Au nanoparticle. Numerous other studies have reported the MXene-based composites for the detection of gases such as CO and CH_4. A study by Yu et al. [54] presented the adsorption of number of gases such as NH_3, H_2, CH_4, CO, CO_2, N_2, NO_2, and O_2 on the monolayer Ti_2CO_2 using the first principal simulations. It was observed that the adsorption energy of NH_3 on Ti_2CO_2 decreased significantly due to biaxial strains. When 3% biaxial strain is applied the adsorption, energy reaches to −0.51 eV indicating an effective capture on Ti_2CO_2 layer. In contrast, the adsorption energies of other gases on Ti_2CO_2 showed minor changes when biaxial strains were applied indicating weak interaction. The work discloses preferable NH_3 adsorption over other gases. The overall investigation revealed that Ti_2CO_2 could be an excellent material for detection, capture, separation, and storage of NH_3 gas. MXene composites showed high sensitivity to a varied range of physical and chemical stimuli due to their excellent physical, electrical, and optoelectronic properties. Therefore, an expected surge and significant advancements are anticipated in the research area of MXene composites as sensors.

4.3.2.3 MXene Composites for Water Pollutant Remediation

MXene composites have emerged as a promising material for the catalytic and destructive treatment of numerous conventional and recalcitrant water pollutants [55]. As discussed in previous sections, the exceptional properties of MXenes such as high specific surface area, electrical conductivity, and strong adsorption capabilities enable their possible use in conventional and non-conventional wastewater treatment techniques. They exhibit excellent adsorption capacity towards heavy metals, organic compounds, and dyes. As discussed, MXene composite showed a strong response in different photocatalytic reactions, hence, they have a tendency to be utilized in advanced oxidation processes for efficient pollutant degradation.

For example, MXene composites have been used as photocatalysts for the degradation of organic pollutants in water. The composites were able to efficiently absorb visible light and produce reactive oxygen species, which can degrade a variety of pollutants. Liu et al. [56] reported a sandwich-like Co_3O_4/MXene composite, synthesized using a hydrothermal method. The authors reported that the composite exhibited enhanced catalytic performance for the degradation of Bisphenol A in water, highlighting its potential as a heterogeneous catalyst in advanced oxidation processes. The study found that the highest catalytic activity was achieved when the Co_3O_4 loading was about 20%, outperforming both pure MXene and Co_3O_4. The catalyst also exhibited excellent stability, recoverability, and good catalytic performance across a wide pH range and in the presence of anions (NO_3^- and Cl^-). EPR and quenching tests confirmed the presence of both SO_4^{*-} and OH^* radicals in the CMs/PMS system, with SO_4^{*-} being the major oxidizing species. This discovery highlights the potential of MXene composite materials in the removal of organic pollutants and further expands the application of MXene in environmental science. Similarly, Yin et al. [57] reported novel ternary MXene composite catalyst material $Cu_2O/TiO_2/Ti_3C_2$, which exhibited efficient and sustainable catalytic activity.

Furthermore, due to their higher specific surface area, MXene composites can act as efficient adsorbents for the removal of different organic and inorganic contaminants from water or direct sewage treatment. Mashtalir et al. [58] reported a high adsorption capacity of $Ti_3C_2T_x$ for cationic methylene blue dye preferentially due to surface OH^- and hydrophilic nature. Some authors had specifically functionalized the surface of MXene to enhance its adsorption capacity, alkalization with LiOH, NaOH, and KOH was reported to take the adsorption capacity of Ti_3C_5 up to 189 mg/g. The higher capacity was attributed to intercalation of methylene blue dye into alkalized MXene.

MXene composites have shown a significant response for the treatment of different types of pollutants. Their unique properties combined with versatility in electrical and optical properties allow their usage for the efficient removal of targeted pollutants such as PFAS from water. However, continued research and development in the MXene research is important to lower the cost of production and strategies to develop new thermodynamically viable composites to enhance the removal efficiency in different water purification processes.

4.4 Conclusion

In summary, the MXene composites have been utilized as catalytic and support material in various types of reactions. These 2D materials have an excellent tendency to act as storage and electrode materials in supercapacitor and battery applications due to higher specific surface area, distinctive surface characteristics, and higher conduction-based effective ion mobility. Moreover, in light-driven reaction, the MXene composites can thrive to execute numerous uphill reactions such as water splitting and CO_2 reduction. Their role as co-catalysts is worthy of being explored in different biosensing applications. Due to the high specific surface area and excellent surface characteristics, MXene composite are suitable options to replace the conventional water treatment materials. In addition, these material composites could prove productive in destructive removal of recalcitrant pollutant from wastewater through simultaneous adsorption and advanced oxidation processes. Furthermore, for future progress, a key focus is needed on the synthesis and optimization of MXene composites to lower the cost of production and increase their catalytic efficiency.

References

1. D. Chen, W. Chen, L. Ma, G. Ji, K. Chang, J.Y. Lee, Mater. Today **17**, 184–193 (2014)
2. N. Dwivedi, C. Dhand, P. Kumar, A. Srivastava, Materials Advances **2**, 2892–2905 (2021)
3. A. Bhat, S. Anwer, K.S. Bhat, M.I.H. Mohideen, K. Liao, A. Qurashi, NPJ 2D Mater. Appl. **5**, 1–21 (2021)
4. M. Naguib, M.W. Barsoum, Y. Gogotsi, Adv. Mater. **33**, 2103393 (2021)
5. M. Naguib, M. Kurtoglu, V. Presser, J. Lu, J. Niu, M. Heon, L. Hultman, Y. Gogotsi, M.W. Barsoum, Adv. Mater. **23**, 4248–4253 (2011)
6. R.M. Ronchi, J.T. Arantes, S.F. Santos, Ceram. Int. **45**, 18167–18188 (2019)
7. A. Szuplewska, D. Kulpińska, A. Dybko, M. Chudy, A.M. Jastrzębska, A. Olszyna, Z. Brzózka, Trends Biotechnol. **38**, 264–279 (2020)
8. J. Yang, W. Bao, P. Jaumaux, S. Zhang, C. Wang, G. Wang, Adv. Mater. Interfaces **6**, 1802004 (2019)
9. A. Chroneos, D. Horlait, W. Lee, S. Middleburgh
10. Y. Luo, G.-F. Chen, L. Ding, X. Chen, L.-X. Ding, H. Wang, Joule **3**, 279–289 (2019)
11. Y. Dall'Agnese, P.-L. Taberna, Y. Gogotsi, P. Simon, J. Phys. Chem. Lett. **6**, 2305–2309 (2015)
12. S.M. George, B. Kandasubramanian, Ceram. Int. **46**, 8522–8535 (2020)
13. X. Yu, T. Wang, W. Yin, Y. Zhang, Int. J. Hydrogen Energy **44**, 2704–2710 (2019)
14. A.E. Allah, J. Wang, Y.V. Kaneti, T. Li, A.A. Farghali, M.H. Khedr, A.K. Nanjundan, B. Ding, H. Dou, X. Zhang, Nano Energy **65**, 103991 (2019)
15. Y. Liu, J. Yu, D. Guo, Z. Li, Y. Su, J. Alloy. Compd. **815**, 152403 (2020)
16. L. He, J. Liu, Y. Liu, B. Cui, B. Hu, M. Wang, K. Tian, Y. Song, S. Wu, Z. Zhang, Appl. Catal. B **248**, 366–379 (2019)
17. M.Q. Zhao, C.E. Ren, Z. Ling, M.R. Lukatskaya, C. Zhang, K.L. Van Aken, M.W. Barsoum, Y. Gogotsi, Adv. Mater. **27**, 339–345 (2015)
18. X. Liu, F. Xu, Z. Li, Z. Liu, W. Yang, Y. Zhang, H. Fan, H.Y. Yang, Coord. Chem. Rev. **464**, 214544 (2022)
19. B.C. Wyatt, B. Anasori, Appl. Mater. Today **27**, 101451 (2022)
20. H. Li, Y. Hou, F. Wang, M.R. Lohe, X. Zhuang, L. Niu, X. Feng, Adv. Energy Mater. **7**, 1601847 (2017)

21. P. Yan, R. Zhang, J. Jia, C. Wu, A. Zhou, J. Xu, X. Zhang, J. Power. Sources **284**, 38–43 (2015)
22. S. Xu, G. Wei, J. Li, W. Han, Y. Gogotsi, J. Mater. Chem. A **5**, 17442–17451 (2017)
23. T. Zhou, C. Wu, Y. Wang, A.P. Tomsia, M. Li, E. Saiz, S. Fang, R.H. Baughman, L. Jiang, Q. Cheng, Nat. Commun. **11**, 2077 (2020)
24. X. Gao, X. Du, T.S. Mathis, M. Zhang, X. Wang, J. Shui, Y. Gogotsi, M. Xu, Nat. Commun. **11**, 6160 (2020)
25. M. Carey, M. Barsoum, Mater. Today Adv. **9**, 100120 (2021)
26. J. Jimmy, B. Kandasubramanian, Eur. Polym. J. **122**, 109367 (2020)
27. D. Parajuli, N. Murali, D. KC, B. Karki, K. Samatha, A.A. Kim, M. Park, B. Pant, Polymers **14**, 3433 (2022)
28. G.J. Adekoya, O.C. Adekoya, R.E. Sadiku, Y. Hamam, S.S. Ray, ACS Omega **7**, 39498–39519 (2022)
29. X. Li, J. Hao, R. Liu, H. He, Y. Wang, G. Liang, Y. Liu, G. Yuan, Z. Guo, Energy Storage Mater. **33**, 62–70 (2020)
30. M.S. Javed, A. Mateen, I. Hussain, A. Ahmad, M. Mubashir, S. Khan, M.A. Assiri, S.M. Eldin, S.S.A. Shah, W. Han, Energy Storage Mater. (2022)
31. Y. Wang, Y. Zheng, J. Zhao, Y. Li, Energy Storage Mater. **33**, 82–87 (2020)
32. X. Wang, W. Chen, Y. Liao, Q. Xiang, Y. Li, T. Wen, Z. Zhong, J. Mater. Sci. **56**, 2486–2496 (2021)
33. W. Luo, Y. Liu, F. Li, J. Huo, D. Zhao, J. Zhu, S. Guo, Appl. Surf. Sci. **523**, 146387 (2020)
34. V.-H. Nguyen, B.-S. Nguyen, C. Hu, C.C. Nguyen, D.L.T. Nguyen, M.T. Nguyen Dinh, D.-V.N. Vo, Q.T. Trinh, M. Shokouhimehr, A. Hasani, Nanomaterials **10**, 602 (2020)
35. K. Li, T. Jiao, R. Xing, G. Zou, J. Zhou, L. Zhang, Q. Peng, Sci. China Mater. **5**, 728–736 (2018)
36. R.A. Rather, A. Mehta, Y. Lu, M. Valant, M. Fang, W. Liu, Int. J. Hydrogen Energy **46**(2021), 21866–21872 (1888)
37. R.A. Rather, M. Khan, I.M. Lo, J. Catal. **366**, 28–36 (2018)
38. M. Li, H. Huang, J. Low, C. Gao, R. Long, Y. Xiong, Small Methods **3**, 1800388 (2019)
39. X. An, W. Wang, J. Wang, H. Duan, J. Shi, X. Yu, Phys. Chem. Chem. Phys. **20**, 11405–11411 (2018)
40. J. Ran, G. Gao, F.-T. Li, T.-Y. Ma, A. Du, S.-Z. Qiao, Nat. Commun. **8**, 13907 (2017)
41. W. Yem, ChemSusChem **11**, 1606–1611 (2018)
42. Y. Zhuang, Y. Liu, X. Meng, Appl. Surf. Sci. **496**, 143647 (2019)
43. Y. Li, Z. Yin, G. Ji, Z. Liang, Y. Xue, Y. Guo, J. Tian, X. Wang, H. Cui, Appl. Catal. B **246**, 12–20 (2019)
44. Y. Li, L. Ding, Y. Guo, Z. Liang, H. Cui, J. Tian, ACS Appl. Mater. Interfaces **11**, 41440–41447 (2019)
45. X. Liu, C. Chen, Mater. Lett. **261**, 127127 (2020)
46. V. Ramalingam, P. Varadhan, H.C. Fu, H. Kim, D. Zhang, S. Chen, L. Song, D. Ma, Y. Wang, H.N. Alshareef, Adv. Mater. **31**, 1903841 (2019)
47. L. Cheng, Q. Chen, J. Li, H. Liu, Appl. Catal. B **267**, 118379 (2020)
48. M. Zhang, J. Qin, S. Rajendran, X. Zhang, R. Liu, Chemsuschem **11**, 4226–4236 (2018)
49. Y. Li, L. Ding, Z. Liang, Y. Xue, H. Cui, J. Tian, Chem. Eng. J. **383**, 123178 (2020)
50. C. Peng, P. Wei, X. Li, Y. Liu, Y. Cao, H. Wang, H. Yu, F. Peng, L. Zhang, B. Zhang, Nano Energy **53**, 97–107 (2018)
51. L. Zhang, K. Khan, J. Zou, H. Zhang, Y. Li, Adv. Mater. Interfaces **6**, 1901329 (2019)
52. D.H. Ho, Y.Y. Choi, S.B. Jo, J.M. Myoung, J.H. Cho, Adv. Mater. **33**, 2005846 (2021)
53. F. Wang, C. Yang, M. Duan, Y. Tang, J. Zhu, Biosens. Bioelectron. **74**, 1022–1028 (2015)
54. X.-F. Yu, Y.-C. Li, J.-B. Cheng, Z.-B. Liu, Q.-Z. Li, W.-Z. Li, X. Yang, B. Xiao, ACS Appl. Mater. Interfaces **7**, 13707–13713 (2015)
55. F. Dixit, K. Zimmermann, R. Dutta, N.J. Prakash, B. Barbeau, M. Mohseni, B. Kandasubramanian, J. Hazard. Mater. **423**, 127050 (2022)
56. Y. Liu, R. Luo, Y. Li, J. Qi, C. Wang, J. Li, X. Sun, L. Wang, Chem. Eng. J. **347**, 731–740 (2018)

57. J. Yin, B. Ge, T. Jiao, Z. Qin, M. Yu, L. Zhang, Q. Zhang, Q. Peng, Langmuir **37**, 1267–1278 (2021)
58. Q. Zhang, J. Teng, G. Zou, Q. Peng, Q. Du, T. Jiao, J. Xiang, Nanoscale **8**, 7085–7093 (2016)

Chapter 5
MXenes for Energy Harvesting and Storage Applications

Rameez Ahmad Mir, Amardeep Amardeep, and Jian Liu

5.1 Introduction

Energy harvesting plays a crucial role in fabricating and utilizing new clean energy, self-powered electronic, and energy storage devices. Environmental pollution and the energy crisis are the major concerns faced by society. To address these prevailing issues, the need for new, clean, and renewable resources is increasing rapidly for sustainable energy. Hydrogen energy derived from water (a renewable source) is considered as a promising approach exhibiting a high-energy density and zero emission of toxic gases [1]. At the same time, the demand for self-powered microelectronic devices is rapidly increasing for sustainable development and smart living. For it, there is a need to develop micro-size efficient charge storage devices like new-generation batteries and supercapacitors that have attained special attention all over the globe to replace traditional bulk devices. Various materials and strategies have been developed for energy harvesting depending on their chemical and physical features. Among the developed materials, two-dimensional (2D) structures exhibiting idiosyncratic structural, morphological, and surface characteristics appeared as the potential source for energy production and storage devices. 2D materials, referring to sheet-like structures or thin films, exhibit an atomic layer thickness and high lateral dimensions [2]. MXenes, a special class of 2D transition metal carbides/nitrides (TMCs/TMNs), appeared as attractive materials for energy harvesting due to their unique physical and chemical properties. MXenes have the general formula

R. A. Mir (✉) · A. Amardeep · J. Liu
School of Engineering, Faculty of Applied Science, The University of British Columbia, 3333 University Way, Kelowna, BC V1V 1V7, Canada
e-mail: rameez.mir@ubc.ca

A. Amardeep
e-mail: amardeep.amardeep@ubc.ca

J. Liu
e-mail: jian.liu@ubc.ca

© The Author(s), under exclusive license to Springer Nature Singapore Pte Ltd. 2024

O. P. Pandey and P. Sharma (eds.), *MXenes: Emerging 2D Materials*,
https://doi.org/10.1007/978-981-97-4064-2_5

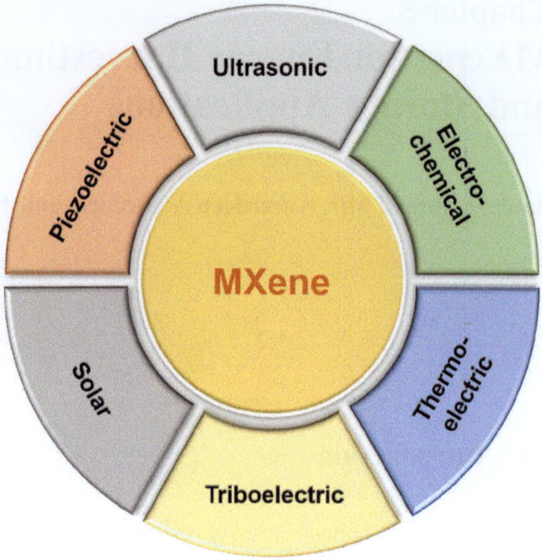

Fig. 5.1 Energy harvesting by MXene-based energy devices

$M_{n+1}X_nT_x$, where M is transition metal, X is carbon or nitrogen, and T is a functional group like fluorine, hydroxyl, or oxygen is a good source of energy harvesting and storage [3, 4]. MXene-based materials are used for harvesting different energy resources, as shown in Fig. 5.1.

Energy harvesting has a broader scope, so this chapter presents a perspective on using MXenes to harvest energy (production and storage) from different environment-friendly renewable energy sources [5]. MXenes emerged as demanding electrocatalyst materials for energy production, such as hydrogen fuel derived from water (water electrolysis) through hydrogen evolution reaction (HER)/oxygen evolution reaction (OER) and electrochemical energy storage (supercapacitors and batteries) due to their excellent electronic conductivity, large specific surface area, tuneable properties, and unique micrographical features [4, 6–8]. The typical electrochemical energy production and storage applications of MXenes are shown in Fig. 5.2.

The different characteristics and structural properties of MXenes, dependent on the synthesis methodologies, which influence energy production and storage abilities and have already been described [10]. Efforts have been made to investigate and improve the performance of 2D materials, especially MXenes, due to potentially prolonging Moore's law in energy production and storage [11]. Along with the unique chemical and physical properties of MXenes, a high population of active electrochemical sites or electrochemical active surface area (ECSA) within a small size and high mechanical stability widens their scope as electrocatalysts for hydrogen production. The essential features of MXenes, responsible for their improved charge storage performance, are high hydrophobicity and unique surface charge distribution [12]. The utilization of nanostructured MXenes gained special attention in

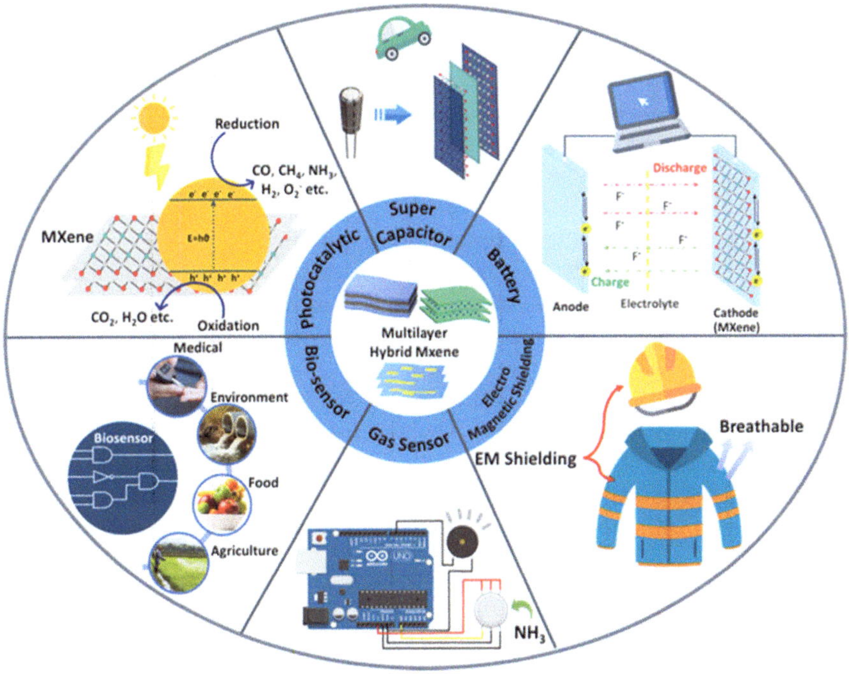

Fig. 5.2 Essential energy harvesting applications of MXenes [with permission Ref. [9] 2022 American Chemical Society]

energy production and storage due to the improvement of the surface features at the nanoscale and rapid operating mechanisms. The role of MXenes in hydrogen production, supercapacitors, and batteries are discussed as follows.

5.2 MXenes for Hydrogen Production (HER/OER)

Currently, the maximum portion of the energy is generated from traditional fossil fuels, and the reservoirs of these fossil fuels are diminishing. Moreover, the utilization of these fossil fuels generates toxic gases, which affect the environment and cause a critical issue of global warming. To address this problem, a global initiative is on the rise seeking the use of natural (renewable) and clean energy alternatives to diminishing traditional resources. Various natural energy resources, such as solar, wind, tidal, and biomass, are being explored for sustainable development. The regional or seasonal factors affect the intermittent availability of these energy sources [13]. Among the available natural resources, water-derived hydrogen (H_2) fuel exhibits different advantages, such as high purity green fuel (zero emission of toxic gases). Water electrolysis is considered as an efficient and suitable method for producing hydrogen either through HER or OER, as shown in Fig. 5.3a. HER at the cathode (water reduction) and OER at the anode (water oxidation) in the water electrolyzer systems are kinetically sluggish and require large overpotentials for water reduction/oxidation, resulting in low energy efficiency. The electrocatalyst is vital in lowering the water-splitting potential and enhancing efficiency. The prominent role of the electrocatalyst in reducing the overpotential and energy barrier for HER and OER is shown in Fig. 5.3b–c [14].

The theoretical potential to generate hydrogen from water is 1.23 V (Fig. 5.3c). However, in practical applications, the required potential is higher than the theoretical value due to the involvement of various resistances in the water electrolyzer systems [14]. Different electrocatalyst species have been developed and tested to lower the water oxidation and reduction potentials [16, 17]. To date, platinum (Pt) or Pt/C and ruthenium oxide (RuO_2), and iridium oxide (IrO_2) exhibited potential results for efficient HER and OER, respectively. The lower abundance and high cost of noble metal electrocatalysis limit their practical/commercial implications. Different metal and non-metal electrocatalysts specie as substitutes to noble metals have been developed, such as transition metal (TM) oxides (TMOs), transition metal dichalcogenides

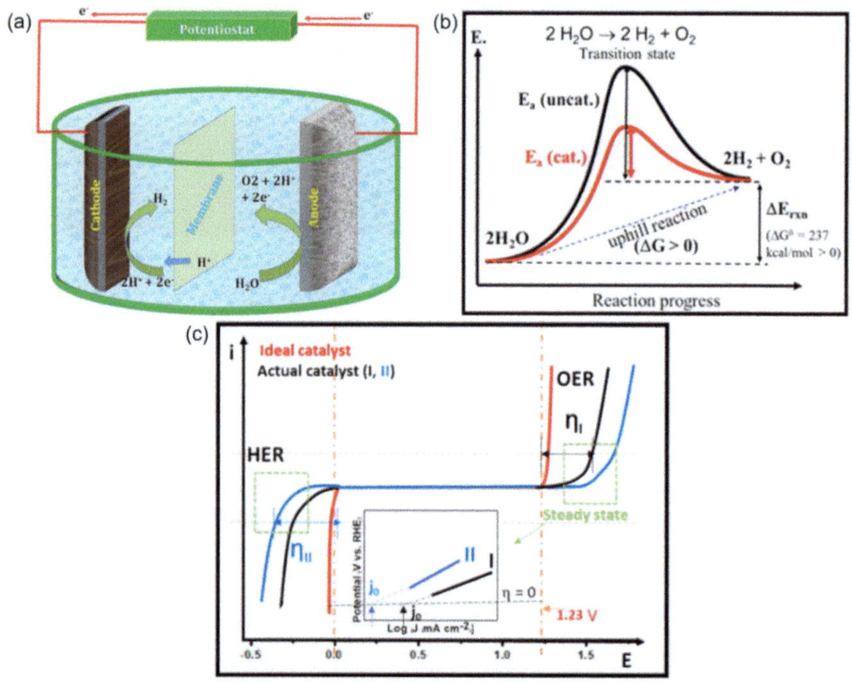

Fig. 5.3 Schematics diagram of water electrolysis, **a** HER/OER processes, [with permission Ref. [14] 2023 Elsevier] and **b–c** role of electrocatalyst in HER and OER [with permission Ref. [15] 2021 Springer]

(TMDCs), TM carbides (TMCs), TM nitrides (TMNs), non-metal supported combinations like graphene species, carbon nitride, and carbon supported electrocatalysts, etc. [14, 18–21]. With advances in 2D material electrocatalysts, especially graphene, graphene-like materials such as TMCs and TMNs showed potential to replace costly noble metal electrocatalysts due to their higher carrier mobility and intrinsic layered structure. One of the essential properties of TMCs and TMNs, which makes them attractive, is their similar electronic structure to noble metal elements like Pt and are considered as a potential candidate to replace Pt and IrO_2/RuO_2 electrocatalysts in HER and OER, respectively [22, 23]. To enhance the HER and OER activity and stability of the TMCs/TMNs, the 2D MXenes attained attention due to their exceptional intrinsic and extrinsic characteristics. The band structure of MXenes similar to Pt (due to carbon incorporation in a metal lattice), a higher number of edge sites and improved active sites due to attached functional groups on the surface, improved the water-splitting performance. Different MXene species based on TM (Ti, Mo, and V) have been developed and utilized as electrocatalyst candidates for HER and OER [24–26]. Modifying the terminal ends of MXenes enhances the HER performance by improving the conductivity and surface edge sites. For practical applications and higher efficiency, the long-term stability of the electrocatalyst is the need of time. Lim et al. [27] developed Mo_2CT_x MXene as a low cost, highly active, and stable electrocatalyst for HER. The hybrid of Mo_2CT_X/2H-MoS_2 MXene revealed higher stability and performance in an aqueous electrolyte. The synthesized hybrid electrocatalyst requires an overpotential of 119 and 182 mV to yield a current density of 10 and 100 mA cm^{-2}, respectively. The strong interfacial adhesion in the hybrid structure reduces the HER overpotential by tailoring the MXene sulfidation. Moreover, the 2H-MoS_2 layer over MXene suppresses the oxidation of the MXene layers. It restricts the excessive coverage or blockade of the active interfaces that improve the charge transfer rate (high current density) without deterioration. The hybrid phase exhibits exceptionally longer durability, up to 10 days/100,000 CV cycles without degradation. The results revealed that in-situ sulfidation is a promising strategy for creating intimate interfaces among different materials and TMDCs for widening their scope in energy conversion systems. Anasori et al. [28] developed ordered double-transition metal MXene $Mo_{2+\alpha}Nb_{2-\alpha}C_3T_x$ MAX phases ($0 \leq \alpha \leq 0.3$) with different compositions as a suitable electrocatalyst for HER, as shown in Fig. 5.4.

Ordering bimetals (Fig. 5.4b–c) such as Mo (outer layer) and Nb (inner layer) affects the HER activity. The synthesized MXene phase exhibits improved HER activity (higher cathodic current) with respect to other MXene phases due to the ordering of double TMs (Fig. 5.5a). The $Mo_{2+\alpha}Nb_{2-\alpha}C_3T_x$ phase requires a lower overpotential than other MXene phases with varying Nb content, as shown in Fig. 5.5b. The said phase also outperforms the other electrocatalysts due to its lower Tafel slope of 91 mV dec^{-1} (Fig. 5.5c), indicating improved charge transfer kinetics.

The influence of Mo and Nb content on conductivity and flake size (Fig. 5.5d) revealed that the HER performance is not likely affected by the flake size, but the occupancy of Mo and Nb (metal) sites plays a key role. The HER performance hasn't yet reached the level determined by costly Pt/C electrocatalysts. The availability of the MXene precursors (transition metals and carbon) on earth determines

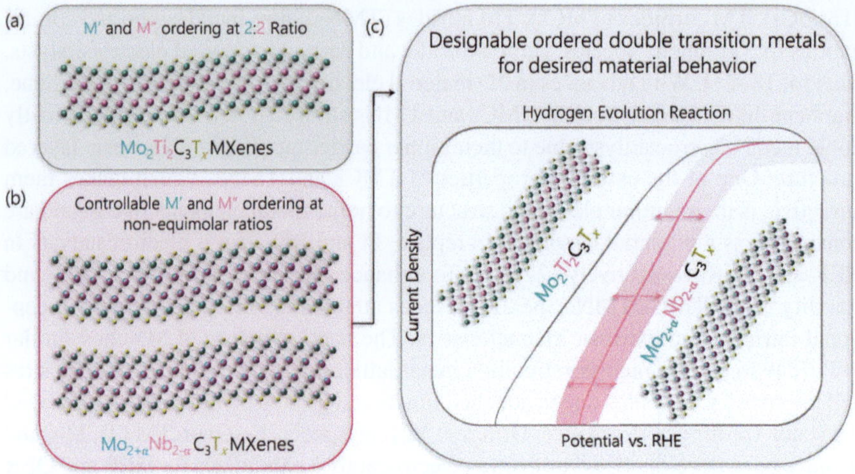

Fig. 5.4 **a** Ordering of $Mo_2Ti_2C_3T_x$ MXenes, where TMs are segregated into separate M atomic layers. **b** $Mo_{2+\alpha}Nb_{2-\alpha}C_3T_x$ MXenes allow controllable ordering in the M layers, and **c** The gained control in $Mo_{2+\alpha}Nb_{2-\alpha}C_3T_x$ MXenes, permits designable material behavior, as demonstrated in HER electrocatalytic applications [with permission Ref. [28] 2023 American Chemical Society]

the potential use of MXenes as a cost-effective electrocatalyst in water electrolyzers. Further modifications are required at both intrinsic and extrinsic levels to boost the activity and stability of MXenes compared to noble metal electrocatalysts. Along with HER in acidic electrolytes, electrocatalytic water splitting is preferred in alkaline electrolyzers. However, the need for suitable electrocatalysts is still in demand due to sluggish OER kinetics, a four-electron transfer reaction [29]. The hydrophobic nature of the MXenes allows the coverage of OH^- ions on the surface-active sites, which improves the formation of O_2, and the structural characteristics enhance the ability of O_2 evolution [30]. Yan et al. [31] synthesized nine atomic layered reactive surface modified (RSM) $V_4C_3T_x$ MXene. The surface modifications with the different amorphous metal hydroxides (Ni, Fe, and V hydroxides) formed during the in-situ process improve the inner skeleton's stability ($V_4C_3T_x$) without disrupting it. The synthesized nanohybrids exhibited higher OER stability for more than 70 h. The exceptional HER/OER activity and stability of the MXene nanohybrids present a significant step forward in exploring the MXenes as non-noble electrocatalysts for water splitting. Various categories of MXenes have been developed and explored as suitable electrocatalysts for HER and OER. The corresponding HER/OER performance of these MXenes is given in Table 5.1, exhibiting better electrocatalytic properties than other metal/non-metal electrocatalysts. However, the relatively higher cost due to typical synthesis procedures and a slightly higher overpotential than Pt/C still restricts their potential applications in commercial systems. Efforts are being made to synthesize them with easy and manageable strategies to improve their activity higher than Pt/C. The further exploration and design modification of the MXene

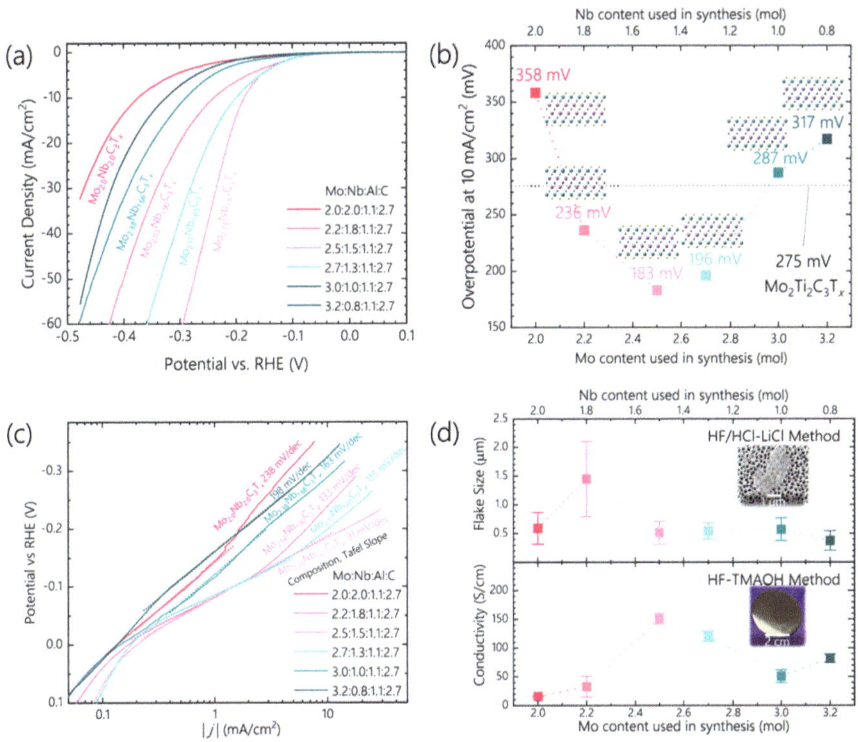

Fig. 5.5 HER activity of HF/HCl-LiCl-produced $Mo_{2+\alpha}Nb_{2-\alpha}C_3T_x$ MXenes. **a** I–V plots of $Mo_{2+\alpha}Nb_{2-\alpha}C_3T_x$ MXenes (Mo:Nb:Al:C composition in legend, actual Rietveld-derived composition next to data) used to derive overpotentials at 10 mA/cm^2 and Tafel slopes are shown in **b** and **c**, respectively. **d** Conductivity measurements of HF-TMAOH-synthesized $Mo_{2+\alpha}Nb_{2-\alpha}C_3T_x$ MXenes compared to flake sizes of HF/HCl-LiCl-synthesized $Mo_{2+\alpha}Nb_{2-\alpha}C_3T_x$ MXenes [with permission Ref. [28] 2023 American Chemical Society]

family contributes to the energy production application. Along with energy production applications, the MXenes also revealed exceptional energy storage performance in both supercapacitors as well in batteries, as discussed below.

5.3 MXenes for Supercapacitors

Supercapacitors, also known as ultracapacitors or electrochemical supercapacitors (ESs), are special systems for energy storage due to their rapid charging characteristics and high storage ability compared to conventional capacitors [10, 56, 57]. The fast adsorption and desorption of electrolyte ions on the electrode surface exhibited a higher efficiency than conventional capacitors. Figure 5.6 shows the basic design of the supercapacitors.

Table 5.1 Performance of different MXenes in HER and OER

Sample	Reaction	Tafel slope (mV dec^{-1})	Overpotential at 10 mA cm^{-2} (mV)	Electrolyte	References
MXene electrocatalysts for HER					
Mo_2CT_x	HER	74.0	189	0.5 mol/L H_2SO_4	[32]
$Ti_3C_2T_x$	HER	128.0	538	0.5 mol/L H_2SO_4	[33]
F-terminated Ti_2CT_x	HER	100.0	170	0.5 mol/L H_2SO_4	[34]
O-terminated $Ti_3C_2T_x$	HER	60.7	190	0.5 mol/L H_2SO_4	[35]
$V_4C_3T_x$	HER	168.0	200	0.5 mol/L H_2SO_4	[36]
N-Ti_2CT_x	HER	67.0	215	0.5 mol/L H_2SO_4	[37]
$Ti_3C_2T_x$ nanofibers	HER	97.0	169	0.5 mol/L H_2SO_4	[38]
$Ni_{0.9}Co_{0.1}$@Nb-$Ti_3C_2T_x$	HER	116.0	43.4	1.0 mol/L KOH	[39]
Pt/$Ti_3C_2T_x$-550	HER	32.3	32.7	0.1 mol/L $HClO_4$	[40]
TBA-$Ti_3C_2T_x$-Pt-20	HER	85.0	70	0.5 mol/L H_2SO_4	[41]
Co^{3+}-Cr_2CT_x	HER	112.0	404	1.0 mol/L KOH	[42]
Co^{3+}-V_2CT_x	HER	98.1	460	1.0 mol/L KOH	
Co^{3+}-Ti_2CT_x	HER	103.0	458	1.0 mol/L KOH	
Co-MoS_2@Mo_2CT_x	HER	82.0	112	1.0 mol/L KOH	[43]
PtNPs/$Ti_3C_2T_x$	HER	59.54	226	0.5 mol/L H_2SO_4	[44]
MoS_2-Ti_3C_2	HER	40.0	110	0.5 mol/L H_2SO_4	[45]
PtO_aPdO_bNPs@$Ti_3C_2T_x$	HER	70.0	57	0.5 mol/L H_2SO_4	[46]
MoS_2/Ti_3C_2 MXene@C	HER	45.0	135	0.5 mol/L H_2SO_4	[47]
MoS_2/$Ti_3C_2T_x$ nanoroll	HER	70.0	168	0.5 mol/L H_2SO_4	[48]

(continued)

Table 5.1 (continued)

Sample	Reaction	Tafel slope (mV dec^{-1})	Overpotential at 10 mA cm^{-2} (mV)	Electrolyte	References
CoP@3D $Ti_3C_2T_x$	HER	58.0	168	1.0 mol/L KOH	[49]
MXene electrocatalysts for OER					
FeNi-LDH/$Ti_3C_2T_x$	OER	43.0	298	1.0 mol/L KOH	[50]
$Ti_3C_2T_x$-CoBDC	OER	48.2	410	0.1 mol/L KOH	[51]
g-C_3N_4/$Ti_3C_2T_x$ film	OER	74.6	420	0.1 mol/L KOH	[52]
Co-Bi/$Ti_3C_2T_x$	OER	53.0	250	1.0 mol/L KOH	[25]
S-$NiFe_2O_4$@Ti_3C_2@NF	OER	46.8	270	1.0 mol/L KOH	[53]
NiCoS/$Ti_3C_2T_x$	OER	58.2	365	1.0 mol/L KOH	[54]
Co^{3+}-Ti_2CT_x	OER	63.2	420	1.0 mol/L KOH	[42]
PtO_aPdO_bNPs@$Ti_3C_2T_x$	OER	78.0	310	0.1 mol/L KOH	[46]
$Ni_{0.7}Fe_{0.3}PS_3$@ $Ti_3C_2T_x$	OER	36.5	282	1.0 mol/L KOH	[47]
Co/N-CNTs@$Ti_3C_2T_x$	OER	79.1	411	0.1 mol/L KOH	[55]
CoP@3D $Ti_3C_2T_x$	OER	51.0	280	1.0 mol/L KOH	[49]

The essential features of supercapacitors are that they exhibit high power density and long-term cyclic stability than traditional capacitors and batteries. Progress has been made to improve the performance, especially the energy density of the ESs, by developing suitable and stable electrode materials that enable the electron transfer and ion diffusion process to occur rapidly and reversibly. On account of the energy storage mechanism, ESs have been classified as electrochemical or electrical double-layer capacitors (EDLCs), pseudocapacitors, and newly introduced hybrid supercapacitors that store and release energy based on non-Faradic and Faradic interactions at the electrode–electrolyte interface, respectively [59–61]. In the case of EDLCs, a double-layer is formed between the ions in the electrolyte and the attracted electrons on the electrode surface. In contrast, the charge is stored in pseudocapacitors by the rapid, reversible Faradic reactions on the electrode surface. The adsorption of electrolyte ions is very fast in EDLCs and the Faradic charge transfer during the pseudocapacitive

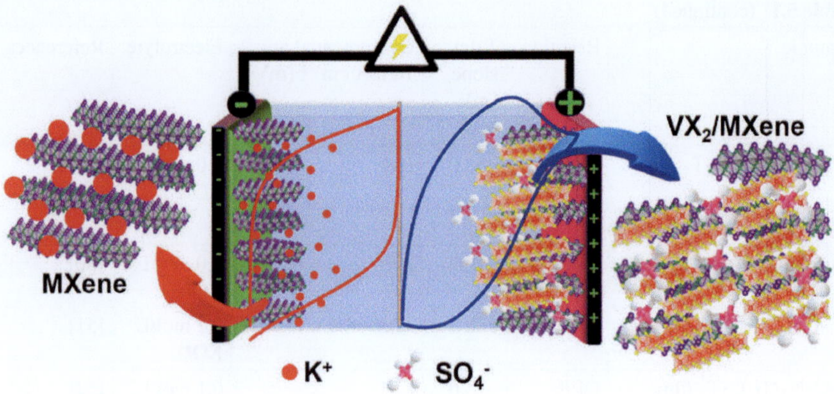

Fig. 5.6 Schematic diagram of electrochemical MXene-based supercapacitor [with permission ref. [58] 2021 American Chemical Society]

process is sluggish but stores more energy than EDLCs. The major drawback of the ESs based on aqueous electrolytes is the small voltage window restricted by the thermodynamic stability of water [62]. The prevailing issues with charge storage devices are the lower energy and power density exhibited by supercapacitors and batteries. To overcome these drawbacks, hybrid supercapacitors (HSCs) combine the energy storage mechanism revealed by the capacitors and batteries. HSCs consist of different electrode materials, a battery or pseudocapacitors type anode and a capacitor type cathode to improve energy and power density, respectively. Carbon-based electrodes are common in EDLCs, pseudocapacitors, and HSCs. Carbon materials exhibiting higher surface area act as electrodes for EDLCs revealing high power density, but the energy density is still not comparable to batteries. Efforts have been made to improve the energy density of ESs. The pseudocapacitors exhibited higher charge transfer kinetics based on the redox reaction mechanism and enhanced the energy density of the ESs. Advanced electrode materials are being searched and developed with a high specific area, high conductivity, and rich redox sites that are highly desired for developing high-performance supercapacitors [10]. To date, different categories of electrodes have been developed, like carbon species, CNTs, mesoporous carbon, carbon shells, transition metal oxides (TMOs), 2D materials like graphene species, transition metal dichalcogenides (TMDCs), transition metal carbides (TMCs) and transition metal nitrides (TMNs). The electrochemical charge storage ability of the well-known 2D material (graphene) is finite due to the absence of redox reactions and the main portion of capacitance arises due to EDLC mechanism. TMOs, especially like MnO_2 and RuO_2 being pseudocapacitive materials, exhibited higher capacitance and energy densities, but the cyclic stability is still lower than the carbon electrodes. Among the developed materials, the 2D TMCs/TMDs, especially the MXenes, have attained special attention as electrode materials for ESs due to a particular layered

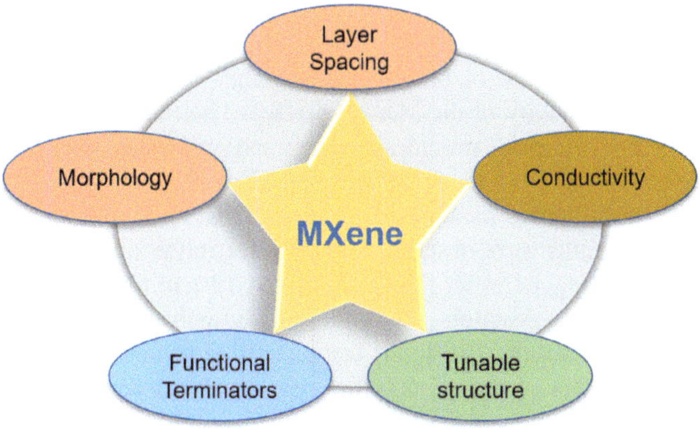

Fig. 5.7 Required features for MXene-based electrodes for energy storage

structure, faster electrolyte ion transport kinetics, and higher number of surface-active redox sites [10]. Figure 5.7 shows the main characteristics of MXenes that enhance the charge storage performance.

The distance between the layers of the MXene electrode forms the channels to improve the ion transport favorable for EDLCs. However, the outermost TMO or the functional groups on the surface layer provides redox-rich sites necessary for making pseudocapacitive energy storage possible. The inner layer of TMCs in MXenes exhibited excellent electron conductivity [57]. Various strategies have been followed to improve the charge storage ability of MXenes, such as redesigning electrodes, regulating surface functional groups, and controllable synthesis of the MXene phase [63]. Moreover, the typical synthesis procedure of the MXenes still restricts their potential applications in supercapacitor technology compared to carbon species and other metal compounds. Different MXene species have been suggested based on transition metals, but only a few have been practically synthesized [63–66]. Due to the complex chemical composition and structure, the electrolyte ion interaction of MXenes with different electrolytes (a combination of organic and aqueous electrolytes) has not been explored yet. Among all MXenes, Ti_3C_2 MXene is mostly studied due to its higher stability and ease of fabrication [67]. Most of the studies revealed that the supercapacitor performance of MXenes is highly dependent on the layered structure and the functional groups on the surface [63]. For improved energy storage in MXene-based supercapacitors, the O-terminated groups provide more electrochemical reaction sites than the –OH and –F groups [68]. Xie et al. [69] confirmed that the Li-ion storage ability of 2D MXenes depends on the surface functional groups. The –OH groups are deliberately removed through heat treatment or electrochemical reactions during the initial cycling (lithiation cycle). The lithiated oxygen-terminated surfaces of MXenes are capable of absorbing more Li-ions and provide a way to increase their capacity. The corresponding effects were observed in the delaminated MXenes, through XAS and were also revealed by the DFT studies.

The capacitance of the MXenes is mainly affected by the surface functional group that influences the conductivity and hydrophilicity. The development of the surface functional groups during the etching of the MAX phase brings hydrophilicity and affects the electronegativity of the MXene surfaces [70]. The electrolyte ion interaction with MXene affects its structural features and performance. Lukatskaya et al. [57] observed a downshift in the XRD pattern of Ti_3C_2Tx, which can be attributed to an increase in the lattice parameter (c-axis of $Ti_3C_2T_x$) in the KOH electrolyte solution. Similar results were observed in other electrolyte solutions due to cation intercalation [71]. Gogotsi et al. [72] synthesized 13-μm thick $Ti_3C_2T_x$ MXene-based electrochemical capacitors and achieved a gravimetric capacitance of 310 Fg^{-1} in 3 M H_2SO_4 solution. The highly ion-accessible MXene electrodes exhibited higher capacitance than graphene and other commercial capacitors [73–77]. Modifying the structural features of the MXene electrodes enhances the accessibility to the redox-active sites. Mesoporous $Ti_3C_2T_x$ MXene revealed a capacitance of 210 F g^{-1} even at 10 mV s^{-1}. The MXenes can deliver a ~1500 F cm^{-3} capacitance higher than state-of-the-art supercapacitor electrodes [73]. Agnese et al. [78] also studied the influence of surface modifications on the supercapacitor performance of Ti_3C_2 MXene in an H_2SO_4 solution. The two-way chemically surface modification of the Ti_3C_2 MXene is given in Fig. 5.8. The resultant product d-Ti_3C_2 (Fig. 5.8b) is due to the delamination of layers and the functional group containing Ti_3C_2 MXene (Fig. 5.8c–d) resulting from the K^+ intercalation mechanism.

The enhancement of capacitance performance in the case of MXenes revealed by the electrochemical testing was attributed to the higher number of oxygen-containing

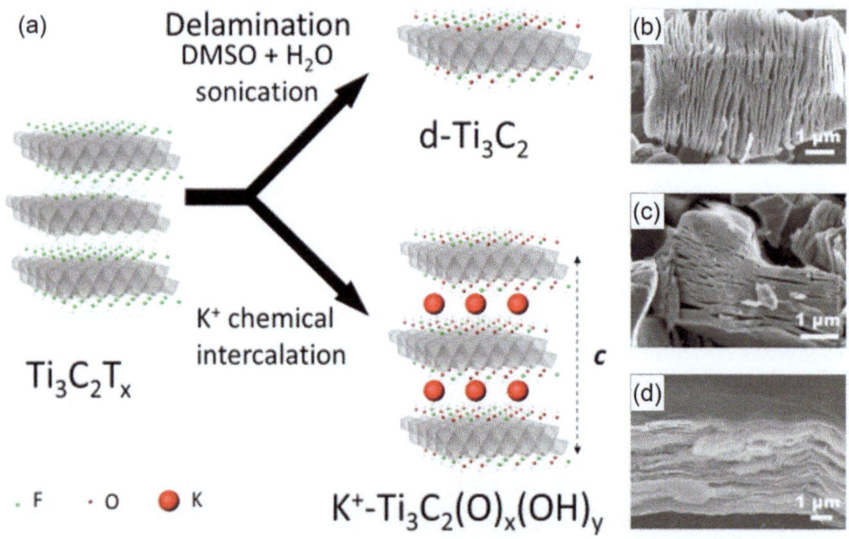

Fig. 5.8 **a** Schematic representation of the modifications of $Ti_3C_2T_x$ MXene, scanning electron microscope (SEM) images of **b** $Ti_3C_2T_x$, **c** KOH-Ti_3C_2, and **d** d-Ti_3C_2 [with permission [78] 2014 Elsevier]

functional groups. The CV plots of the synthesized material tested at 10 mV s^{-1} are shown in Fig. 5.9a. The concentrated electrolyte solution (1 M H_2SO_4 electrolyte) does not affect the lattice parameters of MXene due to the smaller size of H^+ ions compared to K^+ ions. The delaminated d-Ti_3C_2 exhibited a higher capacitance of 520 F cm^{-3} due to its thin electrode size and better charge transfer kinetics. However, the d-Ti_3C_2 MXene shows more dependency on the scan rate (Fig. 5.9b) due to the alignment of flakes parallel to the current collector.

The functional group attached to $Ti_3C_2T_x$, responsible for redox reaction (pseudocapacitive) in acidic electrolytes, was also confirmed by broad peaks in the CV curves. The transition metal (Ti in this case) in MXene can change its oxidation states from 3^+ to 4^+ due to surface redox reactions. A long-term stability test (GCD test) performed for MXenes (d-Ti_3C_2 and KOH-Ti_3C_2) shows no degradation even after 10,000 cycles (Fig. 5.9c–d). The results revealed that the surface modification of 2D MXenes is essential for improving the energy storage performance of MXene-based ESs. Zhou et al. [65] applied the DFT criteria to explore, design, and assemble the Mo_2CT_x MXenes for supercapacitors. The supercapacitor performance of the Mo_2CT_x MXene was investigated in various electrolytes such as 1 M H_2SO_4, KOH, and $MgSO_4$. The adsorption of hydrogen (H), potassium (K), and magnesium (Mg)

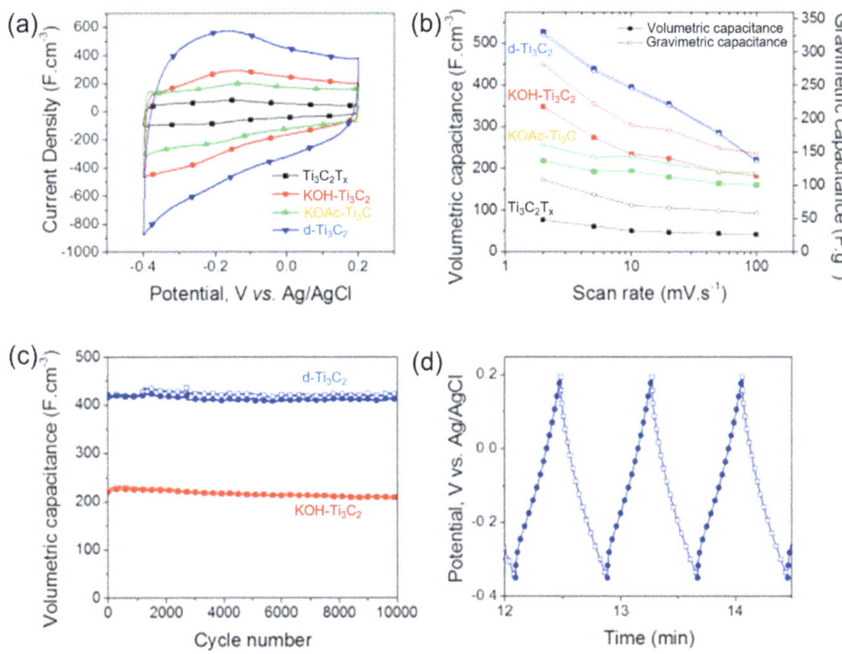

Fig. 5.9 Electrochemical activity of Ti_3C_2 MXene **a** CV profiles at 10 mV s^{-1}. **b** Summary of rate performances. **c** Charge and discharge volumetric capacitance versus cycle number of KOH–Ti_3C_2 and d-Ti_3C_2 electrodes from galvanostatic cycling at 5 A g^{-1}. **d** Galvanostatic charge–discharge profile of d-Ti_3C_2 [with permission Ref. [78] 2014 Elsevier]

on the surface of Mo_2CT_x MXene has been explored. The specific capacitance (70.14 Fg^{-1}) obtained in 1 M H_2SO_4 was more than that of KOH (11.27 Fg^{-1}) and $MgSO_4$ (18.97 Fg^{-1}). The Mo_2CT_x MXene shows a high retention rate of 89.9% (KOH), 96% ($MgSO_4$), and 98% (H_2SO_4) over 5000 cycles. After that, different MXene materials have been developed as supercapacitors based on the transition metal, electrode type, and electrolyte [73, 79]. Along with pristine MXenes, the nanocomposite of the MXenes has gained research attention as supercapacitor candidates due to the improvement in the surface properties necessary for enhancing the storage performance [79]. Luo et al. [64] explored that the tailoring of the interlayer spacing was controlled in accordance with the intercalated agent. The morphological features of the Ti_3C_2 MXene before and after the pillaring process are shown in Fig. 5.10.

The SEM micrograph of CTAB@Ti_3C_2 (Fig. 5.10a) showed the presence of some irregular particles on the Ti_3C_2 matrix, particularly the anchored CTAB. Almost no particles were observed in the Sn(IV)@Ti_3C_2 composites synthesized without the CTAB prepillaring process (Fig. 5.10b). No clear particles can be detected in the Sn(IV) nano complex assisted by CTAB prepillaring process (Fig. 5.10c–d). The TEM image of CTAB–Sn(IV)@Ti_3C_2 (Fig. 5.10e–f) shows that the Sn(IV) nanocomplexes (dot-shaped) are uniformly distributed in the CTAB@Ti_3C_2 matrix. The particle size was reduced to 2–5 nm (Fig. 5.10f) due CTAB effect, acting as the surfactant and the prepillaring agent [64, 80]. Figure 5.10g shows a clear layered structure of the MXene complex, along with the uniform distribution of constituents as observed in the mapping pattern (Fig. 5.10h–i). The remarkable energy storage performance of the developed MXene complex (CTAB–Sn(IV)@Ti_3C_2) was tested as an anode in the hybrid capacitor with activated carbon (AC) as a cathode, as shown in Fig. 5.11a.

The CV (Fig. 5.11b) plots performed at different scan rates in a wide voltage window of 1.0–4.0 V show no electrolyte decomposition. The deflection from the regular rectangular CV curve attributes to the redox reactions on the terminated surface functional groups. The fabricated hybrid capacitor revealed an energy density of 239.50 Wh kg^{-1} due to the pillaring effect. The results revealed that incorporation of the CTAB–Sn(IV)@Ti_3C_2 MXene anode with commercial AC cathode showed a better rate capability necessary for higher energy and power density supercapacitors (Fig. 5.11c). The capacitance retention is outstanding till 4000 cycles with negligible loss as shown in Fig. 5.11d. Moreover, along with the superior capacitance performance of MXenes, the other advantages such as atomic layer thickness, exceptional hydrophilic properties, micro/nanoscale dimensions, and special mechanical characteristics enable them (MXenes) to be converted into flexible electrodes. The high conductivity of MXenes attributes to higher power density than conventional semiconductor metal oxides and also provides a possible way to fulfill the need for rapid charging materials. This led to the development of semiconductor-type MXene-based supercapacitors. Qin et al. [81] developed semitransparent flexible photovoltaic supercapacitor (PSC) and transparent MXene supercapacitors by incorporating a flexible organic photovoltaic (OPV) with $Ti_3C_2T_x$ and bare $Ti_3C_2T_x$ MXene as the electrodes, respectively. The synthesis, morphological features, and size of the developed flakes are shown in Fig. 5.12. MXene integration with a flexible

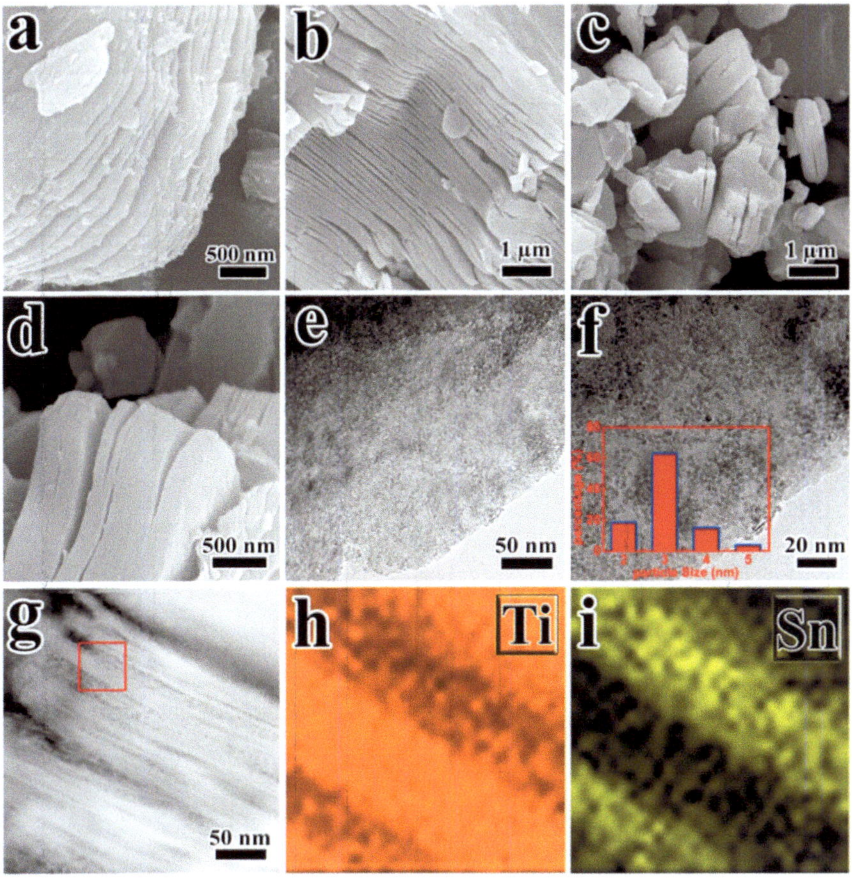

Fig. 5.10 **a** SEM image of CTAB@Ti_3C_2. **b** SEM image of Sn(IV)@Ti_3C_2 without cetyltrimethylammonium bromide (CTAB) prepillaring process. **c–f** SEM and TEM images of CTAB–Sn(IV)@Ti_3C_2. Inset in (**f**): lateral size distribution of the anchored Sn(IV) nano complex. **g–i** STEM image of CTAB–Sn(IV)@Ti_3C_2 and corresponding elemental mapping of Ti and Sn [with permission Ref. [64] 2017 American Chemical Society]

OPV photoelectrode as PSC exhibited an enhanced power conversion efficiency of 13.6%, higher than commercial indium tin oxide (ITO).

The transparent MXene-based supercapacitor exhibited a high volumetric capacitance of 502 F cm^{-3} with improved stability up to 10,000 cycles. The supercapacitor performance of the developed MXene at varying scan rates and applied currents is shown in Fig. 5.13. The cyclic voltammetry (Fig. 5.13b) plots show the retention of a nearly rectangular shape even at a higher scan rate (500 mV s^{-1}), which revealed the excellent reversible capacitive behavior of the electrode material. The results also revealed the potential applicability of the developed material as efficient electrodes for energy storage devices even at higher scan rates (>10 mV s^{-1}). The symmetric shape of the GCD curve (Fig. 5.13c) revealed that the charge/discharge process agrees

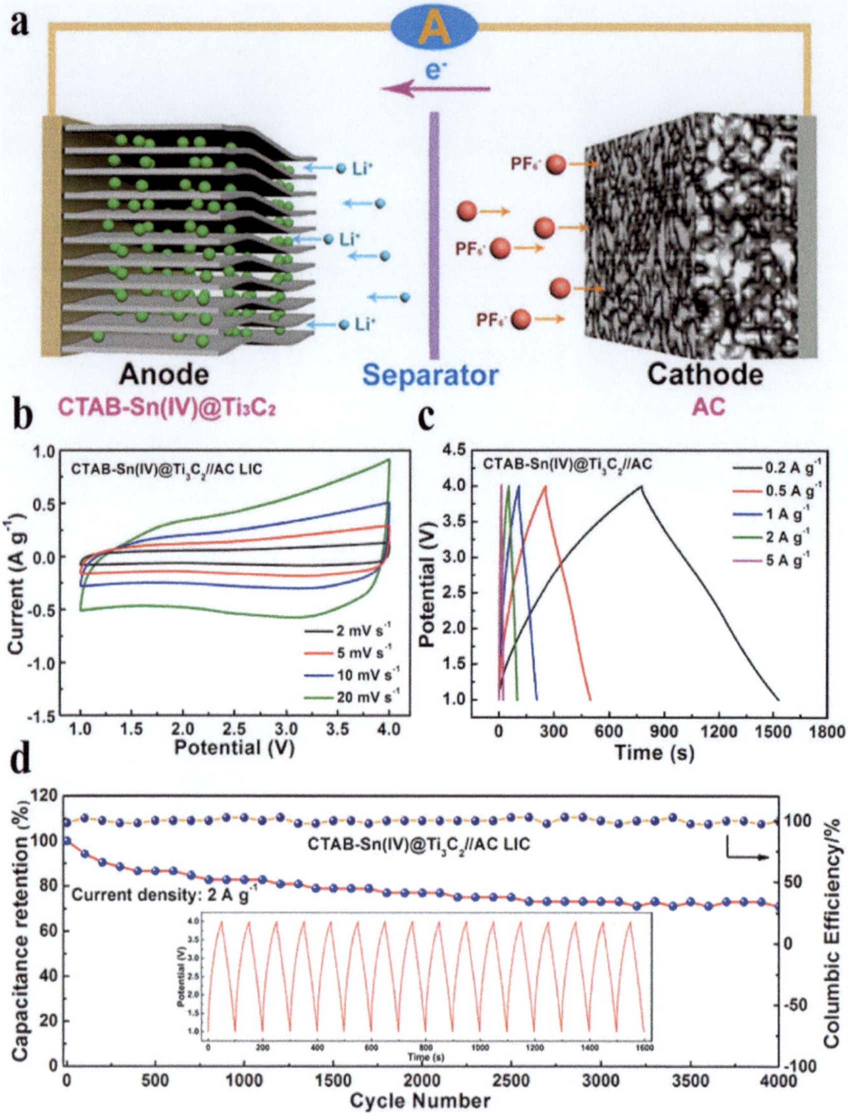

Fig. 5.11 **a** The charging process of CTAB–Sn(IV)@Ti_3C_2//AC lithium ion capacitor (LIC). **b** CV curves of CTAB–Sn(IV)@Ti_3C_2//AC LIC at different scan rates. **c** Typical charge–discharge curves of CTAB–Sn(IV)@Ti_3C_2//AC LIC at different current densities. **d** Long-term cycling performance of LIC at 2 A g^{-1}. Inset in (**d**): charge–discharge curves at 2 A g^{-1} [with permission Ref. [64] 2017 American Chemical Society]

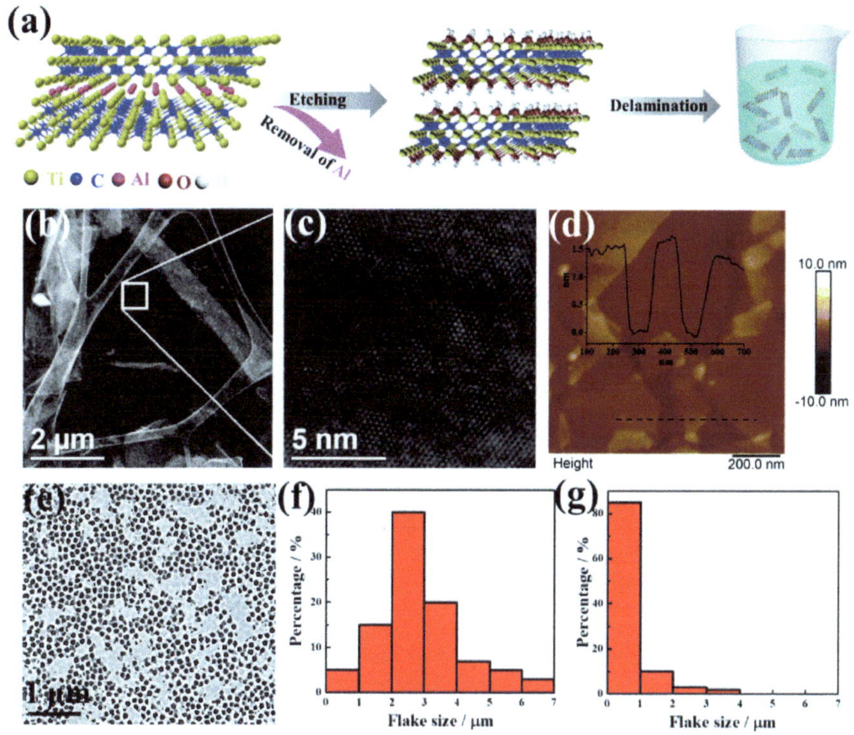

Fig. 5.12 Synthesis of the $Ti_3C_2T_x$ MXene colloidal solution. **a** preparation of MXene from the Ti_3AlC_2 precursor using the minimally intensive layer delamination (MILD) method. **b** Low-magnification and **c** high-magnification STEM images of single $Ti_3C_2T_x$ sheets dispersed onto a lacy carbon grid. **d** AFM image of a $Ti_3C_2T_x$ single layer. **e** Top view SEM image of small-sized $Ti_3C_2T_x$. The $Ti_3C_2T_x$ flake size distributions **f** before and **g** after sonication [with permission Ref. [81] 2017 (open access) Royal Society of Chemistry (RSC)]

with the CV results. The corresponding capacitance values dependent on the applied current and scan rate are given in Fig. 5.13d.

The stability of the MXene supercapacitors was analyzed by performing GCD measurements till 10,000 cycles, and the device shows a retention of 95% as that of initial capacitance. The long-term durability (Fig. 5.13e) of the fabricated supercapacitor is attributed to the non-aqueous organic solid ionogel electrolyte. This type of electrolyte overcomes the leakage issues faced in aqueous electrolytes. In addition, the MXenes can form a flexible supercapacitor, as shown in Fig. 5.13f. The obtained features like high-capacitance, long-term stability, and flexibility as different bending angles show the promising applications of $Ti_3C_2T_x$ Mxene as commercial supercapacitors. The developed PSC exhibited a high transmittance of over 33.5% and an exceptional storage capacity of 88%. This strategy widens the scope of developing the MXene-based, highly active, and stable flexible PSCs essential for printable electronics for future technologies.

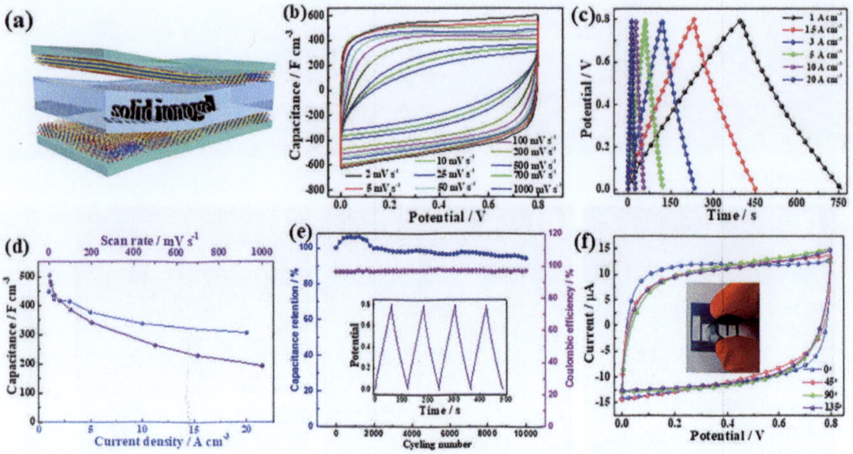

Fig. 5.13 Electrochemical performances of transparent, flexible, solid-state supercapacitors based on $Ti_3C_2T_x$ films. **a** Schematic illustration of a flexible solid-state supercapacitor based on spin coating $Ti_3C_2T_x$ MXene on polyethylene terephthalate (PET) sheets as electrodes and solid ionogel as the electrolyte. **b** CV curves at various scan rates and **c** galvanostatic charge/discharge (GCD) curves at different current densities for the flexible solid-state device. **d** The volumetric capacitances obtained from GCD and CV. **e** Long-term charge–discharge cycling performance and Coulombic efficiency of the $Ti_3C_2T_x$-based flexible solid-state device. Inset in (**e**): the typical GCD curves upon cycling. **f** CV curves of the $Ti_3C_2T_x$ device bent to different angles at 50 mV s^{-1}. Inset in (**f**): optical image of the transparent flexible ultra-thin supercapacitor [with permission Ref. [81] 2017 (open access) Royal Society of Chemistry (RSC)]

5.4 MXenes for Batteries

The exploitation of fossil fuel reservoirs due to the expanded energy usage has forced researchers to develop alternate energy generation technologies, such as solar, water, and wind. Although the aforementioned promising renewable technologies remarkably contribute toward energy production; however, they are geographically intermittent. Thus, the urgency of the energy storage devices was realized, which can efficiently function and power the required gadgets irrespective of the geography and surrounding atmosphere. When discussing energy storage devices, batteries, especially lithium batteries, have demonstrated superior electrochemical performance over other alternatives. Despite the supremacy of lithium batteries, they suffer from low natural abundance and safety aspects due to the hazardous undesirable Li-dendrite formation. Thus, limiting their usage in high-energy density lithium metal batteries. Therefore, the alternatives with huge profuse in earth's crust, such as monovalent, sodium (Na), potassium (K) batteries, and multivalent, magnesium (Mg), calcium (Ca), zinc (Zn), aluminum (Al) batteries, etc., required to be extensively explored to cope with the pressing global demand.

In the above context, advanced material design is essential, which can store the corresponding charged species, such as Li^+, Na^+, K^+, Mg^{2+}, Ca^{2+}, Zn^{2+}, and Al^{3+}, within its structure for respective metal-ion batteries. Graphite, carbonaceous, and 2D layered materials have been of great interest that are capable of reversibly storing specific metal ions. For instance, graphite and its derivatives (graphene oxide, reduced graphene oxide, graphene) are suitable for Li and K-ion batteries; however, they cannot intercalate Na-ions due to specific layer spacing and higher activation energy. Similarly, hard carbon is ideal for Na-ion batteries compared to the alternatives. Therefore, significant efforts were devoted toward extensive research innovation to develop a new class of 2D materials capable of intercalating the above-mentioned ions for respective metal-ion batteries. Eventually, perseverance was attained in the last decade by introducing a family of different 2D layered structured materials like MXenes. The structure of the MXene phase can be tuned during the transformation step of the etching of the MAX phase, which plays a vital role in its application type. For example, smaller interlayer spacing suffices for intercalating smaller ions, i.e., Al^{3+}, Li^{+}, Mg^{2+}, Zn^{2+}, etc., for respective metal-ion batteries; in contrast, larger ions, i.e., Na^{+}, K^{+}, Ca^{2+}, etc. demand a larger interlayer spacing. Therefore, the same MXene is not employable for each kind of metal-ion battery, which necessitates the interlayer-cum-structure tuning of MXenes for applicability in distinct metal-ion batteries due to their inherently intercalating demand. The MXene's ($M_{n+1}X_nT_x$) structure can be tuned by introducing a combination of suitable early transition metal (M = Ti, V, Sc, Mo, etc.), carbon or nitrogen (X), the functional groups (T_x = O, F, OH, Cl, etc.), and the process adopted, where M and X support the 2D skeleton, and n and functional groups participate in tuning the interlayer spacing. The process depends on structural and/or layer spacing modification of MXenes for battery application, which is described in details in references [82–84].

In the above structural and morphological context of MXenes, Greaves et al. [85] have summarized the various shapes of MXenes, which can be transformed into crumpled flocs, delaminated sheets, hollow spheres, nanoribbons, nanopillars, and composites, according to the required applications for different metal-ion batteries. Moreover, Ming et al. [84] have discussed the individual usage of modified MXenes via structural engineering, regulating interlayer spacing, substitution, and surface modification, forming their derivatives, and making hybrid composites for employing them in suitable battery applications, such as single/alkali-ion batteries (i.e., Li^{+}, Na^{+}, K^{+}, etc.), multivalent-ion batteries (i.e., Al^{3+}, Mg^{2+}, Zn^{2+}, etc.), and metal batteries. The flexibility of interlayer tuning via employing different synthesis routes facilitates suitable ionic diffusion within its layered structure sans structural deterioration, resulting in superior rate performance and better electrochemical cyclic stability. The previously mentioned manipulatable property of MXenes makes them more suitable electrodes, especially intercalation-type, than commercially used graphite or other 2D carbon materials for metal batteries. Another major difference between graphite and MXene is their ion-storing potential. In graphite, the lithium-ion gets intercalated close to 0.1 V versus Li/Li^{+}, which is vulnerable to Li-plating, raising serious safety concerns. Whereas MXenes store lithium ions at around 0.5 V versus Li/Li^{+}, avoiding the possibility of lithium plating, thus, eliminating the safety risks. Therefore, employing MXenes instead of graphite in lithium batteries improves the safety aspects; however, the cell voltage gets sacrificed due to the higher storing potential of MXenes.

Electronic properties of electrode materials are also a key requirement or beneficial factor and play a critical role in enhancing the electrochemical performance of energy storage devices. In addition to higher ionic diffusion, MXenes exhibit excellent electronic conductivity similar to its competitor graphene due to the available pool of electrons. Researchers have revealed that most of the MXenes have metal-like electronic properties; however, they can be significantly tuned by surface terminations or surface functional groups [86]. For example, $Ti_3C_2(OH)_2$, i.e., OH functional group, exhibits a narrow band gap of 0.05 eV, whereas $Ti_3C_2F_2$, i.e., F functional group, exhibits a wide band gap of 0.1 eV [87]. Not only the surface terminations but orbital hybridization of functional groups, viz., *p* orbitals and transitions metals' *d* orbital could lead MXenes toward semiconductor behavior from metallic behavior. Moreover, the transition metal *M* in MXenes is crucial in delivering the necessary electronic conductivity. For instance, substituting Ti with Mo in titanium carbide-based MXenes transforms the metallic behavior into semiconducting behavior. Therefore, the surface functional groups and their hybridization, transition metals, and their arrangements could vastly alter the electronic conductivity and related properties. The superior electronic conductivity was attributed to the highly aligned coplanar arrangement of MXene flakes/layers possessing metallic free-electron density. The higher electronic conductivity of MXenes facilitates swift electron transport and charge transfer within the electrodes and provides good electrical connectivity, resulting in improved electrochemical performance, proving to be a potential electrodes for electrochemical energy storage devices, i.e., batteries.

As per the above discussion regarding the suitability of MXenes in batteries, major research focuses on employing MXenes via compositing with active anodes due to the possibility of self-discharge from MXenes' layers having larger interlayer spacing. In this regard, synthesized MXene-anode composites are used in monovalent lithium-ion and sodium-ion batteries, as shown in Fig. 5.14 [88, 89]. Figure 5.14a reveals the SEM image of the as-prepared $Li_4Ti_5O_{12}$-MXene ($Ti_3C_2T_x$) composite, revealing a clear laminated 2D layered flake-shaped morphology corresponding to $Ti_3C_2T_x$ MXene structure. $Li_4Ti_5O_{12}$ has nucleated from the defect sites in the MXene, which can be observed on the sides of the MXene layers. Such composite morphology was attained by etching the MAX (Ti_3AlC_2) phase with HF resulting in MXene ($Ti_3C_2T_x$), followed by treating MXene with LiOH in the presence of H_2O_2 to obtain $Li_4Ti_5O_{12}$-MXene composite. A similar synthesis process was adopted by Huang et al. [89] to achieve sandwich-type $Na_{0.23}TiO_2$-MXene (Ti_3C_2) composite by treating Ti_3C_2 MXene with NaOH and oxidant. The sandwich-type composite morphology is shown in Fig. 5.14b. The synthesized MXene composites were tested in lithium-ion half cells to evaluate their performance. Figure 5.14c shows the cyclic voltammogram (CV) of $Li_4Ti_5O_{12}$-MXene composite against pristine lithium in a half cell at various scan rates viz., 0.2, 0.4, 0.8, 1, and 2 mV s^{-1}. The CV curves demonstrate an excellent rate capability due to negligible distortion of the CV profiles on varying the scan rate. The superior rate performance corresponds to the unique morphological features and the presence of MXene. The $Li_4Ti_5O_{12}$ particles are grown on the surface of MXene, thus, directly attached to them, and MXene is directly in contact with the electrolyte, allowing easy facilitation of Li-ions from the electrolyte to MXene and $Li_4Ti_5O_{12}$ particles. Whereas MXene also facilitates fast electron conduction, resulting in an interconnected path of electrons throughout the structure, enhancing the reaction rate during intercalation/de-intercalation of Li-ions. Therefore, $Li_4Ti_5O_{12}$-Mxene composite exhibits both fast Li-ion diffusion and rapid electronic conduction, proving superior rate performance. Similarly, Fig. 5.14d demonstrates the first three CV cycles of sandwich-type $Na_{0.23}TiO_2$-MXene composite at a scan rate of 1 mV s^{-1}. Except for the first cycle, which includes the formation of solid-electrolyte interphase (SEI), the following cycles overlap with each other, revealing good reversibility. The long galvanostatic cycling of $Li_4Ti_5O_{12}$-Mxene composite was evaluated, as shown in Fig. 5.14e, which revealed a discharge capacity of 178 mAh g^{-1} after 500 cycles at a high current density of 5 A g^{-1} with maintaining the Coulombic efficiency of nearly 100%. This excellent electrochemical cycling stability was assigned to the structural stability, unique morphology, minimal volume expansion, and rapid ionic and electronic diffusion/conduction. The gradual rise in reversible capacity over the cycling is due to the widening of the MXenes' interlayer spacing via trapping Li-ions and infiltration of non-coordinated solvent species. A similar observation was made for sandwich-type $Na_{0.23}TiO_2$-MXene composite at the same current density of 5 A g^{-1}, as shown in Fig. 5.14f. Furthermore, sandwich-type $Na_{0.23}TiO_2$-MXene composite was subjected to galvanostatic cycling against sodium (in a sodium-ion half cell), as shown in Fig. 5.14g. The remarkable cyclic stability was attributed to the above explanation. However, the size of Na-ion is larger than the Li-ions, the expanded MXenes having more interlayer spacing can explain Na-ion storage, which can successfully intercalate/de-intercalate larger Na-ions. Therefore, MXenes or MXene-based composites can easily store monovalent or alkali metal ions for alkali metal(ion) batteries with superior electrochemical performance.

MXenes have demonstrated the capability of storing monovalent alkali-ions and possess the potential to intercalate multivalent ions, such as Al^{3+}, Mg^{2+}, Zn^{2+}, etc., for respective metal-ion batteries. For instance, Xu et al. [90] revealed the storage of Mg^{2+} ions by Ti_3C_2 MXene via pre-storing a cationic surfactant (CTAB). Along with the experimental shreds of evidence, theoretical simulations also verified that the Mg-ion diffusion barrier could be reduced by intercalating CTA^+ cations. Therefore, Mg-ions intercalation/de-intercalation between MXene layers gets significantly improved, resulting in the volumetric capacity of 300 mAh cm^{-3} at a current density of 50 mA g^{-1}. Moreover, Vahid Mohammadi et al. [91] employed MXenes for Al^{3+} storage for high-capacity Al-ion batteries, as presented in Fig. 5.15. Figure 5.15a revealed the typical MXene layered structure of V_2CT_x after etching the corresponding MAX phase (V_2AlC) with concentrated HF for 92 h at room temperature. The galvanostatic potential profile (Fig. 5.15b) indicated a substantial irreversible discharge capacity of about 335 mAh g^{-1} at a current density of 10 mA g^{-1}, which corresponds to chloroaluminate anions' dissolution and SEI formation. Post the first cycle; the second discharge cycle delivered a specific capacity of about 178 mAh g^{-1}, corresponding to $V_2Al_{0.26}CT_x$ intercalated phase. Broad plateaus observed in the potential profile indicated a wide interlayer spacing possessing a range of potential

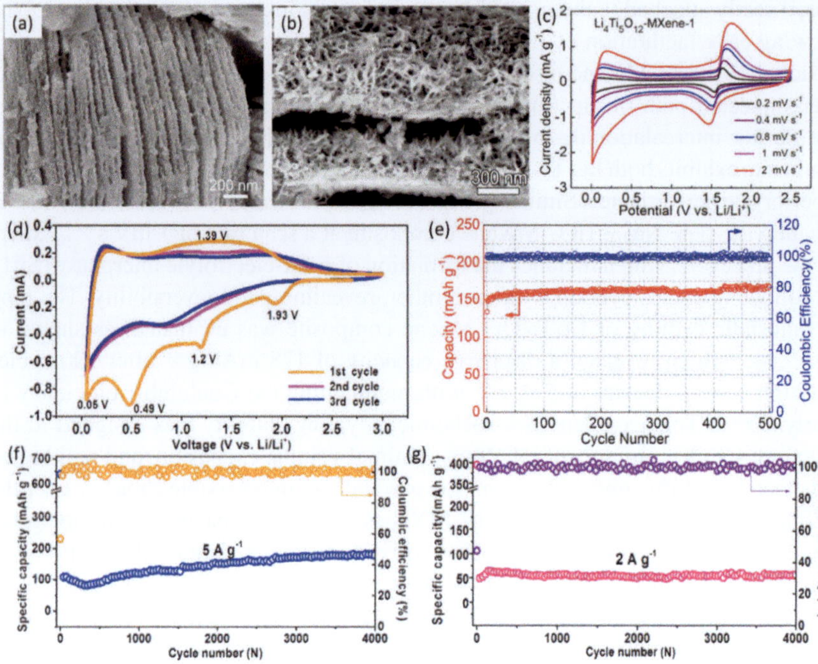

Fig. 5.14 **a** SEM image of $Li_4Ti_5O_{12}$-Mxene ($Ti_3C_2T_x$) composite. **b** SEM image of sandwich-type $Na_{0.23}TiO_2$-MXene (Ti_3C_2) composite. **c** cyclic voltammograms of $Li_4Ti_5O_{12}$-MXene composite at different scan rates, viz., 0.2, 0.4, 0.8, 1, and 2 mV s^{-1} for lithium-ion batteries (LIBs). **d** cyclic voltammograms of sandwich-type $Na_{0.23}TiO_2$-MXene composite at a scan rate of 1 mV s^{-1} for LIBs. **e** galvanostatic cycling of $Li_4Ti_5O_{12}$-MXene composite at a high current density of 5 A g^{-1} for LIBs. **f** galvanostatic cycling of sandwich-type $Na_{0.23}TiO_2$-MXene composite at a high current density of 5 A g^{-1} for LIBs. **g** galvanostatic cycling of sandwich-type $Na_{0.23}TiO_2$-MXene composite at a high current density of 2 A g^{-1} for sodium-ion batteries (SIBs) [with permission from Refs. [88, 89] Copyright 2018 Elsevier]

for intercalation/de-intercalation, which was confirmed with a cyclic voltammogram (as shown in Fig. 5.15c), revealing the corresponding cathodic and anodic peaks. Figure 5.15b clearly shows that there is a severe capacity decay over the cycling for multilayer MXene. Therefore, by further treatment, MXene was modified to a few-layer and subjected to electrochemical cycling (Fig. 5.15d, e). By transforming the multilayer MXene into few-layer MXene, the rapid capacity decay was suppressed; however, a gradual decay was observed, retaining a reversible capacity of 90 mAh g^{-1} after 100 cycles at a higher current density of 100 mA g^{-1}. To evaluate the fast-charging performance of few-layer MXene, a different protocol was employed with a high charge current density of 1000 mA g^{-1} and a low discharge current density of 100 mA g^{-1}. Even at this high charging rate, few-layer MXene was able to retain a specific capacity of 76 mAh g^{-1} after 100 revolutions (Fig. 5.15e). Therefore, the applicability of MXenes is not only limited to alkali metal batteries but

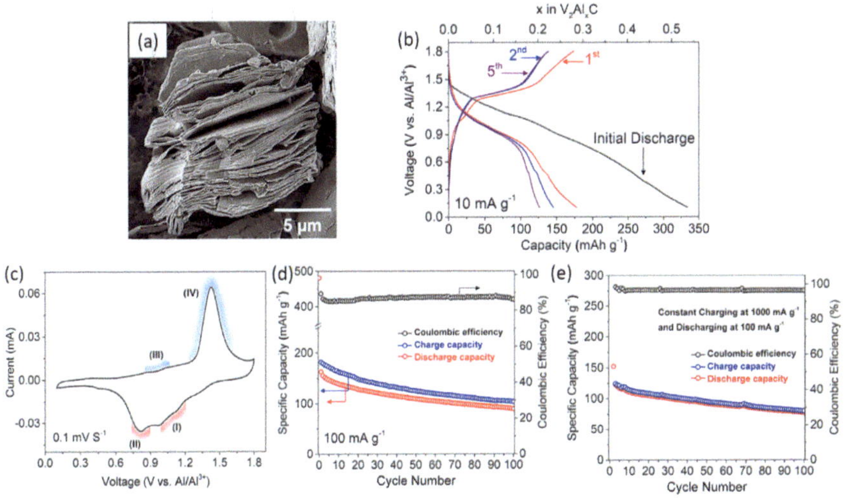

Fig. 5.15 **a** SEM image of MXene (V_2CT_x), obtained by treating V_2AlC MAX phase with 50% HF for 92 h at room temperature. **b** initial galvanostatic potential profiles of multilayer V_2CT_x MXene electrode against Al-metal anode at a current density of 10 mA g^{-1}. **c** cyclic voltammogram of multilayer V_2CT_x MXene electrode at a scan rate of 0.1 mV s^{-1} in a potential window of 0.1–1.8 V (V vs. Al/Al^{3+}). **d** galvanostatic cycling of few-layer V_2CT_x MXene electrode at a current density of 100 mA g^{-1}. **e** galvanostatic cycling of few-layer V_2CT_x MXene electrode at a charge current density of 1000 mA g^{-1} and discharge current density of 100 mA g^{-1}, showing fast-charging capability [with permission from Ref. [91] Copyright 2017 American Chemical Society]

also multivalent-ion batteries, metal-sulfur batteries, and metal batteries. However, the research progress on multivalent-ion batteries, metal-sulfur batteries, and metal batteries is limited. According to the current research progress, structure modification, interlayer spacing tuning, surface or functional group substitution, transition metal doping, and choice of synthesis process are a few strategies to enhance the properties of MXenes for an improvement in electrochemical performance for various metal(ion) battery applications.

5.5 Conclusion

The material composition and architecture of MXenes widen the scope of their utilization in energy conversion and storage devices. Their performance improvement procedure (energy and power density) depends on the surface features. The edge sites and the electronic band structure of MXenes enhance their water-splitting performance to replace precious metal (noble metal) electrocatalysts in water electrolyzer systems. The vertically aligned nanosheets of MXenes decrease the diffusion resistance to enhance the supercapacitor performance. Moreover, the reversible intercalation of various electrolyte ions in the expanded layers of MXenes makes them potential candidates for new-generation supercapacitors. The terminated functional groups on the MXenes play an essential role in energy storage regarding ion adsorption and ion diffusion barriers. With the good qualities of flexible morphologies, wide and tunable interlayer spacing, superior ionic diffusivity, higher electronic conductivity, and reversible volume expansion, MXenes are capable of intercalating/ de-intercalate alkali-ions and multivalent ions due to their inherent large interlayer spacing. The surface groups on the MXenes control the in-situ growth of active species responsible for decreasing the charge–discharge process.

Despite the extensive efforts of researchers and promising qualities with excellent electrochemical performance in supercapacitors and batteries, certain issues are still prevailing in commercializing the MXenes, such as

1. Understanding the intercalation mechanism of multivalent ions. One has to explore the relationship between the interlayer spacing and diffusion barrier.
2. Requirement of facile and innovative synthesis strategies and etching methods. For instance, mostly HF is employed in etching the MAX phase to obtain MXenes, which is unsafe and harmful for humans and the environment.
3. To explore the doping and defect engineering of MXene electrocatalyst for water-splitting application. The potential doping (metal species) as redox sites in energy storage systems is yet to be explored. The theoretical and experimental results may provide a breakthrough energy and power density of MXene-based energy conversion and storage devices.

References

1. L. Huo, B. Liu, G. Zhang, J. Zhang, Universal strategy to fabricate a two-dimensional layered mesoporous Mo_2C electrocatalyst hybridized on graphene sheets with high activity and durability for hydrogen generation. ACS Appl. Mater. Interfaces **8**, 18107–18118 (2016). https://doi.org/10.1021/acsami.6b05007
2. H. Tao, Q. Fan, T. Ma, S. Liu, H. Gysling, J. Texter et al., Two-dimensional materials for energy conversion and storage. Prog. Mater. Sci. **111**, 100637 (2020). https://doi.org/10.1016/j.pmatsci.2020.100637
3. M. Naguib, J. Halim, J. Lu, K.M. Cook, L. Hultman, Y. Gogotsi et al., New twodimensional niobium and vanadium carbides as promising materials for Liion batteries partners 1–2 (2016). https://doi.org/10.1021/ja405735d
4. Y. Wang, T. Guo, Z. Tian, K. Bibi, Y.Z. Zhang, H.N. Alshareef, MXenes for energy harvesting. Adv. Mater. **34**, 1–22 (2022). https://doi.org/10.1002/adma.202108560
5. J. Li, F. Zeng, J.K. El-Demellawi, Q. Lin, S. Xi, J. Wu et al., Nb_2CT_xMXene cathode for high-capacity rechargeable aluminum batteries with prolonged cycle lifetime. ACS Appl. Mater. Interfaces **14**, 45254–45262 (2022). https://doi.org/10.1021/acsami.2c09765
6. J. Liu, W. Peng, Y. Li, F. Zhang, X. Fan, 2D MXene-based materials for electrocatalysis. Trans. Tianjin Univ. **26**, 149–171 (2020). https://doi.org/10.1007/s12209-020-00235-x
7. R. Garg, A. Agarwal, M. Agarwal, A review on MXene for energy storage application: effect of interlayer distance. Mater. Res. Express **7** (2020). https://doi.org/10.1088/2053-1591/ab750d
8. R.M. Ronchi, J.T. Arantes, S.F. Santos, Synthesis, structure, properties and applications of MXenes: current status and perspectives. Ceram. Int. **45**, 18167–18188 (2019). https://doi.org/10.1016/j.ceramint.2019.06.114

9. S.T. Mahmud, M.M. Hasan, S. Bain, S.T. Rahman, M. Rhaman, M.M. Hossain et al., Multilayer MXene heterostructures and nanohybrids for multifunctional applications: a review. ACS Mater. Lett. **4**, 1174–1206 (2022). https://doi.org/10.1021/acsmaterialslett.2c00175
10. P. Das, Z.S. Wu, MXene for energy storage: present status and future perspectives. J. Phys. Energy **2** (2020). https://doi.org/10.1088/2515-7655/ab9b1d
11. F.R. Fan, W. Wu, Emerging devices based on two-dimensional monolayer materials for energy harvesting. Research **2019**, 1–16 (2019). https://doi.org/10.34133/2019/7367828
12. Z. Fu, N. Wang, D. Legut, C. Si, Q. Zhang, S. Du et al., Rational design of flexible two-dimensional MXenes with multiple functionalities. Chem. Rev. **119**, 11980–12031 (2019). https://doi.org/10.1021/acs.chemrev.9b00348
13. J. Song, C. Wei, Z.F. Huang, C. Liu, L. Zeng, X. Wang et al., A review on fundamentals for designing oxygen evolution electrocatalysts. Chem. Soc. Rev. **49**, 2196–2214 (2020). https://doi.org/10.1039/c9cs00607a
14. R. Ahmad, S. Upadhyay, O.P. Pandey, A review on recent advances and progress in Mo_2C@C: a suitable and stable electrocatalyst for HER. Int. J. Hydrog. Energy (2022). https://doi.org/10.1016/j.ijhydene.2022.12.179
15. S. Wang, A. Lu, C.J. Zhong, Hydrogen production from water electrolysis: role of catalysts. Nano Converge **8** (2021). https://doi.org/10.1186/s40580-021-00254-x
16. N. Yuan, Q. Jiang, J. Li, J. Tang, A review on non-noble metal based electrocatalysis for the oxygen evolution reaction. Arab. J. Chem. **13**, 4294–4309 (2020). https://doi.org/10.1016/j.arabjc.2019.08.006
17. S.M. Ibn Shamsah, Earth-abundant electrocatalysts for water splitting: current and future directions. Catalysts **11** (2021). https://doi.org/10.3390/catal11040429
18. H. Yuan, L. Kong, T. Li, Q. Zhang, A review of transition metal chalcogenide/graphene nanocomposites for energy storage and conversion. Chin. Chem. Lett. **28**, 2180–2194 (2017). https://doi.org/10.1016/j.cclet.2017.11.038
19. Y.C. Kimmel, X. Xu, W. Yu, X. Yang, J.G. Chen, Trends in electrochemical stability of transition metal carbides and their potential use as supports for low-cost electrocatalysts. ACS Catal. **4**, 1558–1562 (2014). https://doi.org/10.1021/cs500182h
20. D.J. Ham, J.S. Lee, Transition metal carbides and nitrides as electrode materials for low temperature fuel cells. Energies **2**, 873–899 (2009). https://doi.org/10.3390/en20400873
21. H. Yan, Y. Xie, Y. Jiao, A. Wu, C. Tian, X. Zhang et al., Holey reduced graphene oxide coupled with an Mo_2N–Mo_2C heterojunction for efficient hydrogen evolution. Adv. Mater. **30** (2018). https://doi.org/10.1002/adma.201704156
22. R.B. Levy, M. Boudart, Platinum-like behavior of tungsten carbide in surface catalysis. Science **181**(80), 547–549 (1973)
23. J.S. Lee, S.T. Oyama, M. Boudart, Molybdenum carbide catalysts. J. Catal. **106**, 125–133 (1987). https://doi.org/10.1016/0021-9517(87)90218-1
24. M. Naguib, O. Mashtalir, J. Carle, V. Presser, J. Lu, L. Hultman et al., Two-dimensional transition metal carbides. ACS Nano **6**, 1322–1331 (2012). https://doi.org/10.1021/nn204153h
25. J. Liu, T. Chen, P. Juan, W. Peng, Y. Li, F. Zhang et al., Hierarchical cobalt borate/MXenes hybrid with extraordinary electrocatalytic performance in oxygen evolution reaction. Chemsuschem **11**, 3758–3765 (2018). https://doi.org/10.1002/cssc.201802098
26. Z. Kang, M.A. Khan, Y. Gong, R. Javed, Y. Xu, D. Ye et al., Recent progress of MXenes and MXene-based nanomaterials for the electrocatalytic hydrogen evolution reaction. J. Mater. Chem. A **9**, 6089–6108 (2021). https://doi.org/10.1039/d0ta11735h
27. K.R.G. Lim, A.D. Handoko, L.R. Johnson, X. Meng, M. Lin, G.S. Subramanian et al., 2H-MoS_2on Mo_2CT_xMXene nanohybrid for efficient and durable electrocatalytic hydrogen evolution. ACS Nano **14**, 16140–16155 (2020). https://doi.org/10.1021/acsnano.0c08671
28. B.C. Wyatt, A. Thakur, K. Nykiel, Z.D. Hood, S.P. Adhikari, K.K. Pulley et al., Design of atomic ordering in $Mo_2Nb_2C_3T_x$ MXenes for hydrogen evolution electrocatalysis. Nano Lett. (2023). https://doi.org/10.1021/acs.nanolett.2c04287
29. J. Abed, S. Ahmadi, L. Laverdure, A. Abdellah, C.P. O'Brien, K. Cole et al., In situ formation of nano Ni–Co oxyhydroxide enables water oxidation electrocatalysts durable at high current densities. Adv. Mater. **33** (2021). https://doi.org/10.1002/adma.202103812

30. M.P. Browne, D. Tyndall, V. Nicolosi, The potential of MXene materials as a component in the catalyst layer for the oxygen evolution reaction. Curr. Opin. Electrochem. **34**, 101021 (2022). https://doi.org/10.1016/j.coelec.2022.101021
31. C.F. Du, X. Sun, H. Yu, W. Fang, Y. Jing, Y. Wang et al., $V_4C_3T_x$ MXene: a promising active substrate for reactive surface modification and the enhanced electrocatalytic oxygen evolution activity. InfoMat **2**, 950–959 (2020). https://doi.org/10.1002/inf2.12078
32. Z.W. Seh, K.D. Fredrickson, B. Anasori, J. Kibsgaard, A.L. Strickler, M.R. Lukatskaya et al., Two-dimensional molybdenum carbide (MXene) as an efficient electrocatalyst for hydrogen evolution. ACS Energy Lett. **1**, 589–594 (2016). https://doi.org/10.1021/acsenergylett.6b00247
33. A.D. Handoko, K.D. Fredrickson, B. Anasori, K.W. Convey, L.R. Johnson, Y. Gogotsi et al., Tuning the basal plane functionalization of two-dimensional metal carbides (MXenes) to control hydrogen evolution activity. ACS Appl. Energy Mater. **1**, 173–180 (2018). https://doi.org/10.1021/acsaem.7b00054
34. S. Li, P. Tuo, J. Xie, X. Zhang, J. Xu, J. Bao et al., Ultrathin MXene nanosheets with rich fluorine termination groups realizing efficient electrocatalytic hydrogen evolution. Nano Energy **47**, 512–518 (2018). https://doi.org/10.1016/j.nanoen.2018.03.022
35. Y. Jiang, T. Sun, X. Xie, W. Jiang, J. Li, B. Tian et al., Oxygen-functionalized ultrathin $Ti_3C_2T_x$ MXene for enhanced electrocatalytic hydrogen evolution. Chemsuschem **12**, 1368–1373 (2019). https://doi.org/10.1002/cssc.201803032
36. M.H. Tran, T. Schäfer, A. Shahraei, M. Dürrschnabel, L. Molina-Luna, U.I. Kramm et al., Adding a new member to the MXene family: synthesis, structure, and electrocatalytic activity for the hydrogen evolution reaction of $V_4C_3T_x$. ACS Appl. Energy Mater. **1**, 3908–3914 (2018). https://doi.org/10.1021/acsaem.8b00652
37. Y. Yoon, A.P. Tiwari, M. Lee, M. Choi, W. Song, J. Im et al., Enhanced electrocatalytic activity by chemical nitridation of two-dimensional titanium carbide MXene for hydrogen evolution. J. Mater. Chem. A **6**, 20869–20877 (2018). https://doi.org/10.1039/C8TA08197B
38. W. Yuan, L. Cheng, Y. An, H. Wu, N. Yao, X. Fan et al., MXene nanofibers as highly active catalysts for hydrogen evolution reaction. ACS Sustain. Chem. Eng. **6**, 8976–8982 (2018). https://doi.org/10.1021/acssuschemeng.8b01348
39. C.F. Du, X. Sun, H. Yu, Q. Liang, K.N. Dinh, Y. Zheng et al., Synergy of Nb doping and surface alloy enhanced on water-Alkali electrocatalytic hydrogen generation performance in Ti-based MXene. Adv. Sci. **6**, 1–7 (2019). https://doi.org/10.1002/advs.201900116
40. Z. Li, Z. Qi, S. Wang, T. Ma, L. Zhou, Z. Wu et al., In situ formed Pt_3Ti nanoparticles on a two-dimensional transition metal carbide (MXene) used as efficient catalysts for hydrogen evolution reactions. Nano Lett. **19**, 5102–5108 (2019). https://doi.org/10.1021/acs.nanolett.9b01381
41. Y. Yuan, H. Li, L. Wang, L. Zhang, D. Shi, Y. Hong et al., Achieving highly efficient catalysts for hydrogen evolution reaction by electronic state modification of platinum on versatile $Ti_3C_2T_x$ (MXene). ACS Sustain. Chem. Eng. **7**, 4266–4273 (2019). https://doi.org/10.1021/acssuschemeng.8b06045
42. S.Y. Pang, Y.T. Wong, S. Yuan, Y. Liu, M.K. Tsang, Z. Yang et al., Universal strategy for HF-free facile and rapid synthesis of two-dimensional MXenes as multifunctional energy materials. J. Am. Chem. Soc. **141**, 9610–9616 (2019). https://doi.org/10.1021/jacs.9b02578
43. J. Liang, C. Ding, J. Liu, T. Chen, W. Peng, Y. Li et al., Heterostructure engineering of Co-doped MoS_2 coupled with Mo_2CT_X MXene for enhanced hydrogen evolution in alkaline media. Nanoscale **11**, 10992–11000 (2019). https://doi.org/10.1039/c9nr02085c
44. J. Filip, S. Zavahir, L. Lorencova, T. Bertok, A.B. Yousaf, K.A. Mahmoud et al. Tailoring electrocatalytic properties of Pt nanoparticles grown on $Ti_3C_2T_X$ MXene surface. J. Electrochem. Soc. **166**, H54–H62 (2019). https://doi.org/10.1149/2.0991902jes
45. N.H. Attanayake, S.C. Abeyweera, A.C. Thenuwara, B. Anasori, Y. Gogotsi, Y. Sun et al., Vertically aligned MoS_2 on Ti_3C_2 (MXene) as an improved HER catalyst. J. Mater. Chem. A **6**, 16882–16889 (2018). https://doi.org/10.1039/c8ta05033c
46. B. Cui, B. Hu, J. Liu, M. Wang, Y. Song, K. Tian et al., Solution-plasma-assisted bimetallic oxide alloy nanoparticles of Pt and Pd embedded within two-dimensional $Ti_3C_2T_x$ nanosheets

as highly active electrocatalysts for overall water splitting. ACS Appl. Mater. Interfaces **10**, 23858–23873 (2018). https://doi.org/10.1021/acsami.8b06568

47. C.F. Du, K.N. Dinh, Q. Liang, Y. Zheng, Y. Luo, J. Zhang et al., Self-assemble and in situ formation of $Ni_{1-x}Fe_xPS_3$ nanomosaic-decorated MXene hybrids for overall water splitting. Adv. Energy Mater. **8**, 1–9 (2018). https://doi.org/10.1002/aenm.201801127
48. J. Liu, Y. Liu, D. Xu, Y. Zhu, W. Peng, Y. Li et al., Hierarchical "nanoroll" like $MoS_2/Ti_3C_2T_x$ hybrid with high electrocatalytic hydrogen evolution activity. Appl. Catal. B Environ. **241**, 89–94 (2019). https://doi.org/10.1016/j.apcatb.2018.08.083
49. L. Xiu, Z. Wang, M. Yu, X. Wu, J. Qiu, Aggregation-resistant 3D MXene-based architecture as efficient bifunctional electrocatalyst for overall water splitting. ACS Nano **12**, 8017–8028 (2018). https://doi.org/10.1021/acsnano.8b02849
50. M. Yu, S. Zhou, Z. Wang, J. Zhao, J. Qiu, Boosting electrocatalytic oxygen evolution by synergistically coupling layered double hydroxide with MXene. Nano Energy **44**, 181–190 (2018). https://doi.org/10.1016/j.nanoen.2017.12.003
51. L. Zhao, B. Dong, S. Li, L. Zhou, L. Lai, Z. Wang et al., Interdiffusion reaction-assisted hybridization of two-dimensional metal-organic frameworks and $Ti_3C_2T_x$ nanosheets for electrocatalytic oxygen evolution. ACS Nano **11**, 5800–5807 (2017). https://doi.org/10.1021/acsnano.7b01409
52. T.Y. Ma, J.L. Cao, M. Jaroniec, S.Z. Qiao, Interacting carbon nitride and titanium carbide nanosheets for high-performance oxygen evolution. Angew. Chem. Int. Ed. **55**, 1138–1142 (2016). https://doi.org/10.1002/anie.201509758
53. Y. Tang, C. Yang, Y. Yang, X. Yin, W. Que, J. Zhu, Three dimensional hierarchical network structure of S-$NiFe_2O_4$ modified few-layer titanium carbides (MXene) flakes on nickel foam as a high efficient electrocatalyst for oxygen evolution. Electrochim. Acta **296**, 762–770 (2019). https://doi.org/10.1016/j.electacta.2018.11.083
54. H. Zou, B. He, P. Kuang, J. Yu, K. Fan, Metal-organic framework-derived nickel-cobalt sulfide on ultrathin MXene nanosheets for electrocatalytic oxygen evolution. ACS Appl. Mater. Interfaces **10**, 22311–22319 (2018). https://doi.org/10.1021/acsami.8b06272
55. Y. Zhang, H. Jiang, Y. Lin, H. Liu, Q. He, C. Wu et al., In situ growth of cobalt nanoparticles encapsulated nitrogen-doped carbon nanotubes among $Ti_3C_2T_x$ (MXene) matrix for oxygen reduction and evolution. Adv. Mater. Interfaces **5**, 1–9 (2018). https://doi.org/10.1002/admi.201800392
56. T. Brousse, D. Bélanger, J.W. Long, To be or not to be pseudocapacitive? J. Electrochem. Soc. **162**, A5185–A5189 (2015). https://doi.org/10.1149/2.0201505jes
57. M.R. Lukatskaya, S. Kota, Z. Lin, M.Q. Zhao, N. Shpigel, M.D. Levi et al., Ultra-high-rate pseudocapacitive energy storage in two-dimensional transition metal carbides. Nat. Energy **6**, 1–6 (2017). https://doi.org/10.1038/nenergy.2017.105
58. A. Sharma, P. Mane, B. Chakraborty, C.S. Rout, 1T-VS2/MXene hybrid as a superior electrode material for asymmetric supercapacitors: experimental and theoretical investigations. ACS Appl. Energy Mater. **4**, 14198–14209 (2021). https://doi.org/10.1021/acsaem.1c02959
59. Z.S. Iroc, C. Subramani, S.S. Dash, A brief review on electrode materials for supercapacitor **11**, 10628–10643 (2016). https://doi.org/10.20964/2016.12.50
60. J.R. Miller, Engineering electrochemical capacitor applications. J. Power. Sources **326**, 726–735 (2016). https://doi.org/10.1016/j.jpowsour.2016.04.020
61. S.A. Thomas, J. Cherusseri, A review of Nb_2CT_x MXene as an emerging 2D material: synthesis, applications in rechargeable batteries and supercapacitors, progress, and outlook. Energy Fuels (2023). https://doi.org/10.1021/acs.energyfuels.3c00503
62. M. He, K. Fic, E. Frackowiak, P. Novák, E.J. Berg, Ageing phenomena in high-voltage aqueous supercapacitors investigated by in situ gas analysis. Energy Environ. Sci. **9**, 623–633 (2016). https://doi.org/10.1039/c5ee02875b
63. C. Wang, X. Wang, L. Zhang, L. Yin, MXene for high energy and power density: a perspective. J. Phys. Energy **2** (2020). https://doi.org/10.1088/2515-7655/abab66
64. J. Luo, W. Zhang, H. Yuan, C. Jin, L. Zhang, H. Huang et al., Pillared structure design of MXene with ultralarge interlayer spacing for high-performance lithium-ion capacitors. ACS Nano **11**, 2459–2469 (2017). https://doi.org/10.1021/acsnano.6b07668

65. H. He, J. Wang, Q. Xia, L. Wang, Q. Hu, A. Zhou, Effect of electrolyte on supercapacitor performance of two-dimensional molybdenum carbide (Mo_2CT_x) MXene prepared by hydrothermal etching. Appl. Surf. Sci. 568 (2021). https://doi.org/10.1016/j.apsusc.2021.150971
66. Q. Zhu, J. Li, P. Simon, B. Xu, Two-dimensional MXenes for electrochemical capacitor applications: progress, challenges and perspectives. Energy Storage Mater. **35**, 630–660 (2021). https://doi.org/10.1016/j.ensm.2020.11.035
67. Y. Gogotsi, B. Anasori, The rise of MXenes. ACS Nano **13**, 8491–8494 (2019). https://doi.org/10.1021/acsnano.9b06394
68. M. Hu, T. Hu, Z. Li, Y. Yang, R. Cheng, J. Yang et al., Surface functional groups and interlayer water determine the electrochemical capacitance of $Ti_3C_2T_x$ MXene. ACS Nano **12**, 3578–3586 (2018). https://doi.org/10.1021/acsnano.8b00676
69. Y. Xie, M. Naguib, V.N. Mochalin, M.W. Barsoum, Y. Gogotsi, X. Yu et al., Role of surface structure on li-ion energy storage capacity of two-dimensional transition-metal carbides. J. Am. Chem. Soc. **136**, 6385–6394 (2014). https://doi.org/10.1021/ja501520b
70. K. Maleski, V.N. Mochalin, Y. Gogotsi, Dispersions of two-dimensional titanium carbide MXene in organic solvents. Chem. Mater. **29**, 1632–1640 (2017). https://doi.org/10.1021/acs.chemmater.6b04830
71. M.R. Lukatskaya, O. Mashtalir, C.E. Ren, Y. Dall'Agnese, P. Rozier, P.L. Taberna et al. Cation intercalation and high volumetric capacitance of two-dimensional titanium carbide. Science **341**(80), 1502–1505 (2013). https://doi.org/10.1126/science.1241488
72. M. Naguib, V.N. Mochalin, M.W. Barsoum, Y. Gogotsi, 25th anniversary article: MXenes: a new family of two-dimensional materials. Adv. Mater. **26**, 992–1005 (2014). https://doi.org/10.1002/adma.201304138
73. J. Pang, R.G. Mendes, A. Bachmatiuk, L. Zhao, H.Q. Ta, T. Gemming et al., Applications of 2D MXenes in energy conversion and storage systems. Chem. Soc. Rev. **48**, 72–133 (2019). https://doi.org/10.1039/c8cs00324f
74. Y. Sun, Q. Wu, Y. Xu, H. Bai, C. Li, G. Shi, Highly conductive and flexible mesoporous graphitic films prepared by graphitizing the composites of graphene oxide and nanodiamond. J. Mater. Chem. **21**, 7154–7160 (2011). https://doi.org/10.1039/c0jm04434b
75. A. Burke, R&D considerations for the performance and application of electrochemical capacitors. Electrochim. Acta **53**, 1083–1091 (2007). https://doi.org/10.1016/j.electacta.2007.01.011
76. D. Pech, M. Brunet, H. Durou, P. Huang, V. Mochalin, Y. Gogotsi et al., Ultrahigh-power micrometre-sized supercapacitors based on onion-like carbon. Nat. Nanotechnol. **5**, 651–654 (2010). https://doi.org/10.1038/nnano.2010.162
77. M. Zhu, Y. Huang, Q. Deng, J. Zhou, Z. Pei, Q. Xue et al., Highly flexible, freestanding supercapacitor electrode with enhanced performance obtained by hybridizing polypyrrole chains with MXene. Adv. Energy Mater. **6** (2016). https://doi.org/10.1002/aenm.201600969
78. Y. Dall'Agnese, M.R. Lukatskaya, K.M Cook, P.L. Taberna, Y. Gogotsi, P. Simon, High capacitance of surface-modified 2D titanium carbide in acidic electrolyte. Electrochem. Commun. **48**, 118–122 (2014). https://doi.org/10.1016/j.elecom.2014.09.002
79. A. Zhang, R. Liu, J. Tian, W. Huang, J. Liu, MXene-based nanocomposites for energy conversion and storage applications. Chem. A Eur. J. **26**, 6342–6359 (2020). https://doi.org/10.1002/chem.202000191
80. M. Pramanik, V. Malgras, J. Lin, S.M. Alshehri, T. Ahamad, J.H. Kim et al., Electrochemical property of mesoporous crystalline iron phosphonate anode in li-ion rechargeable battery. J. Nanosci. Nanotechnol. **16**, 9180–9185 (2016). https://doi.org/10.1166/jnn.2016.12883
81. L. Qin, J. Jiang, Q. Tao, C. Wang, I. Persson, M. Fahlman et al., A flexible semitransparent photovoltaic supercapacitor based on water-processed MXene electrodes. J. Mater. Chem. A **8**, 5467–5475 (2020). https://doi.org/10.1039/d0ta00687d
82. L. Verger, C. Xu, V. Natu, H.M. Cheng, W. Ren, M.W. Barsoum, Overview of the synthesis of MXenes and other ultrathin 2D transition metal carbides and nitrides. Curr. Opin. Solid State Mater. Sci. **23**, 149–163 (2019). https://doi.org/10.1016/j.cossms.2019.02.001

83. S. Liu, Z. Song, X. Jin, R. Mao, T. Zhang, F. Hu, MXenes for metal-ion and metal-sulfur batteries: synthesis, properties, and electrochemistry. Mater. Rep. Energy **2**, 100077 (2022). https://doi.org/10.1016/j.matre.2021.100077
84. F. Ming, H. Liang, G. Huang, Z. Bayhan, H.N. Alshareef, MXenes for rechargeable batteries beyond the lithium-ion. Adv. Mater. **33**, 1–35 (2021). https://doi.org/10.1002/adma.202004039
85. M. Greaves, S. Barg, M.A. Bissett, MXene-based anodes for metal-ion batteries. Batter Supercaps **3**, 214–235 (2020). https://doi.org/10.1002/batt.201900165
86. Q. Tang, Z. Zhou, P. Shen, Are MXenes promising anode materials for Li ion batteries? Computational studies on electronic properties and Li storage capability of Ti_3C_2 and $Ti_3C_2X_2$ (X = F, OH) monolayer. J. Am. Chem. Soc. **134**, 16909–16916 (2012). https://doi.org/10.1021/ja308463r
87. M. Naguib, M. Kurtoglu, V. Presser, J. Lu, J. Niu, M. Heon et al., Two-dimensional nanocrystals produced by exfoliation of Ti_3AlC_2. Adv. Mater. **23**, 4248–4253 (2011). https://doi.org/10.1002/adma.201102306
88. J. Wang, S. Dong, H. Li, Z. Chen, S. Jiang, L. Wu et al., Facile synthesis of layered $Li_4Ti_5O_{12}$-Ti_3C_2Tx (MXene) composite for high-performance lithium ion battery. J. Electroanal. Chem. **810**, 27–33 (2018). https://doi.org/10.1016/j.jelechem.2017.12.079
89. J. Huang, R. Meng, L. Zu, Z. Wang, N. Feng, Z. Yang et al., Sandwich-like $Na_{0.23}TiO_2$ nanobelt/Ti_3C_2 MXene composites from a scalable in situ transformation reaction for long-life high-rate lithium/sodium-ion batteries. Nano Energy **46**, 20–28 (2018). https://doi.org/10.1016/j.nanoen.2018.01.030
90. M. Xu, S. Lei, J. Qi, Q. Dou, L. Liu, Y. Lu et al., Opening magnesium storage capability of two-dimensional MXene by intercalation of cationic surfactant. ACS Nano **12**, 3733–3740 (2018). https://doi.org/10.1021/acsnano.8b00959
91. A. Vahidmohammadi, A. Hadjikhani, S. Shahbazmohamadi, M. Beidaghi, Two-dimensional vanadium carbide (MXene) as a high-capacity cathode material for rechargeable aluminum batteries. ACS Nano **11**, 11135–11144 (2017). https://doi.org/10.1021/acsnano.7b05350

Chapter 6
Implementation of MXenes for Water Treatment

Aadil Bathla

6.1 Introduction

Rapid industrialization and irrigation consume a lot of water. It is accompanied by the release of various toxic organic pollutants (e.g., dyes, pesticides, pharmaceuticals and personal care products (PCPs), and other hydrocarbons) [1]. To date, numerous techniques have been employed (e.g., adsorption, membrane separation, coagulation, and catalytic oxidation) to treat wastewater. A wide variety of catalytic materials like carbon (graphene oxide (GO), carbon nanotubes, C-dots), metal oxides (e.g., TiO_2, ZnO, and MnO_2), zeolites, and polymers have been utilized for the wastewater treatment [2, 3]. However, the practical application of these materials has not generally been acknowledged due to their operational challenges (e.g., a broad bandgap (3.2 eV) and rapid recombination of e^-/h^+ pairs) and deactivation issues (e.g., partial oxidation and deposition of coke and intermediates on active catalytic sites) [2, 4].

MXenes, a newly discovered family of 2D materials, have recently attracted much attention due to their fascinating physicochemical features [5, 6]. The exfoliation of the MAX phase precursor typically results in the formation of MXenes. The standard chemical formula for the MAX phase is $M_{n+1}AX_n$, where M corresponds to early transition metals (e.g., Ti, Mo, and Cr), X might be carbon (C), nitrogen (N), or a mixture of two and A stands for elements from group 3A or 4A (e.g., Al, Si, B, Se) [7, 8]. The distinctive structural and compositional characteristics of MXenes make them potential candidates for diverse applications e.g., degradation [9], water splitting [10], carbon capture [11], sensing [12], CO_2 reduction [13], and N_2 fixation [14].

The present book chapter highlights the recent advances made in the MXene-based material for the treatment of water along with the discussion on their catalytic

A. Bathla (✉)
Department of Civil and Environmental Engineering, Hanyang University, 222 Wangsimni-Ro, Seoul 04763, South Korea
e-mail: aadil93@hanyang.ac.kr

© The Author(s), under exclusive license to Springer Nature Singapore Pte Ltd. 2024

O. P. Pandey and P. Sharma (eds.), *MXenes: Emerging 2D Materials*,
https://doi.org/10.1007/978-981-97-4064-2_6

mechanisms. Lastly, the current challenges and prospects in the treatment of wastewater have been discussed with respect to their scalability for real-world applications. Accordingly, this chapter is anticipated to explore the way forward in developing an effective MXene-based catalytic system for water purification.

6.2 Utilization of MXenes in Water Treatment

6.2.1 Removal of Heavy Metal Ions

The heavy metal ions (e.g., $Hg^{2+}$and Pb^{2+}) released from various industries (e.g., paint, mining, and electronic industries) pose serious threats to the surface and subsurface environments [8]. Due to their high toxicity and non-biodegradability, heavy metal ions can be quickly accumulated with living organisms, causing long-term damage to humans and other species [15]. In this regard, the utility of magnetic MXene (MGMX) nanocomposite synthesized via hydrothermal approach using iron oxide (e.g., Fe_2O_3) nanoparticles and $Ti_3C_2T_x$ is vividly explored [16]. The MGMX showcased high adsorption capacity of 1128.4 mg/g for Hg^{2+} ions, attributed to the existence of negatively charged surface functional groups (e.g., =O and OH groups) [16] (Fig. 6.1). Similarly, molybdenum disulfide (MoS_2)-functionalized MXene demonstrated enhanced adsorption capacity (1435.2 mg/g) for Hg^{2+} ions [17]. The adsorption of Hg^{2+} onto MoS_2/MX-II can occur via two distinct ways: (1) the development of a Hg–O complex with O_2; and (2) direct adsorption of Hg^{2+} onto MoS_2 surfaces through Hg–S ionic interactions [17]. The silane-modified MXene ($Ti_3C_2T_x$–KH570) demonstrated higher adsorption capacity (147.29 mg/g) for Pb^{2+} relative to pristine $Ti_3C_2T_x$ (48.28 mg/g). The higher adsorption of Pb^{2+} is attributed to the presence of hydroxyl and carbonyl groups in $Ti_3C_2T_x$ and KH570 (methacryloxypropyl chain) [18, 19]. Additionally, the introduction of KH570 into the multilayered framework of $Ti_3C_2T_x$ increases the separation between different layers to enhance the adsorption of Pb^{2+} ions [18]. Similarly, the utility of amine (NH_2) functionalized MXene was evaluated for the adsorption of Pb^{2+} ions (384.63 mg/g). The improved adsorption for Pb^{2+} ions arises from the surface complexation of amine and hydroxyl groups [18].

6.2.2 Anion Removal

In addition to the adsorption of heavy metal ions, the utility of MXene-based material has been investigated for the removal of negatively charged ions (e.g., Cr (VI)). For instance, Ti_3C_2 coupled with TiO_2 demonstrated ~99% removal efficiency for Cr (VI) in acidic conditions (pH = 2) [20]. At low pH, the higher concentration of H^+ ions significantly promotes the reduction of Cr (VI) to less hazardous Cr (III) onto

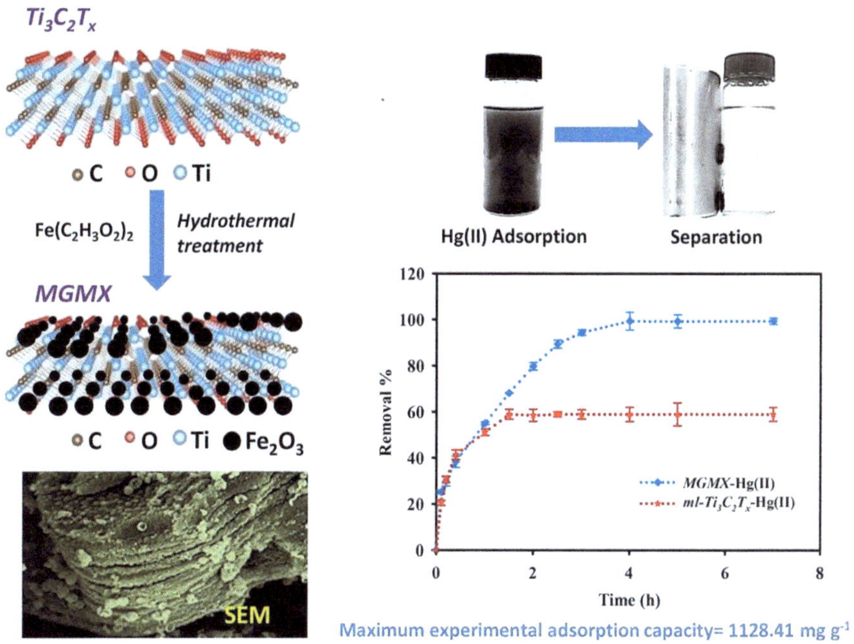

Fig. 6.1 Schematic for the synthesis and Hg^+ ion removal using MXene. Reprinted with permission © Elsevier 2018 [16]

the Ti_3C_2/TiO_2 surface through the formation of Ti-C-Cr (III) covalent bond [20] (Fig. 6.2). Similarly, $Ti_3C_2T_x$ nanosheets demonstrated higher efficiency (250 mg/g) for the removal to toxic $Cr_2O_7^{2-}$ anion. The higher removal efficiency is attributed to the presence of surface hydroxyl group on $Ti_3C_2T_x$ nanosheets, resulting in the effective electrostatic interactions between $Cr_2O_7^{2-}$ and $Ti_3C_2T_x$ nanosheets [21]. Similarly, the MXene-based polyelectrolyte nanocomposite (PDDA) showcased higher efficiency (363 mg/g) for the removal of perrhenate anion (Re (VII)) [22]. The introduction of poly(diallyldimethylammonium chloride, PDDA) into the framework of Ti_2CT_x enhances the adsorption of (Re (VII)) onto the surface of $Ti_3C_2T_x$ to accelerate its reduction to (Re(VII)O^{2-}) amorphous ReO_2 (Fig. 6.3) [22]. In addition, the utility of $Ti_3C_2T_x$ modified with a glassy carbon electrode was evaluated for the detection and reduction of bromate ions (BrO_3^-) to Br^- [8].

6.2.3 Dye Removal

A good number of MXene-based composites have been explored for the photocatalytic degradation of dyes. MXenes, due to their advantageous plasmonic properties (e.g., light harvesting efficacy, high electron density, and effective separation of

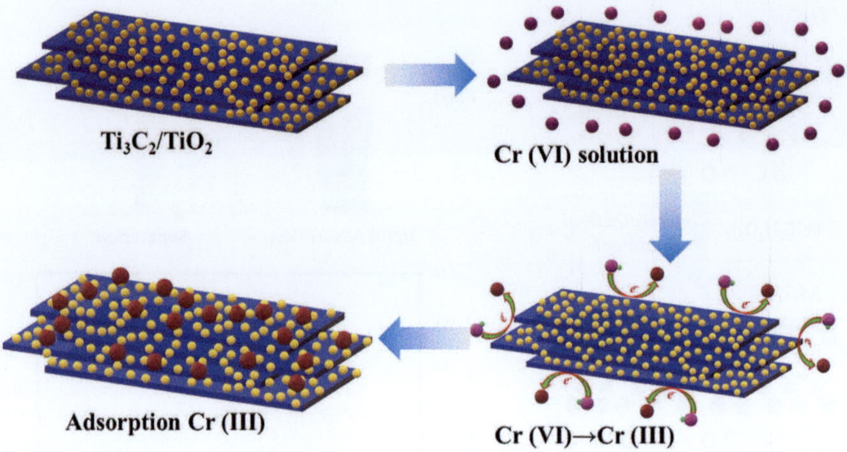

Fig. 6.2 Schematic for the removal of Cr(VI) ion using Ti_3C_2/TiO_2 MXene. Reprinted with permission © Elsevier 2020 [20]

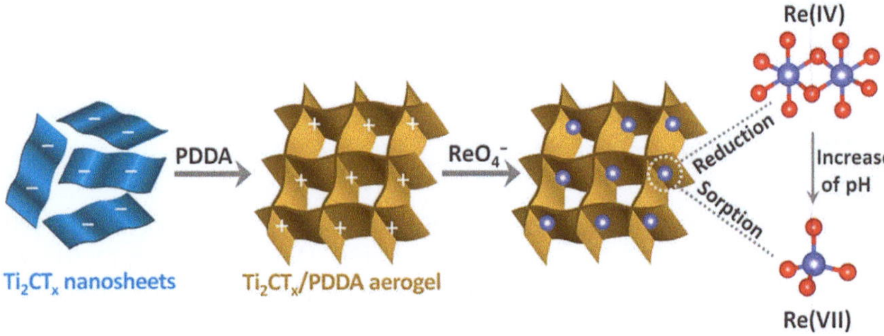

Fig. 6.3 Schematic for the removal of Re(IV) ion using Ti_2CT_x modified with PDDA aerogel. Reprinted with permission © ACS 2019 [22]

charge carry and easy combination with another photocatalyst can develop efficient heterojunction systems for the photodegradation of dyes [8]. For instance, the Ti_3C_2 coupled with 001 faceted TiO_2 exhibited enhanced photocatalytic degradation of methyl orange (MO) dye with the kinetic rate of 18.8 min^{-1}/g [23]. Under visible-light exposure, Ti_3C_2 served as electron tapping sites to prolong the separation of photogenerated charge carriers (e^-–h^+). As a result, the electron-rich TiO_2 interacts with surface adsorbed O_2 and H_2O molecules to form reactive oxygen species (ROS, OH^- and $O_2{}^-$) (Fig. 6.4) [23]. These ROS species react with adsorbed dye molecules to mineralize them into CO_2 and H_2O. The catalytic efficiency of $Ti_3C_2T_x$–TiO_2 decorated with black phosphorus was investigated for the removal of rhodamine-B (RhB) (99%) and tetracycline hydrochloride (93%) under visible light [24]. The synergy interactions between black phosphorus and $Ti_3C_2T_x$–TiO_2 significantly suppress the

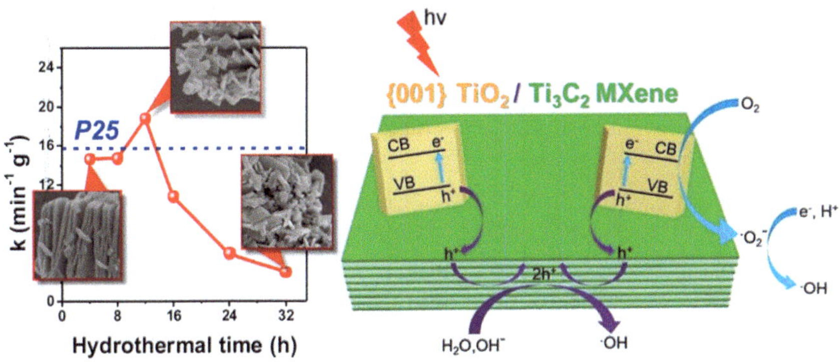

Fig. 6.4 Morphology and mechanism for photodegradation of organic dye using Ti_3C_2 modified with {001}TiO_2. Reprinted with permission © ACS 2016 [23]

recombination of photo-excited charge carries to enhance the photodegradation efficiency of black phosphorus-coated $Ti_3C_2T_x$–TiO_2 [24]. Similarly, the introduction of CeO_2 onto Ti_3C_2 nanosheet significantly improves the photodegradation efficiency (75%) relative to pristine Ti_3C_2 (63%) and CeO_2 (24%) [25]. The utility of Fe_2O_3 coupled with Ti_3C_2 was investigated for the removal of different organic dyes, e.g., MO, methylene blue (MB), and congo red (CR). The Fe_2O_3/Ti_3C_2 showcased higher degradation efficiency (94%) even after five successive cycles [26].

6.2.4 *Removal of Organic Pollutants*

In addition to the degradation of dyes, the MXene-based materials have been utilized for the removal of different pharmaceuticals and PCPs. For instance, $CoFe_2O_4$ coupled with @$Ti_3C_2T_x$ was examined for the degradation of naproxen (99%) [27]. This resulted in higher degradation efficiency of $CoFe_2O_4/Ti_3C_2T_x$. The unique accordion-like structure of $Ti_3C_2T_x$ provides an efficient area for the immobilization and dispersion of $CoFe_2O_4$ NPs. As a result, the incorporation of $CoFe_2O_4$ significantly boosted the degradation efficacy of $Ti_3C_2T_x$ through the formation of active sulfate and hydroxyl radicals [27]. Similarly, the catalytic efficiency of single atom Co supported over 2D Ti_2AlN substrate was investigated for the removal of different organic pollutants (e.g., carbamazepine, sulfamethoxazole, and 2,4-dichlorophen) [28]. The Ti atoms in Ti_2AlN accelerate the Co^{3+}/Co^{2+} reduction cycle to enhance the degradation of organic pollutants through peroxymonosulfate activation (Fig. 6.5). The CuO decorated MXene was evaluated for the photocatalytic degradation of carbamazepine [28]. Introducing CuO onto MXene considerably improves the degradation efficiency of MXene by promoting the formation of sulfate and hydroxyl radicals [28].

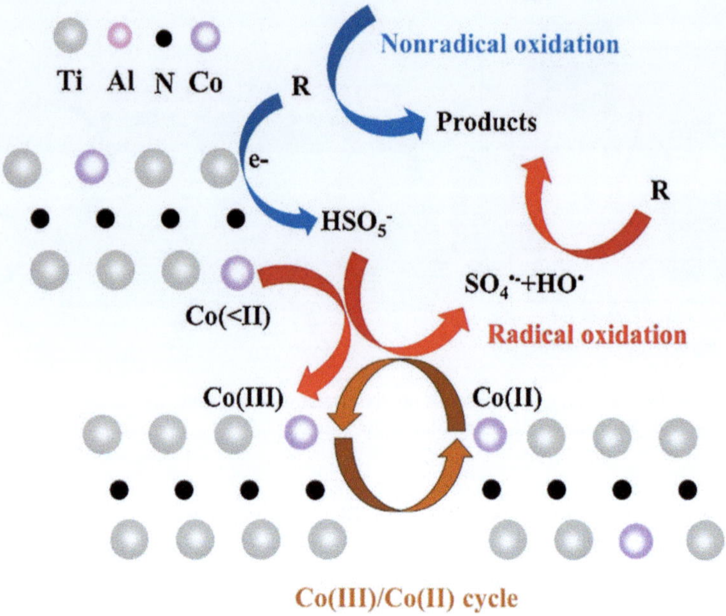

Fig. 6.5 Schematic of the reaction mechanism of PMS/Co@$Ti_{2-x}N$. Reprinted with permission © Elsevier 2021 [28]

6.3 Factors Affecting the Performance of MXenes

The performance of MXene-based materials can be significantly influenced by several operational parameters such as pH, catalyst dosage, oxidant concentration, reaction temperature, intensity of light, co-existing ions, and stability [29]. Therefore, the present section discusses the effect of different operational parameters on the catalytic proficiency of MXene-based materials.

6.3.1 *Effect of pH*

The pH of the solution has a significant role in the catalytic degradation of organic pollutants present in the wastewater. The pH of the solution considerably influences the formation of active surface radicals, adsorption, and dissociation of pollutant molecules onto the surface of the MXene-based photocatalyst [30]. In persulfate-based systems, the primary reactive species are $SO_4^{\bullet}$ (pH 2–7), $SO_4^{\bullet}$ and $HO^{\bullet}$ (pH 7–10), and $HO^{\bullet}$ (pH > 10) [31]. It has been observed that the rate of decomposition of peroxydisulfate varies slightly at pH 3–10, but it accelerates at pH < 3 or pH > 10 [32]. Under acidic conditions, the protonation of surface active groups increases, which in turn enhances the adsorption of anionic pollutants toward the positively charged

MXene-based catalyst. The low pH of the solution also facilitates the leaching of metal ions to enhance the catalytic process. However, this process increases the instability of the catalyst [29]. The pH of the solution also affects the dissociation of organic pollutants. For instance, the equilibrium dissociation constant (pK_a) of different pollutants was observed as tetracycline (3.3 (pK_{a1}), 7.7 (pK_{a2}), and 9.7 (pK_{a3})), sulfamoxazole (1.6 (pK_{a1}) and 5.7 (pK_{a2})), and rhodamine-B (3.1 (pK_{a1})) [29].

6.3.2 *Effect of Catalyst Amount*

Generally, the photocatalytic degradation of organic pollutants is directly proportional to the catalyst dosage. The increase in the amount of catalyst provides more exposure to the active sites to facilitate the absorption of photons [29, 33]. As a result, under exposure to light irradiations, the formation of ROS (OH^- and O_2^- radicals) increases, which in turn accelerates the mineralization of organic pollutants. For instance, the increase in dosage of catalyst (MXene-based photocatalyst) from 0.3 to 1.2 g/L, significantly increases the reaction rate from 10.04×10^{-2} to 26.60×10^{-2} min^{-1}, respectively [34]. However, it has been observed that the reaction rate is proportional to catalyst dosage up to a certain limit. The excessive increase in catalyst dosage results in the consumption of oxidants to generate ROS and a reduction in the catalytic performance is observed [29].

6.3.3 *Effect of Oxidant Concentration*

The oxidant concentration considerably affects the catalytic efficiency of MXene-based materials for the degradation of organic pollutants in wastewater [29]. For instance, the catalytic efficiency of Fe_3O_4-loaded Ti_3C_2 MXene for the degradation of organic dyes increases with an increase in the concentration of H_2O_2 from 10 to 90 mmol/L [35]. Similarly, the Ti_3C_2 decorated with spinel cobalt ferrite showcased enhancement in the catalytic efficiency for naproxen when the concentration of persulfate increased from 0.4 to 0.6 mM attributed to the increased formation of ROS [27]. However, the excessive oxidant concentration also endows a negative impact on the catalytic efficiency of MXenes. Therefore, it is important to determine the optimum concentration of catalyst and oxidant for the efficient treatment of wastewater using MXene-based materials [36].

6.3.4 *Effect of Temperature*

The reaction temperature has a considerable impact on the catalytic efficiency of MXene-based material utilized in the Fenton-like process [29, 37]. For instance, nZVI@Ti_3C_2/H_2O_2 was investigated for the degradation of ranitidine. It was observed that the reaction rate increased from 3.89×10^{-2} to 7.68×10^{-2} as the reaction temperature increased from 15 to 35 °C, respectively [38]. The higher temperature boosted up the formation of ROS through the decomposition of $S_2O_8{}^{2-}$ and $HSO_5{}^{-}$, resulting in improved catalytic efficiency for ranitidine degradation [39].

6.3.5 *Effect of Co-existing Ions*

The presence of different co-existing ions in the wastewater significantly influences the catalytic performance of MXene-based materials. In addition, humic acid is the major component of natural existing organic matter in the water sources [29]. The presence of Cl^-, $HPO4^{2-}$, $HCO_3{}^-$, and humic acid is reported to have a significant effect on the catalytic efficiency of Ti_3C_2-coated polyurethane sponge for the degradation of carbamazepine [40]. Similarly, $HCO_3{}^-$ and humic acid considerably retreaded the catalytic efficiency of $Fe_2CoTi_3O_{10}$-MXene for the degradation of 2,4-dichlorophenoxyacetic acid [33]. The inhibiting effect of co-existing ions and organic matters is attributable to the (i) competitive adsorption onto the active sites, (ii) interactions between hydroxyl radicals and co-existing ions, (iii) slow decomposition of persulfate to active species [41, 42].

6.4 Conclusion and Future Prospects

MXenes have a remarkable potential for use in environmental applications, particularly in wastewater treatment. The tunable structures, exciting physiochemical properties, distinctive morphology, and compositional diversity characteristics make them capable of removing various contaminants, such as heavy metals, inorganic ions, and organic dyes. The present book chapter highlights the recent advances achieved in the applicability of diverse MXene-based materials. A large number of studies have been carried out that account for the applications of MXenes in wastewater treatment. However, some constraints still need to be solved to enhance the utility of MXene-based material for real-world applications.

Several by-products are produced during the catalytic elimination of organic dyes. Sometimes these by-products are considerably more toxic than the original pollution. In addition, the production of steady by-products can help speed up the deactivation of the MXene-based catalyst. Consequently, in the future, a larger emphasis should be given to the research of the actual degradation mechanism for MXene-based materials for the real-time measurement of intermediates and by-products formed during the catalytic processes.

Further research into the rational design of catalysts based on MXene is needed. The catalytic characteristics of MXenes can be enhanced through careful control of their surface shape and surface terminations. In this regard, effective strategies should be built to boost the catalytic efficiency through the diverse routes of modification (e.g., coupling with different semiconductors, facet modification, impurity doping, and coupling with the porous organic framework). The introduction of iron and copper on the MXenes surface or inside the layered structure promises to improve catalytic performance. The advantages of low-cost and high-efficiency transition metal-based catalysts can make them a popular choice for heterogeneous catalysis. The anchoring hetero component efficiently restrains the stacking of MXenes nanosheets, elemental doping and anchoring active components like metal oxides are also useful ways for enhancing catalytic activity.

The lab-scale catalytic systems are beneficial to understand the basic mechanism involved in the oxidation of VOCs. However, it is important to scale up the catalytic operations under harsh real-world conditions (e.g., the mixed phase of diverse VOCs, moisture presence, and photon energy variation). Furthermore, theoretical modeling (e.g., density functional theory (DFT) simulation) of the catalyst's surface and structure can be used to investigate the potential interactions between the targeted pollutants and the material. The catalytic reactor should be effectively planned and constructed to reduce environmental impact and to positively leverage the mass transfer mechanisms and reaction kinetics.

References

1. A. Bathla, S.A. Younis, B. Pal, K.-H. Kim, Recent progress in bimetallic nanostructure impregnated metal-organic framework for photodegradation of organic pollutants. Appl. Mater. Today **24**, 101105 (2021)
2. A. Bathla, K. Vikrant, D. Kukkar, K.-H. Kim, Photocatalytic degradation of gaseous benzene using metal oxide nanocomposites. Adv. Colloid Interface Sci. 102696 (2022).
3. S. Zhu, D. Wang, Photocatalysis: basic principles, diverse forms of implementations and emerging scientific opportunities. Adv. Energy Mater. **7**, 1700841 (2017)
4. A. Bathla, S.A. Younis, K.-H. Kim, X. Li, TiO_2-based catalytic systems for the treatment of airborne aromatic hydrocarbons. Mater. Horiz. (2023)
5. Y. Ibrahim, M. Meslam, K. Eid, B. Salah, A.M. Abdullah, K.I. Ozoemena, A. Elzatahry, M.A. Sharaf, M. Sillanpää, A review of MXenes as emergent materials for dye removal from wastewater. Sep. Purif. Technol. **282**, 120083 (2022)
6. Y. Sheth, S. Dharaskar, V. Chaudhary, M. Khalid, R. Walvekar, Prospects of titanium carbide-based MXene in heavy metal ion and radionuclide adsorption for wastewater remediation: a review. Chemosphere 133563 (2022)
7. L.-F. Hong, R.-T. Guo, Y. Yuan, X.-Y. Ji, Z.-S. Li, Z.-D. Lin, W.-G. Pan, Recent progress of two-dimensional MXenes in photocatalytic applications: a review. Mater. Today Energy **18**, 100521 (2020)
8. A. Yaqub, Q. Shafiq, A.R. Khan, S.M. Husnain, F. Shahzad, Recent advances in the adsorptive remediation of wastewater using two-dimensional transition metal carbides (MXenes): a review. New J. Chem. **45**, 9721–9742 (2021)

9. N.H. Solangi, R.R. Karri, S.A. Mazari, N.M. Mubarak, A.S. Jatoi, G. Malafaia, A.K. Azad, MXene as emerging material for photocatalytic degradation of environmental pollutants. Coord. Chem. Rev. **477**, 214965 (2023)
10. J. Qiao, L. Kong, S. Xu, K. Lin, W. He, M. Ni, Q. Ruan, P. Zhang, Y. Liu, W. Zhang, Research progress of MXene-based catalysts for electrochemical water-splitting and metal-air batteries. Energy Storage Mater. **43**, 509–530 (2021)
11. S.P. Sreenilayam, I.U. Ahad, V. Nicolosi, D. Brabazon, Mxene materials based printed flexible devices for healthcare, biomedical and energy storage applications. Mater. Today **43**, 99–131 (2021)
12. T.P. Nguyen, D.M.T. Nguyen, H.K. Le, D.-V.N. Vo, S.S. Lam, R.S. Varma, M. Shokouhimehr, C.C. Nguyen, Q. Van Le, MXenes: applications in electrocatalytic, photocatalytic hydrogen evolution reaction and CO_2 reduction. Mol. Catal. **486**, 110850 (2020)
13. J. Shen, Z. Wu, C. Li, C. Zhang, A. Genest, G. Rupprechter, L. He, Emerging applications of MXene materials in CO_2 photocatalysis. FlatChem **28**, 100252 (2021)
14. D.L.T. Nguyen, M.A. Tekalgne, T.H.C. Nguyen, M.T.N. Dinh, S.S. Sana, A.N. Grace, M. Shokouhimehr, D.-V.N. Vo, C.K. Cheng, C.C. Nguyen, Recent development of high-performance photocatalysts for N_2 fixation: a review. J. Environ. Chem. Eng. **9**, 104997 (2021)
15. L. Ling, M. Fan, B. Wang, R. Zhang, Application of computational chemistry in understanding the mechanisms of mercury removal technologies: a review. Energy Environ. Sci. **8**, 3109–3133 (2015)
16. A. Shahzad, K. Rasool, W. Miran, M. Nawaz, J. Jang, K.A. Mahmoud, D.S. Lee, Mercuric ion capturing by recoverable titanium carbide magnetic nanocomposite. J. Hazard. Mater. **344**, 811–818 (2018)
17. A. Shahzad, J. Jang, S.-R. Lim, D.S. Lee, Unique selectivity and rapid uptake of molybdenum-disulfide-functionalized MXene nanocomposite for mercury adsorption. Environ. Res. **182**, 109005 (2020)
18. Y. Du, B. Yu, L. Wei, Y. Wang, X. Zhang, S. Ye, Efficient removal of Pb (II) by $Ti_3C_2T_x$ powder modified with a silane coupling agent. J. Mater. Sci. **54**, 13283–13297 (2019)
19. S. Luo, X. Xu, G. Zhou, C. Liu, Y. Tang, Y. Liu, Amino siloxane oligomer-linked graphene oxide as an efficient adsorbent for removal of Pb (II) from wastewater. J. Hazard. Mater. **274**, 145–155 (2014)
20. H. Wang, H. Cui, X. Song, R. Xu, N. Wei, J. Tian, H. Niu, Facile synthesis of heterojunction of MXenes/TiO_2 nanoparticles towards enhanced hexavalent chromium removal. J. Colloid Interface Sci. **561**, 46–57 (2020)
21. Y. Ying, Y. Liu, X. Wang, Y. Mao, W. Cao, P. Hu, X. Peng, Two-dimensional titanium carbide for efficiently reductive removal of highly toxic chromium (VI) from water. ACS Appl. Mater. Interfaces **7**, 1795–1803 (2015)
22. L. Wang, H. Song, L. Yuan, Z. Li, P. Zhang, J.K. Gibson, L. Zheng, H. Wang, Z. Chai, W. Shi, Effective removal of anionic Re (VII) by surface-modified Ti_2CT_x MXene nanocomposites: implications for Tc (VII) sequestration. Environ. Sci. Technol. **53**, 3739–3747 (2019)
23. C. Peng, X. Yang, Y. Li, H. Yu, H. Wang, F. Peng, Hybrids of two-dimensional Ti_3C_2 and TiO_2 exposing 001 facets toward enhanced photocatalytic activity. ACS Appl. Mater. Interfaces **8**, 6051–6060 (2016)
24. B. Shao, J. Wang, Z. Liu, G. Zeng, L. Tang, Q. Liang, Q. He, T. Wu, Y. Liu, X. Yuan, $Ti_3C_2T_x$ MXene decorated black phosphorus nanosheets with improved visible-light photocatalytic activity: experimental and theoretical studies. J. Mater. Chem. A **8**, 5171–5185 (2020)
25. W. Zhou, J. Zhu, F. Wang, M. Cao, T. Zhao, One-step synthesis of Ceria/Ti_3C_2 nanocomposites with enhanced photocatalytic activity. Mater. Lett. **206**, 237–240 (2017)
26. V.Q. Hieu, T.K. Phung, T.-Q. Nguyen, A. Khan, V.D. Doan, V.A. Tran, Photocatalytic degradation of methyl orange dye by Ti_3C_2–TiO_2 heterojunction under solar light. Chemosphere **276**, 130154 (2021)
27. A. Fayyaz, K. Saravanakumar, K. Talukdar, Y. Kim, Y. Yoon, C.M. Park, Catalytic oxidation of naproxen in cobalt spinel ferrite decorated $Ti_3C_2T_x$ MXene activated persulfate system: mechanisms and pathways. Chem. Eng. J. **407**, 127842 (2021)

28. H. Song, R. Du, Y. Wang, D. Zu, R. Zhou, Y. Cai, F. Wang, Z. Li, Y. Shen, C. Li, Anchoring single atom cobalt on two-dimensional MXene for activation of peroxymonosulfate. Appl. Catal. B **286**, 119898 (2021)
29. L. Wang, H. Jiang, H. Wang, P.L. Show, A. Ivanets, D. Luo, C. Wang, MXenes as heterogeneous fenton-like catalysts for removal of organic pollutants: a review. J. Environ. Chem. Eng. 108954 (2022)
30. U. Ushani, X. Lu, J. Wang, Z. Zhang, J. Dai, Y. Tan, S. Wang, W. Li, C. Niu, T. Cai, Sulfate radicals-based advanced oxidation technology in various environmental remediation: a state-of-the–art review. Chem. Eng. J. **402**, 126232 (2020)
31. F. Ghanbari, M. Moradi, Application of peroxymonosulfate and its activation methods for degradation of environmental organic pollutants. Chem. Eng. J. **310**, 41–62 (2017)
32. Z.A. Alothman, N. AlMasoud, X.Y. Mbianda, I. Ali, Synthesis and characterization of γ-cyclodextrin-graphene oxide nanocomposite: sorption, kinetics, thermodynamics and simulation studies of tetracycline and chlortetracycline antibiotics removal in water. J. Mol. Liq. **345**, 116993 (2022)
33. M. Ding, W. Chen, H. Xu, Z. Shen, T. Lin, K. Hu, Q. Kong, G. Yang, Z. Xie, Heterogeneous $Fe_2CoTi_3O_{10}$-MXene composite catalysts: synergistic effect of the ternary transition metals in the degradation of 2,4-dichlorophenoxyacetic acid based on peroxymonosulfate activation. Chem. Eng. J. **378**, 122177 (2019)
34. P. Yang, S. Li, L. Xiaofu, A. Xiaojing, D. Liu, W. Huang, Singlet oxygen-dominated activation of peroxymonosulfate by CuO/MXene nanocomposites for efficient decontamination of carbamazepine under high salinity conditions: Performance and singlet oxygen evolution mechanism. Sep. Purif. Technol. **285**, 120288 (2022)
35. Y. Cui, M. Liu, H. Huang, D. Zhang, J. Chen, L. Mao, N. Zhou, F. Deng, X. Zhang, Y. Wei, A novel one-step strategy for preparation of Fe_3O_4-loaded Ti_3C_2 MXenes with high efficiency for removal organic dyes. Ceram. Int. **46**, 11593–11601 (2020)
36. M. Ding, W. Chen, H. Xu, C. Lu, T. Lin, Z. Shen, H. Tao, K. Zhang, Synergistic features of superoxide molecule anchoring and charge transfer on two-dimensional $Ti_3C_2T_x$ MXene for efficient peroxymonosulfate activation. ACS Appl. Mater. Interfaces **12**, 9209–9218 (2020)
37. Y. Liu, R. Luo, Y. Li, J. Qi, C. Wang, J. Li, X. Sun, L. Wang, Sandwich-like Co_3O_4/MXene composite with enhanced catalytic performance for Bisphenol A degradation. Chem. Eng. J. **347**, 731–740 (2018)
38. Y. Ma, X. Lv, D. Xiong, X. Zhao, Z. Zhang, Catalytic degradation of ranitidine using novel magnetic Ti_3C_2-based MXene nanosheets modified with nanoscale zero-valent iron particles. Appl. Catal. B **284**, 119720 (2021)
39. X. Liu, F. Huang, Y. Yu, P. Zhao, Y. Zhou, Y. He, Y. Xu, Y. Zhang, Ofloxacin degradation over Cu–Ce tyre carbon catalysts by the microwave assisted persulfate process. Appl. Catal. B **253**, 149–159 (2019)
40. R. Zhou, H. Song, D. Zu, S. Pan, Y. Wang, F. Wang, Z. Li, Y. Shen, C. Li, Self-floating Ti_3C_2 MXene-coated polyurethane sponge with excellent photothermal conversion performance for peroxydisulfate activation and clean water production. Sep. Purif. Technol. **282**, 119990 (2022)
41. Y. Gao, Q. Wang, G. Ji, A. Li, Degradation of antibiotic pollutants by persulfate activated with various carbon materials. Chem. Eng. J. **429**, 132387 (2022)
42. Y. Ma, D. Xiong, X. Lv, X. Zhao, C. Meng, H. Xie, Z. Zhang, Rapid and long-lasting acceleration of zero-valent iron nanoparticles @Ti_3C_2-based MXene/peroxymonosulfate oxidation with bi-active centers toward ranitidine removal. J. Mater. Chem. A **9**, 19817–19833 (2021)

Chapter 7
Emergence and Recent Advances in MXenes for Diverse Sensing Applications

B. Sheetal Priyadarshini, Rahul Mitra, and Unnikrishnan Manju

7.1 Introduction

The demographic, social, and economic evolution has resulted in an accelerating and relentless shift in the urban lifestyle from a reactive to a proactive model. Major changes in the quality of life in recent years have galvanized technological developments in the arena of digital systems, like sensors, that can not only provide actionable information related to health and factors affecting it but also present and process the data intelligently to sustain behavioral change in society. Traditional sensors have been around for more than a century, but smart sensors integrated with information and communication technologies have aged a little more over three decades. Phenomenal developments in computational sciences and cloud networking endow sensors with pervasive connectivity between the physical world and the digital world of the internet of things (IoT) that has been pivotal in revolutionizing how we look at and perceive data [1].

Sensors play a key role in enabling innovative solutions in the digital ecosystem by employing a wide spectrum of signal acquisition and transducing corresponding to various technological domains. They are driven by the insight of having a tangible connection with human end users while accommodating the critical need to adapt to a constantly shrinking form factor space (as in smartphones). Further, the rapid commercialization of low-power micromechanical systems (MEMS) has catalyzed new arenas of ambient sensing like air quality sensing and humidity level sensing that are evolving into crowdsourced maps of environmental parameters over a large geographical area [2]. Consequently, research domains have witnessed a rocketing

B. S. Priyadarshini · R. Mitra · U. Manju (✉)
Materials Chemistry Department, CSIR-Institute of Minerals and Materials Technology, Bhubaneswar, Odisha 751013, India
e-mail: manju@immt.res.in

Academy of Scientific and Innovative Research (AcSIR), Uttar Pradesh, Ghaziabad 201002, India

© The Author(s), under exclusive license to Springer Nature Singapore Pte Ltd. 2024

O. P. Pandey and P. Sharma (eds.), *MXenes: Emerging 2D Materials*,
https://doi.org/10.1007/978-981-97-4064-2_7

interest in developing sensors enabled by information and communication technology (ICT) while catering to the evolving need for miniaturization and biocompatibility.

However, the presence of noise in the ambient environment may interfere with the ability of the sensing material to resolve the signal, thus limiting the efficiency of a commercial sensor [3, 4]. Two-dimensional materials have emerged as versatile materials in the field of sensors due to their distinct electronic and structural properties that can be tailored to our advantage. The sensing performance of 2D materials relies upon the composition, thickness, and quality of the atomic layers which can be optimized through different synthesis procedures like chemical exfoliation, physical vapor deposition, and wet chemical approaches [5]. The highly tuneable surface, electrical, and morphological properties make 2D materials an outstanding choice for next-generation sensors [6–8].

Among the widely studied 2D compounds like graphene, transition metal chalcogenides, and transition carbides, MXenes are a relatively new candidate in the domain of sensing systems after their recent discovery in 2011. The performance of any sensor is highly dependent on the structural defects, surface-terminated groups, hydrophilicity, and dopants introduced into the sensing materials. MXenes are gaining interest as an efficient sensing material due to their highly tunable bandgap from semiconductor to metallic, hydrophobic surface properties with abundant functional groups, excellent thermal conductivity, mechanical robustness, and thermal and chemical stability. This group of 2D materials is highly favored for stress, pressure, gas, and electrochemical sensors due to the high surface area to volume ratio, biocompatibility, and adjustable lateral size [9].

7.2 Basic Principles and Fundamentals of a Sensor

A sensor is basically a device that processes a physical stimulus of interest and produces a recordable and quantified output that can be interpreted by the user. The words sensor and transducer have been used interchangeably in research articles, but they differ in their working principles. While transducers typically convert one form of energy to another, sensors process the information from the real world into electrical form only, as visualized in Fig. 7.1. We can say that the inputs of the sensor are modified by the transducer into functional and usable formats. An ideal sensor has high precision and sensitivity, a quick and linear transfer function, lower fabrication cost, prolonged durability, low hysteresis, and fast response time for hassle-free repeatability.

For a given parameter, there can be more than one sensing method depending on smooth data acquisition, personal feasibility, and maximum economic benefits. Each sensor has varying levels of sensitivity, specificity, and accuracy that are highly specific to the working environment. Sensing modalities can be broadly classified into three main categories [10]:

(i) Contact sensors:

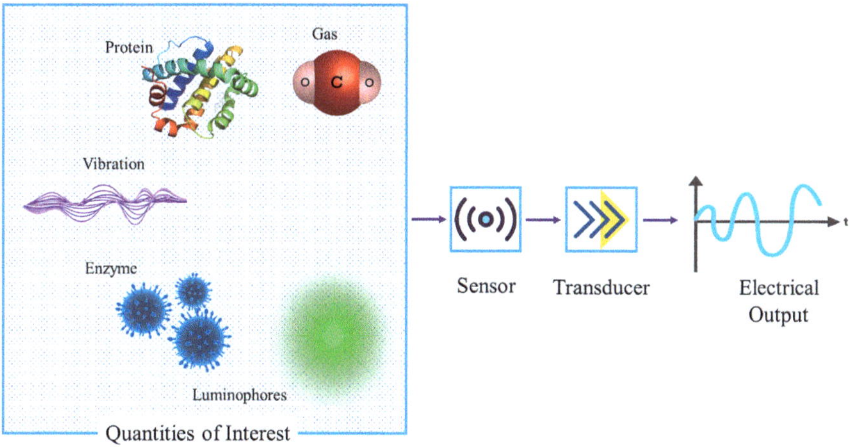

Fig. 7.1 The mechanism of sensing

This approach necessitates physical contact of the subject (e.g., human body) or sample (e.g., liquid, gas, or equipment) with the concerned sensor. This method makes signal acquisition easier, but it may perturb the state of the subject. For on-body sensing, comfort and biocompatibility are essential considerations to ensure an ergonomic design. The on-body sensors are further classified into invasive sensors (implanted inside the body), like in-vivo blood pressure monitoring, and non-invasive sensors (attached to the skin as a patch), like pulse oximetry.

(ii) Non-contact sensors:

This form of sensing does not need physical contact with the quantity of interest. This is generally employed for ambient sensing applications like passive infrared sensors and tracking the motion of individuals in a house. They offer the added advantage of causing minimum perturbation to the subject while signal acquisition along with faster response time as compared to contact sensors.

(iii) Sample analysis sensors:

This method involves the collection of a representative sample by a human or automated sampling system which is further processed by the sensors to provide a measurable output. This procedure is primarily adopted in the healthcare and environmental sectors for monitoring body fluids and bacteria growth in water, respectively.

Depending on the parameter of interest, sensors are broadly classified as mechanical, electrochemical, optical, biological, and ambient sensors. Mechanical sensors transduce inputs like velocity, displacement, acceleration, or motion into electrical output [11, 12]. A strain gauge is the most commonly used mechanical sensor that detects changes in the resistance of the sensing element when exposed to varying levels of strain. Optical sensors detect the photons of light in the visible, infrared, and ultraviolet regions and operate by measuring the changes in wavelength, the intensity of emission or absorption, and phase changes in the beams due to interference. These sensors have been widely used in industries for automated doors and measuring levels of liquids in tanks.

Electrochemical sensors comprise a sensing electrode, a reference electrode, a counter electrode, and a solid or liquid electrolyte. When operated at a lower temperature (<150 °C), they sense changes in pH, conductivity, dissolved ions, and gases in the electrolyte, and at higher temperatures (>500 °C), they sense molten metals using solid electrolytes. Biological sensors or biosensors employ an immobilized biochemical sensing material (enzyme, antibody, nucleic acid, or enzyme) which, when it comes in contact with the analyte of interest, produces certain products like gases, ions, or heat that can be detected to give quantified outputs. Biosensors are now available as over-the-counter blood glucose and cholesterol sensors which play a significant role in day-to-day healthcare monitoring [1, 10].

7.3 Sensor Fabrication Techniques

MXene-based sensors are either fabricated by blending MXenes into a variety of polymer matrices or developed into multidimensional heterostructures of MXenes with other 0D, 1D, or 2D nanomaterials. Real-world sensing demands highly flexible and efficient sensors with ergonomic design to ensure minimum hindrance and maximum comfort. The most commonly employed method for fabricating MXene-based polymer composite is dip coating, spray coating, electrospinning, and wet spinning, schematically represented in Fig. 7.2. In dip coating method, the substrate is dipped into the solution containing the coating material for a while and pulled out at a pre-determined speed to tailor the thickness and uniformity of the deposition. Yue et al. demonstrated a piezoresistive sensor by dipping a sponge into MXene of different concentrations along with polyvinyl alcohol (PVA) nanofiber as spacers to fabricate a piezoresistive sensor [13].

Electrospinning is a relatively controlled synthesis procedure where the MXene/polymer melt solution is fed into a spinneret fitted with a syringe pump. The activation of the pump results in the ejection of pendant-shaped droplets that form into a Taylor cone due to AC or DC voltage. When stretched over longer distances, the jet extends to form a straight line and undergoes whipping motion due to bending instabilities. The fibers, thus formed, settled, and solidified on a drum collector that can be tailored to have different diameters and rotating speeds to control the deposition of the nanofiber. Sharma et al. fabricated a capacitive sensor with MXene/polyvinylidene fluoride-trifluoroethylene (TrFE) nanofibrous scaffolds as the active layer through electrospinning. The spinning time of the collector drum was varied and monitored to reach the optimal thickness of the scaffolds [14].

Spray coating involves compression of the coating solution, which is then dispensed out in the form of aerosols. The tendency of a pressurized liquid to turn into mist by forming small droplets when the liquid is suddenly expelled out of the can form the basic principle of spray coating. The thickness and uniformity of

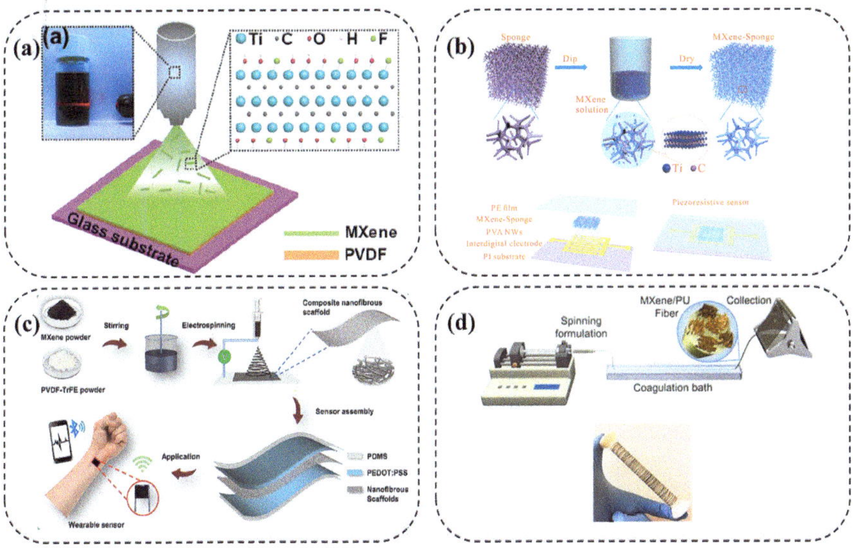

Fig. 7.2 Schematic of the fabrication process of **a** MXene-based sensor via spray coating method (inset shows the structure of MXene layer and optical image of the solution). Adapted with permission from Ref. [15]. Copyright 2019, The Royal Society of Chemistry **b** MXene sponge. Reproduced with permission from Ref. [13]. Copyright 2018, Elsevier. **c** MXene-based sensor via electrospinning. Adapted with permission from Ref. [14]. Copyright 2020, American Chemical Society, and **d** wet-spinning setup for MXene fibrillar nanocomposite sensor. Adapted with permission from Ref. [16]. Copyright 2020, John Wiley and Sons

the deposited layer can be controlled by altering the distance between the nozzle and the substrate and changing the spray pressure. Li et al. [15] fabricated a multi-layer MXene/polyvinylidene fluoride (PVDF) composite using spray coating, spin coating, and hot-pressing technique. They found that spray coating produced the most efficient multi-layer structure with a larger interlayer spacing, increasing the dielectric constant and high breakdown strength.

The wet spinning process has also been employed to fabricate MXene-based polymer nanocomposite sensors. Seyedein et al. used polyurethane (PU) elastomer and a dispersion solution of MXene in an organic solvent like dimethylformamide or dimethyl sulfoxide to synthesize a strain sensor through a wet spinning procedure. The MXene/PU. sensor exhibited an all-time high gauge factor of 12,900 and a considerable sensing strain of approximately 152%. They produced fibers over a length of 1 km, which vouches for the large-scale production of the sensor [16].

However, the major challenge in fabricating an effective sensor is deciding the optimum amount of MXene that should be added to a polymer to form a conductive network within the matrix which should also be able to break within the desired strain range due to increasing distance between the MXene sheets. When the sensor approaches the threshold value of strain, the MXene flakes that form a conductive network move apart, which causes the collapsing of the network resulting in no electric current passing through the sensor, this marks the breakdown of the sensor and corresponds to the amount of strain that the sensor can detect. If the concentration of MXene is higher than the optimal value, even at higher strain, the MXene sheets interact with each other and cannot detect changes in resistivity or conductivity. Thus, the concentration of MXene plays a crucial role in the fabrication of a sensor with improved sensitivity and a higher gauge factor.

7.4 Comparison of MXenes with Other Sensing Elements

Various 2D materials like graphene oxide (GO), reduced graphene oxide (RGO), and transition metal dichalcogenide (TMDC) have been widely studied in relation to developing commercial mechanical, electrochemical, biological, and gas sensors. Nevertheless, the highly tuneable surface and electrical properties of MXenes have attracted researchers to play around with the pristine and composite of MXene in the field of sensing. A major pre-requisite for fabricating functional sensors is superior mechanical strength with high flexibility. The theoretically calculated Young's modulus of Ti_3C_2 is 502 GPa. The elasticity of single layer $Ti_3C_2T_x$ obtained from force-indentation experiments based on force–deflection relationship, using Poisson's ratio of $\upsilon = 0.227$, ranges from 278 to 393 Nm^{-1}. The average elasticity of double-layer MXene was 655 ± 19 Nm^{-1}, whereas, for a monolayer sheet, it averaged 326 ± 29 Nm^{-1}, indicating strong interlayer interaction of surface groups. The nanoindentation studies reveal that $Ti_3C_2T_x$ possessed a higher elasticity than GO, RGO, and MoS_2, thus reinforcing its competency in the sensing arena. Consequently, MXene sheets showed higher foldability and were easily rolled into a cone shape (radius <20 nm) and airplane shape without any significant damage to the structure, which caters to the design flexibility of the sensors. The $Ti_3C_2T_x$ thin films with a thickness of 3.3 microns showed a high mechanical strength of 22 MPa, which was comparable to RGO and MoS_2 thin films [17, 18].

The unique electronic structure of MXenes severely depends on the surface terminations, and DFT calculations have shown that surface functionalization with –F and –OH transforms the metallic Ti_3C_2 into semiconductors with reduced carrier density and conductivity. When considering multi-layer MXenes, intercalation by different moieties enlarges the intercalation spaces and thus decreases the resistance by order of magnitude, thus enhancing conductivity. The layered hierarchical architecture, with rich interlayer interactions and intercalations, endows MXene-based heterogeneous structures with ultra-high device sensitivity and superior noise immunity, which is imperative for developing commercial sensors. A nacre-mimetic bioinspired "brick–mortar" architectural strain sensor was designed by the amalgamation of 2D titanium carbide (MXene) $Ti_3C_2T_x$/1D silver nanowire, which exhibited high gauge factor over 8700 within a strain region of 76–83%. Thus, surface functionalization and hierarchical layering assist in tailoring the property of the MXene suited to the mode of sensing, which makes MXenes a versatile category of compounds that can be used in a multitude of sensing platforms [19].

7.5 Surface Functionalization and Modification in MXenes

As mentioned in the previous section, the surface functionalization of MXenes is an easy way to tailor the functional properties of MXenes suited to sense the physical quantity of interest. The surface properties of $Ti_3C_2T_x$ can be governed by adjusting the type and fraction of the T part in the compound during synthesis procedures via covalent and non-covalent modifications. The non-covalent surface modification is achieved through a synergistic effect among van der Waal's forces, electrostatic attraction, and hydrogen bonding. The covalent modifications, however, are categorized into three groups depending on the type of atoms, namely (i) small molecule functionalization involving alkali metal hydroxides, acid halide or acid anhydride, organic amines, and epoxy compounds, (ii) single heteroatoms functionalization involving vanadium, sulfur, nitrogen or phosphorous, and (iii) surface-initiated functionalization involving polymers like polystyrene (PS) or polyvinyl alcohol (PVA) [20].

The synthesis of sensors necessitates the need to have ex-situ blended or in-situ polymerized MXene composite with the polymer. The hydroxyl groups on the surface of MXenes are highly reactive, and surface modification can improve the affinity between MXene and the polymeric phase. For a real-world working sensor, more extended durability and repeatability are required, which demands a self-healing MXene polymer composite with necessary surface modifications using a reactive elastomer as the host matrix. Guo et al. demonstrated the esterification of surface hydroxyl groups in MXenes with carboxyl acid groups of serin. This functionalized MXene was added to a reactive elastomer, epoxidized natural rubber in this case, to improve the self-healing property and lifespan of the MXene/polymer composite. The addition of the functionalized MXene to the polymer resulted in the formation of hydrogen bonds between the serin-modified MXene and the hydroxyl or carboxylic groups of the reactive elastomer. This bonding consequently decreases the percolation threshold of this MXene-based nanocomposite sensor due to the formation of well-organized MXene networks within the composite film.

This also agrees with another research that argues that a randomly oriented network of pure, unfunctionalized MXenes in a non-reactive polymer exhibited a percolation threshold of around 40 wt%. This percentage was drastically reduced to 6 wt% in the case of serine-functionalized MXene incorporated into reactive epoxidized natural rubber, which is correlated to the establishment of hydrogen bonds due to functionalization, as shown in Fig. 7.3a. The study also revealed that an utterly cut MXene/epoxidized natural rubber self-healed to its 100% initial mechanical strength after 90 min. A marked ease in the twisting, bending, and stretching of the nanocomposite sensor was also observed after self-healing. So, the hydrogen bonds enriched the mechanical properties of the sensor and enhanced the flexibility, twistability, and self-healing properties as they formed bridges of stress transfer between the polymer and MXene [21].

In another study by Chen et al., surface functionalization of $Ti_3C_2T_x$ with a representative of the single assembled monolayer (SAM) group, namely, 1H, 1H, 2H,

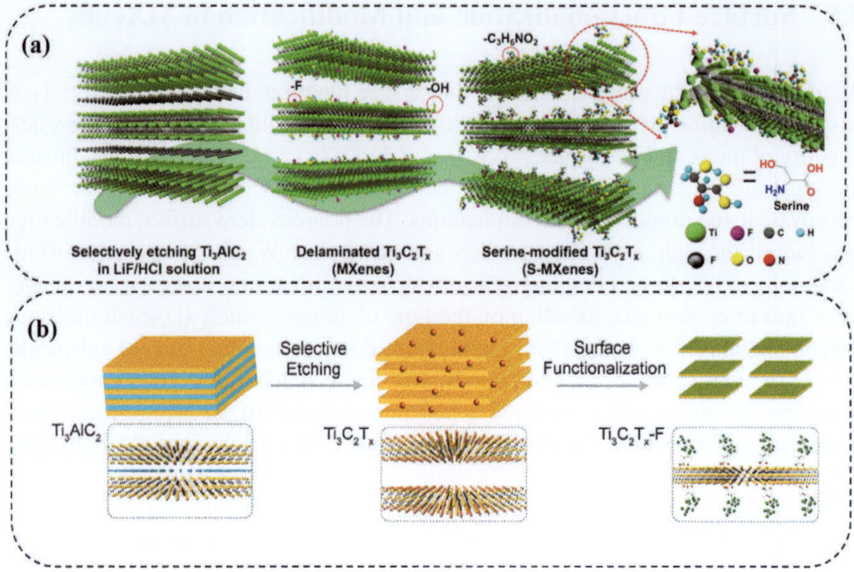

Fig. 7.3 Surface functionalization mechanism of MXenes **a** with serine due to esterification reaction between the hydroxyl group of MXene and a carboxyl group on serine. Adapted with permission from Ref. [21]. Copyright 2020, American Chemical Society. **b** with 1H, 1H, 2H, 2H-perfluorooctyltriethoxysilane (FOTS). Adapted with permission from Ref. [22]. Copyright 2020, American Chemical Society

2H-fluorooctyltriethoxyalkylsilane (FOTS), depicted in Fig. 7.3b, resulted in the formation of a superhydrophobic layer over the surface –OH and –O functional groups which significantly reduced the environmental degradation of the MXene when exposed to air or aqueous environments. MXenes are prone to rapid degradation in the air or humid environments due to their hydrophilic nature stemming from the –OH, –F, or –O terminal groups that have a high affinity for oxidation. This molecular protection layer mitigates the problem of oxidative reduction and helps in dealing with the reduced shelf-life of the MXene-based sensors. The flexible FOTS molecules are intercalated between the MXene layers increasing the interlayer spacing, which enhances the sensing abilities of the functionalized sensor. In contrast to pristine $Ti_3C_2T_x$, showing a 6.6% response, the $Ti_3C_2T_x$–F (FOTS functionalized MXene) shows a 14% response toward 120 ppm of ethanol. The surface functionalized sensor portrays higher selectivity, sensitivity, response, and recovery time in the detection of oxygen-based volatile organic compounds [22].

7.6 MXene-Based Sensors

7.6.1 Pressure and Strain-Based Sensors

Flexible and portable pressure sensors have distinguished themselves amid a lot of interest for a multitude of different uses, such as electronic skins, human–machine interfaces (HMI), flexible touch displays, and intelligent robotics [23–26]. This is because of their proficient adaptability, sustaining electrical sensor behavior concerning human tasks without interfering with or limiting the motions of the person wearing the sensor, high integration prospects, flexibility, and ease of use. By adopting a variety of sensing techniques, such as piezoresistive, capacitive, piezoelectric, and triboelectric phenomena, the majority of reported pressure sensors have currently achieved good sensing performance [11, 27–29].

Typically, conducting nanostructured material networks and flexible plastic substrates are combined to create flexible wearable pressure sensors. Due to their exceptional mechanical flexibility and electrical conductivity, remarkable nanostructured materials, like metal nanoparticles, carbon nanotubes, metal nanowires, polymer nanofibers, carbonized nanofibers, graphene, and their nanocomposites, have recently been used to fabricate a variety of flexible pressure sensors [30]. MXenes, which are two-dimensional layered nanomaterials with a large specific surface area, superior hydrophilicity, and high conductivity, are frequently employed in transparent electrode materials, nanocomposites, and electrochemical energy storage [31].

Hierarchical MXene composite meshes have not been used very often to create flexible pressure sensors, despite having a huge specific surface area and robust conductivity. Guo et al. have made a wearable transient pressure sensor using MXene nanosheets that could be used for broad-range human–machine interfacing (HMI) [32]. The simple device construction procedure is depicted in Fig. 7.4. First, by selectively etching the precursor, conductive MXene ($Ti_3C_2T_x$) nanosheets were produced. MXene sheets with a lateral dimension of a few micrometers were present in the resultant colloidal solution. The pores of the anodized aluminum oxide membrane were visible through the well-exfoliated ultrathin MXene sheets. The sensitivity (S) of the pressure sensor can be defined as $S = \delta(\Delta I/I_0)/\delta P$, where ΔI is the relative change in current, I_0 is the sensor current under no load, and P is the pressure under load. It is generally defined as the ratio of relative change in current to the applied pressure. Therefore, the sensitivity of a pressure sensor can be extracted from the slope of current curves plotted against different applied pressures. Representative current graphs ($\Delta I/I_0$) of the sensor for various external pressures are shown in Fig. 7.4a, b. To determine the average and typical sensing performance with a standard variation of around 5%, a total of five samples were tested. The sensitivity of flexible pressure sensors was divided into three zones, as shown in Fig. 7.4c. S_1 in the low-pressure range (23–982 Pa), S_2 in the pressure range from 982 to 10 kPa, and S_3 in the pressure range from 10 to 30 kPa. S_1, S_2, and S_3 had sensitivities of 0.55, 3.81, and 2.52 kPa^{-1}, respectively. The MXene/tissue-paper-based pressure sensor with its repeatable sensing capability is displayed in Fig. 7.4d. The sensing response was sustained during 10,000 compressive loading/unloading experiments between 0 and 3036 Pa, showing the possibility for reliable and consistent human motion biomonitoring. Additionally, Fig. 7.4e displays the MXene/tissue-paper-based pressure sensor's reaction time (11 ms) and recovery time (25 ms) at a pressure of 3036 Pa, guaranteeing a real-time detection sensitivity to an instant human–machine interface. The I–V curves, as shown in Fig. 7.4f, demonstrated that the resistance (slope of the I–V curves) remained constant for each pressure and that the pressure-sensitive sensing response of the sensor was stable at various external pressures.

In order to demonstrate its performance for strain sensing, Yang et al. coated single/few-layer MXene sheets onto a network-structured polyurethane (PU) mat (network-P mat) [33, 34]. Numerous surface functional groups on MXene nanosheets allow for their attachment onto the surface of PU mats via hydrogen bonds or electrostatic interactions. The highly interlocking network structure of the stretchy network-P mat, which was created using electrospinning, allowed the MXene sheets to readily penetrate into it for ornamentation. The network-structured MXene/polyurethane mat

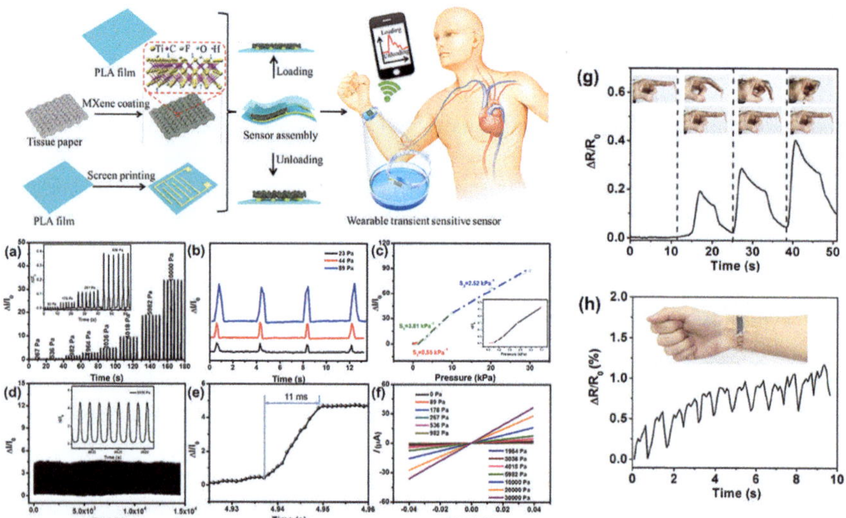

Fig. 7.4 Representation of a schematic showing how flexible and wearable pressure sensors are made using MXene nanosheets. **a** Current–Pressure curves between loading and unloading, with the current response in the low-pressure range of 89−536 Pa in the inset. **b** Current–Pressure curves under loading and unloading at 23, 44, and 89 Pa. **c** Sensing sensitivity of the flexible pressure sensor to pressure. **d** The durability test under a pressure of 3036 Pa. The inset shows a zoomed view. **e** Response time of the pressure sensor. **f** I−V curves of the sensor under various applied pressures. Adapted with permission from Ref. [32]. Copyright 2019, American Chemical Society. **g** Applications of the network-MXene/polyurethane mat strain sensor for tracking physiological signals and detecting physical motion. **h** Pulse beat monitoring using the network-M/P mat strain sensor. Adapted with permission from Ref. [33]. Copyright 2019, The Royal Society of Chemistry

(network-M/P mat)-based strain sensor as-prepared demonstrated extremely high sensitivity (with a gauge factor of 228), extremely low limit of detection (0.1%), a wide and adjustable sensing range (up to 150%), and outstanding mechanical durability (over 3200 cycles). Additionally, simultaneous realization of multifunctional detection, including lateral stretch, vertical pressure, bending, and subtle vibration monitoring, was possible as shown in Fig. 7.4g, h. The network-M/P mat-based strain sensor can thus monitor a wide spectrum of bodily motions, from delicate physiological signals, such as pulse waves, to forceful physical actions, such as joint and muscle movement, proving its considerable potential for practical applications.

Furthermore, $Ti_3C_2T_x$ MXenes paired with CNTs have been investigated for producing adaptable and scalable strain sensors by Cai et al. [35]. By layer-by-layer (LBL) spray coating delaminated $Ti_3C_2T_x$ MXene flakes and hydrophilic single-walled carbon nanotubes (SWNTs), sandwich-like $Ti_3C_2T_x$ MXene/CNT sensing layers (1–2 μm) were created. Owing to the water-based sandwich-like microstructure, the ultrathin devices possessed low limits of detection of 0.1%, great sensitivity with a gauge factor up to 772.6, configurable sensing ranges of 30–130% strain, and thin device dimensions of 2 μm. Only a few alterations in the loading and unloading signal were seen after 5000 cycles. Due to the remarkable sensing capabilities, both subtle deformations like speech, and noticeable gestures, such as walking, running, and leaping, were detected.

7.6.2 Electrochemical Sensors

Ti_3C_2 is the most common form of MXene and has distinct advantages over other 2D materials that are driving its application in the development of electrochemical sensors. These distinct qualities can be summed up as follows.

As compared to other 2D materials, MXene has the clear benefit of high electrical conductivity, which is essential for speeding the pace of heterogeneous electron transport. One of the most crucial reasons for its use in electrochemical sensors is its outstanding conductivity feature. MXenes show adequate stability and dispersibility in solutions. The most common method for creating customized electrodes is drop-casting, which calls for pre-treatment before applying the well-dispersed coating solution. This is vital for the creation of electrochemical sensors. MXene can be a potent substrate for the creation of flexible conductive platforms that can be utilized to create wearable electrochemical sensors, which are critically needed for clinical analysis and health care monitoring. This is due to its great stretchability and biocompatibility. MXene has a significant potential to be coupled with a range of functional materials or biomolecules for various analytical applications because of its 2D layered structure and distinctive surface with numerous chemical groups.

In a study by Shahzad et al. [36], they created a straightforward and reasonably priced electrochemical sensor for dopamine (DA) detection. In the study, a well-dispersed $Ti_3C_2T_x$ solution was drop casted on a glassy carbon electrode (GCE) along with a conducting binder, i.e., Nafion. This helped to strongly immobilize

$Ti_3C_2T_x$ on the electrode surface, resulting in the formation of a stable film that could accumulate positively charged species in a targeted manner. Nafion deprotonates at physiological pH, and it is likely that a negatively charged ion trapped on the surface of the MXene sensor electrode further prevents negatively charged anions (ascorbic acid—AA and uric acid—UA) from diffusing to the electrode. The XRD patterns of Ti_3AlC_2 and $Ti_3C_2T_x$ are shown in Fig. 7.5a. The crystalline Ti_3AlC_2 peaks with (002), (004), (101), (104), and (105) planes were represented by the diffraction peaks at 9.58°, 19.2°, 34.02°, 39.04°, and 41.8°, respectively [35]. Following etching, new diffraction peaks of the (001) series were seen in place of the Ti_3AlC_2 characteristic diffraction peaks. With an interplanar spacing of 13.52, which was substantially bigger than the 9.29 from the (002) plane of Ti_3AlC_2, the primary peak at 6.28° was assigned to the (002) plane of $Ti_3C_2T_x$. Successful etching and delamination were suggested by this shift in the (002) peak, as well as an increase in the d-spacing and peak widening. These findings from SEM and TEM, depicted in Fig. 7.5b, d, and e demonstrated that the etching procedure was gentle enough to delaminate the layers and that it took place along a certain crystal orientation so as to preserve the parent MAX phase's regular hexagonal symmetry with the 2D layers connected by weak van der Waals interaction of hydrogen bonds.

The CV characteristics for the three electrodes are displayed in Fig. 7.5c. A control scan was performed with just the GCE in the buffer. All three electrodes had a distinct redox peak when DA was present. With greater current responsiveness and a lower peak-to-peak separation, $Ti_3C_2T_x$/GCE clearly offered a better sensing performance. With its excellent repeatability and stability performances (Fig. 7.5f,

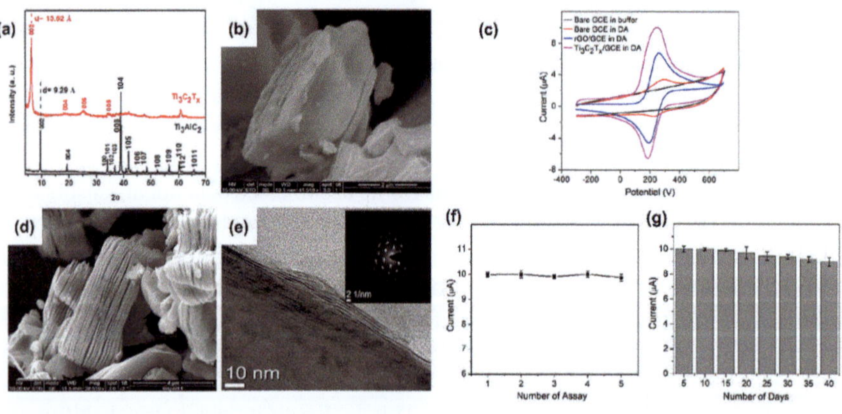

Fig. 7.5 **a** X-ray diffraction (XRD) pattern of Ti_3AlC_2 and $Ti_3C_2T_x$. **b** SEM image of Ti_3AlC_2. **c** Cyclic voltammograms (CVs) of bare and modified glassy carbon electrodes (GCE) in 1 μM Dopamine solutions of 0.1 M phosphate-buffered saline (PBS) buffer solution with pH 7.2. **d** SEM image of $Ti_3C_2T_x$. **e** Bright-field T.E.M. image of $Ti_3C_2T_x$ with inset corresponding to its selected area electron diffraction (SAED) pattern. **f** Repeatability and **g** stability tests of DA sensing (1 μM) with $Ti_3C_2T_x$/GCE. Adapted with permission from Ref. [36]. Copyright 2019, The Korean Society of Industrial and Engineering Chemistry. Published by Elsevier B.V.

g), MXene/GCE-based electrochemical sensors can be a very promising alternative for DA sensing.

Without immobilizing enzymes, MXene might potentially be used to create electrochemical non-biosensors. In order to detect the drinking water contaminant BrO_3^-, the first electrochemical non-biosensor based on pure Ti_3C_2 MXene was created in 2018 [37]. Ti_3C_2 MXene, which served as both a signal-enhancing matrix and a reducing agent, demonstrated exceptional electrocatalytic properties for efficient BrO_3^- reduction. The MXenes were also included in other electrode systems, such as the graphite composite paste electrode (GCPE) [38] and the screen-printed electrode (SPE) [39]. To detect adrenaline, the MXene/GCPE electrochemical sensor was created. Additionally, Ti_2C MXene has been included in the electrochemical sensing system for the first time. The constructed sensor attained a line of detection (LOD) of 9.5 nM and had 99.2–100.8% recoveries when used to detect adrenaline in pharmaceutical samples. Acetaminophen (ACOP) and isoniazid were determined by cyclic voltammetry (CV) using the MXene/SPE electrochemical sensor (INZ). In comparison to bare SPE in 0.1 M H_2SO_4, the Ti_3C_2 MXene had shown superior electrocatalytic activity toward the oxidation of ACOP and INZ, and the separated oxidation peak potentials ensured simultaneous detection of the targets. Arranging nanomaterials in a specific location on an electrode with regulated architecture, shape, and thickness is a particularly difficult problem in the fabrication of different electrochemical sensing devices [40]. Sharifuzzaman et al. [40] developed a green, simple, and economical one-pot deposition technique of Ti_3C_2 nanosheets for producing large-scale and high-quality MXene nanosheets thin films. They showed that by utilizing an electroplating process with a DMSO solution containing colloidal Ti_3C_2 nanosheets as the electrolyte, Ti_3C_2 nanosheets could be accurately and evenly deposited on various sensing electrodes at room temperature in a matter of minutes. In DMSO solvent, Ti_3C_2 nanosheets totally dispersed and were very stable. The electric field created by the redox process in the colloidal Ti_3C_2 nanosheet solution under the influence of a constant applied voltage drove the nanoparticles toward a particular electrode interface, where they were cathodically electroplated.

Electrochemical sensors have demonstrated considerable application potential for MXene, and a variety of electrochemical sensors, including electrochemical non-biosensors, electrochemical biosensors, electrochemiluminescence sensors, and photoelectrochemical sensors, have been created based on MXene. It also shows significant potential in the production of conductive substrates for different electrode systems, particularly for patterned electrodes. This opens the door to the development of a wide range of advanced electrochemical sensing devices, such as wearable, flexible sensors for non-invasive body fluid monitoring and portable detection instruments.

7.6.3 Photoluminescent-Type Sensors

When the excited state decays to the ground state, the absorbed energy is released as light. A phenomenon known as PL Quantum dots (QDs) contains separate energy levels due to the quantum confinement phenomenon; as a result, the excited states can primarily decay by radiation. MXene QDs (MQDs) exhibit vivid PL with quantum yields of 10–20% while maintaining a high density of active surface sites, making them an excellent material for PL sensors. Bulk Ti_3AlC was used to create amino-functionalized Ti_3C_2 quantum dots in a study by Ai et al. [41]. On the one hand, the usage of Ti_3C_2 MXene quantum dots (MQDs) doped with nitrogen atoms and the addition of amino groups can both boost the FL intensity of MQDs. On the other hand, after changing the MQD surface properties, the aminated functional groups are advantageous for the association of Ni^{2+} with MQDs. They created a simple, high-output approach for fabricating Ti_3C_2 M.Q.D.s with a small particle size of just 2.73 nm, a high photoluminescence yield of 13.55%, and outstanding photostability. The experimental model posits a fluorescence turn-on sensor based on the quenching of MQD. FL by Ni^{2+} and the recovery of MQD. FL for the precise detection of histidine (His), as shown in Fig. 7.6a. After etching the Al layer off of Ti_3AlC_2 with HF acid to produce Ti_3C_2 MXene, M.Q.D.s were made by hydrothermally and ultrasonically cutting huge layers of Ti_3C_2 MXene. The XPS spectra in Fig. 7.6b, c demonstrated typical Ti_3C_2 MXene nanosheet characteristic peaks with Ti-C $2p^3$ and Ti-C $2p^1$ double peaks, demonstrating that MQDs continue to preserve the chemical composition of the nanosheets. In order to remove the remaining Al and the original bulk material Ti_3AlC_2, the Al layer from the Ti_3AlC_2 phase was etched, and the result was the Ti_3C_2 MXene precursor, which can be seen in Fig. 7.6d. Scanning electron microscopy clearly reveals the accordion-like structures of delaminated Ti_3C_2. The homogeneous distribution of the elements Ti and C is visible in the energy dispersive spectra of Ti_3C_2 (Fig. 7.6e, f). Energy-dispersive X-ray spectroscopy (EDX) also demonstrated the production of Ti_3C_2 by confirming the existence of Ti, C, and O elements. The FL of MQDs was rapidly suppressed, and equilibrium was attained in approximately 10 min, as illustrated in Fig. 7.6g. Histidine (His) was introduced after the solution reached total stability, and the FL of MQDs recovered fast. After approximately 15 min, it also reached complete stability. Figure 7.6g clearly shows how FL changes with the addition of various compounds, clearly indicating how FL of MQDs changes from quenching to recovery. Figure 7.6h illustrates the FL intensity recovery of MQDs fluctuating as the concentration of His increases. The restored photoluminescence intensity increased linearly with the concentration of His in the range of 100–1000 nM under ideal circumstances ($R = 0.996$). As a result, the study presented a viable method for using novel MXene materials in photoluminescent-based sensing in addition to developing a sensitive assay for His activity sensing.

The PL-type sensing process may be anticipated when these low-dimensional nanoscale materials are used as PL sensors since the quantum confinement phenomena are relevant to QDs and nanoscale materials with different dimensions.

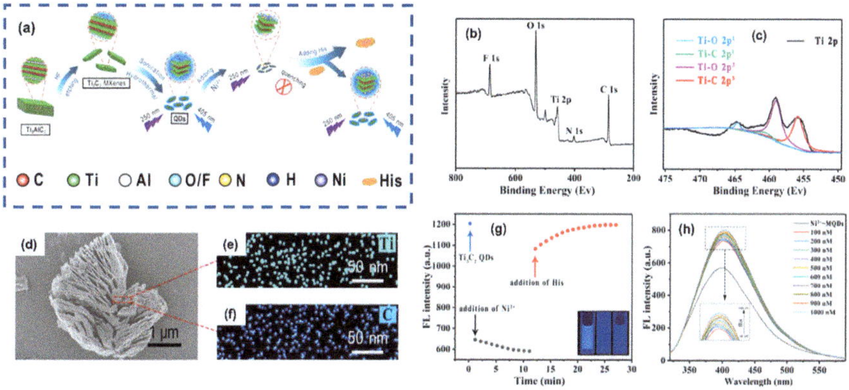

Fig. 7.6 **a** Mechanism of transducing the signal using the Histidine (His)-based Ni^{2+}-modulated MXene quantum dot (MQD) photoluminescent (PL) sensor. **b** XPS survey spectra of Ti_3C_2 M.Q.D.s. **c** High-resolution XPS spectra of Ti 2p region with deconvoluted peaks of the M.Q.D.s. **d** SEM images of the parent bulk material Ti_3AlC_2 and the as-obtained layered Ti_3C_2 MXene. **e, f** Elemental mapping images of the pristine Ti_3C_2 MXene nanosheets using energy-dispersive spectroscopy (EDS). **g** Time-dependent Fluorescence (FL) intensity of MQDs in the aqueous solution at 410 nm after adding Histidine and Ni^{2+} (insets from left to right are M.Q.D.s, Ni^{2+}−MQDs, and His@Ni^{2+}−MQDs solution under 365 nm UV lamp). **h** FL response of Ni^{2+}−MQDs after adding various His concentrations. Adapted with permission from Ref. [41]. Copyright 2021, American Chemical Society

Using MXene nanosheets (MNSs), Desai et al. created a sensor for Ag and Mn ions [42]. Tetramethylammonium hydroxide (TMAOH) was added to Ti_3C_2 MXene solution before being ultrasonically fragmented to produce the MNSs. The synthesized MNSs displayed a PL emission peak at 461 nm through the quantum confinement effect after absorbing 381 nm UV light. Chemical sensors of the electrical and optical types get benefit from the characteristics of MXenes. The strong sensing performance of chemical sensors is primarily due to the low noise response of MXenes and the high adsorption energy of gas molecules on the MXene surface, which are ascribed to the high conductivity and surface functional group in MXene. Future research will benefit from the excellent sensor performance of MXenes have achieved.

7.6.4 Biological Sensors

As indications for the metabolic, physiological, and immunological processes of biological systems, several biomaterials and biomarkers are used. These molecules may be detected qualitatively and quantitatively, which can be used to identify metabolic disorders, gauge the effectiveness of therapeutic interventions, and assess environmental risk factors, including toxins and infections. Therefore, a key need for biosensors is the capacity to accurately and selectively detect tiny quantities of analytes in multicomponent systems. As a result, great effort has been put into improving the specificity and LOD of biosensors. Since biosensors operate on the same principles as chemical sensors, it is possible to enhance the LOD of MXene-based biosensors by guaranteeing an abundance of active surface sites, managing MXene geometry, and utilizing MXene-based composites. The formation and anchoring of suitable bioreceptors that may trigger certain interactions, such as antibody-antigen and enzyme-ligand interactions, is sometimes necessary to achieve high selectivity.

In order to simultaneously and very selectively screen a panel of biomarkers in sweat, the work by Lei et al. [43] offered a stretchy, wearable, modular multi-functional biosensing device. The first use of 2D MXene ($Ti_3C_2T_x$) nanosheets is in an oxygen-rich enzyme biosensor. The solid-liquid-air tri-phase interface of the working electrode is based on the utilization of a $Ti_3C_2T_x$/Prussian Blue (PB) composite in combination with a hydrophobic substrate. $Ti_3C_2T_x$/PB composites significantly outperformed graphene/Prussian blue (PB) and carbon nanotubes (CNTs)/PB composites in terms of electrochemical performance for the detection of hydrogen peroxide due to the outstanding conductivity and electrochemical activity of exfoliated $Ti_3C_2T_x$. A sweat-uptake layer, a sensor layer (glucose, lactate, and pH sensors), and a cover layer made up the three modules that made up the integrated system, as represented in Fig. 7.7a. A porous cloth for collecting sweat and serpentine tunnels made up the sweat-uptake layer. For the purpose of installing active sensors, five sensor spaces and serpentine tunnels were etched into the sensor layer. Monitoring perspiration throughout 10–30 min of vigorous cycling action showed the wearable multifunctional biosensor patch's promise (Fig. 7.7c). The patch was powered by a portable electrochemical analyzer with an internal power supply that was attached to the wristband sweat sensor, as shown in Fig. 7.7b. Following wireless transmission to a distant mobile device, such as a smartphone or tablet computer, real-time analysis was then shown via a specially created cell-phone application. The sensor array performance of the glucose, pH, and lactate was appraised repetitively before and after meals, as shown in Fig. 7.7d, e, and f. The development of ultra-sensitive enzymatic wearable biosensors for early illness detection and individually tailored healthcare was therefore made possible by this study.

To create a medium-free biosensor for detecting H_2O_2, Wang et al. [44] immobilized hemoglobin (Hb) on the surface of a multi-layer Ti_3C_2 material that resembled an accordion. Favorable enzyme immobilization properties were investigated in the MXene-Ti_3C_2 system. The constructed biosensor performed well at detecting H_2O_2, with a linear range of 0.1–260 M and a lower detection limit of 20 nM. Additionally, they created Ti_3C_2 MXene nanocomposites modified with TiO_2 nanoparticles using in-situ hydrolysis and hydrothermal methods, and they applied them to create a mediator-free biosensor for hydrogen peroxide detection. Nanocomposites containing a lot of TiO_2 nanoparticles supported on Ti_3C_2 layered substrates have multiple benefits when utilized as carriers for enzyme immobilization.

Cholesterol is a crucial lipid in the human body and an essential component of cell membranes, helping to maintain the permeability and cohesiveness of cell membranes. A number of disorders have cholesterol as a significant biomarker. Low

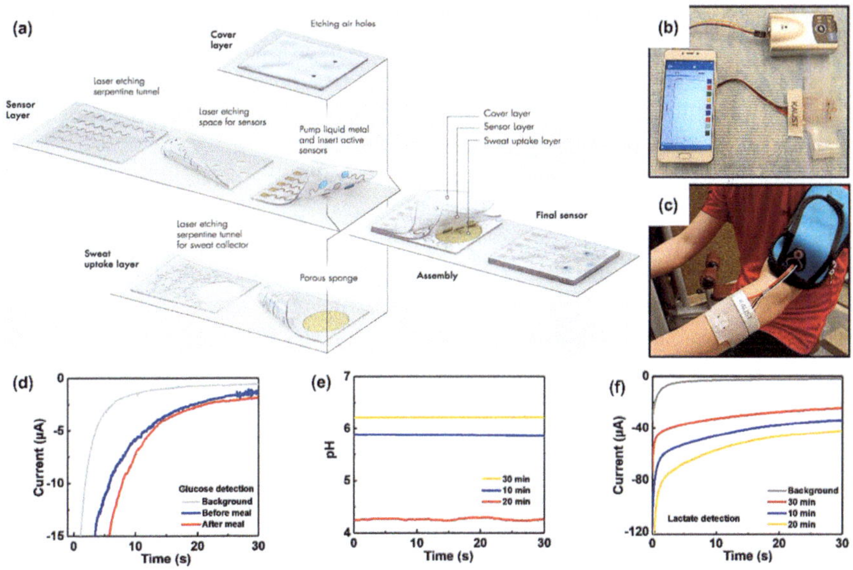

Fig. 7.7 **a** The final assembly of the sensor patch system, which consists of a sweat-uptake layer, a sensor layer, and a cover layer is shown schematically. **b** The wearable sweat-monitoring patch is coupled to a small electrochemical analyzer that enables a wireless interface with Bluetooth-enabled mobile devices and charges the patch as well as regulates it. **c** A skin-mounted, portable electrochemical analyzer is linked to the wearable sweat-monitoring patch. **d** Three different glucose sensors were used to measure the chronoamperometric responses of glucose sensors and pH changes before and after meals. **e** Throughout the activity, the pH sensor readings were recorded many times. **f** At various points throughout the workout, the chronoamperometric responses of the lactate sensor were measured. Adapted with permission from Ref. [43]. Copyright 2019, John Wiley and Sons

cholesterol levels can result in hypolipoproteinemia, malnutrition, sepsis, malnutrition, and anemia. 2.83–5.20 mM is the typical range for maintaining cholesterol levels in the body. As a result, precise, sensitive methods to track cholesterol levels are needed in the academic and medical domains. Ti_3AlC_2 was etched with HF and LiF by Xia et al. to create $Ti_3C_2T_x$ [45]. High specific surface area, biocompatibility, strong electrical conductivity, and dispersibility are all characteristics of the synthesized $Ti_3C_2T_x$ that support electron transport.

7.6.5 Gas and Humidity Detection Sensors

Gas sensing is essential for illness diagnosis, air pollution monitoring, explosive detection, and chemical process control. Chemiresistive gas sensors have garnered a lot of interest among the many gas-sensing techniques due to their better performance and affordable price. They have low selectivity, a high working temperature, and high-power consumption, however. MXenes are attractive candidates for chemiresistive gas sensor applications due to their high surface-to-volume ratio, excellent surface conductivity, surface-terminated capabilities, and hydrophilicity. It should be noted that while hydrophilic absorbents like MXenes may successfully absorb hydrophilic gas molecules, their application for the identification of hydrophobic gas molecules (also known as nonpolar molecules) may be constrained.

Engineering the structure and surface chemistry of sensors based on 2D $Ti_3C_2T_x$ MXenes may improve their ability to detect gases. According to Yuan et al. [46], a versatile and highly effective volatile organic compounds (VOC) sensor was made using a three-dimensional (3D) $Ti_3C_2T_x$ MXene framework. Electrospinning was used to create a 3D polymer (PVA/PEI mixture) framework with cross-linked fibers, which was subsequently transferred to electrodes with Au interdigitation and submerged in the MXene dispersion. The surface of the fibers was functionalized by MXene by self-assembly as a consequence of electrostatic interactions among the negatively charged MXene (induced by functional groups) and positively charged 3D polymeric framework (originated from the PEI component). The sensor provided better sensitivity toward VOCs, such as acetone, methanol, and ethanol in parts per billion level, in contrast to pure MXene, because of the densely interconnected porous structure.

Additionally, the sensor responded less to polar inorganic gas molecules (NO_2, NH_3, and H_2O) and barely at all to hydrocarbons (toluene and cyclohexane). Charge transfer occurs as a result of interactions between VOCs and polar inorganic gas molecules and the functional groups of MXene (mostly –OH and –F). The strength of the hydrogen bond produced can be correlated with the gas response amplitude. Intriguingly, the metallic conductivity of the MXene caused the resistance of the sensor to rise independent of the nature of the molecule.

MXene-based devices can potentially be improved in terms of gas detection by hybridization with other nanomaterials. According to Chen et al. [48], liquid-phase exfoliation and inject printing were used to create a flexible nanohybrid room-temperature sensor made of TMD (WSe_2) and MXene ($Ti_3C_2T_x$). Numerous hetero-junction contacts were produced by evenly decorating the 300 nm MXene scaffolds with WSe_2 flakes that typically had a size of less than 30 nm. After being hybridized with WSe_2, $Ti_3C_2T_x$ developed a greater sensitivity to gas molecules. A $Ti_3C_2T_x$/WSe_2 sensor was successful in selectively and sensitively detecting O-containing V.O.C.s (ethanol, methanol, and acetone), but it was less successful in detecting hydrocarbons. More specifically, $Ti_3C_2T_x/WSe_2$ was 12 times more sensitive to ethanol than $Ti_3C_2T_x$.

Furthermore, Li et al. [47] exhibited a very sensitive Ti_3C_2/TiO_2 humidity sensor. By performing an in-situ alkali oxidation of Ti_3C_2 in a KOH solution, the Ti_3C_2/TiO_2 composite was demonstrated. The optimal structure–response relationship for the Ti_3C_2/TiO_2 composite was investigated, and the potential humidity–sensing method was arrived at. Additionally, an array of flexible humidity sensors based on an ultrasensitive composite was created, and they explored their potential for touch-less control of electrical devices as well as their use in the non-contact detection of liquids. 100 mg of the produced 2D Ti_3C_2 MXene powder was added gradually to 60 mL of KOH solutions with varied KOH concentrations (6, 12, and 24 mol/L). The

KOH solutions were then agitated vigorously at 50 °C for 2, 10, and 20 h. The reaction products were washed in ethanol and deionized water until the pH of the mixture reached 7. The samples were retrieved and vacuum heated (80 °C, 24 h) in preparation for further evaluation and humidity sensor assembly, as shown in Fig. 7.8a. The Ti_3C_2/TiO_2 sensor, with its time-dependent response to stepped changes in the RH, is depicted in Fig. 7.8b. When the sensor was loaded in a dry atmosphere (7% RH), the capacitance varied appropriately as the RH changed and quickly returned to its starting condition. The estimated response and recovery durations were 2.0 and 0.5 s, respectively. This performance is primarily related to the suitable structure, substantial surface area, and plenty of flaws and adsorption sites, which can accelerate the absorption and desorption processes in the composite. The developed Ti_3C_2/TiO_2 sensor was then utilized to measure the RH fluctuation on a fingertip near-surface over distances ranging from 1 to 10 mm. The findings are displayed in Fig. 7.8c. On a flexible PET substrate, they created a 5 × 5 pixel array of humidity sensors with separately linked Au electrodes that were synthesized using the sputter-coating technique. Finally, a switcher with a 20 ms switching time was used to monitor each sensor separately. Figure 7.8d, e shows a 3D positional mapping of three fingertips and how far away they are from the sensor array's surface, demonstrating the sensor potential for non-contact human–machine interaction.

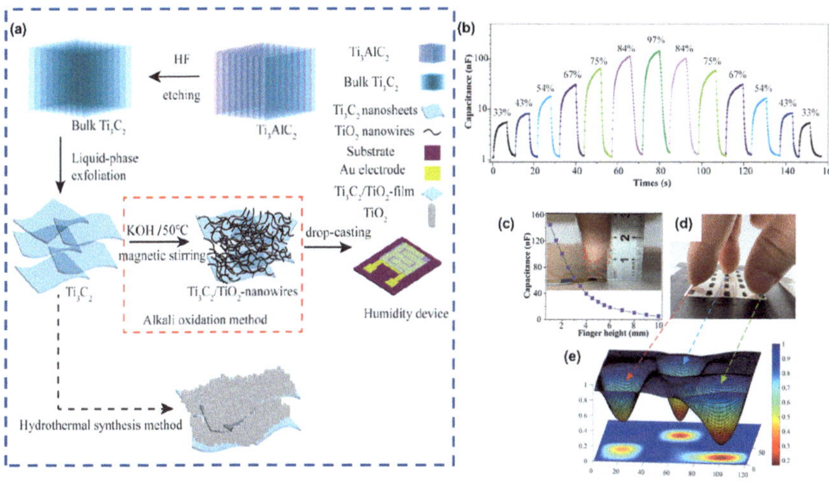

Fig. 7.8 **a** MXene-based humidity sensor preparation process: Hydrofluoric acid (HF) etching, liquid-phase exfoliation (LPE), alkali oxidation, and other methods for material synthesis. **b** Response and recovery characteristics in terms of capacitance (nF) of sensor A5 (A5 synthesis conditions—10 h reaction time and KOH concentration of 12 mol/L) exposed to a series of relative humidity (RH) conditions. **c** Fingertip near-surface assessed by the established humidity sensor using a height-resolved capacitance diagram. **d** A photo showing the three fingertips above the 5 × 5 grid. **e** 3D mapping of three fingers approaching. Adapted with permission from Ref. [47]. Copyright 2019, American Ceramic Society

7.6.6 Capacitive Sensors

The operation of MXene-based sensors can alternatively be used for capacitive measurements. Capacitive sensors, as opposed to electrochemical ones, assess the induced polarization or ionic conduction of adsorbed solutes under AC bias and fields, which can result in extremely quick and sensitive responses.

MXene and PVDF-TrFE were combined to create the composite nanofibrous scaffolds (CNS) by Sharma et al. [14], which were then synthesized using the electrospinning technique. Two types of impacts were observed during the mixing and spinning processes. First, the electrostatic attraction between functional groups and H or F atoms that are found in the PVDF-TrFE polymer causes the van der Waals attractive force to exist in the MXene. The second is the change from α-phase to β-phase crystallization, which synergistically alters the mechanical characteristics of the fiber as a whole. The aforementioned mechanism has been detailed in Fig. 7.9e. The sensor with the 5% MXene concentration had the highest relative change in capacitance ($\Delta C/C_0$) of all the concentrations, which may be a result of the composite's increased sensitivity in the high-pressure and low-pressure regions. Due to its improved performance, the 5% CNS was taken into consideration for more electromechanical characterization and focused applications. The sensor device was dynamically loaded and unloaded at various pressures to examine its wide detection range and dependability, as observed in Fig. 7.9a. Additionally, the capacitance response of the sensor was markedly linear with pressure. The response changed abruptly with each loading and unloading of pressure, but it remained constant when there was no discernible transition. The results supported the idea that the conceived sensor can clearly identify various pressure levels and tiny to large deformations with a stable and hysteresis-free response. According to Fig. 7.9b, the response and relaxation durations of the constructed sensor were assessed at a 1.5 kPa pressure, or 1 mm of compression. As examined further in Fig. 7.9a, b, which show that similar dynamic response and relaxation periods (0.15 s) were observed, the CNS's elastic recovery time is equal to its compression time, showing no hysteresis.

As seen in Fig. 7.9c, several loading cycles were applied to the CNS-based sensor in the low-pressure range. The input pressure had a strong correlation with the capacitive response of the sensor ($\Delta C/C_0$) in the low-pressure range. It was thereby confirmed that the sensor could detect minute pressure changes since it could work smoothly in a very low-pressure range without producing transients or hysteresis. The developed sensor was able to detect 1.5 Pa, which is comparable to one long grain of rice (38 mg), on the sensor surface area. Figure 7.9d shows the ultra-lightweight object detection capacity of the CNS-based sensor. Figure 7.9d also depicts the sensor response for the successive loading of the rice grains, which supports the accuracy of the detection. As shown in Fig. 7.9f, the CNS-based sensor demonstrated good stability and durability throughout a 10,000 load and unload testing at a high pressure of 167 kPa. The magnified views of a few chosen cycles taken from the beginning and conclusion of the test are shown in the left and right insets, respectively. The

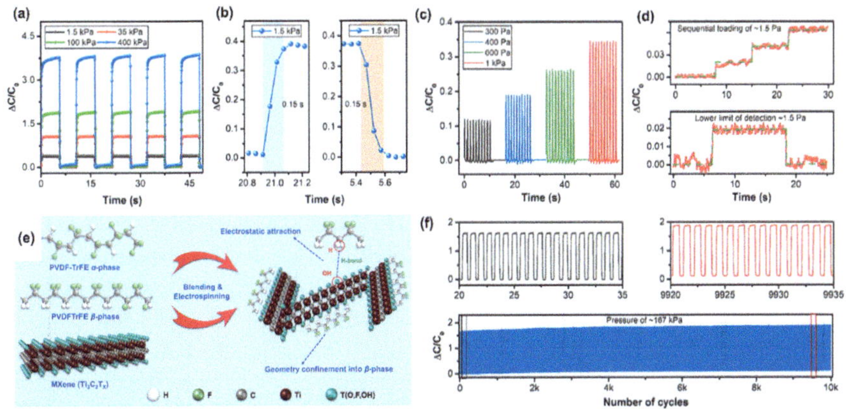

Fig. 7.9 **a** The MXene composite nanofibrous scaffolds (CNS)-based sensor's cyclic capacitive response to a 5 wt% MXene concentration under various loading/unloading pressure levels. **b** Response and relaxation periods for a 1.5 kPa loading/unloading cycle. **c** Capacitance response in terms of its relative change during low-pressure loading and unloading cycles. **d** Sequential loading and unloading of a long kernel of rice amounting to 38 mg to demonstrate the lower limit of detection. **e** Chemical representation of the synergistic outcome when MXene is added to the polymer matrix of PVDF-TrFE in the CNS. **f** 10,000 loading and unloading cycles of the CNS-based pressure sensor under (>40% compression) a high-pressure of 167 kPa. Selected cycles from the start and finish of the stability test are shown in the insets. Adapted with permission from Ref. [14]. Copyright 2020, American Ceramic Society

insets show that even in a high-pressure situation, the CNS's real degeneration and hysteresis are bare minimal.

Due to their low hysteresis and good linearity, capacitive sensors may also play a significant and decisive role in electronic skin (E-skin). However, a capacitive sensor has to be stretchy and self-healing in order to accurately replicate the functions of human skin. The development of such a sensor is constrained by electrode materials, which are frequently neither stretchable nor self-healing. For application in capacitive strain sensors for E-skin, a highly stretchy and self-healing MXene ($Ti_3C_2T_X$)/ polyvinyl alcohol (PVA) hydrogel electrode was created by Zhang et al. [49]. The hydrogel's conductivity and capacity to self-heal were improved by the addition of MXene to PVA. The electrode had a high stretchability at break (~1200%) and self-healing that occurs instantly (~0.15 s). A capacitive sensor based on these electrodes had strong mechanical durability (a 5.8% drop in relative capacitance change after 10,000 cycles), great linearity (up to 200%), minimal hysteresis, and a sensitivity of around 0.40. Additionally, this sensor retained its functionality during a self-healing test, demonstrating its potential for tracking human mobility.

7.6.7 Other Miscellaneous Sensors

Apart from the multitude of sensing modalities, MXenes have also been used for sensing variations in temperature. Inspired by the potential of tailoring the properties of MXenes through application-specific surface modifications, Tran et al. [50] fabricated a smart hybrid titanium-based MXene (Ti_2CT_x) with poly(2-(dimethylamino)ethyl methacrylate) (PDMAEMA) through surface-initiated controlled photoiniferter polymerization (SIPIP). The thermo-responsive tuning of conductivity observed in the hybrid MXene-PDMAEMA structure is more pronounced than that exhibited by pristine MXenes due to the switchable conductive behavior of PDMAEMA below and above lower solution critical temperature (LCST). The prepared thin films exhibit conductivity switching as the temperature cycles above and below the LCST of PDMAEMA. This is attributed to the shrinking of PDMAEMA, which is covalently bonded to the MXene sheets. This reduces the interlayer distance, thus facilitating electron transport across the sheets. At temperatures below LCST, the PDMAEMA moieties expand and push away the sheets increasing the interlayer distance. This results in reduced electron transport which is reflected in the conductivity curves. This assembly acts as an efficient temperature sensor that can be used to gauge thermal variations during chemical reactions and in microelectronics.

Optical sensors have gradually assumed importance as an auxiliary branch of biochemical sensing utilizing surface plasmon resonance (SPR) and evanescent surface waves as the driving factors. For the first time, Chen et al. [51] investigated the effect of MXene on the refractive index (RI) sensors and reported a sensitivity boost of ~30% with MXene integration. MXene nanosheets (Ns) were the ideal choice of filler material due to their highly hydrophilic surface groups and large surface area that can enhance overall sensitivity. Additionally, the characteristic large-band optical absorption facilitates its use as an optical sensor over a broad spectrum of wavelengths. The group utilized a fiber optic SPR sensor and RI sensor and deposited the MXene Ns onto the sensing region. In the case of SPR sensor, a significant dip in the resonance curve with an apparent red shift was observed that was dependent on the concentration of MXene. This indicates that the sensitivity of directly proportional to the concentration of MXene layers. This also reveals that MXene can be used to tailor the working window of the sensor by adjusting the resonance dip in the region of interest.

For the RI sensor, the evanescent surface waves act as stimuli to detect changes in the RI of the environment. To enhance these waves on the surface of RI sensors, various complicated micro-nano processing methods are employed that add to the sensor's fragility, making them difficult to use. In this work, it was demonstrated that the light intensity sensitivity, i.e., the interaction of the evanescent waves on the surface of the fiber optic sensor, was dramatically enhanced on depositing MXene of different concentrations on the surface of the sensor. At a wavelength of 500 nm, the sensor witnessed an over eight-fold enhancement in sensitivity due to the characteristic wide-band absorption spectrum of MXene. This suggests that by depositing MXenes onto the surface of optical sensors, the sensitivity and detection spectrum of the sensors can be significantly enhanced and precisely tailored to the application-specific wavelength.

7.7 Key Challenges and Future Outlook

The unique structure, metal-like conductivity, and rich surface properties of MXenes make it a versatile material for the development of new-age sensors and wearable electronics. The hydrophilic properties due to the oxygen, hydroxyl, and fluorine surface terminations in MXenes facilitate the formation of composites with polymers to develop free-standing films. The principal mechanism behind the sensing abilities of these composites lies in the linear relationship between conductivity and an external stimulus like strain, pressure, humidity, and temperature. The first and foremost challenge in the development of efficient sensors is the synthesis of MXene precursor. MAX phase is prepared by milling highly combustible materials like titanium and aluminum at higher temperatures. This necessitates extreme precaution as several rounds of milling are required to produce superior-grade MAX phase powder. Proper handling and prompt waste disposal are crucial post-etching by HF.

The large-scale industrial production of MXene sensors is also restricted due to the inadequate mechanical stability of MXene. To this end, MXene has been incorporated into polymer matrices to form composites that are robust and also exhibit overall enhancement in stretchability which is a pre-requisite for flexible electronics and sensors. However, the type of polymer (glassy or rubbery) used and interactions between the polymer and MXene require further research and understanding for the two phases to work in harmony and maximize the efficiency of the sensors. The most viable approach is using first-principles calculations and machine-learning models to analyze the synergistic interactions and develop self-healing and durable sensors. Another bottleneck is assessing the optimal MXene concentration that should be incorporated in the composite for high-performance sensors. The levels of MXene are critical to the sensing abilities as it dictates the change in conductivity which forms the basic mechanism responsible for sensing. The concentration of MXene in a sensor should be high enough to form conductive channels that exhibit a perceivable change in conductivity on being exposed to a stimulus, while a still higher concentration results in the formation of stable conductive channels that are resistant to changes in conductivity, thus affecting the sensing performance.

Further, the oxidation of MXenes is also a growing concern among researchers as it affects the shelf-life and durability of the sensors. This has been overcome to some extent by encapsulation of the sensor with a hydrophobic layer and thermally insulating polymers. The surface modification of MXenes also holds great potential as it can significantly alter the surface property, band gap, and plasmonic behavior of MXene. This has the potential to tap into the semi-conducting properties of MXenes, which can have future implications in transistor-based devices.

Wearable electronics based on MXenes have prospects in diverse fields as they provide low-cost remote health monitoring and enhanced wearing experience. However, coherent signal processing modules are essential parts of a dynamic sensor that can seamlessly acquire bodily signals. The integration of a polymer-based sensing layer with a flexible signal conditioning circuit will bridge the gap between signal acquisition, processing, and output. The flexible signal conditioning circuit comprises a silicon integrated circuit on a flexible circuit board that can be easily attached to the skin. Additionally, Bluetooth modules can also be employed to communicate and store the output in smartphones or display devices for personal healthcare monitoring.

7.8 Conclusions

MXenes are a novel family of 2D materials that have been thoroughly researched due to their exceptional properties, which are highly advantageous in a bevy of applications. Due to the exotic 2D structure of MXenes and their naturally high specific surface area and adjustable surface-mediated features, MXene-based sensors have attracted the greatest attention and curiosity in research. Through methods including surface functionalization, macro structuring, and composite creation, investigations on MXene-based sensors have thus been enhancing the selectivity and sensitivity of many types of chemical, biological, and physical sensors.

One of the most encouraging results of research on these sensors is that the outstanding electrical conductivity apprised to pure MXenes is not sacrificed when the MXene surface is modified. The development of MXene-based sensors is still in its early innings; therefore, there are still issues that need to be resolved before they can be used commercially. There is a need for clarification of these requirements since the performance standards and other requirements for the commercialization of these sensors have not yet been systematically developed. By classifying MXene-based sensors according to the types of target stimuli and analytes and outlining the methodologies designed for diverse sensing applications, this book chapter offers a thorough overview of the research area of MXene-based sensors. Based on this overview, we were able to identify interesting research areas for the future and the likelihood that they would succeed in each classed area of MXene-based sensors.

Research in this area is currently mostly focused on evaluating the full potential of MXenes and capabilities for transferring their advantageous properties to sensors to attain remarkable performance in terms of LOD and sensitivity. These elements are essential for the widespread use of MXene-based sensors. Their uses could be broadly classified into a plethora of sensors which include pressure and strain sensors, electrochemical sensors, photoluminescent-type sensors, biosensors, capacitive sensors, gas and humidity sensors, and other miscellaneous ones. Therefore, in order to enhance sensor efficacy in terms of parameters like stability, reliability, multimodal applicability, and productivity for conformity with industrial requirements, future research should widen their concentrated focus. The development of next-generation sensor technologies based on MXenes will be sped up by addressing the aforementioned commercialization criteria. This might eventually result in the creation of low-cost, high-performance, and multimodal sensors for soft-electronics applications. We anticipate that this analysis will provide a clear picture of the future commercialization of MXene-based sensors and inspire a wide range of in-depth and useful investigations.

References

1. J. Vetelino, A. Reghu, *Introduction to Sensors* (CRC Press, 2017). https://doi.org/10.1201/9781315218274
2. A.S. Algamili, M.H.M. Khir, J.O. Dennis, A.Y. Ahmed, S.S. Alabsi, S.S. Ba Hashwan, M.M. Junaid, A review of actuation and sensing mechanisms in MEMS-based sensor devices. Nanoscale Res. Lett. **16**, 16 (2021). https://doi.org/10.1186/s11671-021-03481-7
3. A.S. Rajamani, M. Divagar, V.V.R. Sai, Plastic fiber optic sensor for continuous liquid level monitoring. Sens. Actuators A Phys. **296**, 192–199 (2019). https://doi.org/10.1016/j.sna.2019.07.021
4. E.D. Scott, C.T. Hayward, R.F. Kubichek, J.C. Hamann, J.W. Pierre, B. Comey, T. Mendenhall, Single and multiple sensor identification of avalanche-generated infrasound. Cold Reg. Sci. Technol. **47**, 159–170 (2007). https://doi.org/10.1016/j.coldregions.2006.08.005
5. R. Mitra, A. Prusty, U. Manju, Ferroelectric perovskites as electro-optic switching devices, modulators and optical memory, in *Perovskite Metal Oxides* (Elsevier, 2023), pp. 617–643. https://doi.org/10.1016/B978-0-323-99529-0.00022-9
6. H. Jiang, L. Zheng, Z. Liu, X. Wang, Two-dimensional materials: from mechanical properties to flexible mechanical sensors. InfoMat. **2**, 1077–1094 (2020). https://doi.org/10.1002/inf2.12072
7. L. Zhang, K. Khan, J. Zou, H. Zhang, Y. Li, Recent advances in emerging 2D material-based gas sensors: potential in disease diagnosis. Adv. Mater. Interfaces **6**, 1901329 (2019). https://doi.org/10.1002/admi.201901329
8. R. Mitra, U. Manju, Negative capacitance and intrinsic ferroelectric behavior in α-MoO_3 culminating as a robust piezoelectric energy harvester. ACS Appl. Electron. Mater. **5**, 3130–3143 (2023). https://doi.org/10.1021/acsaelm.3c00203
9. H. Riazi, G. Taghizadeh, M. Soroush, MXene-based nanocomposite sensors. ACS Omega **6**, 11103–11112 (2021). https://doi.org/10.1021/acsomega.0c05828
10. M.J. McGrath, C.N. Scanaill, Sensing and sensor fundamentals, in *Sensor Technologies* (Apress, Berkeley, CA, 2013), pp. 15–50. https://doi.org/10.1007/978-1-4302-6014-1_2
11. R. Mitra, B. Sheetal Priyadarshini, A. Ramadoss, U. Manju, Stretchable polymer-modulated PVDF-HFP/TiO_2 nanoparticles-based piezoelectric nanogenerators for energy harvesting and sensing applications. Mater. Sci. Eng. B. **286**, 116029 (2022). https://doi.org/10.1016/j.mseb.2022.116029
12. S. Das Mahapatra, P.C. Mohapatra, A.I. Aria, G. Christie, Y.K. Mishra, S. Hofmann, V.K. Thakur, Piezoelectric materials for energy harvesting and sensing applications: roadmap for future smart materials. Adv. Sci. **8** (2021). https://doi.org/10.1002/advs.202100864
13. Y. Yue, N. Liu, W. Liu, M. Li, Y. Ma, C. Luo, S. Wang, J. Rao, X. Hu, J. Su, Z. Zhang, Q. Huang, Y. Gao, 3D hybrid porous Mxene-sponge network and its application in piezoresistive sensor. Nano Energy **50**, 79–87 (2018). https://doi.org/10.1016/j.nanoen.2018.05.020
14. S. Sharma, A. Chhetry, M. Sharifuzzaman, H. Yoon, J.Y. Park, Wearable capacitive pressure sensor based on MXene composite nanofibrous scaffolds for reliable human physiological signal acquisition. ACS Appl. Mater. Interfaces **12**, 22212–22224 (2020). https://doi.org/10.1021/acsami.0c05819

15. W. Li, Z. Song, J. Zhong, J. Qian, Z. Tan, X. Wu, H. Chu, W. Nie, X. Ran, Multilayer-structured transparent MXene/PVDF film with excellent dielectric and energy storage performance. J. Mater. Chem. C. **7**, 10371–10378 (2019). https://doi.org/10.1039/c9tc02715g
16. S. Seyedin, S. Uzun, A. Levitt, B. Anasori, G. Dion, Y. Gogotsi, J.M. Razal, MXene composite and coaxial fibers with high stretchability and conductivity for wearable strain sensing textiles. Adv. Funct. Mater. **30**, 1910504 (2020). https://doi.org/10.1002/adfm.201910504
17. Y. Pei, X. Zhang, Z. Hui, J. Zhou, X. Huang, G. Sun, W. Huang, $Ti_3C_2T_X$ MXene for sensing applications: recent progress, design principles, and future perspectives. ACS Nano **15**, 3996–4017 (2021). https://doi.org/10.1021/acsnano.1c00248
18. Z. Ling, C.E. Ren, M.-Q. Zhao, J. Yang, J.M. Giammarco, J. Qiu, M.W. Barsoum, Y. Gogotsi, Flexible and conductive MXene films and nanocomposites with high capacitance. Proc. Natl. Acad. Sci. **111**, 16676–16681 (2014). https://doi.org/10.1073/pnas.1414215111
19. X. Shi, H. Wang, X. Xie, Q. Xue, J. Zhang, S. Kang, C. Wang, J. Liang, Y. Chen, Bioinspired ultrasensitive and stretchable MXene-based strain sensor via nacre-mimetic microscale "brick-and-Mortar" architecture. ACS Nano **13**, 649–659 (2019). https://doi.org/10.1021/acsnano.8b07805
20. V. Kedambaimoole, K. Harsh, K. Rajanna, P. Sen, M.M. Nayak, S. Kumar, MXene wearables: properties, fabrication strategies, sensing mechanism and applications, Mater. Adv. 3784–3808 (2022). https://doi.org/10.1039/d1ma01170g
21. Q. Guo, X. Zhang, F. Zhao, Q. Song, G. Su, Y. Tan, Q. Tao, T. Zhou, Y. Yu, Z. Zhou, C. Lu, Protein-inspired self-healable Ti_3C_2 MXenes/rubber-based supramolecular elastomer for intelligent sensing. ACS Nano **14**, 2788–2797 (2020). https://doi.org/10.1021/acsnano.9b09802
22. W.Y. Chen, S.N. Lai, C.C. Yen, X. Jiang, D. Peroulis, L.A. Stanciu, Surface functionalization of $Ti_3C_2T_x$ MXene with highly reliable superhydrophobic protection for volatile organic compounds sensing. ACS Nano **14**, 11490–11501 (2020). https://doi.org/10.1021/acsnano.0c03896
23. P.-K. Yang, C.-P. Lee, 2D-layered nanomaterials for energy harvesting and sensing applications, in *Applied Electromechanical Devices and Machines for Electric Mobility Solutions* (2020), pp. 1–14. https://doi.org/10.5772/intechopen.85791
24. H. Yuan, T. Lei, Y. Qin, R. Yang, Flexible electronic skins based on piezoelectric nanogenerators and piezotronics. Nano Energy **59**, 84–90 (2019). https://doi.org/10.1016/j.nanoen.2019.01.072
25. X. Gao, J. Yang, J. Wu, X. Xin, Z. Li, X. Yuan, X. Shen, S. Dong, Piezoelectric actuators and motors: materials, designs, and applications. Adv. Mater. Technol. **5**, 1–26 (2020). https://doi.org/10.1002/admt.201900716
26. J. Tichý, J. Erhart, E. Kittinger, J. Přívratská, *Fundamentals of Piezoelectric Sensorics* (2010). https://doi.org/10.1007/978-3-540-68427-5
27. M. Lallart, S. Pruvost, D. Guyomar, Electrostatic energy harvesting enhancement using variable equivalent permittivity. Phys. Lett. Sect. A Gen. At. Solid State Phys. **375**, 3921–3924 (2011). https://doi.org/10.1016/j.physleta.2011.09.043
28. A.A. Khan, A. Mahmud, D. Ban, Evolution from single to hybrid nanogenerator: a contemporary review on multimode energy harvesting for self-powered electronics. IEEE Trans. Nanotechnol. **18**, 21–36 (2019). https://doi.org/10.1109/TNANO.2018.2876824
29. S. Bhatta, R. Mitra, A. Ramadoss, U. Manju, Enhanced voltage response in TiO_2 nanoparticle-embedded piezoelectric nanogenerator. Nanotechnology **33**, 335402 (2022). https://doi.org/10.1088/1361-6528/ac6df5
30. S. Li, Z. Zhao, J. Zhao, Z. Zhang, X. Li, J. Zhang, Recent advances of ferro-, piezo-, and pyroelectric nanomaterials for catalytic applications. ACS Appl. Nano Mater. **3**, 1063–1079 (2020). https://doi.org/10.1021/acsanm.0c00039
31. L. Zhang, C. Tang, C. Zhang, A. Du, First-principles screening of novel ferroelectric MXene phases with a large piezoelectric response and unusual auxeticity. Nanoscale **12**, 21291–21298 (2020). https://doi.org/10.1039/d0nr06609e

32. Y. Guo, M. Zhong, Z. Fang, P. Wan, G. Yu, A wearable transient pressure sensor made with MXene nanosheets for sensitive broad-range human-machine interfacing. Nano Lett. **19**, 1143–1150 (2019). https://doi.org/10.1021/acs.nanolett.8b04514
33. K. Yang, F. Yin, D. Xia, H. Peng, J. Yang, W. Yuan, A highly flexible and multifunctional strain sensor based on a network-structured MXene/polyurethane mat with ultra-high sensitivity and a broad sensing range. Nanoscale **11**, 9949–9957 (2019). https://doi.org/10.1039/C9NR00488B
34. H. Yuan, T. Lei, Y. Qin, R. Yang, Author's accepted manuscript. Nano Energy (2019). https://doi.org/10.1016/j.nanoen.2019.01.072
35. Y. Cai, J. Shen, G. Ge, Y. Zhang, W. Jin, W. Huang, J. Shao, J. Yang, X. Dong, Stretchable $Ti_3C_2T_x$ MXene/carbon nanotube composite based strain sensor with ultrahigh sensitivity and tunable sensing range. ACS Nano **12**, 56–62 (2018). https://doi.org/10.1021/acsnano.7b06251
36. F. Shahzad, A. Iqbal, S.A. Zaidi, S.-W. Hwang, C.M. Koo, Nafion-stabilized two-dimensional transition metal carbide ($Ti_3C_2T_x$ MXene) as a high-performance electrochemical sensor for neurotransmitter. J. Ind. Eng. Chem. **79**, 338–344 (2019). https://doi.org/10.1016/j.jiec.2019.03.061
37. P.A. Rasheed, R.P. Pandey, K. Rasool, K.A. Mahmoud, Ultra-sensitive electrocatalytic detection of bromate in drinking water based on Nafion/$Ti_3C_2T_x$ (MXene) modified glassy carbon electrode. Sens. Actuators B Chem. **265**, 652–659 (2018). https://doi.org/10.1016/j.snb.2018.03.103
38. S.S. Shankar, R.M. Shereema, R.B. Rakhi, Electrochemical determination of adrenaline using MXene/graphite composite paste electrodes. ACS Appl. Mater. Interfaces **10**, 43343–43351 (2018). https://doi.org/10.1021/acsami.8b11741
39. Y. Zhang, X. Jiang, J. Zhang, H. Zhang, Y. Li, Simultaneous voltammetric determination of acetaminophen and isoniazid using MXene modified screen-printed electrode. Biosens. Bioelectron. **130**, 315–321 (2019). https://doi.org/10.1016/j.bios.2019.01.043
40. M. Sharifuzzaman, S.C. Barman, M.A. Zahed, S. Sharma, H. Yoon, J.S. Nah, H. Kim, J.Y. Park, An electrodeposited MXene-$Ti_3C_2T_x$ nanosheets functionalized by task-specific ionic liquid for simultaneous and multiplexed detection of bladder cancer biomarkers. Small **16**, 2002517 (2020). https://doi.org/10.1002/smll.202002517
41. F. Ai, C. Fu, G. Cheng, H. Zhang, Y. Feng, X. Yan, X. Zheng, Amino-functionalized Ti_3C_2 MXene quantum dots as photoluminescent sensors for diagnosing histidine in human serum. ACS Appl. Nano Mater. **4**, 8192–8199 (2021). https://doi.org/10.1021/acsanm.1c01425
42. M.L. Desai, H. Basu, R.K. Singhal, S. Saha, S.K. Kailasa, Ultra-small two dimensional MXene nanosheets for selective and sensitive fluorescence detection of Ag^+ and Mn^{2+} ions. Colloids Surfaces A Physicochem. Eng. Asp. **565**, 70–77 (2019). https://doi.org/10.1016/j.colsurfa.2018.12.051
43. Y. Lei, W. Zhao, Y. Zhang, Q. Jiang, J. He, A.J. Baeumner, O.S. Wolfbeis, Z.L. Wang, K.N. Salama, H.N. Alshareef, A MXene-based wearable biosensor system for high-performance in vitro perspiration analysis. Small **15**, 1901190 (2019). https://doi.org/10.1002/smll.201901190
44. F. Wang, C. Yang, C. Duan, D. Xiao, Y. Tang, J. Zhu, An organ-like titanium carbide material (MXene) with multilayer structure encapsulating hemoglobin for a mediator-free biosensor. J. Electrochem. Soc. **162**, B16–B21 (2015). https://doi.org/10.1149/2.0371501jes
45. T. Xia, G. Liu, J. Wang, S. Hou, S. Hou, MXene-based enzymatic sensor for highly sensitive and selective detection of cholesterol. Biosens. Bioelectron. **183**, 113243 (2021). https://doi.org/10.1016/j.bios.2021.113243
46. W. Yuan, K. Yang, H. Peng, F. Li, F. Yin, A flexible VOCs sensor based on a 3D Mxene framework with a high sensing performance. J. Mater. Chem. A. **6**, 18116–18124 (2018). https://doi.org/10.1039/C8TA06928J
47. N. Li, Y. Jiang, C. Zhou, Y. Xiao, B. Meng, Z. Wang, D. Huang, C. Xing, Z. Peng, High-performance humidity sensor based on urchin-like composite of Ti_3C_2 MXene-derived TiO_2 nanowires. ACS Appl. Mater. Interfaces **11**, 38116–38125 (2019). https://doi.org/10.1021/acsami.9b12168

48. W.Y. Chen, X. Jiang, S.-N. Lai, D. Peroulis, L. Stanciu, Nanohybrids of a MXene and transition metal dichalcogenide for selective detection of volatile organic compounds. Nat. Commun. **11**, 1302 (2020). https://doi.org/10.1038/s41467-020-15092-4
49. J. Zhang, L. Wan, Y. Gao, X. Fang, T. Lu, L. Pan, F. Xuan, Highly stretchable and self-healable MXene/polyvinyl alcohol hydrogel electrode for wearable capacitive electronic skin. Adv. Electron. Mater. **5**, 1900285 (2019). https://doi.org/10.1002/aelm.201900285
50. M.H. Tran, R. Brilmayer, L. Liu, H. Zhuang, C. Hess, A. Andrieu-Brunsen, C.S. Birkel, Synthesis of a smart hybrid MXene with switchable conductivity for temperature sensing. ACS Appl. Nano Mater. **3**, 4069–4076 (2020). https://doi.org/10.1021/acsanm.0c00118
51. Y. Chen, Y. Ge, W. Huang, Z. Li, L. Wu, H. Zhang, X. Li, Refractive index sensors based on $Ti_3C_2T_x$ MXene fibers. ACS Appl. Nano Mater. **3**, 303–311 (2020). https://doi.org/10.1021/acsanm.9b01889

Chapter 8
Progress of MXenes in Biomedical Sciences

Namita, Arti, Naushad Alam, and Jamilur R. Ansari

8.1 Introduction

MXenes are derived from the MAX phases having the general formulation $M_{m+1}AX_m$ with M stands for transition (group B) element, A for group A elements of the periodic table preferably third and fourth group elements and X for C, N or combination of both. Here 'm' holds for only three values 1, 2 and 3. MXene so obtained has the general formula $M_{m+1}X_mT_x$ where 'm' has the value from 1 to 3 [1].

MXenes are emerging as a base component in the multiple domains of magnetic, electrical, capacitive, optical, semiconductor, and catalytic fields along with sensing and biomedicines on a large scale. The compactness of the atomic sheets in MXenes led to diversification in the biomedical sciences [2, 3]. MXenes are associated with plenty of functional groups on their surface, which strongly assist in the promotion of MXenes bio-Engineering. MXenes are best suited in the biomedical fields owing to their properties like atomic stacking features, modifiable conformation and water solubility characteristics. MXenes are specifically exploited in photothermal treatment due to their immense optical absorbance as well as their capacity to convert under the near infra-red region. It can undergo tumor ablation, penetrates organ-tissues and treats cancerous cells, so, specifically that it does not cause any damage to the other organs. In the bioimaging procedures, MXenes assist remarkably during chemotherapy and real-time monitoring [4]. MXenes also have applications in energy storage [5]. For the ion storage capacity of MXenes, the oxygen terminations are highly preferable in comparison to those of hydroxyl and fluorine terminations [6]. This is because the use of fluorine and hydroxyl

Namita · Arti · N. Alam
Department of Physics, Lalit Narayan Mithila University, Darbhanga, Bihar 846004, India

J. R. Ansari (✉)
Department of Packaging & Logistics, College of Science and Technology, Yonsei University, 1 Yonseidae-Gil, Wonju-Si, Gangwon-do 26493, South Korea
e-mail: jransari.phd@gmail.com

© The Author(s), under exclusive license to Springer Nature Singapore Pte Ltd. 2024
O. P. Pandey and P. Sharma (eds.), *MXenes: Emerging 2D Materials*,
https://doi.org/10.1007/978-981-97-4064-2_8

terminations leads to reduced capacity and restricted transportation of lithium ions [7]. V_2CT_x demonstrated an excellent capacity of Li^+ ion storage amongst all MXenes under identical conditions [8].

In the general formula of MXene $M_{m+1}X_mT_x$, 'M' stands for the transition metals like Hf, Zr, Cr, Ti, Sc, V, Nb, Ta, Mo etc. [9, 10]. 'X' stands for carbon or nitrogen whereas T_x indicates the type of surface functional groups like –O, –F or –OH groups. There are various MXenes which are non-magnetic in nature, such as Hf_2CO_2, $ZrCO_2$, $Cr_2C(OH)_2$, Ti_2CO_2, Cr_2CF_2, Sc_2CO_2, $Sc_2G(OH)_2$ and Sc_2CF_2 [11]. But MXenes with ferromagnetic and antiferromagnetic properties are also available like Cr_2C exhibits ferro- and Ti_3C_2, $Cr_2TiC_2F_2$ exhibit anti-ferro magnetic behavior [12, 13]. Even MXenes like Ti_3C_2 with so many structural defects are utilized and find applications in sonodynamic therapy (SDT). Since, such structure are capable of generating reactive oxygen species (ROS), which are essential for numerous medical treatments.

The hybridization [14] of MXenes with different functional groups led to the diversification of MXenes in the broad field of Biomedicines. The intrinsic properties of MXenes have been harnessed continuously to extract more information in the field. MXenes containing multiple transition metals have also been discovered. The bonding between M and A, in MAX phases, is relatively more chemically active, though weaker than the bonding between M and X. This property makes the selectivity in etching the layers of A element easier. Consequently, the synthesis of MXenes from MAX phases is assisted effectively. MXenes are not only obtained from the MAX phases, but non-MAX-phase precursors also participate in the synthesis of MXenes [15]. Mo_2Ga_2C and $Zr_3Al_3C_5$ are such non-MAX phase precursors [16] which yielded Mo_2CT_x and $Zr_3C_2T_x$ MXenes, respectively, by etching layers of Ga from the former one and Aluminum carbide (Al_3C_3) from the latter. An illustration of Biomedical applications of MXenes from Nanomedicine to Biomaterials is shown in Fig. 8.1.

In these non-MAX phases, metal layers [17] are separated by the layers of carbon. It has been revealed in the previous study that MXenes should be synthesized and annealed below 800 °C within a controlled atmosphere to obtain 2D instead of getting a cubic phase. For the treatment of cancer and tumors, MXenes' surface chemistry has been utilized in which superparamagnetic two-dimensional Ti_3C_2 MXenes were fabricated. The superparamagnetic crystals of Fe_3O_4 were subjected to the in-situ growth over the surface of Ti_3C_2 MXenes. This enhanced the magnetic resonance imaging of cell analytes as well as promoted the photothermal destruction of the cancerous cells and ablated the tumor tissues.

The synthesis of MXenes includes a wet etching mechanism also. This category involves etching using hydrofluoric acid (HF), fluoride salt and the process of delamination, which discloses many properties in the two-dimensional state. The MXene has an overall crystalline structure with hexagonal close-packed stacking and sometimes face-centered cubic stacking. Surface terminations are largely affected by the conditions of delamination as well as the etching atmosphere. The transition metals,

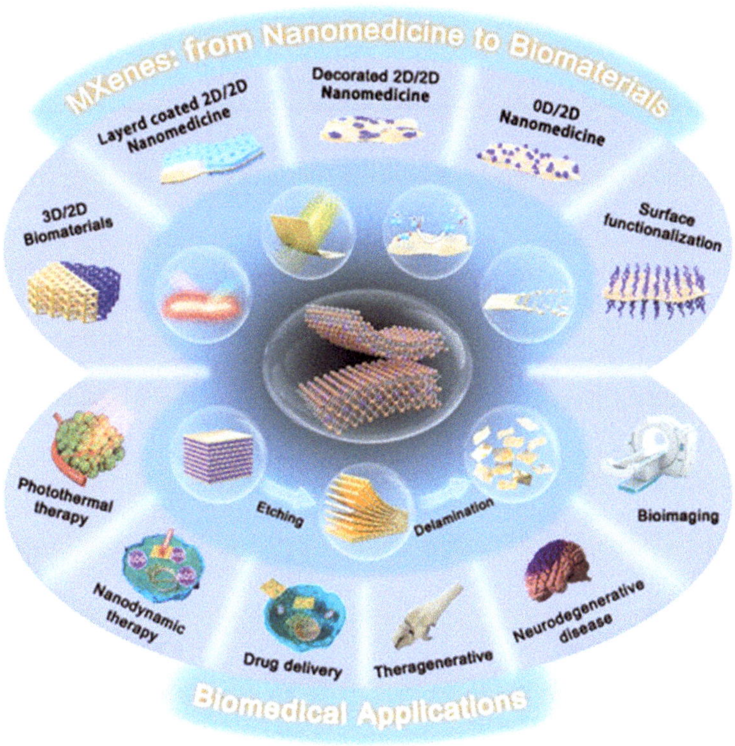

Fig. 8.1 An illustration of biomedical applications of MXenes from nanomedicine to biomaterials. Reproduced with permission from L. Chen, X. Dai, W. Feng, Y. Chen, Biomedical Applications of MXenes: From Nanomedicine to Biomaterials, Acc. Mater. Res. 3 (2022) 785–798, American Chemical Society. https://doi.org/10.1021/accountsmr.2c00025 [19]

the after-treatments of synthesis and the storage factor all influence the surface functional groups of MXenes. Surface chemistry [18] of MXenes, on the other hand, depends on the types of etching solutions involved.

8.2 Structure of MXenes

8.2.1 Classification on the Basis of Arrangements of Atoms

MXenes are classified into four categories on the basis of their structural arrangements. They are

(i) **Mono-transition metal MXenes**
Examples—Ti_2C and Nb_4C_3.

(ii) **Solid solution MXenes**: M layers consist of two transition elements with randomly arranged patterns. These are comparatively less stable than ordered arrangements with the same composition.
Examples—$(Ti, V)_3C_2$ and $(Ti, V)_2C$.

(iii) **Ordered double-transition metal (M) MXenes**: These are the solids where one or two layers of one M element are adjusted between the layers of another M element.
Examples—$(Cr_2V)\ C_2$ and $(Mo_2Ti_2)\ C_3$.
The forms of ordered double transition metals MXenes are $M'_2M''C_2$ and $M'_2M''_2C_3$, where M' and M'' are the two different transition metals. These are often denoted as M′M″Xene. At 0 K, most of them are found to be stable. Their stability could be varied when surface termination and entropy are considered.

(iv) **Ordered divacancy MXenes**
Examples—$Mo_{1.33}C$ and $W_{1.33}C$ [20].

With the atomic etching of Al and Sc from the atomically laminated [21] 3D Max phase parent $(Mo_{2/3}Sc_{1/3})_2AlC$, two-dimensional sheets of $Mo_{1.33}C$ having ordered divacancies with enhanced electrical conductivity are evolved. Such vacancies are not present in its counterpart Mo_2C.

8.2.2 Classification on the Basis of Dimensions

(i) **1D-MXenes**
They are not still extensively studied so far. $Ti_{n+1}C_n (n = 1, 2)$, V_2C nanoribbons, and armchair-shaped and zigzag nanoribbons were chosen to explore the characteristics of 1D-MXenes.

(ii) **2D-MXenes**
The assembly formation, as well as the self-piling up of layers, are quite difficult in 2D-MXenes, though a wide range of studies can be seen in this structural configuration.

(iii) **3D-MXenes**
This structure has manifold applications like energy storage in the form of supercapacitors [22], lithium-ion batteries and electrocatalysis.

8.3 Synthesis of MXenes

Top-down approach is easy to access for the synthesis of MXenes instead of Bottom-up approach.

8.3.1 Top-Down Method

There are several etching techniques, which are constituted under the Top-down approach, such as HF etching method, Bifluoride (NH_4HF_2) method, LiF/HCl etching method, molten salt etching and minimally intensive layer delamination (MILD) etching method. All these methods are used for etching Al from several Al-based MXenes like Ti_3AlC_2, Ti_4AlN_3, Ti_2AlC, etc. The etching responses are alike when other fluoride salts are used in place of LiF for the LiF/HCl method. Similarly, H_2SO_4 can be the alternative to HCl in the process and so on. The MILD etching method produced an exceptionally superb MXene over other previously discussed etching methods. Since fluorine is an environmentally hazardous element, so without exploiting the fluoride base, several MXenes, such as Nb_2CT_x are synthesized, employing the etching technique with the electrochemical exfoliation method. Several properties, like the transfer of electrons and enzymatic performance, are improved by utilizing fluorine-excluded MXenes. Not only aluminum (Al) was used as a unit, but rather $AlCl_3$ unit was also involved, which are called non-MAX phases, constituting YAl_3C_3 and $Zr_3Al_3C_5$. The MAX phases as well as non-MAX phases both are used to classify various etching methods during the synthesis of MXenes. UV inhalation [23] physical approach is also adopted for the etching mechanism. A surface chemical reaction is also motivated using UV radiation. It also promotes the green synthesis of MXenes. Ga can be etched out of Mo_2Ga_2C precursors using UV radiation when combined with a mild acid. Mo_2Ga_2C is reported to respond efficiently to UV radiation by absorbing effectively. MXenes with ultra-pure graphene-like two-dimensional architecture evolved using this method.

8.3.2 Bottom-Up Method

Since the internal structure, configuration and many such features of MXenes have not been studied and revealed so deeply, the synthesis of MXenes via Bottom-up routes becomes quite inaccessible.

Three methods dominate under Bottom-Up routes.

(i) **Chemical vapor deposition (CVD)**—CVD process, carried away using two steps only, yielded many MXenes like Mo_2C, WC and their hybrid structures [25] like Mo_2C/graphene heterostructures, Mo_2C-graphene hybrid [26] films and many more.
(ii) **Template method**—Using this method, several two-dimensional structures like MoO_3 and h-MoN nanosheets were obtained [27].
(iii) **Plasma-enhanced pulsed-laser deposition (PE-PLD)**—This process proceeds with the conjugation of two mechanisms—Plasma-enhanced CVD (PE CVD) and pulsed-laser deposition (PLD).

According to the composition, MXenes can be divided into three different formulas: M_2X, M_3X_2, M_4X_3. On the basis of structure, MXenes can also be divided:

mono-transition metal MXenes, solid solution MXenes, ordered double-transition metal (M) MXenes, and ordered divacancy MXenes, as shown in Fig. 8.2.

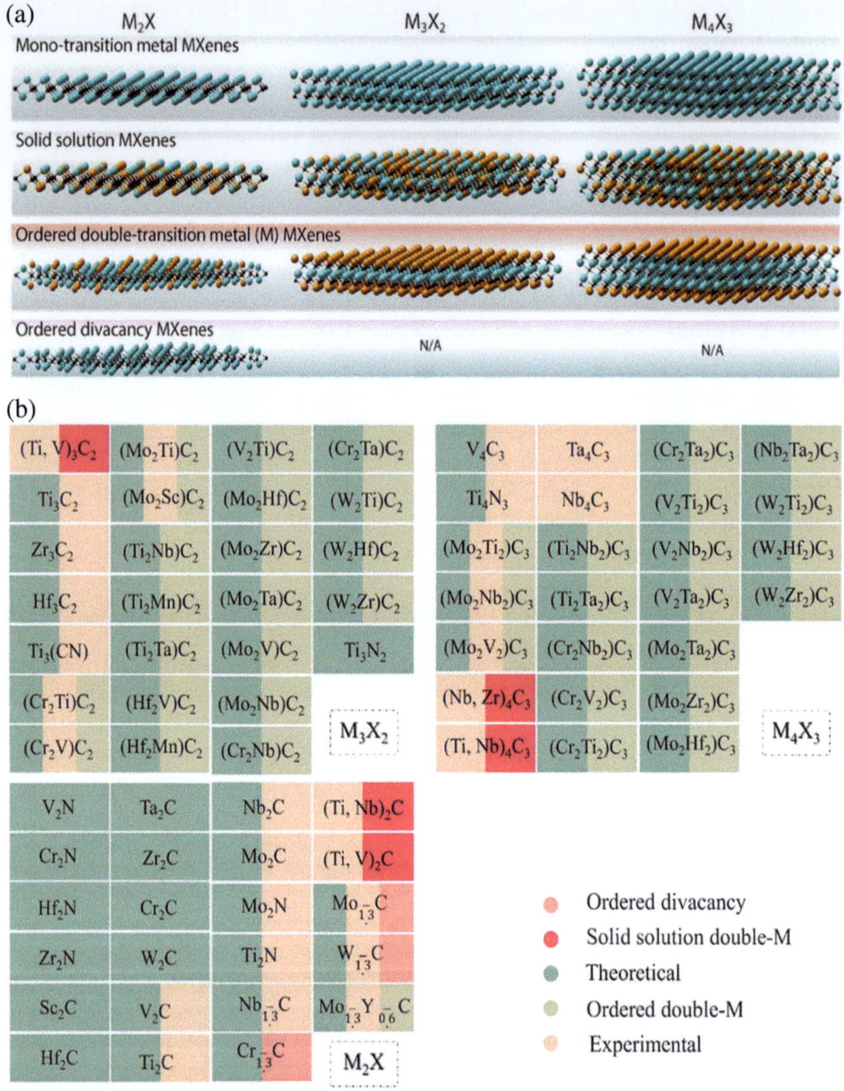

Fig. 8.2 **a** According to the composition, MXenes can be divided into three different formulas: $M_2 X$, $M_3 X_2$, $M_4 X_3$. On the basis of structure, MXenes can also be divided: mono-transition metal MXenes, solid solution MXenes, ordered double-transition metal (M) MXenes, ordered divacancy MXenes. Reproduced with permission from J. Huang, Z. Li, Y. Mao, Z. Li, Progress and biomedical applications of MXenes, Nano Sel. 2 (2021) 1480–1508. https://doi.org/10.1002/nano.202000309 [24]

8.3.3 Bottom-Up Approach Over Top-Down Technique

There are a few factors that are responsible for the choice of the synthesis method being adopted for MXenes, which provide an ease during preparation procedures.

(i) Easy to manage the structure
(ii) The composition of elements
(iii) Morphological modification.

8.3.4 Various Synthesis Mechanisms of MXenes

Elimination of 'A' part from the primary MAX phases results in the yielding of MXenes. For example, when Al is removed from Ti_3AlC_2, Ti_3C_2 MXenes are collected.

(i) **Alkali Etching Strategy**—In the method, conc. NaOH solution is treated with Ti_3AlC_2 to yield Ti_3C_2 MXene free from fluorine.
(ii) **Molten Salt Etching Route**—In this strategy, molten salt of Lewis acid is used. The cation of the salt and A element undergoes a direct redox coupling reaction. This method is useful for obtaining MXenes from MAX phases containing Al or even having no Al content (like Zn, Ga or Si).
(iii) **Molten Cadmium bromide salt route**—Here, cadmium bromide is exploited as an etchant to produce MXenes with bromide termination.
(iv) **Halogen-assisted etching strategy**—Here, halogens are used during synthesis.
(v) **Ultrafast acoustic-assisted synthesis**—Acoustic waves are utilized.
(vi) **Water-free etching methods**—This method has also been used recently to fabricate MXene-based composites.
(vii) **HF and in situ HF etching strategy**—This is one of the strategies generally employed for synthesizing biomedical-consistent MXenes.

MXenes are not only obtained from the MAX phases, but non-MAX-phase precursors also participate in the synthesis of MXenes. Mo_2Ga_2C and $Zr_3Al_3C_5$ are such non-MAX phase precursors which yielded Mo_2CT_x and $Zr_3C_2T_x$ MXenes, respectively, by, etching layers of Ga from the former one and Aluminum carbide (Al_3C_3) from the latter. In these non-MAX phases, metal layers are separated by the layers of carbon. The etching time and etching temperature are both very significant as these can compel the MAX phase [28] to convert completely to MXenes. But, this results in the least quality yield of the flakes because of the presence of more defects [29]. MXenes prepared using V, Nb, Ta and Mo need a highly aggressive atmosphere while etching.

8.3.5 *Freeze-Drying Treatment*

The freeze-drying method was employed to enhance the electrochemical activity and ease the self-stacking mechanism during the synthesis of MXene film electrodes.

With the help of the sublimation process, the solvent molecules, which have been frozen, were eliminated. Hence the unfavorable conditions led by Van der Waals forces were relieved. The interlayer distance also increased. The freeze-dried MXene, also called f-MXene films, depict the ultimate structural porosity with a fascinating and effective ion-diffusion activity as well as a transport channel [30]. Figure 8.3 shows 0D/2D MXene-based nanomedicines.

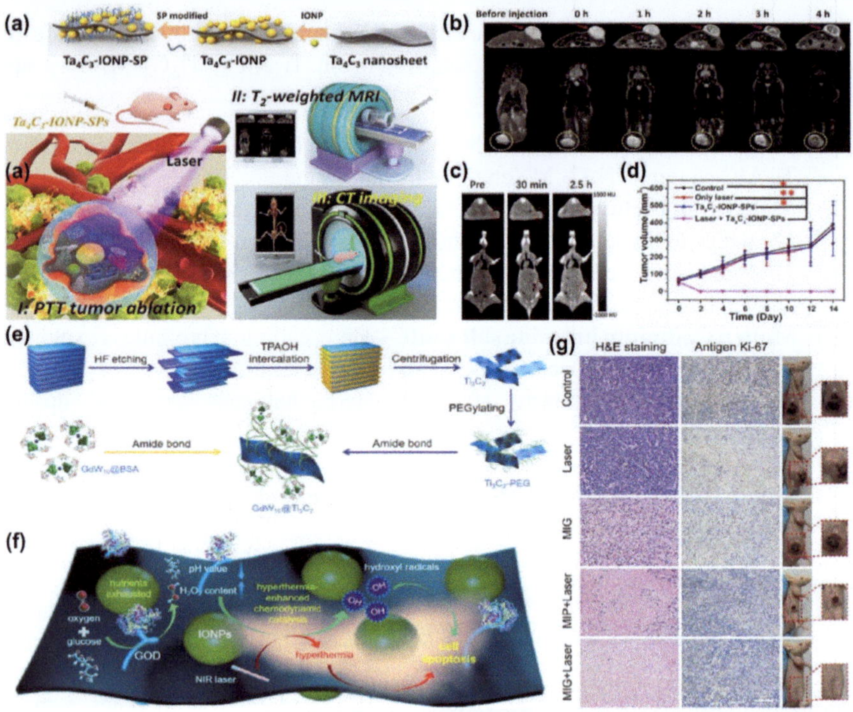

Fig. 8.3 0D/2D MXene-based nanomedicine. **a** Schematic illustration of the preparation of superparamagnetic composite Ta_4C_3 for multimodal imaging-guided PTT. **b** MR images and **c** CT images of tumor-bearing mice after injection of superparamagnetic composite Ta_4C_3. **d** Tumor volume curve of mice under different treatments. **e** Schematic illustration of the preparation of POM-functionalized Ti_3C_2 MXenes. **f** Schematic illustration of IONP-functionalized Ti_3C_2 MXenes for hyperthermia-enhanced sequential catalytic cancer therapy. **g** Immunohistochemical images of tumor slices and corresponding photos of tumor-bearing mice after different treatments. Reproduced with permission from L. Chen, X. Dai, W. Feng, Y. Chen, Biomedical Applications of MXenes: From Nanomedicine to Biomaterials, Acc. Mater. Res. 3 (2022) 785–798, American Chemical Society. https://doi.org/10.1021/accountsmr.2c00025 [19]

8.3.6 MXenes Surface Engineering

The surface of MXenes constitutes a larger area, some functional groups like –OH, –O and –F, high-density electric charges [31]. So, the surface engineering is done with the help of processes like electrostatic adsorption, Van der Waals forces and hydrogen bonding [32]. Undresses MXenes are unstable and react quickly with the particles present in the environment, so without disturbing the morphology, MXenes undergo various surface alteration mechanisms.

The surface of MXenes can be modified in three ways:

(a) Compound-based strategy
(b) Inorganic nanoparticle-based strategy
(c) Metal-nanoparticle-based strategy.

8.3.6.1 Compound-Based Strategy

It is also called the polymer-based method, in which polymers like polyethylene glycol (PEG), soybean phospholipid (SP), hyaluronic acid (HA), polyvinyl pyrrolidone (PVP) etc. and even their mixture is engineered in order to alter the surface properties of MXenes. The strategies attained to manage the polymers on the surface using self-initiated photo-grafting and photo-polymerization (SIPGP).

Physical adsorption interaction methods and non-covalent bonds are some common methods deployed for the surface-adjusting techniques.

8.3.6.2 Inorganic Nanoparticle-Based Strategy

The inorganic nanoparticle used for altering the surface properties are mesoporous [33] silica nanoparticles (MSN), ZnO [34], SnO_2, SnS, CoS, $NiFe_2O_4$ and Mn_3O_4. The combination of nanoparticles and MXenes is used in biosensors also.

The modified features of the surface of MXenes now provide enhanced hydrophilicity and dispersity, mobilizing surface particles and promoting excellent bioactivity and stability [35].

8.3.6.3 Metal Nanoparticle-Based Strategies

Noble metals are included under this strategy Ag, Si and Pt are the most commonly used metals under this category. Biofouling resistance comes into play, where the electrochemical signal is received efficiently and speedily. Besides Au, Ag and Pt, Cobalt and Bismuth nanoparticles are also used in the mechanism where cobalt can upgrade the photoelectric catalysis process along with the charge transfer responses, while Bismuth can boost the sensing capabilities.

8.4 MXenes-Derived Quantum Dots (MQDs)

Primarily, MQDs are plenteous with multiple features like brilliant electrical conductivity [36], abundant active sites, highly dispersible, easily manageable morphology, effective water solubility [37], optically active and variability in functionalization. Owing to these features, they are remarkably applied as sensors, catalysts, and store energy, used in optoelectronic devices [38] and in various biomedical fields.

8.4.1 *Synthesis of MQDs*

MQDs [39], such as Ti_3C_2 QDs, have been synthesized using the hydrothermal method from Ti_3C_2 MXenes. They have an excellent capability of fluorescent emission, due to which they are recognized as effective nanoprobes [40] for diagnosing cancer biomarkers [41, 42]. Likewise, V_2C [43] QDs prepared using V_2C MXenes proved to possess an extraordinary feature of magnetic resonance imaging.

8.4.2 *Properties of MQDs*

(i) The as-prepared Ti_3C_2 QDs, in an aqueous solution, exhibit unique salt-tolerant features, anti-photobleaching capacity and consistent dispersibility.
(ii) They can detect alkaline phosphatase (ALP) activity as well as monitor the enzymatic behavior accurately to determine embryonic stem cells (ESC).
(iii) MQDs show brilliant photoluminescence effects over two-dimensional MXenes nanosheets.
(iv) Monolayered hydrophilic MQDs Ti_3C_2 were reported to possess an excellent quantum confinement effect with excitation-based PL spectra and 10% quantum yield.

8.4.3 *Difficulties with MQDs*

(i) Synthesis techniques and mechanisms
(ii) Functionalization of morphology
(iii) Biocompatibility and cytotoxicity.

8.5 Current Assistance of Mxenes in Biomedical Areas

MXenes are currently being utilized in several biomedical fields.

(i) Therapeutic Operations
(ii) Bioimaging with contrasting media
(iii) Nanozymatic strategies
(iv) Biosafety, biodegradability and biocompatibility
(v) Photothermal therapy
(vi) Photoacoustic imaging
(vii) CT imaging
(viii) Reactive Oxygen Species (ROS) scavenging
(ix) Nano-dynamic therapy
(x) Nano-catalytic therapy.

8.6 Features of MXenes Useful for Biomedical Extension

(i) **Strongly absorbable in the near-infrared region (NIR)**—This property of MXene helps in the photothermal treatments in an efficient manner, particularly in the NIR I or NIR II bio-window.
(ii) **Surface Plasmon Resonance Effect and tunable bandgap**—This feature enables MXenes to produce reactive oxygen species (ROS) for Nano-dynamic treatments upon illumination of ultrasonic waves.
(iii) **Ultrasmall dimension and ample defects**—Such MXenes are incorporated in imaging and sensing biomolecules.
(iv) **Presence of metallic components in MXenes**—The atoms of heavy metals present in MXenes led them to be utilized as unique contrasting media during magnetic resonance imaging (MRI) and computed tomography (CT).
(v) **Atomic Redox activities**—The atoms of the metallic component of MXenes show redox performance which contributes to the enhanced catalytic ability towards multiple enzymatic behaviors.
(vi) **Ease of Compatibility and degradation**—This feature facilitates the removal of unwanted elements from the body and is hence utilized for biomedical theranostic applications [44].
(vii) **The ability of Impersonating Enzymatic Behavior**—MXenes serve as nanozymes in biomedical zones. The term "MXenzyme" was also introduced in recent studies, which account for the interdependence between V_2C MXenes as well as enzymes and so-named "V_2C MXenzymes". It was reported that V_2C MXenzymes exactly resemble the behavior of many enzymes effectively exploited in biochemical reactions [45]. The enzyme-mimicking capability of MXenes is shown in Fig. 8.4.

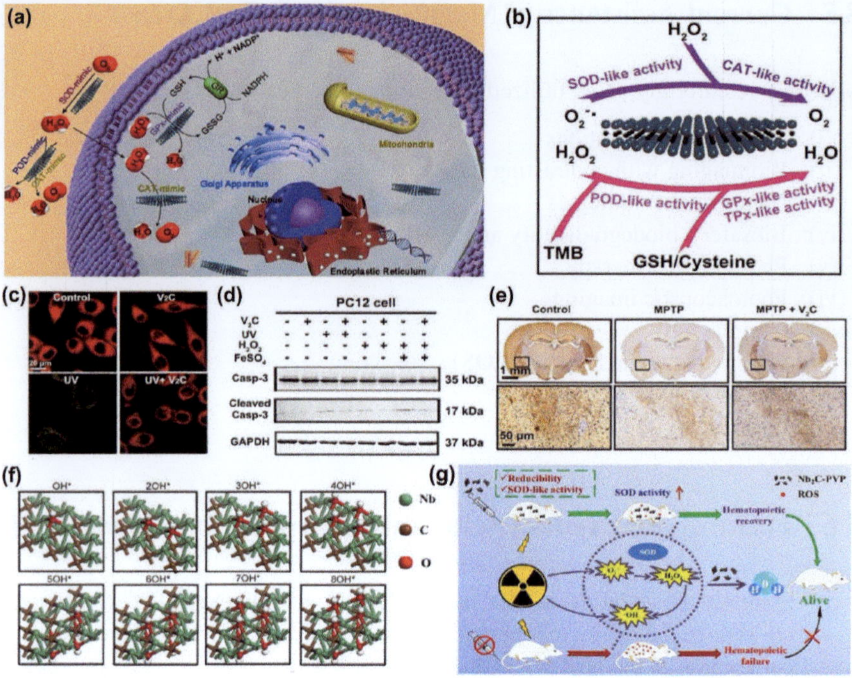

Fig. 8.4 Enzyme-mimicking capability of MXenes. **a, b** Schematic illustration of the multiple enzyme-mimicking capabilities of V_2C MXenzyme for ROS scavenging. **c** Confocal laser scanning microscope (CLSM) images of C11-BODIPY581/591-stained L929 cells under corresponding treatments to assess lipid peroxidation. **d** Western blot results of specific enzyme expression in PC12 cells under the corresponding treatments. **e** Immunohistochemical images of TH expression in brains after the corresponding treatments (magnified images of squares below). **f** Simulated structures of Nb_2C under attack of $^{\bullet}OH$. **g** Schematic illustration of the enzyme-mimicking activity of Nb_2C MXenes to promote hematopoietic recovery after IR irradiation. Reproduced with permission from L. Chen, X. Dai, W. Feng, Y. Chen, Biomedical Applications of MXenes: From Nanomedicine to Biomaterials, Acc. Mater. Res. 3 (2022) 785–798, American Chemical Society. https://doi.org/10.1021/accountsmr.2c00025 [19]

8.7 Properties

8.7.1 Colloidal Stability

Colloidal Stability of MXenes can be achieved by the functionalization of the surface with natural or synthetic macromolecules through non-covalent interactions. The fundamental properties of MXenes are shown in Table 8.1.

Table 8.1 MXenes fundamental properties

	Electrical properties	Optical properties	Magnetic properties	Mechanical properties
Affecting factors	• Synthesis Method • Constituting material • Surface Properties • Internal structural arrangement • External factors like temperature, pH, humidity etc. • Energy density	• Energy band structures which further depends upon: – Surface functional group – Electric field – Electronic localization – Doping	• Composition of 'M' in Mxene • Surface termination and surface functional groups • External electric field	• Thickness of atomic layers • Surface termination • Surface functionalization • Number of atomic layers
Fixing of problem	• Adopting methods yielding less number of defects, for example fluorine-free synthesis • Varying surface functional groups to modify bandgaps • Increasing interlayer separation • Modifying previous architectures	• By optimum calibration of layer thickness • Through variation in ion-intercalation and molecular intercalation techniques • By tuning bandgaps through modification in surface termination, the phenomenon of catalysis is enhanced • By altering 'M' contained in MXenes	• Rise in external electric field value, magnetic moment and hence the effective mass reduces • Thus, the conductivity enhances	To form composites using different materials like 2D MXenes having micro/ nano scaling when combined with 1D nanofibers of cellulose having strong bonding between H-atom to produce soft-layered actuator

8.7.2 Optical Properties

Due to excellent photoluminescence optical response, an effective participation of MXenes in photocatalytic reactions, biosensors and in many biomedical areas like photoacoustic imaging, photothermal therapy (PTT), photodynamic treatments (PDT), and helps in limiting the undesirable drug leakage.

8.7.3 Magnetic Properties

8.7.3.1 Effect of External Electric Field

Due to the presence of an external electric field [46], few MXenes, such as Ti_2C get converted from antiferromagnetic (AFM) semiconductor to Ferrimagnetic (FIM) semiconductor, magnetic as well as non-magnetic metals, non-magnetic semiconductor and half-metal.

8.7.3.2 Effect of Surface Functional Groups

The magnetic properties of MXenes start vanishing due to the existence of surface terminals. There exist some p-d bonds between M atoms and T groups which reduces the number of atoms near fermi levels. This leads to the disappearance of magnetism from the sites.

8.7.4 Hydrophilic Properties

The properties of hydrophilicity and biocompatibility, both are closely related to each other. Hydrophilicity arises due to surface termination having functional groups like hydroxyl (–OH), (–O) and (–F).

Owing to hydrophilic activities of MXenes, many advantages can be observed.

(i) Enhance the interaction with the polymer matrix.
   - So, can be deployed for the synthesis of nanocomposites.

(ii) Its combination with electrical conductivity and 2D layered atomic structures [47] led to the discovery of biocompatible FET.
   - So, the response becomes faster and easier to collect.

(iii) This property is effective in MXenes surface functionalization [48] as well as antibacterial performances [49].
   - So, obstruct the bacterial growth during direct assimilation.

(iv) Facilitates with large surface area.
   - So, enhances drug-holding capacity of MXenes.

(v) Polar or ionic molecule's adsorption becomes possible.
   - So, helpful for environment also.

8.8 MXenes Approach in Biomedical Sciences

8.8.1 Sensing Devices

MXenes find a multiple application in the field of biomedical sciences when employed as sensing devices. These are utilized in bioimaging, biosensing, therapeutics, as drug carriers etc. using optical and electrical signals.

8.8.2 Biological Sensors

A biosensor consists of a sensing component—It responds to its complementary analyzing element and a transducer—It converts biochemical signals to electrical and optical signals. An analyzer—It is used to interpret and analyze the collected information.

MXenes have become a recognizing candidate in the evolution of extraordinary biosensors [50, 51] owing to their high surface-to-volume ratio, unique electrical and optical features, fascinating availability of large number of surface functional groups along with brilliant hydrophilic behavior.

(a) **Electrical Sensing**—A biosensor can sense several compounds like H_2O_2 [24], nitride [52, 53], glutathione [54], glucose [55–58], dopamine [59, 60], epinephrine[61], organophosphate [62] pesticides [63–66] using electrical signals generated from the surface modified by introducing different analytes constituting proteins and amino acids.

(b) **Optical Sensing**—Sensing devices can sense optical signals based on which they are classified into two types—electrochemiluminescence (ECL) biosensor and surface plasmon resonance (SPR) biosensor.

- ECL biosensors—This biosensor is utilized for sensing proteins, enzymes and DNA. These biosensors are cost-effective, respond very quickly, manifest extraordinary controllability and sensitivity [67].
- SPR biosensors—This biosensor was used for sensing carcinoembryonic antigen (CEA) in human serum for detection [68] and treatment of cancer cells. These biosensors exhibit excellent and real-time and quick response, support molecular motion, can work with very little sample quantity, broadly applicable, best performance analysis, enhanced sensitivity with brilliant monitoring and analysis processes.

(c) **Physical Sensor**—MXenes possess electronic [69] as well as mechanical features which hold the ability to modify the nature of electrical signals to respond to some physical conditions like pressure and tension [70–73]. Several human body functions like muscular strains/ Physiological detection processes are carried out by Physical sensors, fabricated using MXenes.

Synthesis of Physical (or Pressure) sensor:

MXene tissue paper is first prepared using the solution of MXene nanosheets in which normal tissue papers are dipped. The tissue paper so obtained is then subjected over the layers of PLA using interdigitated electrodes. At last, PLA sheet which is black in color and of extremely thin dimension is used to wrap the developed structure [74]. This Pressure sensor [2] showed several astonishing features like excellent sensitivity, least detection limit, negligible power dissipation, and unique reproducibility even after 10 K repeated cycles. The feature which cannot be overlooked is its easy degradability.

8.8.3 Bioimaging

This category includes several types of imaging related to the biomedical fields—Imaging based on photo-acoustics, X-ray computation and luminescence property of the materials. Medical diagnosis using bioimaging has become possible on account of quantum confinement, photothermal effects and several contrasting agents (Cas).

(a) **Bioimaging based on Photo-acoustics**—This is also termed as Photoacoustic imaging (PAI). The radiation photons scatter remarkably in the body tissue system which control many diagnosis procedures in the biomedical fields. The photo-acoustics based on light are capable of solving this issue by transforming optical energy into sound energy and hence in-depth tissue imaging could become possible. In this technique, biomolecules absorb photons to produce ultrasonic waves, which in turn, are utilized by the detector for obtaining tissue images [75].
    MXene being a semi-metallic possess efficient absorption and conversion ability as well as having localized surface plasmon resonance (LSPR) property due to which it could participate in the bioimaging process based on acoustics. MXenes also serve as an efficient contrast agent for photoacoustic imaging.

(b) **Bioimaging based on Magnetic Resonance effect**—This is also termed as MRI. Due to the presence of heavy metals [76], MXenes are utilized as contrasting agents (CA) for MR imaging. The techniques applying magnetic resonance effect provide high resolution, enhanced sensitivity response with negligible ionization effect. This imaging technology contributes in the treatment of cancer and tumors.
    For the treatment of cancer and tumors, MXenes' surface chemistry has also been utilized in which superparamagnetic two dimensional Ti_3C_2 MXenes were fabricated. The superparamagnetic crystals [77] of Fe_3O_4 were subjected to the in-situ growth over the surface of Ti_3C_2 MXenes. This enhanced the magnetic resonance imaging of cell analytes as well as promoted the photothermal destruction of the cancerous cells and ablated the tumor tissues. MXenes for bioimaging are shown in Fig. 8.5.

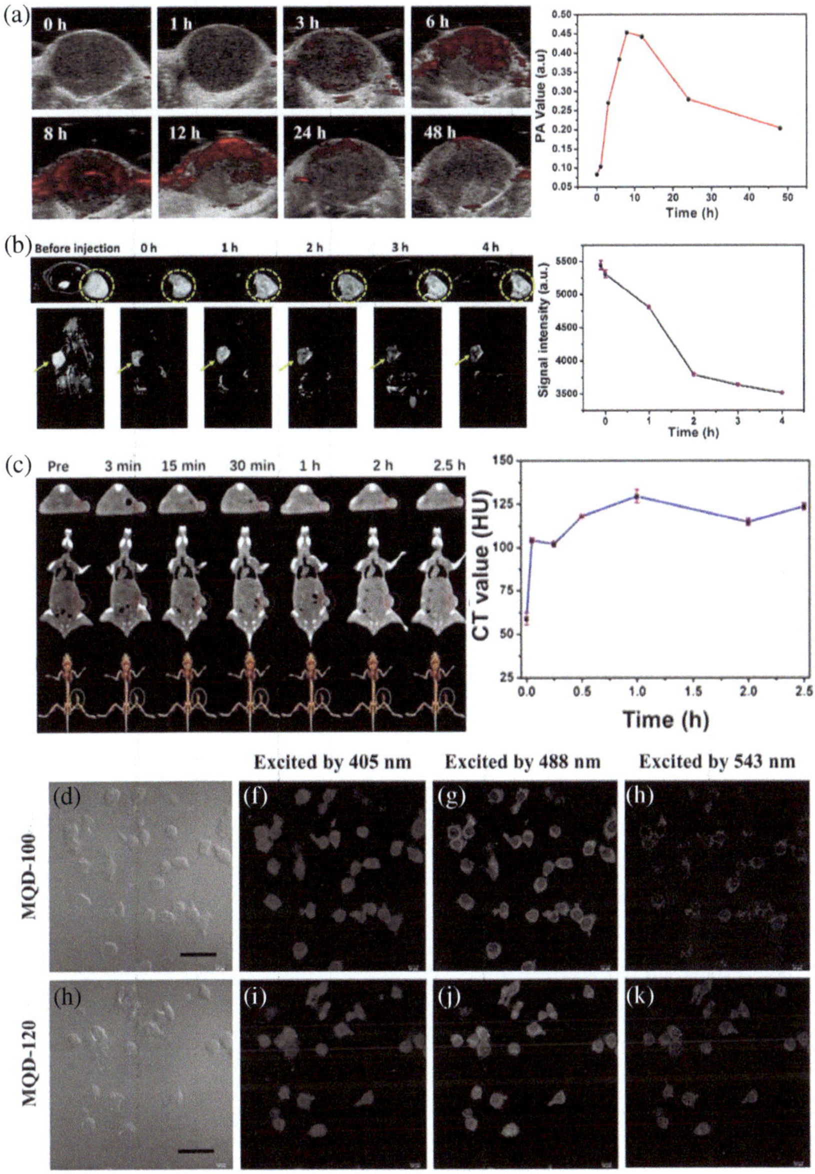

Fig. 8.5 MXenes for bioimaging. **a** After intravenous injection with Nb_2C–MSNs–SNO, PA images in vivo and corresponding signal intensities of the tumor were shown. **b** Before and after intravenously injection with Ti_3C_2–IONPs–SPs, transverse, coronal section and corresponding signal intensities of the T2-weighted MRI of 4T1 tumor-bearing mice were showed. **c** The left picture showed transverse, coronal and 3D-rendering CT images of 4T1 tumor-bearing mouse, while the right picture demonstrated time-dependent CT values after intravenously injecting with Ta_4C_3-IONP-SPs. In which the red circles mark the tumor sites. **d**–**k** Bright-field and confocal imaging of RAW264.7 cells were shown. Reproduced with permission from J. Huang, Z. Li, Y. Mao, Z. Li, Progress and biomedical applications of MXenes, Nano Sel. 2 (2021) 1480–1508. https://doi.org/10.1002/nano.202000309 [24]

(c) **Tomography based on X-Ray computation**—This is also known by the term CT imaging. Computed tomography (CT) is remarkably used for tracing lesions very distinctly. This technique is executed through the variation in absorbed X-Rays at the tissue sites during bioimaging. MXenes are used as a contrasting-agents in the form of nanoparticles which consists of an element with a large charge number. Ta_4C_3 MXene based nanocomposite [78] Ta_4C_3-IONP-SPs (iron oxide nanoparticles) were reported to exhibit an efficient photothermal conversion capacity of 32.5% with an outstanding success in the 100% removal of the tumor cells.

(d) **Bioimaging based on Luminescence**—MXenes exhibit tunable luminescence property which enables them to be utilized in bioimaging. The photoluminescence of MXene reveals about its least dimension and plentiful defects which encourage their applications in bioimaging, biosensing and molecular imaging. Monolayered hydrophilic MQDs Ti_3C_2 were reported to possess an excellent quantum confinement effect with excitation-based PL spectra and 10% quantum yield (QY). The as-synthesized Ti_3C_2 MQDs are capable of sensing Zn^{2+} ions with large selectivity.

The functionalization of the two dimensional sheet of Ti_3C_2 MXene with GdW_{10}-based polyoxometalates (POMs) [79] yields a multifunctional hybrid nanocomposite $GdW_{10}@Ti_3C_2$ which incorporates in the hydrothermal treatment to destroy tumor cells by root. Not only this, both MR and CT imaging become possible with this structure. The hybrid structure acts like a contrast agent for the bioimaging based on magnetic resonance and computed tomography. Many previous studies demonstrated that $GdW_{10}@Ti_3C_2$ nano-bio-composite has an excellent biocompatibility, so can be efficiently utilized in biomedical fields. The functionalization of the two dimensional sheet of Ti_3C_2 MXene with GdW_{10}-based polyoxometalates (POMs) is shown in Fig. 8.6.

8.8.4 Tissue Engineering

MXenes are widely used in tissue engineering which constitutes engineering with bones, skin, nerves, heart and kidney tissues.

(a) **Bone-based Engineering**—$Ti_3C_2T_z$ MXenes are used to modify the tensile strength of nanocomposites. In the form of nanosheets, $Ti_3C_2T_z$ MXenes become highly biocompatible, regenerate guided bone [80] structure, improves osteogenic differentiation and promote osteo-inducibility. MXenes are biocompatible both in vivo as well as vitro [81, 82].

(b) **Skin-based Engineering**—Skin can be regenerated and repaired when skin tissues are engineered using MXenes. Bacterial cellulose (BC)/ Ti_3C_2 hydrogel was reported to utilize the process of cell co-culture and electrical simulation coupling for performing skin tissue engineering.

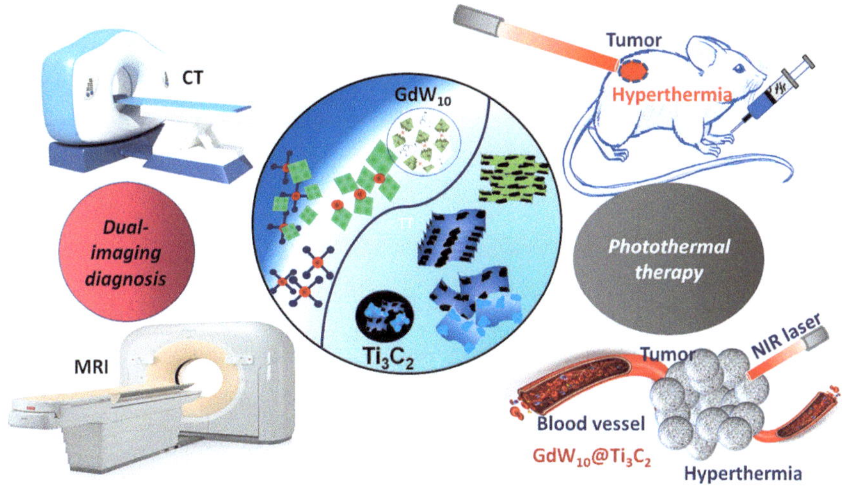

Fig. 8.6 The functionalization of two dimensional sheet of Ti_3C_2 MXene with GdW_{10}-based polyoxometalates (POMs)

(c) **Nerve-based Engineering**—Ti_3C_2 MXene neural [83] microelectrode has been reported to capture neuronal activities very accurately with high sensitivity. These microelectrodes have outstanding flexibility, extremely small impedance and are highly conductive. Neuroelectronic devices based on Ti_3C_2 have also been fabricated. Ti_3C_2 micro-neural electrodes [84] have some advantages like:

- The interface impedance is about four times less than that of gold microelectrodes of equal dimension.
- They have more signal-to-noise ratio and reduced baseline noise.
- Neurons based on Ti_3C_2 are more sustainable, can develop axons and construct functional networks.

(d) **Cardiac-based Engineering**—MXene Ti_2C was used to fabricate engineered cardiac patches (ECP) capable of fixing myocardial infarction (MI) [85]. It has been found that when Ti_2C-8-cryogel ECP was transplanted to the infarcted heart, the heart started working remarkably well, the size of the infarct reduced and further inflammation was obstructed. A new arteriole has also been evolved along with the vasculation process.

(e) **Kidney-based Engineering**—Ti_3C_2-PVP nanosheets (TPNS) [86] demonstrated an excellent anti-oxidative and perfectly biocompatible properties. TPNS provides the cytoprotective effects against ROS destruction as well as against cell apoptosis. The concentration of TPNS largely affects the rate of cell death or the protection effect. The treatment of Acute Kidney Injury (AKI) and many diseases associated with ROS are confirmed by the non-enzymatic antioxidant MXene assisted by ultrathin TPNS and mediated by redox reaction.

(f) **Three-dimensional Bioprinting**
3D bioprinting has been revealed to be an efficient technique for the regeneration of impaired tissues. MXene Ti_3C_2 was used in the form of nanosheet to process an electroconductive cell-laden bioink [87]. Ti_3C_2 is completely dissolved into hyaluronic acid/alginate (HA/Alg) hydrogels and yields MXene-based hydrogel bioinks. The as-prepared bioinks demonstrated excellent electrical conductivity and rheological properties. This technology played a significant part in the advancement of tissue as well as neural engineering [88].

8.8.5 Antimicrobial Biomedical Applications

$Ti_3C_2T_x$ MXene nanosheets interact directly with the cell membrane due to which in the course of time, the cell structure destroys completely. High reducing ability, surface reactivity feature and nano-scale size all make the direct penetration and contact to the cell membranes possible. The MXene nanosheet antimicrobial exercises greatly depend on the dimension and time to which they have been exposed. Smaller the size of the nanosheets, higher will be antibacterial functions. The direct contact between the sharp edges of the nanosheet and membrane surface of the bacteria play an important role in the antibacterial mechanism. Titanium carbide @ cuprous oxide (Ag@ Ti_3C_2@Cu_2O) nanocomposite [89] was also prepared with enhanced photocatalytic antimicrobial performance.

8.8.6 Therapeutics

(a) **Photothermal Therapy (PTT)**—PTT is the best alternative to the traditional treatment of cancer cells as well as chemotherapy which have many side-effects. PTT helps in tumor ablation in which NIR light is absorbed from the tumor-center transformed into heat energy. MXene is the most suitable aspirant for PTT due to its highly efficient photothermal conversion ability, abundant surface functional groups and hydrophilic property. PTT response varies with the depth to which the laser beam [90] penetrates or tumor is located, photothermal conversion efficiency, the time of optical excitation and optical power density. PTT is also utilized to energize and activate bacteria resistance to multiple drugs.
(b) **Therapeutic microneedle patch**—For the photothermal ablation of tumor cells, a microneedle made of polyvinylpyrrolidone (PVP) assisted by Nb_2C MXene nanosheets [91, 92] was developed. This microneedle [93] has a special feature of good solubility in the presence of NIR biowindow (NIR-II). When this needle is penetrated into the tumor location, it starts dissolving and releases Nb_2C Nanosheets [94].

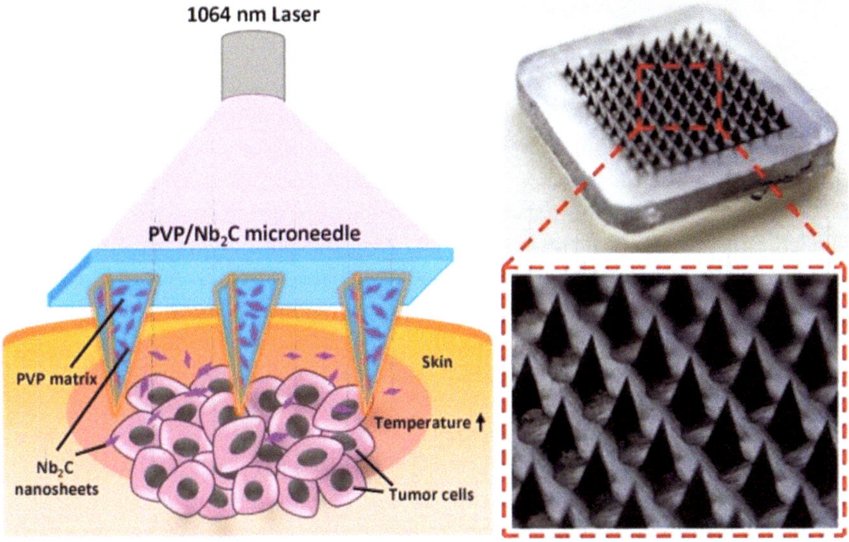

Fig. 8.7 Therapeutic microneedle patch functions for photonic implantable medicines. Reproduced with permission from S. Lin, H. Lin, M. Yang, M. Ge, Y. Chen, Y. Zhu, A two-dimensional MXene potentiates a therapeutic microneedle patch for photonic implantable medicine in the second NIR biowindow, Nano-scale. 12 (2020) 10,265–10,276. https://doi.org/10.1039/D0NR01444C [93]

This process restricts the further growth of tumor cells and enhances the rate of survival. The MXene microneedle was tested to be biocompatible with negligible toxic effects and functions for photonic implantable medicinesis shown in Fig. 8.7 [95].

(c) **Photodynamic therapy (PDT)**—MXenes are used to sensitize reactive oxygen species (ROS) in the PDT mechanism. The tumor cells are efficiently eradicated by employing sensitized ROS. So, ROS generation has become very significant in the biomedical fields. Ti_3C_2-DOX was found to generate ROS by degrading the HA cell to reveal the active sites of Ti_3C_2 nanosheets.
Previous studies enlighten that energy transfer mechanism of photo-excited electrons and LSPR effects are responsible for the generation of $^1O_2$, by utilizing Ti_3C_2 and other MXenes in the techniques.

(d) **Thermodynamic Therapy (TDT)**—MXenes' surface is rich in such compositions that enable their bandgaps to be modified and enhance surface plasmon resonance (SPR) effect. Due to this, MXenes comfortably generate ROS when irradiated with ultrasonic waves.

The tumor sites have lack of oxygen molecules, so it is quite difficult to generate ROS at those sites in order to destroy the tumor and cancer cells. Knowing this, the thermodynamic therapy was introduced which employs free radicals to damage cancer cells. The treatment mechanism was analyzed using multiple peak fluorescence, photoacoustic and photothermal bioimaging.

8.8.7 As Drug Carriers

MXenes are emerging as effective drug carriers because of the availability of large functional groups over the surface, their planar structure, presence of excessive surface negative charges and excellent biocompatible property. Ti_3C_2 was the earliest MXene used for the drug delivery mechanism, whose surface needed to be altered and modified for the better performance. The tumor sites constitute small pH values due to which the electrostatic force experienced by drug and MXenes is obstructed. Ti_3C_2-DOX responds to heat, light, pH-values and enzymatic activities. A thin film [96] of therapeutic mesopore was grown over the surface of Nb_2C MXene nanosheet using the sol–gel method. This nano-therapeutic Nb_2C MXene nanosheet demonstrated photothermal conversion efficiency (PTCE-light to heat) of 28.6% that removes cancer cells by adopting conjugation of chemotherapy and photothermal treatment (PTT).

When DOX is carried on by cellulose/MXene hydrogel [97], DOX releasing is found to be accelerated when illuminated by 808 nm wavelength radiation. This accelerated release is observed because of the expansion of holes on gaining energy.

8.8.8 Surface Engineering for the Colloidal Stability

Colloidal Stability of MXenes can be achieved by the functionalization of the surface with natural or synthetic macromolecules through non-covalent interactions.

Soyabean phospholipid (SP), a natural macromolecule was reported to provide colloidal stability to several composites of MXenes like MnO_x/Ti_3C_2, MnO_x/Ta_4C_3 and Ta_4C_3 via physical absorption technique. Another macromolecule Bovine serum albumin (BSA) was also found to contribute for the colloidal Stability to $W_{1.33}C$ *i*-MXenes using hydrogen bonding and Van der Waals forces.

There are few synthetic biocompatible macromolecules of polymers like poly vinyl alcohol (PVA), Poly vinyl pyrrolidone (PVP) and poly ethylene glycol (PEG) which also participate in managing colloidal stability to the ultrathin MXenes [98]. Biomedical properties and applications of MXenes in shown in Fig. 8.8 and Table 8.2.

8.8.9 Compatibility and Degradability in Biomedical Science

The biodegradation of Nb_2C MXenes was reported which explained that their decomposition took place in just a single day with human myeloperoxidase (hMPO) and hydrogen peroxide. It employed the enzyme-assisted degradation route. Later Ti_3C_2 MXenes underwent degradation using enzyme/ROS-triggered route. Mo_2C MXenes degraded biologically using pH-dependent degradation route and it took about two days to decompose. The biodegradation of MXenes generally takes a long time, and

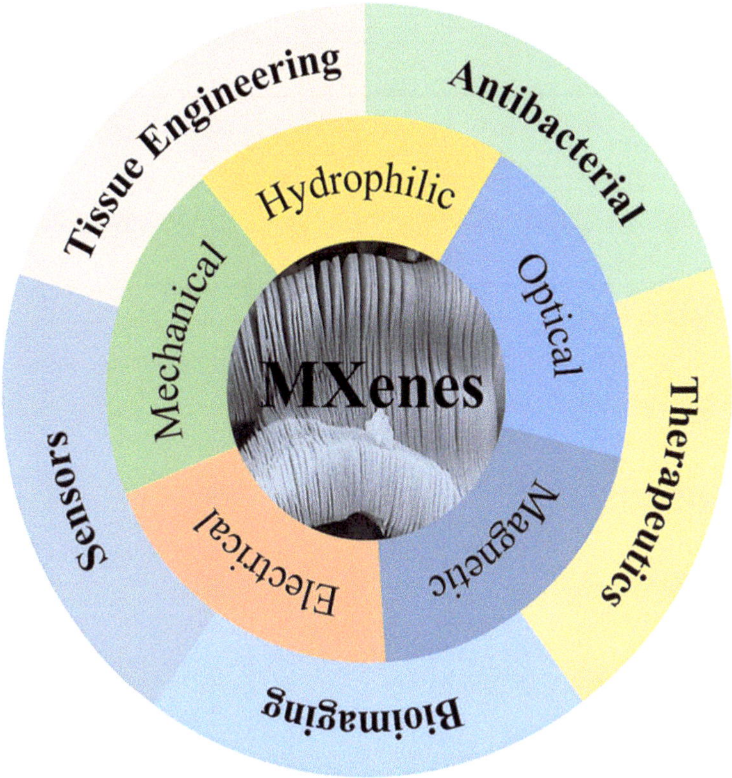

Fig. 8.8 Biomedical properties of MXenes in summary, Reproduced with permission from J. Huang, Z. Li, Y. Mao, Z. Li, Progress and biomedical applications of MXenes, Nano Sel. 2 (2021) 1480–1508. https://doi.org/10.1002/nano.202000309 [24]

when they are adversely etched and delaminated, it becomes chemically unstable, which can be then decomposed through oxidation reactions [99]. The process of degradation can be accelerated in the presence of abundance of enzymes and strong ionic environment. The biological compatibility of biomaterials based on MXenes is largely influenced by the morphology, structural configuration as well as surface chemistry.

Table 8.2 Properties in biomedical application

	Photo-acoustic imaging (PAI)	Magnetic resonance imaging (MRI)	Computed tomography (CT)	Al^{3+} added Ti_3C_2 Mxene	V_2C Mxene NSs with an algae extraction layer	Nanothera peutic Nb_2C Mxene NSs	Mo_2C MXenes
Photothermal conversion efficiency (PTCE)	Very high	48.60%	32.50%	Nearly 58.3%	48%	28.60%	Nearly 43.3%
Energy transformation	Light energy to sound energy	Light energy to heat energy	Light energy to heat energy	Light energy to heat energy	Excellent absorption in NIR region	Light energy to heat energy	Light energy to heat energy

8.9 Necessary Demands for Advancement in Biomedical Areas

(a) **Stability and Control Over Techniques**
Injection of MXenes for curing tumor effects may lead to an uncontrolled leakage of drugs and many such instabilities. More work needs to be done to enhance the stability factor of MXenes.

(b) **Hybridization of MXenes**
Single MXenes, without any hybridization could not carry on the diversity in therapeutic and rehabilitation effects. MXenes should be decorated with structures having diverse dimensional features such as 0D, 2D and 3D layered biomaterials.

(c) **Extension in Multiple Fields**
Biomedical fields provide MXenes to participate in analyzing sensing, imaging, photothermal, photodynamic and thermodynamic operations as well as drug-delivery systems. Biosensing strategies are attained using electrical and optical signals.

(d) **Combination of Additional Elements**
MXenes need to be combined with some additional functional elements to broaden its applicability. MXene-based biomaterials should be fabricated.

(e) **Solution to the Problems**

- Oxygen-deficient microenvironment of tumor-sites.
- Biased heat-resistance of cancer cells.
- MXenes are limited to the cancer and tumor cells treatments only.

(f) **Ecofriendly Synthesis Mechanism**
Ecofriendly approaches are essentially to be attained in order to fabricate MXene nanosheets. The inexpensive and scalable synthesis methods significantly need to be discovered.

8.9.1 Subsequent Future-Possibilities

(i) Investigation of long-term toxicity
(ii) Pharmacokinetics
(iii) Biodistribution of MXenes in small animal models as well as large animal models
(iv) Explore more synthetic methods of top-down and bottom-up results in an uncontrolled effort over experiments
(v) To develop more innovative surface modification strategy
(vi) Researches on electrical and optical properties [100] are more but less on magnetic properties.

(vii) To develop unique structures to improve electron-transmission capacity and increase the drug-carrying area.

8.9.2 Living Cell Toxicity in Biomedical Applications

The study of living cell toxicity becomes very relevant when one makes a selection for working on biomedical sciences. The Ti-based MXene $Ti_3C_2T_x$ absorption [101] largely depends on the concentration of titanium. The inductively-coupled plasma mass spectrometry (ICP-MS) is used for calculating and evaluating the toxicity content of MXene. It was reported in several studies related to embryo's cell membrane that $Ti_3C_2T_x$ solution up to 100 μM concentration has negligible toxicity but above 100 μM, it will become fatal. There may be severe blocking of capillaries, specifically in heart that may result in unexpected life risk.

The cytotoxicity depends on the types of cells to be treated. The cancer cells were found to have adverse toxic effects than any other. When the small particles of size less than 44 μm are made to interact with various forms of $Ti_3C_2T_x$ like TiC, Ti_2AlC, and Ti_3AlC having high concentration greater than or equal to 400 μg/mL, results found to be more hazardous. $Ti_3C_2T_x$ forms are not much harmful to $MSU_{1.1}$ cells even with large concentration [102].

8.10 Conclusions

MXenes are emerging and developing as nanomedicines for the remediation of cancerous cells, for bioimaging and biosensing as well as for the antibacterial activities. Mechanical properties of MXenes enable the manufacturing of wearable sensing devices with high flexibility. The actuator, so obtained, for improving the mechanical properties of MXenes, possesses a remarkably high stiffness, elevated tensile strength and larger values of Young's modulus. These features enhance the fast and direct absorption of moisture into a body. The MXene concentration largely affects the survival rate of biological cells. The larger the concentration, more is the risk to the cancerous cells along with those of the normal cells. Below 25 mg/L concentration of MXene, normal cells are found to have the survival rate greater than 70%. TiC, Ti_2AlC, and Ti_3AlC_2 with concentration greater that 400 μg/mL and size less than 44 μm, the cell membranes start destroying. Medical diagnosis using bioimaging has become possible on account of MXenes quantum confinement, photothermal effects and several contrasting agents (CAs). Due to the presence of heavy metals, MXenes are utilized as contrasting agents (CA) for MR imaging. MXenes are widely recognized to be applied as nanomedicines for various theranostic treatments. Mxenes, as two dimensional V_2C MXenzymes, find its application in catalytic therapy for the treatment of neurogenerative diseases. Nb_2C MXenes also demonstrate enzyme-like properties for providing protection from radioactive contaminants. Non-covalent interactions using hydrogen bonds and Van der Waals forces for the surface engineering are least stable, so some innovative and feasible methods should be discovered in order to synthesize biocompatible MXenes. MXenes are now extensively utilized not only in theranostic areas but in various multifunctional therapies. Biosafety is also one of the factors to be considered while applying for fabricating diverse functionalized MXenes. The cytotoxicity depends on the types of cells to be treated. The cancer cells were found to have adverse toxic effects than any other. When the small particles of size less than 44 μm are made to interact with various forms of $Ti_3C_2T_x$ like TiC, Ti_2AlC, and Ti_3AlC having high concentration greater than or equal to 400 μg/mL, results are found to be more hazardous. $Ti_3C_2T_x$ forms are not much harmful to $MSU_{1.1}$ cells even with large concentration. For the treatment of cancer and tumors, MXenes surface chemistry has been utilized in which a superparamagnetic two dimensional Ti_3C_2 MXenes were fabricated. The superparamagnetic crystals of Fe_3O_4 were subjected to the in-situ growth over the surface of Ti_3C_2 MXenes. This enhanced the magnetic resonance imaging of cell analytes as well as promoted the photothermal destruction of the cancerous cells and ablated the tumor tissues.

Acknowledgements J.R. Ansari gratefully acknowledges the financial support by BP Fellowship of National Research Foundation of Korea at Department of Packaging and Logistics, College of Science and Technology, Yonsei University, South Korea. NA gratefully acknowledges LNMU, Darbhanga for the support in writing this Chapter. Namita and Arti are thankful to the Department of Physics, LNMU in carrying out the work in this chapter.

Conflicts of Interest The authors declare that they have no conflict of interest.

References

1. Z. Liu, M. Zhao, H. Lin, C. Dai, C. Ren, S. Zhang, W. Peng, Y. Chen, 2D magnetic titanium carbide MXene for cancer theranostics. J. Mater. Chem. B. **6**, 3541–3548 (2018). https://doi.org/10.1039/C8TB00754C
2. B. Anasori, Y. Xie, M. Beidaghi, J. Lu, B.C. Hosler, L. Hultman, P.R.C. Kent, Y. Gogotsi, M.W. Barsoum, Two-dimensional, ordered, double transition metals carbides (MXenes). ACS Nano **9**, 9507–9516 (2015). https://doi.org/10.1021/acsnano.5b03591
3. B. Anasori, M.R. Lukatskaya, Y. Gogotsi, 2D metal carbides and nitrides (MXenes) for energy storage. Nat. Rev. Mater. **2**, 16098 (2017). https://doi.org/10.1038/natrevmats.2016.98
4. X. Lin, Z. Li, J. Qiu, Q. Wang, J. Wang, H. Zhang, T. Chen, Fascinating MXene nanomaterials: emerging opportunities in the biomedical field. Biomater. Sci. **9**, 5437–5471 (2021). https://doi.org/10.1039/D1BM00526J
5. M. Tebyetekerwa, I. Marriam, Z. Xu, S. Yang, H. Zhang, F. Zabihi, R. Jose, S. Peng, M. Zhu, S. Ramakrishna, Critical insight: challenges and requirements of fibre electrodes for wearable electrochemical energy storage. Energy Environ. Sci. **12**, 2148–2160 (2019). https://doi.org/10.1039/C8EE02607F
6. Y. Ding, J. Zhang, A. Guan, Q. Wang, S. Li, A.M. Al-Enizi, L. Qian, L. Zhang, G. Zheng, Promoting N_2 electroreduction to ammonia by fluorine-terminating $Ti_3C_2T_x$ MXene. Nano Converg. **8**, 14 (2021). https://doi.org/10.1186/s40580-021-00264-9

7. S. Chiluwal, A.M. Rao, R. Podila, Strategies for improving rechargeable lithium-ion batteries: from active materials to CO_2 emissions. Nanotechnol. Rev. **10**, 1993–2026 (2021). https://doi.org/10.1515/ntrev-2021-0114
8. J.R. Ansari, C.A. Sunilbhai, K.K. Sadasivuni, MXenes and their composites for energy storage and conversion, in *Mxenes Their Composites* (Elsevier, 2022), pp. 201–240. https://doi.org/10.1016/B978-0-12-823361-0.00021-6
9. R. Momeni Feili, M. Dadsetani, R. Nejatipour, A. Ebrahimian, Electron energy loss structures of terminated scandium and hafnium MXene monolayers from first-principles calculations. J. Electron. Mater. **49**, 2502–2520 (2020). https://doi.org/10.1007/s11664-020-07946-w
10. H. Hadipour, Y. Yekta, Ab initio study of the effective Coulomb interactions and Stoner ferromagnetism in M_2C and $M_2CO_2M_X$−enes (M = Sc, Ti, V, Cr, Fe, Zr, Nb, Mo, Hf, Ta). Phys. Rev. B **100**, 195118 (2019). https://doi.org/10.1103/PhysRevB.100.195118
11. J. Zhou, X. Zha, X. Zhou, F. Chen, G. Gao, S. Wang, C. Shen, T. Chen, C. Zhi, P. Eklund, S. Du, J. Xue, W. Shi, Z. Chai, Q. Huang, Synthesis and electrochemical properties of two-dimensional hafnium carbide. ACS Nano **11**, 3841–3850 (2017). https://doi.org/10.1021/acsnano.7b00030
12. J. Yang, X. Luo, S. Zhang, L. Chen, Investigation of magnetic and electronic properties of transition metal doped Sc_2CT_2 (T = O, OH or F) using a first principles study. Phys. Chem. Chem. Phys. **18**, 12914–12919 (2016). https://doi.org/10.1039/C6CP00138F
13. Q. Wu, Y. Zhang, Q. Zhou, J. Wang, X.C. Zeng, Transition-metal dihydride monolayers: a new family of two-dimensional ferromagnetic materials with intrinsic room-temperature half-metallicity. J. Phys. Chem. Lett. **9**, 4260–4266 (2018). https://doi.org/10.1021/acs.jpclett.8b01976
14. A. VahidMohammadi, J. Rosen, Y. Gogotsi, The world of two-dimensional carbides and nitrides (MXenes). Science **372**, eabf1581 (2021). https://doi.org/10.1126/science.abf1581
15. R.M. Ronchi, J.T. Arantes, S.F. Santos, Synthesis, structure, properties and applications of MXenes: current status and perspectives. Ceram. Int. **45**, 18167–18188 (2019). https://doi.org/10.1016/j.ceramint.2019.06.114
16. S. Jolly, M.P. Paranthaman, M. Naguib, Synthesis of Ti_3C_2Tz MXene from low-cost and environmentally friendly precursors. Mater. Today Adv. **10**, 100139 (2021). https://doi.org/10.1016/j.mtadv.2021.100139
17. Q. Xue, Z. Pei, Y. Huang, M. Zhu, Z. Tang, H. Li, Y. Huang, N. Li, H. Zhang, C. Zhi, Mn_3O_4 nanoparticles on layer-structured Ti_3C_2 MXene towards the oxygen reduction reaction and zinc–air batteries. J. Mater. Chem. A. **5**, 20818–20823 (2017). https://doi.org/10.1039/C7TA04532H
18. H. Huang, R. Jiang, Y. Feng, H. Ouyang, N. Zhou, X. Zhang, Y. Wei, Recent development and prospects of surface modification and biomedical applications of MXenes. Nano-scale **12**, 1325–1338 (2020). https://doi.org/10.1039/C9NR07616F
19. L. Chen, X. Dai, W. Feng, Y. Chen, Biomedical applications of MXenes: from nanomedicine to biomaterials. Acc. Mater. Res. **3**, 785–798 (2022). https://doi.org/10.1021/accountsmr.2c00025
20. Q. Tao, M. Dahlqvist, J. Lu, S. Kota, R. Meshkian, J. Halim, J. Palisaitis, L. Hultman, M.W. Barsoum, P.O.Å. Persson, J. Rosen, Two-dimensional Mo1.33C MXene with divacancy ordering prepared from parent 3D laminate with in-plane chemical ordering. Nat. Commun. **8**, 14949 (2017). https://doi.org/10.1038/ncomms14949
21. J. Zhou, X. Zha, F.Y. Chen, Q. Ye, P. Eklund, S. Du, Q. Huang, A two-dimensional zirconium carbide by selective etching of Al_3C_3 from nanolaminated $Zr_3Al_3C_5$. Angew. Chem. Int. Ed. **55**, 5008–5013 (2016). https://doi.org/10.1002/anie.201510432
22. S. Zheng, X. Shi, P. Das, Z. Wu, X. Bao, The road towards planar microbatteries and micro-supercapacitors: from 2D to 3D device geometries. Adv. Mater. **31**, 1900583 (2019). https://doi.org/10.1002/adma.201900583
23. L. Wang, H. Song, L. Yuan, Z. Li, Y. Zhang, J.K. Gibson, L. Zheng, Z. Chai, W. Shi, Efficient U(VI) reduction and sequestration by Ti_2CT_x MXene. Environ. Sci. Technol. **52**, 10748–10756 (2018). https://doi.org/10.1021/acs.est.8b03711

24. J. Huang, Z. Li, Y. Mao, Z. Li, Progress and biomedical applications of MXenes. Nano Sel. **2**, 1480–1508 (2021). https://doi.org/10.1002/nano.202000309
25. L. Liu, Y. Wei, S. Jiao, S. Zhu, X. Liu, A novel label-free strategy for the ultrasensitive miRNA-182 detection based on MoS_2/Ti_3C_2 nanohybrids. Biosens. Bioelectron. **137**, 45–51 (2019). https://doi.org/10.1016/j.bios.2019.04.059
26. Z. Kang, Z. Zheng, H. Wei, Z. Zhang, X. Tan, L. Xiong, T. Zhai, Y. Gao, Controlled growth of an Mo_2C—graphene hybrid film as an electrode in self-powered two-sided Mo_2C—graphene/$Sb_2S_{0.42}Se_{2.58}/TiO_2$ photodetectors. Sensors **19**, 1099 (2019). https://doi.org/10.3390/s19051099
27. X. Xiao, H. Yu, H. Jin, M. Wu, Y. Fang, J. Sun, Z. Hu, T. Li, J. Wu, L. Huang, Y. Gogotsi, J. Zhou, Salt-templated synthesis of 2D metallic MoN and other nitrides. ACS Nano **11**, 2180–2186 (2017). https://doi.org/10.1021/acsnano.6b08534
28. J. Chen, K. Chen, D. Tong, Y. Huang, J. Zhang, J. Xue, Q. Huang, T. Chen, CO_2 and temperature dual responsive "Smart" MXene phases. Chem. Commun. **51**, 314–317 (2015). https://doi.org/10.1039/C4CC07220K
29. X. Sang, Y. Xie, M.-W. Lin, M. Alhabeb, K.L. Van Aken, Y. Gogotsi, P.R.C. Kent, K. Xiao, R.R. Unocic, Atomic defects in monolayer titanium carbide ($Ti_3C_2T_x$) MXene. ACS Nano **10**, 9193–9200 (2016). https://doi.org/10.1021/acsnano.6b05240
30. R. Liang, Y. Li, M. Huo, H. Lin, Y. Chen, Triggering sequential catalytic fenton reaction on 2D MXenes for hyperthermia-augmented synergistic nanocatalytic cancer therapy. ACS Appl. Mater. Interfaces **11**, 42917–42931 (2019). https://doi.org/10.1021/acsami.9b13598
31. Z. Guo, J. Zhou, C. Si, Z. Sun, Flexible two-dimensional $Ti_{n+1}C_n$ (n = 1, 2 and 3) and their functionalized MXenes predicted by density functional theories. Phys. Chem. Chem. Phys. **17**, 15348–15354 (2015). https://doi.org/10.1039/C5CP00775E
32. H. Zhang, S. Zuo, M. Qiu, S. Wang, Y. Zhang, J. Zhang, X.W. (David) Lou, Direct probing of atomically dispersed Ru species over multi-edged TiO_2 for highly efficient photocatalytic hydrogen evolution. Sci. Adv. **6**, eabb9823 (2020). https://doi.org/10.1126/sciadv.abb9823
33. J. Cao, Z. Zhou, Q. Song, K. Chen, G. Su, T. Zhou, Z. Zheng, C. Lu, X. Zhang, Ultrarobust $Ti_3C_2T_x$ MXene-based soft actuators via bamboo-inspired mesoscale assembly of hybrid nanostructures. ACS Nano **14**, 7055–7065 (2020). https://doi.org/10.1021/acsnano.0c01779
34. F. Wang, M. Cao, Y. Qin, J. Zhu, L. Wang, Y. Tang, ZnO nanoparticle-decorated two-dimensional titanium carbide with enhanced supercapacitive performance. RSC Adv. **6**, 88934–88942 (2016). https://doi.org/10.1039/C6RA15384D
35. Q. Yang, H. Yin, T. Xu, D. Zhu, J. Yin, Y. Chen, X. Yu, J. Gao, C. Zhang, Y. Chen, Y. Gao, Engineering 2D Mesoporous Silica@MXene-integrated 3d-printing scaffolds for combinatory osteosarcoma therapy and NO-augmented bone regeneration. Small **16**, 1906814 (2020). https://doi.org/10.1002/smll.201906814
36. M. Ghidiu, M.R. Lukatskaya, M.-Q. Zhao, Y. Gogotsi, M.W. Barsoum, Conductive two-dimensional titanium carbide 'clay' with high volumetric capacitance. Nature **516**, 78–81 (2014). https://doi.org/10.1038/nature13970
37. Z. Guo, J. Zhou, L. Zhu, Z. Sun, MXene: a promising photocatalyst for water splitting. J. Mater. Chem. A. **4**, 11446–11452 (2016). https://doi.org/10.1039/C6TA04414J
38. K. Hantanasirisakul, M. Zhao, P. Urbankowski, J. Halim, B. Anasori, S. Kota, C.E. Ren, M.W. Barsoum, Y. Gogotsi, Fabrication of $Ti_3C_2T_x$ MXene transparent thin films with tunable optoelectronic properties. Adv. Electron. Mater. **2**, 1600050 (2016). https://doi.org/10.1002/aelm.201600050
39. X. Li, F. Liu, D. Huang, N. Xue, Y. Dang, M. Zhang, L. Zhang, B. Li, D. Liu, L. Wang, H. Liu, X. Tao, Nonoxidized MXene quantum dots prepared by microexplosion method for cancer catalytic therapy. Adv. Funct. Mater. **30**, 2000308 (2020). https://doi.org/10.1002/adfm.202000308
40. H. Wang, J. Sun, L. Lu, X. Yang, J. Xia, F. Zhang, Z. Wang, Competitive electrochemical aptasensor based on a cDNA-ferrocene/MXene probe for detection of breast cancer marker Mucin1. Anal. Chim. Acta **1094**, 18–25 (2020). https://doi.org/10.1016/j.aca.2019.10.003

41. M. El Aamri, G. Yammouri, H. Mohammadi, A. Amine, H. Korri-Youssoufi, Electrochemical biosensors for detection of microRNA as a cancer biomarker: pros and cons. Biosensors **10**, 186 (2020). https://doi.org/10.3390/bios10110186
42. S. Hroncekova, T. Bertok, M. Hires, E. Jane, L. Lorencova, A. Vikartovska, A. Tanvir, P. Kasak, J. Tkac, Ultrasensitive $Ti_3C_2T_X$ MXene/Chitosan nanocomposite-based amperometric biosensor for detection of potential prostate cancer marker in urine samples. Processes **8**, 580 (2020). https://doi.org/10.3390/pr8050580
43. S. Zhong, B. Xu, A. Cui, S. Li, S. Liao, G. Wang, G. Liu, B. Sun, Robust net magnetic moment in janus V-based nitride MXenes: insight from first-principles calculations. ACS Omega **5**, 864–870 (2020). https://doi.org/10.1021/acsomega.9b03779
44. Z. Liu, H. Lin, M. Zhao, C. Dai, S. Zhang, W. Peng, Y. Chen, 2D superparamagnetic tantalum carbide composite MXenes for efficient breast-cancer theranostics. Theranostics **8**, 1648–1664 (2018). https://doi.org/10.7150/thno.23369
45. W. Feng, X. Han, H. Hu, M. Chang, L. Ding, H. Xiang, Y. Chen, Y. Li, 2D vanadium carbide MXenzyme to alleviate ROS-mediated inflammatory and neurodegenerative diseases. Nat. Commun. **12**, 2203 (2021). https://doi.org/10.1038/s41467-021-22278-x
46. P. Lv, Y.-L. Li, J.-F. Wang, Monolayer Ti_2C MXene: manipulating magnetic properties and electronic structures by an electric field. Phys. Chem. Chem. Phys. **22**, 11266–11272 (2020). https://doi.org/10.1039/D0CP00507J
47. R.P. Pandey, K. Rasool, V.E. Madhavan, B. Aïssa, Y. Gogotsi, K.A. Mahmoud, Ultrahigh-flux and fouling-resistant membranes based on layered silver/MXene ($Ti_3C_2T_x$) nanosheets. J. Mater. Chem. A. **6**, 3522–3533 (2018). https://doi.org/10.1039/C7TA10888E
48. S. Kumar, Y. Lei, N.H. Alshareef, M.A. Quevedo-Lopez, K.N. Salama, Biofunctionalized two-dimensional Ti_3C_2 MXenes for ultrasensitive detection of cancer biomarker. Biosens. Bioelectron. **121**, 243–249 (2018). https://doi.org/10.1016/j.bios.2018.08.076
49. K. Rasool, M. Helal, A. Ali, C.E. Ren, Y. Gogotsi, K.A. Mahmoud, Antibacterial activity of $Ti_3C_2T_x$ MXene. ACS Nano **10**, 3674–3684 (2016). https://doi.org/10.1021/acsnano.6b00181
50. F. Wang, C. Yang, M. Duan, Y. Tang, J. Zhu, TiO_2 nanoparticle modified organ-like Ti_3C_2 MXene nanocomposite encapsulating hemoglobin for a mediator-free biosensor with excellent performances. Biosens. Bioelectron. **74**, 1022–1028 (2015). https://doi.org/10.1016/j.bios.2015.08.004
51. F. Wang, C. Yang, C. Duan, D. Xiao, Y. Tang, J. Zhu, An organ-like titanium carbide material (MXene) with multilayer structure encapsulating hemoglobin for a mediator-free biosensor. J. Electrochem. Soc. **162**, B16–B21 (2015). https://doi.org/10.1149/2.0371501jes
52. H. Liu, K. Guo, J. Lv, Y. Gao, C. Duan, L. Deng, Z. Zhu, A novel nitrite biosensor based on the direct electrochemistry of horseradish peroxidase immobilized on porous Co_3O_4 nanosheets and reduced graphene oxide composite modified electrode. Sens. Actuators B Chem. **238**, 249–256 (2017). https://doi.org/10.1016/j.snb.2016.07.073
53. P. Urbankowski, B. Anasori, T. Makaryan, D. Er, S. Kota, P.L. Walsh, M. Zhao, V.B. Shenoy, M.W. Barsoum, Y. Gogotsi, Synthesis of two-dimensional titanium nitride Ti_4N_3 (MXene). Nano-scale **8**, 11385–11391 (2016). https://doi.org/10.1039/C6NR02253G
54. Q. Gao, H. Zhu, L. Dong, W. Shi, R. Chen, Z. Song, C. Huang, J. Li, X. Dong, Y. Zhou, Q. Liu, L. Ma, X. Wang, J. Zhou, Y. Liu, E. Boja, A.I. Robles, W. Ma, P. Wang, Y. Li, L. Ding, B. Wen, B. Zhang, H. Rodriguez, D. Gao, H. Zhou, J. Fan, Integrated proteogenomic characterization of HBV-related hepatocellular carcinoma. Cell **179**, 561-577.e22 (2019). https://doi.org/10.1016/j.cell.2019.08.052
55. R.B. Rakhi, P. Nayak, C. Xia, H.N. Alshareef, Novel amperometric glucose biosensor based on MXene nanocomposite. Sci. Rep. **6**, 36422 (2016). https://doi.org/10.1038/srep36422
56. H.L. Chia, C.C. Mayorga-Martinez, N. Antonatos, Z. Sofer, J.J. Gonzalez-Julian, R.D. Webster, M. Pumera, MXene titanium carbide-based biosensor: strong dependence of exfoliation method on performance. Anal. Chem. **92**, 2452–2459 (2020). https://doi.org/10.1021/acs.analchem.9b03634

57. H. Gu, Y. Xing, P. Xiong, H. Tang, C. Li, S. Chen, R. Zeng, K. Han, G. Shi, Three-dimensional porous $Ti_3C_2T_x$ MXene–graphene hybrid films for glucose biosensing. ACS Appl. Nano Mater. **2**, 6537–6545 (2019). https://doi.org/10.1021/acsanm.9b01465
58. Y. Lei, W. Zhao, Y. Zhang, Q. Jiang, J. He, A.J. Baeumner, O.S. Wolfbeis, Z.L. Wang, K.N. Salama, H.N. Alshareef, A MXene-based wearable biosensor system for high-performance in vitro perspiration analysis. Small **15**, 1901190 (2019). https://doi.org/10.1002/smll.201901190
59. R.G. Fiddian-Green, W. Silen, Mechanisms of disposal of acid and alkali in rabbit duodenum. Am. J. Physiol. **229**, 1641–1648 (1975). https://doi.org/10.1152/ajplegacy.1975.229.6.1641
60. J. Zheng, B. Wang, A. Ding, B. Weng, J. Chen, Synthesis of MXene/DNA/Pd/Pt nanocomposite for sensitive detection of dopamine. J. Electroanal. Chem. **816**, 189–194 (2018). https://doi.org/10.1016/j.jelechem.2018.03.056
61. S.S. Shankar, R.M. Shereema, R.B. Rakhi, Electrochemical determination of adrenaline using MXene/graphite composite paste electrodes. ACS Appl. Mater. Interfaces **10**, 43343–43351 (2018). https://doi.org/10.1021/acsami.8b11741
62. J. Nakazaki, H. Segawa, Evolution of organometal halide solar cells. J. Photochem. Photobiol. C Photochem. Rev. **35**, 74–107 (2018). https://doi.org/10.1016/j.jphotochemrev.2018.02.002
63. C.-H. Jia, M. Li, J. Liu, L. Zhao, J. Lin, P.-L. Lai, X. Zhou, Y. Zhang, Z.-G. Chen, H.-Y. Li, A.-L. Liu, C.-L. Yang, T.-M. Gao, Y. Jiang, X.-C. Bai, IKK-β mediates hydrogen peroxide induced cell death through p85 S6K1. Cell Death Differ. **20**, 248–258 (2013). https://doi.org/10.1038/cdd.2012.115
64. Q. Xu, S. Chen, J. Xu, X. Duan, L. Lu, Q. Tian, X. Zhang, Y. Cai, X. Lu, L. Rao, Y. Yu, Facile synthesis of hierarchical MXene/ZIF-67/CNTs composite for electrochemical sensing of luteolin. J. Electroanal. Chem. **880**, 114765 (2021). https://doi.org/10.1016/j.jelechem.2020.114765
65. F. Zhao, Y. Yao, C. Jiang, Y. Shao, D. Barceló, Y. Ying, J. Ping, Self-reduction bimetallic nanoparticles on ultrathin MXene nanosheets as functional platform for pesticide sensing. J. Hazard. Mater. **384**, 121358 (2020). https://doi.org/10.1016/j.jhazmat.2019.121358
66. H. Wu, X. Zhang, C. Wei, C. Wang, M. Jiang, X. Hong, Z. Xu, D. Chen, X. Huang, Modular assembly of enzyme loaded nanoparticles in 3D hollow fiber electrode for electrochemical sensing. Chem. Eng. J. **421**, 129721 (2021). https://doi.org/10.1016/j.cej.2021.129721
67. M. Song, S. Pang, F. Guo, M. Wong, J. Hao, Fluoride-free 2D niobium carbide MXenes as stable and biocompatible nanoplatforms for electrochemical biosensors with ultrahigh sensitivity. Adv. Sci. **7**, 2001546 (2020). https://doi.org/10.1002/advs.202001546
68. M. Mohammadniaei, A. Koyappayil, Y. Sun, J. Min, M.-H. Lee, Gold nanoparticle/MXene for multiple and sensitive detection of oncomiRs based on synergetic signal amplification. Biosens. Bioelectron. **159**, 112208 (2020). https://doi.org/10.1016/j.bios.2020.112208
69. Y. Wang, Y. Xu, M. Hu, H. Ling, X. Zhu, MXenes: focus on optical and electronic properties and corresponding applications. Nanophotonics **9**, 1601–1620 (2020). https://doi.org/10.1515/nanoph-2019-0556
70. Y. Guo, M. Zhong, Z. Fang, P. Wan, G. Yu, A wearable transient pressure sensor made with MXene nanosheets for sensitive broad-range human-machine interfacing. Nano Lett. **19**, 1143–1150 (2019). https://doi.org/10.1021/acs.nanolett.8b04514
71. Y. Ma, Y. Yue, H. Zhang, F. Cheng, W. Zhao, J. Rao, S. Luo, J. Wang, X. Jiang, Z. Liu, N. Liu, Y. Gao, 3D synergistical MXene/reduced graphene oxide aerogel for a piezoresistive sensor. ACS Nano **12**, 3209–3216 (2018). https://doi.org/10.1021/acsnano.7b06909
72. Y. Yue, N. Liu, W. Liu, M. Li, Y. Ma, C. Luo, S. Wang, J. Rao, X. Hu, J. Su, Z. Zhang, Q. Huang, Y. Gao, 3D hybrid porous Mxene-sponge network and its application in piezoresistive sensor. Nano Energy **50**, 79–87 (2018). https://doi.org/10.1016/j.nanoen.2018.05.020
73. Y. Cai, J. Shen, G. Ge, Y. Zhang, W. Jin, W. Huang, J. Shao, J. Yang, X. Dong, Stretchable $Ti_3C_2T_x$ MXene/carbon nanotube composite based strain sensor with ultrahigh sensitivity and tunable sensing range. ACS Nano **12**, 56–62 (2018). https://doi.org/10.1021/acsnano.7b06251

74. H. Zhang, G. Yang, X. Zuo, H. Tang, Q. Yang, G. Li, Computational studies on the structural, electronic and optical properties of graphene-like MXenes (M_2CT_2, M = Ti, Zr, Hf; T = O, F, OH) and their potential applications as visible-light driven photocatalysts. J. Mater. Chem. A. **4**, 12913–12920 (2016). https://doi.org/10.1039/C6TA04628B
75. S. Iravani, R.S. Varma, MXenes and MXene-based materials for tissue engineering and regenerative medicine: recent advances. Mater. Adv. **2**, 2906–2917 (2021). https://doi.org/10.1039/D1MA00189B
76. Y. He, L. Ma, L. Zhou, G. Liu, Y. Jiang, J. Gao, Preparation and application of Bismuth/MXene nanocomposite as electrochemical sensor for heavy metal ions detection. Nanomaterials **10**, 866 (2020). https://doi.org/10.3390/nano10050866
77. C. Xu, L. Wang, Z. Liu, L. Chen, J. Guo, N. Kang, X.-L. Ma, H.-M. Cheng, W. Ren, Large-area high-quality 2D ultrathin Mo_2C superconducting crystals. Nat. Mater. **14**, 1135–1141 (2015). https://doi.org/10.1038/nmat4374
78. F. He, C. Tang, G. Zhu, Y. Liu, A. Du, Q. Zhang, M. Wu, H. Zhang, Leaf-inspired design of mesoporous Sb_2S_3/N-doped $Ti_3C_2T_x$ composite towards fast sodium storage. Sci. China Chem. **64**, 964–973 (2021). https://doi.org/10.1007/s11426-020-9942-9
79. L. Zong, H. Wu, H. Lin, Y. Chen, A polyoxometalate-functionalized two-dimensional titanium carbide composite MXene for effective cancer theranostics. Nano Res. **11**, 4149–4168 (2018). https://doi.org/10.1007/s12274-018-2002-3
80. J. Zhou, Z. Zhang, J. Joseph, X. Zhang, B.E. Ferdows, D.N. Patel, W. Chen, G. Banfi, R. Molinaro, D. Cosco, N. Kong, N. Joshi, O.C. Farokhzad, C. Corbo, W. Tao, Biomaterials and nanomedicine for bone regeneration: progress and future prospects. Exploration **1**, 20210011 (2021). https://doi.org/10.1002/EXP.20210011
81. X. Zhao, L. Wang, J. Li, L. Peng, C. Tang, X. Zha, K. Ke, M. Yang, B. Su, W. Yang, Redox-mediated artificial non-enzymatic antioxidant MXene nanoplatforms for acute kidney injury alleviation. Adv. Sci. **8**, 2101498 (2021). https://doi.org/10.1002/advs.202101498
82. K. Chen, N. Qiu, Q. Deng, M.-H. Kang, H. Yang, J.-U. Baek, Y.-H. Koh, S. Du, Q. Huang, H.-E. Kim, Cytocompatibility of Ti_3AlC_2, Ti_3SiC_2, and Ti_2AlN: in vitro tests and first-principles calculations. ACS Biomater. Sci. Eng. **3**, 2293–2301 (2017). https://doi.org/10.1021/acsbiomaterials.7b00432.
83. B. Xu, M. Zhu, W. Zhang, X. Zhen, Z. Pei, Q. Xue, C. Zhi, P. Shi, Ultrathin MXene-micropattern-based field-effect transistor for probing neural activity. Adv. Mater. **28**, 3333–3339 (2016). https://doi.org/10.1002/adma.201504657
84. W. Sun, X. Wang, J. Feng, T. Li, Y. Huan, J. Qiao, L. He, D. Ma, Controlled synthesis of 2D Mo_2C/graphene heterostructure on liquid Au substrates as enhanced electrocatalytic electrodes. Nanotechnology **30**, 385601 (2019). https://doi.org/10.1088/1361-6528/ab2c0d
85. G. Ye, Z. Wen, F. Wen, X. Song, L. Wang, C. Li, Y. He, S. Prakash, X. Qiu, Mussel-inspired conductive Ti_2C-cryogel promotes functional maturation of cardiomyocytes and enhances repair of myocardial infarction. Theranostics **10**, 2047–2066 (2020). https://doi.org/10.7150/thno.38876
86. H. Rastin, B. Zhang, A. Mazinani, K. Hassan, J. Bi, T.T. Tung, D. Losic, 3D bioprinting of cell-laden electroconductive MXene nanocomposite bioinks. Nano-scale **12**, 16069–16080 (2020). https://doi.org/10.1039/D0NR02581J
87. W. Cao, C. Ma, D. Mao, J. Zhang, M. Ma, F. Chen, MXene-reinforced cellulose nanofibril inks for 3D-printed smart fibres and textiles. Adv. Funct. Mater. **29**, 1905898 (2019). https://doi.org/10.1002/adfm.201905898
88. M. Naguib, M. Kurtoglu, V. Presser, J. Lu, J. Niu, M. Heon, L. Hultman, Y. Gogotsi, M.W. Barsoum, Two-dimensional nanocrystals produced by exfoliation of Ti_3AlC_2. Adv. Mater. **23**, 4248–4253 (2011). https://doi.org/10.1002/adma.201102306
89. K. Wang, Z. Lou, L. Wang, L. Zhao, S. Zhao, D. Wang, W. Han, K. Jiang, G. Shen, Bioinspired interlocked structure-induced high deformability for two-dimensional titanium carbide (MXene)/natural microcapsule-based flexible pressure sensors. ACS Nano **13**, 9139–9147 (2019). https://doi.org/10.1021/acsnano.9b03454

90. Y. Lei, A.H. Alshareef, W. Zhao, S. Inal, Laser-scribed graphene electrodes derived from lignin for biochemical sensing. ACS Appl. Nano Mater. **3**, 1166–1174 (2020). https://doi.org/10.1021/acsanm.9b01795
91. X. Han, X. Jing, D. Yang, H. Lin, Z. Wang, H. Ran, P. Li, Y. Chen, Therapeutic mesopore construction on 2D Nb_2C MXenes for targeted and enhanced chemo-photothermal cancer therapy in NIR-II biowindow. Theranostics **8**, 4491–4508 (2018). https://doi.org/10.7150/thno.26291
92. Z.U.D. Babar, M.S. Anwar, M. Mumtaz, M. Iqbal, R.-K. Zheng, D. Akinwande, S. Rizwan, Peculiar magnetic behaviour and Meissner effect in two-dimensional layered Nb_2C MXene. 2D Mater. **7**, 035012 (2020). https://doi.org/10.1088/2053-1583/ab86d2
93. S. Lin, H. Lin, M. Yang, M. Ge, Y. Chen, Y. Zhu, A two-dimensional MXene potentiates a therapeutic microneedle patch for photonic implantable medicine in the second NIR biowindow. Nano-scale **12**, 10265–10276 (2020). https://doi.org/10.1039/D0NR01444C
94. H. Kim, Z. Wang, H.N. Alshareef, MXetronics: electronic and photonic applications of MXenes. Nano Energy **60**, 179–197 (2019). https://doi.org/10.1016/j.nanoen.2019.03.020
95. M. Naguib, V.N. Mochalin, M.W. Barsoum, Y. Gogotsi, 25th anniversary article: MXenes: a new family of two-dimensional materials. Adv. Mater. **26**, 992–1005 (2014). https://doi.org/10.1002/adma.201304138
96. J. Halim, M.R. Lukatskaya, K.M. Cook, J. Lu, C.R. Smith, L.-Å. Näslund, S.J. May, L. Hultman, Y. Gogotsi, P. Eklund, M.W. Barsoum, Transparent conductive two-dimensional titanium carbide epitaxial thin films. Chem. Mater. **26**, 2374–2381 (2014). https://doi.org/10.1021/cm500641a
97. W. Cao, C. Ma, S. Tan, M. Ma, P. Wan, F. Chen, Ultrathin and flexible CNTs/MXene/cellulose nanofibrils composite paper for electromagnetic interference shielding. Nano-Micro Lett. **11**, 72 (2019). https://doi.org/10.1007/s40820-019-0304-y
98. S. Cao, B. Shen, T. Tong, J. Fu, J. Yu, 2D/2D heterojunction of ultrathin MXene/Bi_2WO_6 nanosheets for improved photocatalytic CO_2 reduction. Adv. Funct. Mater. **28**, 1800136 (2018). https://doi.org/10.1002/adfm.201800136
99. R. Tang, S. Zhou, C. Li, R. Chen, L. Zhang, Z. Zhang, L. Yin, Janus-structured Co-Ti_3C_2 MXene quantum dots as a Schottky catalyst for high-performance photoelectrochemical water oxidation. Adv. Funct. Mater. **30**, 2000637 (2020). https://doi.org/10.1002/adfm.202000637
100. X. Jiang, A.V. Kuklin, A. Baev, Y. Ge, H. Ågren, H. Zhang, P.N. Prasad, Two-dimensional MXenes: from morphological to optical, electric, and magnetic properties and applications. Phys. Rep. **848**, 1–58 (2020). https://doi.org/10.1016/j.physrep.2019.12.006
101. D. Shan, J. He, L. Deng, S. Yan, H. Luo, S. Huang, Y. Xu, The underlying mechanisms of enhanced microwave absorption performance for the $NiFe_2O_4$-decorated $Ti_3C_2T_x$ MXene. Results Phys. **15**, 102750 (2019). https://doi.org/10.1016/j.rinp.2019.102750
102. K. Keum, J.W. Kim, S.Y. Hong, J.G. Son, S. Lee, J.S. Ha, Flexible/stretchable supercapacitors with novel functionality for wearable electronics. Adv. Mater. **32**, 2002180 (2020). https://doi.org/10.1002/adma.202002180

Chapter 9
MXenes as a Promising Material for Electromagnetic Interference Shielding

Wei Lu and Hongtao Guo

9.1 Introduction

With advances in knowledge and mechanization, the use of different electrical and electronic devices in hospitals, the military, business, and scientific sectors, all of which produce electromagnetic radiation, has grown enormously. These stray radiations interfere with electronic equipment signals, leading to the production of EMI, which is often hazardous to human health, as well as to precision equipment in hospitals and aerospace [1]. With the increasing complexity of the electromagnetic environment, electromagnetic radiation pollution has become an important environmental pollutant, second only to air pollution, noise pollution, and water pollution, and has become the fourth major public hazard. Therefore, high-performance EMI shielding materials are urgently needed to reduce the harm of electromagnetic waves (EMW) [2].

Electrical conductivity plays an important role in the shielding properties of EMI shielding materials. Conventional metallic materials (e.g., silver, copper, iron, etc.) are frequently employed for EMI shielding. However, their vulnerability to corrosion and high density constrains their broader application in compact electronic devices [3]. As a prominent member of the 2D nanomaterial category, MXene exhibits promising potential across various domains due to its excellent conductivity and expansive surface area. In contrast to conventional 2D materials, MXenes have evolved into a diverse material family, with numerous members successfully synthesized to date [4–6]. One of the earliest discovered MXenes is the material featuring a $Ti_3C_2T_x$ structure. This particular material has gained widespread adoption due to its mature synthetic process and exceptional electrical conductivity. Numerous studies have underscored the unique potential of MXenes in the realm of EMI shielding. Nevertheless, a further comprehensive understanding of MXenes as absorbents and

W. Lu (✉) · H. Guo
School of Materials Science and Engineering, Tongji University, Shanghai 201804, China
e-mail: weilu@tongji.edu.cn

© The Author(s), under exclusive license to Springer Nature Singapore Pte Ltd. 2024

O. P. Pandey and P. Sharma (eds.), *MXenes: Emerging 2D Materials*,
https://doi.org/10.1007/978-981-97-4064-2_9

EMI shielding materials for multifunctional applications is warranted [7–9]. Therefore, exploring its absorbing and shielding mechanisms and its extended applications is urgent.

9.2 EMI Shielding Mechanisms

EMI shielding materials serve to obstruct the transmission of EM waves produced by radiation sources, safeguarding the shielded materials by enhancing the reflection and absorption attenuation of electromagnetic (EM) waves. For internal compact refractive index uniform EMI shielding materials, when the EM wave (P_1) reaches the material surface from the air, part of the energy is reflected back to the air in the form of a reflected wave (P_R), and other parts penetrate the material to form transmitted wave (P_T). The residual energy is attenuated or lost and eventually converted into heat to be absorbed [10]. Generally, the EMI shielding capacity of a material is usually measured by electromagnetic shielding efficiency (EMI SE), whose unit is decibels (dB), is shown in Eq. (9.1):

$$\mathrm{EMI\,SE\,(dB)} = 20\ \mathrm{lg}\frac{E_I}{E_T} = 20\ \mathrm{lg}\frac{H_I}{H_T} = 10\ \mathrm{lg}\frac{P_I}{P_T} \tag{9.1}$$

The electric field (E), magnetic field (H), and power intensity (P) are denoted in the equation, while T and I signify the transmitted and incident electromagnetic (EM) waves. Typically, an EMI shielding effectiveness (SE) exceeding 20 dB is deemed a fundamental criterion for commercial use, indicating a 99% reduction in EM wave transmission [11]. According to Schelkunoff's theory, the materials weaken electromagnetic wave energy mainly through reflection, absorption, and multiple reflection loss to realize electromagnetic shielding, as shown in Eqs. (9.2–9.5):

$$\mathrm{SE_T} = \mathrm{SE_R} + \mathrm{SE_A} + \mathrm{SE_M} \tag{9.2}$$

$$\mathrm{SE_R} = 168.2 + 10\log\left(\frac{\sigma_r}{f\mu_r}\right) \tag{9.3}$$

$$\mathrm{SE_A} = 20\log e \times t\sqrt{\frac{\omega\sigma_r\mu_r}{2}} = 20\log e \times \frac{t}{\delta} \tag{9.4}$$

$$\mathrm{SE_M} = 20\log\left(1 - e^{-\frac{2t}{\delta}}\right) = 20\log\left(1 - 10^{\frac{SE_A}{10}}\right) \tag{9.5}$$

Within these equations, the EM waves' angular frequency is denoted by ω, while σ_r stands for the relative electrical conductivity, f represents the frequency of the EM waves, and t signifies the thickness of the shielding layer. The relative magnetic permeability (compared to free space's magnetic permeability, $4\pi \times 10^{-7}$ H/m) is indicated by μ_r. Additionally, δ, known as the skin depth, characterizes the depth below the shield's surface where the EM field strength diminishes to $1/e$ of its original value, as outlined in Eq. (9.6):

$$\delta = \frac{1}{\sqrt{\pi f \mu \sigma}} \tag{9.6}$$

SE_R, SE_A, and SE_M are the attenuation of EM waves caused by reflection, absorption, and multiple reflection loss, respectively. Reflection refers to the phenomenon that an EM wave changes its propagation direction and returns to the original material when propagating at the interface of two materials with different refractive indices. The energy attenuation caused by reflection is related to the refractive index, electrical conductivity, magnetic permeability, and frequency of EM wave. Materials with high conductivity can achieve strong reflection loss. After absorbing electromagnetic radiation, materials will dissipate in the form of heat energy. Generally speaking, thickness and electrical conductivity determine the ability of materials to absorb EM waves, while dielectric constant and permeability are the main factors affecting absorption loss [12].

To assess the overall effectiveness of EMI shielding, it is essential to mitigate the impact stemming from density and thickness variations. This assessment involves the specific shielding effectiveness (SSE), calculated as the ratio of EMI shielding effectiveness to either density or thickness, and the surface-specific shielding effectiveness (SSE/t or SSE/d), determined by dividing the EMI shielding effectiveness by the material's thickness or density. The equations corresponding to these evaluations are presented as follows (Eqs. 9.7–9.9):

$$\text{SSE} = \text{SE/t (dB/mm, usually used for 2D film)} \tag{9.7}$$

$$\text{SSE} = \text{SE}/\rho \text{ (dB cm}^2\text{/g, usually used for 3D aerogel/foam)} \tag{9.8}$$

$$\text{SSE/t} = \text{SE}/\rho\text{t (dB cm}^2\text{/g)} \tag{9.9}$$

The measurement of EMI shielding performance is based on the calculation of the S-parameters (S_{11}, S_{21}, S_{22}, and S_{12}) ascertained by the vector network analyzer. The scattering parameters (S_{11} and S_{21}) can be recorded to calculate the coefficients of reflectance (R), absorbance (A), and transmittance (T) for the SE_T as shown in the following equations.

The regular layered structure, high free electron density, and abundant active functional groups are the unique advantages of MXene materials compared with other conductive nanomaterials. Firstly, the structured layered structure can better attenuate the EM waves in the form of eddy current loss and ohmic loss. Second, the high density of free electrons endows it with unparalleled electrical conductivity. Finally, the rich active functional groups make MXene exhibit good dispersion in water or organic solvents, which creates convenient conditions for subsequent processing

[13]. The specific attenuation process of EM wave in MXene material is as follows Eqs. (9.10–9.15):

$$R = |S_{11}|^2 \tag{9.10}$$

$$T = |S_{21}|^2 \tag{9.11}$$

$$A = 1 - R - T \tag{9.12}$$

$$SE_R = -10\log(1 - R) \tag{9.13}$$

$$SE_A = -10\log\left(\frac{T}{1 - R}\right) \tag{9.14}$$

$$SE_T = 50 + 10\log\left(\frac{\sigma}{f}\right) + 1.7t\sqrt{\sigma f} \tag{9.15}$$

Due to high electrical conductivity, a large number of free electrons exist on the MXene surface. Therefore, when the EM waves reach the surface of MXene material, some parts are reflected immediately, and the remaining EM waves interact with the highly conductive lattice structure of MXene to generate current, leading to ohmic loss and further dissipation. The residual parts enter the second layer of the MXene material and repeat the attenuation process of the first layer. In addition, in alternating electromagnetic fields, the rich active functional groups on the surface of the MXene lamellae will interact with the Ti atoms, resulting in polarization loss. The lamellar gap of MXene is large, and the EM wave passing through the first layer will also be transmitted between the lamellae of MXene [14, 15]. The above attenuation mechanism will be repeated. Finally, these EM waves are absorbed and dissipated, and almost no EM waves pass through, ultimately exhibiting superior EMI shielding performance.

9.3 Recent Progress on MXene-Based EMI Shielding Composites

The excellent EMI shielding performance of MXene is mainly related to its high conductivity and unique multilayer structure. Yin's group reported excellent EMI shielding performance of Ti_2CT_x [16]. The Ti_2CT_x derived from fluoride salts exhibits a wealth of surface functional groups and possesses high electrical conductivity, measuring approximately 0.3 S cm^{-1}. With an average EMI shielding effectiveness (SE) of approximately 61.1 dB, and a maximum SE reaching around 70 dB at a thickness of 0.8 mm, it surpasses the performance of many carbon-based and metal-based materials reported thus far. The unique 2D morphology and abundance of functional groups on the surface of MXenes facilitate their facile hybridization with other nanomaterials, enabling effective modifications. Attaining exceptional EMI shielding performance proves challenging with homogeneous materials alone. Hence, the design of material structure and composition plays a critical role in meeting the requirements for MXene-based materials in the EMI shielding domain [17, 18]. Drawing from existing literature, this section presents various product forms of materials known for their excellent EMI shielding capabilities, such as hydrogels, films, aerogels, foams, sponges, and others. Additionally, a review of additional functional applications is included within this section.

9.3.1 Pure MXene Material

Owing to its outstanding electrical conductivity, pure MXene materials often exhibit excellent EMI shielding performance. In the work by Liu et al. [19], multilayer $Ti_3C_2T_x$ was produced through the etching of Ti_3AlC_2 with 40% HF at room temperature (RT) for 24 h. The resulting multilayer $Ti_3C_2T_x$/wax composite, containing 60 wt% $Ti_3C_2T_x$, demonstrated exceptional EMI shielding performance of 39.1 dB at a thickness of 2 mm. Meanwhile, Hu et al. [20] synthesized multilayer $Ti_3C_2T_x$ by etching Ti_3AlC_2 with a 40% HF solution at 50 °C for only 0.5 h. Despite this brief processing time, the resulting multilayer $Ti_3C_2T_x$ exhibited impressive EMI shielding capabilities. For instance, the $Ti_3C_2T_x$/wax composite with 70 wt% $Ti_3C_2T_x$ achieved an EMI shielding performance of 34 dB at 18 GHz. Previous research has primarily focused on the EMI shielding performance of multilayer $Ti_3C_2T_x$. He et al. [21] delved into the disparities between $Ti_3C_2T_x$ nanosheets and multilayer $Ti_3C_2T_x$ in terms of EMI shielding performance. Varied etchants led to distinct outcomes (Fig. 9.1a–b) and morphologies. Notably, the $Ti_3C_2T_x$ nanosheet exhibited significantly superior EMI shielding performance compared to multilayer $Ti_3C_2T_x$ due to the establishment of local conductive networks (Fig. 9.1c–d). Furthermore, the $Ti_3C_2T_x$-wax matrix with 80 wt% $Ti_3C_2T_x$ loading showcased an impressive EMI shielding performance of 58.1 dB at a mere 1 mm thickness.

The exploration of EMI shielding characteristics through heat treatment is a topic of extensive research. The team led by Yin et al. [22] delved into the EMI shielding properties of multilayer $Ti_3C_2T_x$ subjected to annealing at 800 °C (designated as $Ti_3C_2T_x$-800) in an Ar environment. The cumulative EMI shielding effectiveness of $Ti_3C_2T_x$-800/wax reached 32 dB. Subsequently, $Ti_3C_2T_x$-200 was created by annealing $Ti_3C_2T_x$ at 200 °C for 2 h in an Ar atmosphere with 5% H_2 [23]. When compared to $Ti_3C_2T_x$/epoxy, $Ti_3C_2T_x$-200/epoxy exhibited improved EMI shielding performance. In a separate study, Ji and colleagues [24] explored the EMI shielding capabilities of multilayer $Ti_3C_2T_x$ following annealing at various temperatures (800, 950, 1100, 1250, 1400, and 1550 °C for 1 h) under Ar conditions. Notably, the multilayer $Ti_3C_2T_x$ treated at 1100 °C demonstrated the most remarkable EMI shielding effectiveness, achieving an outstanding result of 76.1 dB.

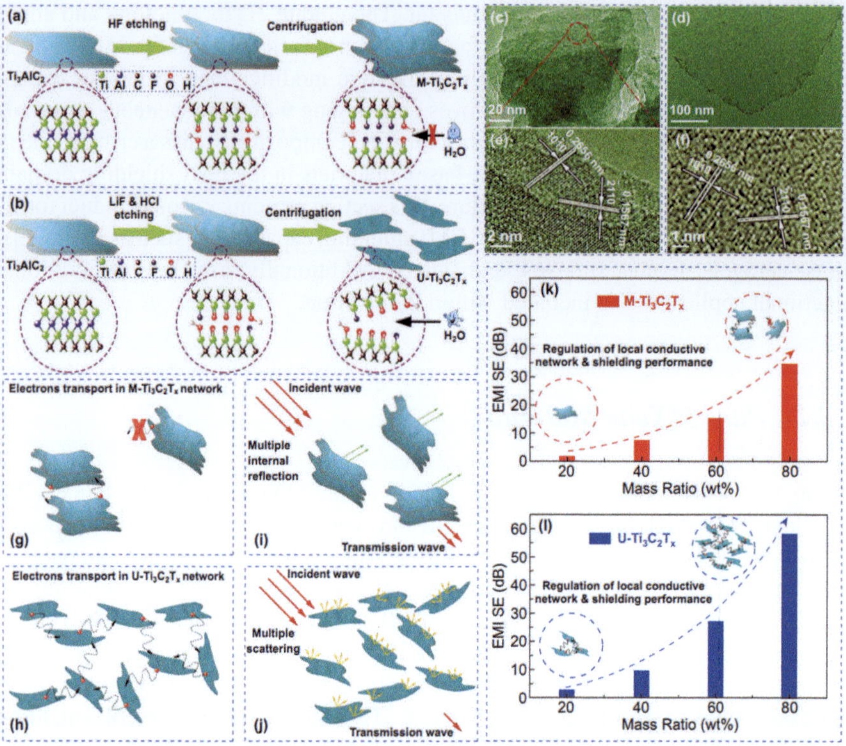

Fig. 9.1 Illustration **a** depicts the synthetic process of multilayer $Ti_3C_2T_x$, while **b** showcases the production of $Ti_3C_2T_x$ nanosheets. The control of the local conductive network and shielding capabilities is demonstrated for **c** multilayer $Ti_3C_2T_x$ and **d** $Ti_3C_2T_x$ nanosheets. Reproduced with permission from Ref. [21]. Copyright 2019, The Royal Society of Chemistry

9.3.2 MXene-Based Hydrogel

Hydrogels are a type of porous engineering material, composed of interconnected hydrophilic building blocks surrounded by water. They possess both a porous structure and a water-enriched interior environment [25]. These characteristics make hydrogels suitable for various applications, including tissue engineering, soft electronics, and electromagnetic interference (EMI) shielding. One advantage of hydrogels is their ability to easily disperse functional nanomaterials, which can form conductive paths. This contributes to the high conductive loss of incident electromagnetic waves (EMWs) [26, 27]. Additionally, the presence of water in hydrogels enables potent polarization loss, as water molecules undergo changes in hydrogen-bond networks and polarization in the gigahertz (GHz) and terahertz (THz) frequency bands.

To construct hydrogels with a stable 3D structure, a network of cross-linked polymer chains is essential. Through dynamic ionic/hydrogen bonds or covalent crosslinking, the dispersed polymer chains in an aqueous solution assemble to form hydrogels with excellent elasticity and flexibility [28].

To enable the integration of conductive fillers into a polymer matrix and produce self-supporting and conductive hydrogels, cross-linked polymer networks play a vital role despite the weak interactions between inorganic nanoparticles [29]. Various methods, including physical crosslinking techniques like crystallization crosslinking and coordination interactions, as well as chemical crosslinking, are employed in the preparation of hydrogel-based EMI shields. Achieving high conductivity is crucial for the development of high-performance EMI shields. However, the non-conductive nature of polymer-based components in hydrogels, such as PVA, polyacrylamide (PAAM or PAM), and cellulose, imposes limitations on their EMI shielding capability. Consequently, the incorporation of conductive nanomaterials into hydrogels becomes imperative to enhance the effectiveness of hydrogel-based EMI shielding materials.

Yu et al. [30] reported the development of MXene-based PVA/PAAM hydrogel by polymerizing acrylamide (AAm) and connecting the polymers through hydrogen bonds (Fig. 9.2). The addition of MXene resulted in increased conductivity of the hydrogel, leading to an increase in EMI shielding effectiveness (SE) from 12 dB to approximately 35 dB with 1.1 wt% MXene. However, a further increase in MXene content to 2.2 wt% caused a decrease in SE to around 30 dB. This non-monotonic change in SE was attributed to the competition between increased electron conduction from MXene and decreased ion conduction. Moreover, higher MXene content led to decreased cell wall thickness, which negatively affected the electrical conductivity of networks and multiple reflections of incident EM waves, resulting in decreased EMI SE. For practical applications, it is important for hydrogel-based EMI shields to have low-temperature tolerance and anti-drying performance. To achieve long-term stability, the researchers performed solvent displacement of the MXene-based hydrogel using a binary solvent of glycerol (Gly) and water. This resulted in the formation of MXene organic hydrogel. After freezing at −25 °C, the EMI SE of the MXene organic hydrogel remained above 30 dB, whereas the SE of the MXene hydrogel decreased to 14 dB. Furthermore, during a 7-day storage period at room temperature, the SE of the MXene hydrogel declined from 30.8 to 1.54 dB, while the SE of the MXene organic hydrogel only decreased to 25.3 dB. However, after 60 min of solvent displacement, the content of Gly in the organic hydrogels increased, leading to a gradual decline in EMI SE below 20 dB due to the weak polarization loss of Gly molecules in the GHz band.

The conductive polymer PEDOT: PSS demonstrates promising potential for EMI shielding applications due to its intrinsic electrical conductivity [31, 32]. Within the PEDOT: PSS molecules, PEDOT contributes to high electrical conductivity, while PSS upholds the polymer's structure and enhances solubility. However, diminishing the quantity of PSS in the polymer boosts the crystallinity of PEDOT and improves conductivity but compromises stability. To further boost the conductivity of PEDOT: PSS, Wang et al. [33] employed an ionic liquid (IL) to modify the polymer and produce conductive hydrogels. Through a dry-annealing process for film formation and subsequent rehydration, they developed conductive polymer hydrogels devoid

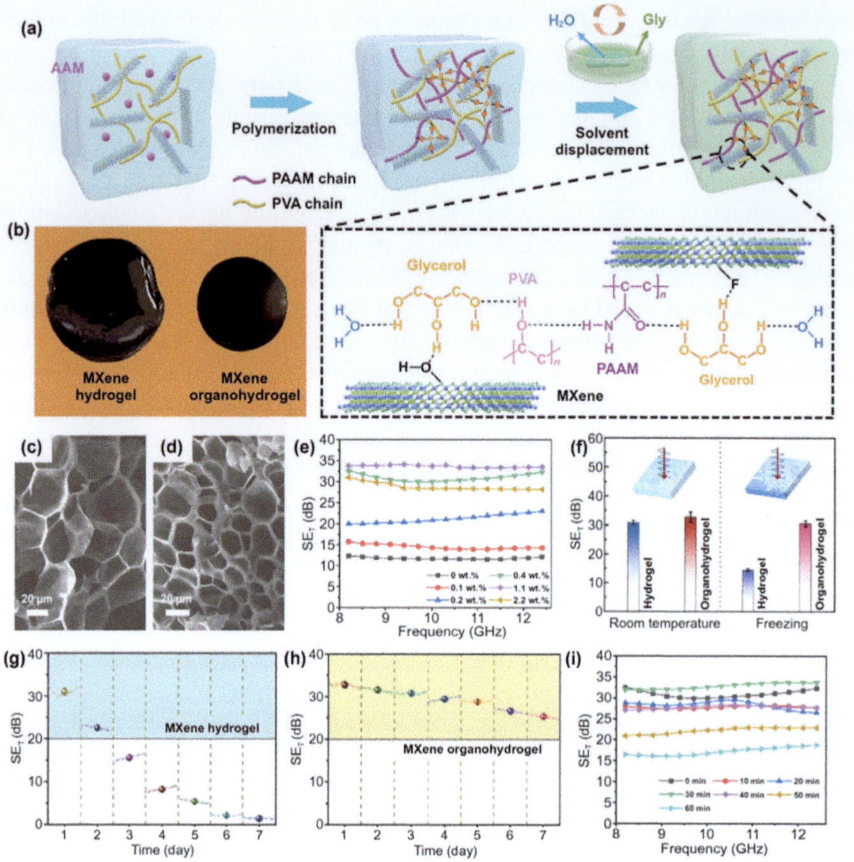

Fig. 9.2 SEM images of porous structures with varying displacing times are presented in (**a**) and (**b**). The EMI shielding effectiveness of MXene-based hydrogels is evaluated in (**c**), while (**d**) illustrates the average shielding effectiveness of hydrogel and organic hydrogel pre- and post-freezing. Additionally, the stability of hydrogels over time is depicted in (**e**) and (**f**), and (**g**) showcases the EMI SE of MXene-based organic hydrogel under water-Gly displacement conditions. Reproduced with permission from Ref. [30], Copyright 2022, Springer

of conductive inorganic additives. The altered PEDOT: PSS hydrogels, doped with 50 wt% ethyl-3-methylimidazolium bis(trifluoromethylsulfonyl)imide (EMIMTFSI) and rehydrated in a Gly/water solution, showcased a conductivity of 30,600 S m^{-1}, significantly surpassing that of standard PEDOT: PSS hydrogels. These modified hydrogels achieved an EMI shielding effectiveness (SE) of 53 dB at a mere thickness of 0.045 mm, with a normalized SE per thickness of 1,182.9 dB mm^{-1} in the X-band. Utilizing additive manufacturing techniques, the customization of hydrogel-based EMI shields was successfully implemented.

In their study, Liu and colleagues [34] introduced Ti_3C_2-MXene-modified ink based on PEDOT: PSS for 3D printing. Employing a freeze–thaw technique, they directly converted the printed structures into resilient and conductive hydrogels with excellent shape fidelity. Through the dispersion of freeze-dried PEDOT: PSS in MXene solution, they developed printable inks by adjusting the solid concentrations of PEDOT: PSS and MXene for optimal printability. The printed structures underwent freezing to establish a framework (depicted in Fig. 9.3a), followed by thawing in an H_2SO_4 solution. This step led to the dissolution of certain PSS components, enhancing the crystallinity of PEDOT and yielding a flexible, stretchable, fatigue-resistant, and high-fidelity Ti_3C_2-MXene modified PEDOT: PSS hydrogel (refer to Fig. 9.3b). These hydrogels showcased notable conductivities ranging from 1525.8 to 5517.2 S m^{-1}, contingent upon the water content. With their enhanced conductivity and varied water content, the hydrogels derived from PEDOT: PSS exhibited an EMI SE of 51.7 dB at a thickness measuring 0.295 mm (as shown in Fig. 9.3c–e).

In a study by Yang et al. [35], a novel hydrogel composed of "trashed" MXene sediment (MS) and biomimetic pores was manufactured using a facile, scalable unidirectional freezing followed by a salting-out approach. These MS-based hydrogels exhibited good EMI shielding performance due to the synergistic effects of the MS-based conductive network, PVA chains, water, and porous structure. The MXene-based hydrogels showed an EMI SE ranging from 31 to 91 dB at varying thicknesses of 1.0 to 7.5 mm, respectively. They also achieved an SE of over 40 dB at a thickness of 2.0 mm in the ultra-broad band GHz band (8.2–40 GHz). The addition of AgNWs further increased the shielding performance of these MS-based hydrogels by enhancing multiple reflection losses caused by the biomimetic-aligned porous structure. This allowed for the achievement of anisotropic EMI shielding

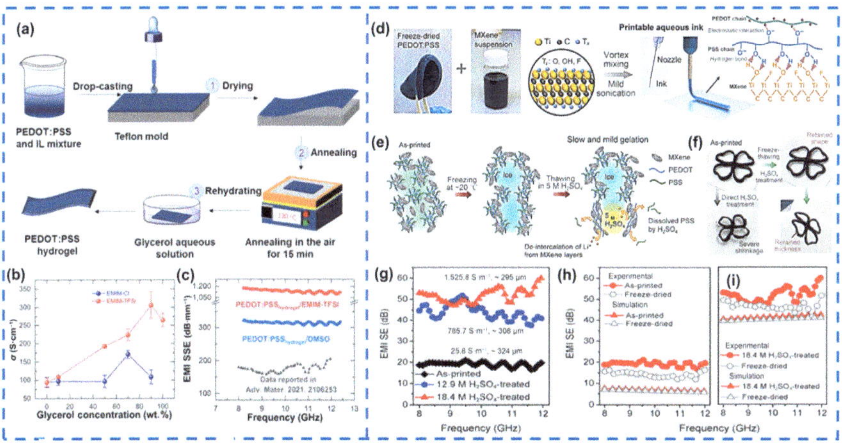

Fig. 9.3 The freeze-thawing treatment mechanism is explored in (**a**), highlighting the exceptional fidelity of the hydrogels in (**b**). Curves demonstrating EMI shielding effectiveness in the X-band are depicted in (**c**), while both simulated and experimental shielding effectiveness of MXene/PEDOT: PSS hydrogels are presented in (**d**) and (**e**). Reproduced with permission from Ref. [36], Copyright 2021, Wiley–VCH GmbH

performance for the hydrogel-based EMI shields, enabling the preparation of hydrogels with controllable EMI shielding performance based on their interior porous structures.

9.3.3 *MXene-Based Film*

To cater to the requirements of intelligent wearable MA materials, it is crucial to develop highly flexible 2D MXene-based films. Additionally, incorporating diverse functionalities into these flexible MXene-based films can effectively address a wider array of practical challenges [37]. There exist two primary approaches for crafting flexible 2D films: (i) utilizing textile-oriented substrates and (ii) arranging nanostructured materials through vacuum filtration. During the production of MXene films, the material tends to self-stack in layers, diminishing the active sites on its surface. This phenomenon impacts its unique characteristics, such as reducing electromagnetic wave attenuation, impeding ion conduction, and constraining the effective incorporation of other functional substances.

9.3.3.1 Pure MXene Film

In a study by Koo and colleagues, a titanium carbide ($Ti_3C_2T_x$) film measuring 45 μm exhibited an EMI SE of 92 dB. Notably, a 2.5 μm thick film demonstrated an EMI SE exceeding 50 dB, marking it as the highest among materials of comparable thicknesses developed to date [38]. The outstanding electrical conductivity of $Ti_3C_2T_x$ films, coupled with multiple internal reflections, contributed significantly to this remarkable performance. Through a comprehensive investigation into the EMI shielding properties of $Ti_3C_2T_x$ MXene-assembled films, they scrutinized the individual contributions of each layer toward the overall shielding effectiveness [39]. Theoretical analysis revealed the growing significance of multiple reflections, surface reflection, and bulk absorption in the shielding mechanism below the skin depth. A film composed of 24 layers, with a thickness of 55 nm, showcased an EMI SE of 20 dB, indicating an exceptionally high absolute shielding effectiveness of 3.89 × 106 dB cm^2 g^{-1}. Furthermore, employing vacuum-assisted filtration, they fabricated Ti_3CNT_x and $Ti_3C_2T_x$ MXene free-standing films with varying thicknesses and evaluated their EMI shielding capabilities at different annealing temperatures [40]. Their observations highlighted the superior EMI SE of the Ti_3CNT_x film compared to the more conductive $Ti_3C_2T_x$ or metal foils of equivalent thicknesses. This enhanced shielding performance of Ti_3CNT_x was achieved through thermal annealing, leading to significantly increased absorption of electromagnetic waves owing to its stratified, metamaterial-esque structure. Han and collaborators conducted a systematic exploration of the shielding properties of 16 distinct MXene films, all demonstrating excellent EMI shielding performance surpassing 20 dB at micrometer-scale thicknesses [41]. Among these films, the $Ti_3C_2T_x$ film exhibited the most impressive

EMI shielding capabilities, with a film measuring approximately 40 nm in thickness showcasing an EMI shielding performance of 21 dB.

9.3.3.2 Porous MXene-Based Film

Utilizing a porous framework provides a viable solution to the self-stacking challenge encountered with MXene sheets. The anticipated superiority of MXene porous film over alternative materials in shielding effectiveness highlights its potential. Moreover, the lightweight nature of the porous film, emphasizing absorption over reflection, presents more desirable attributes when compared to conventional shielding materials. This quality can mitigate secondary contamination resulting from reflected electromagnetic waves (EMW). With the ongoing rapid development of flexible device hardware and portable electronics, there arises a need for ultrathin EMI shielding films that deliver both exceptional flexibility and robust mechanical characteristics.

In a study by Xu et al. [42], a straightforward freeze-drying method was employed to produce f-Ti_2CT_x/PVA foams and films featuring a porous structure. The schematic in Fig. 9.4a delineates the fabrication process for f-Ti_2CT_x/PVA foams and films, with f-Ti_2CT_x nanosheets obtained through etching the Ti_2AlC MAX phase using LiF and HCl. Subsequently, the f-Ti_2CT_x solution was combined with a PVA solution to generate the precursor pure sol, as depicted in Fig. 9.4b. Notably, the f-Ti_2CT_x/PVA foam exhibited low-density characteristics, capable of remaining atop a dandelion without dislodging, boasting a measured density of merely 10.9 mg cm^{-3}. Additionally, f-Ti_2CT_x/PVA-1 showcased an extraordinarily high porosity of 99.3%, positioning it as a standout ultralight aerogel material. The compression process of the f-Ti_2CT_x/PVA composite to produce thin films is illustrated in Fig. 9.4c. The 2D profile of the f-Ti_2CT_x/PVA foam (Fig. 9.4d) revealed an RL_{min} of −18.7 dB and a thickness of 3.9 mm at 8.2 GHz. The average shielding efficiency of f-Ti_2CT_x/PVA $foam^{-1}$ was calculated at 26 dB, with SE_T at 28 dB and SER at 2 dB, as shown in Fig. 9.4e. Upon reaching the surface of the prepared sample, EMWs undergo a combination of reflection, transmission, and absorption processes. Figure 9.4f illustrates the absorption-dominated EMI shielding properties of the MXene/PVA composites. The contribution of absorption or reflection to the shielding properties significantly relies on the specific characteristics of the manufactured product. For MXene/PVA composites with film-like structures and a conductivity of 8.0×10^{-4} S m^{-1}, electromagnetic wave reflection serves as the primary shielding mechanism. In contrast, foamed MXene/PVA composites demonstrate enhanced impedance matching due to their structure, resulting in relatively reduced EM wave reflection. The porous nature of the foam and the layered arrangement of the MXene flakes facilitate internal reflection of the EMW upon entry into the foam, aiding in its dissipation. Dipoles and interface polarization collaborate to maximize the dissipation of the multi-reflected EM wave, leading to absorption-dominated EMI shielding performance.

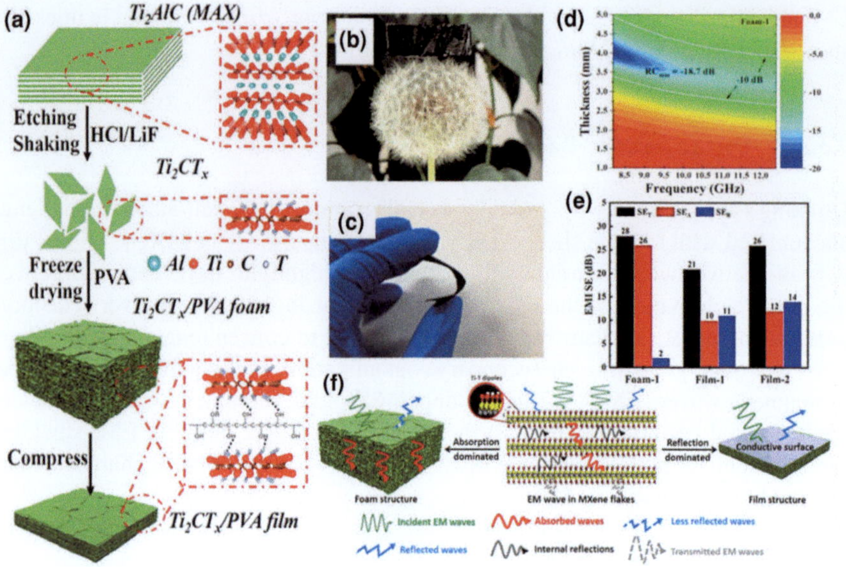

Fig. 9.4 The synthesis of f-Ti_2CT_x/PVA syntactic foams and films is depicted in (**a**). An ultralow-density sheet of f-Ti_2CT_x/PVA foam delicately perched on a dandelion is showcased in (**b**). A flexible f-Ti2CTx/PVA thin film is presented in (**c**). The 2D profile and reflection coefficient curves are provided across different frequencies in (**d**). EMI shielding effectiveness tests, absorption, and reflection of Ti_2CT_x/PVA hybrid materials are detailed in (**e**). The EMI shielding mechanism of MXene/PVA composites with foam or film structure is illustrated schematically in (**f**). Reproduced with permission [42]. Copyright 2019, American Chemical Society

9.3.3.3 Flexible MXene-Based Film

Given the swift progression in flexible hardware and the rising need for ultrathin EMI shielding films that are both highly flexible and mechanically robust, there is a growing demand in the market. To combat EMW pollution concerns, Zhang et al. [43] have devised a sandwich structure film with electrical insulation properties through a vacuum-assisted stepwise filtration method. This film integrates calcium ion-cross-linked sodium alginate montmorillonite and $Ti_3C_2T_x$ MXene. The innovative design not only delivers exceptional EMI shielding effectiveness (50.01 dB), but also serves as a fire retardant, ensuring optimal safeguarding of electronic devices during potential fire incidents. In comparison to alternative $Ti_3C_2T_x$ layers, the interlayer film upholds remarkable EMI shielding capabilities, impressive mechanical strength (84.4 MPa), and superior fire-resistant features. Notably, when evaluated against hybrid films, the interlayer film exhibits a more pronounced EMI shielding impact and performs admirably in extended thermal aging assessments at 80 °C.

Although $Ti_3C_2T_x$ MXene-based materials for EMI shielding display promising potential in mitigating the increasing threat of EM radiation, their utility is restricted by their singular loss mechanism [26, 44]. Liang et al. [45] have tackled this constraint through the fabrication of brick-like NiCo/MX-CNT films by integrating $Ti_3C_2T_x$ MXenes and conductive CNTs. These films demonstrate elevated electrical conductivity, impressive EMA attenuation capability, and exhibit magnetic and elastic characteristics. Achieving an EMI SE of 99.99999991% (90.7 dB), the brick-like NiCo/MX-CNT films stand out as one of the most effective composite films. Even at a mere thickness of 53 mm, the NiCo/MX-CNT composite film excels in both reflecting and absorbing EMW, surpassing the EMI SE of pure CNT films (71 dB) and pure MXene films (61 dB). By varying the film thickness within the range of 9 to 116 mm, the composite film allows for the modulation of EMI shielding performance from 46 to 105 dB. Furthermore, the dense brick-like-layered structure of the film offers exceptional flexibility, foldability, and stable mechanical properties, significantly broadening its potential for practical applications in complex environments.

In response to the growing demand for portable and wearable electronic devices, researchers are compelled to innovate high-performance, flexible EMI shielding materials to combat the escalating issue of EMW pollution [46]. Building upon this premise, these materials exhibit remarkable EMI shielding efficacy and exceptional mechanical pliability. Luo et al. [47] employed a convenient vacuum-assisted filtration approach to develop a cross-linked MXene network within a natural rubber (NR) matrix, resulting in the production of flexible and highly conductive $Ti_3C_2T_x$ MXenes NR nanocomposite films (Fig. 9.5a). The electrostatic repulsion arising from the negative charges of MXene and NR latex facilitates the uniform dispersion of MXene flakes at the NR particle interface, establishing an interconnected network that enhances electron transport efficiency at lower MXene concentrations. With a MXene content of 6.71 vol%, the nanocomposite achieves a conductivity of 1400 S m^{-1} and an EMI shielding effectiveness of -53.6 dB (Fig. 9.5b).

The NR matrix experiences a significant enhancement from the 3D MXene network, leading to a remarkable increase in tensile strength and modulus by 700% and 150% respectively, compared to pure NR. Furthermore, the MXene/NR nanocomposite films demonstrate consistent EMI shielding capabilities and tensile properties during cyclic deformation, positioning them as crucial components for the upcoming generation of flexible and foldable electronic devices. In order to elucidate the EMI shielding mechanism, Fig. 9.5c provides a schematic representation of the potential EMW attenuation mechanism. Previous studies have established that polymer nanocomposites possess extensive surfaces and interfaces with multiple reflections, resulting in more intricate EMI shielding mechanisms compared to homogeneous conductive materials. Within MXene/NR nanocomposites, the porous cross-linked network of MXenes facilitates the penetration of incident EMW through the internal porous structure rather than through direct reflection. Through multiple scattering and intra-interface polarization, the incident wave is effectively attenuated and dissipated across the surface and interface of the porous structure, leading to a notably improved EMA performance. The abundant end groups and numerous charge carriers present on the MXene sheet are capable of interacting with the incident EMW, thereby dissipating it as heat.

Transparent conductive films possessing exceptional EMI shielding capabilities exhibit significant potential within the realm of optoelectronic devices. To fulfill

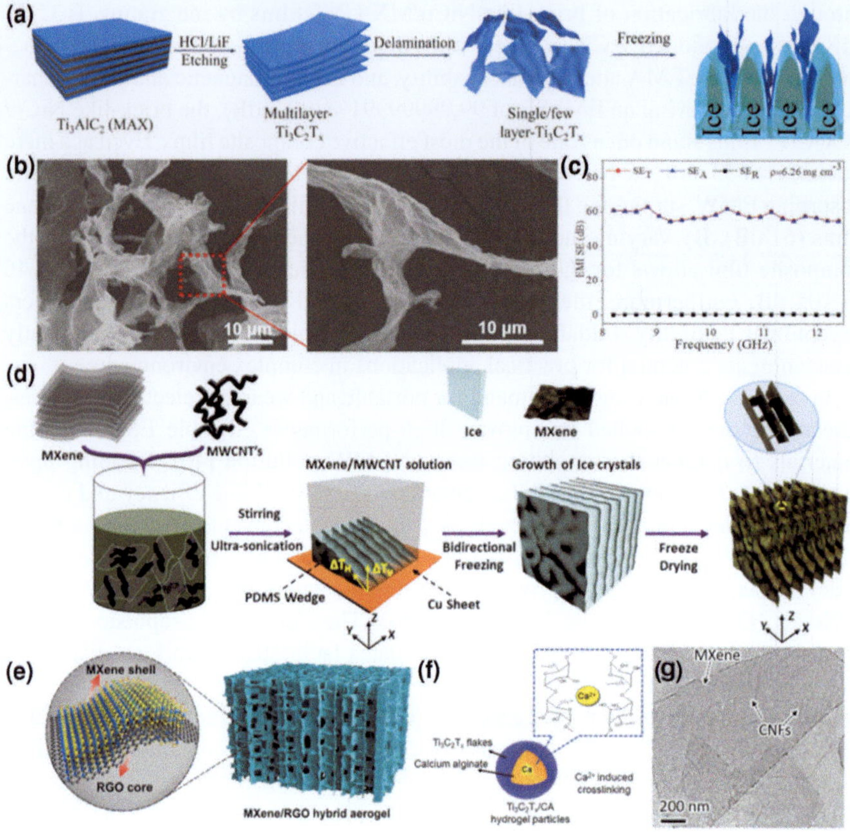

Fig. 9.5 The synthesis schematic of the $Ti_3C_2T_x$/NR film is presented in (**a**). Investigating MXene/rubber composites, the EMI SE under varying MXene loadings is examined in (**b**). A diagram illustrating the potential of nanocomposite films in enhancing EMI shielding properties is depicted in (**c**). Reproduced with permission [47]. Copyright 2019, Elsevier

the stringent requirements of being robust, stable, flexible, and adaptable to harsh conditions, the development of such films remains a challenging endeavor. In their research, Zhou et al. [48] introduced a flexible transparent conductive film comprising a hybrid structure of 2D MXene nanosheets and 1D AgNWs, showcasing strong interfacial adhesion and superior EMI shielding performance alongside an impressive electro-photo-thermal response. The MXene/AgNWx-PVA film demonstrated a minimal electrical resistance of 18.3 Ω sq^{-1} and an optical transmittance of 52.3%, resulting in an outstanding EMI shielding efficiency of 32 dB within the X-band at a thickness of 10 μm. These distinctive characteristics of the multifunctional films underscore their immense potential as high-performance transparent EMI shielding materials tailored for cutting-edge optoelectronic devices operating in harsh environments. In a separate study, Jin et al. [49] engineered transparent, conductive, and flexible MXene grid/silver nanowire hierarchical films, boasting both exceptional optical transmittance and remarkable EMI shielding effectiveness. The MXene grid delivered an average EMI shielding performance of 21.7 dB, a rarity among MXenes and other graphene-based transparent conductive films renowned for their high optical transmittance. Through the incorporation of 1D AgNWs, the MXene grid/AgNW hierarchical structures were meticulously designed and optimized to showcase elevated optical transmittance, enhanced electrical conductivity, superior EMI shielding efficiency, and heightened mechanical durability.

9.3.4 MXene-Based Aerogels, Foams, and Sponges

Due to its diverse surface functional groups such as –OH, –O, and –F, MXene demonstrates exceptional surface activity alongside its notable attributes of maintaining high electrical conductivity, a considerable specific surface area, lightweight nature, and ease of processing. These characteristics enable MXene to seamlessly integrate with a wide range of materials, showcasing great promise as a primary candidate for porous and lightweight EMI shielding structures like aerogels, foams, and sponges within the realm of electromagnetic interference protection [38].

9.3.4.1 MXene-Based Aerogel

Initially, diverse types of pure MXene-based aerogels have been synthesized by researchers to assess their effectiveness in shielding against electromagnetic interference. Han et al. [50] applied a dual-directional freeze-casting method to produce three kinds of porous MXene aerogels ($Ti_3C_2T_x$, Ti_2CT_x, and Ti_3CNT_x). The EMI shielding performance of these aerogels achieved 70.5 dB, 69.2 dB, and 54.1 dB, respectively, at a thickness of 1 mm. Particularly, the Ti_2CT_x aerogel, with a density measuring 5.5 mg cm^{-3} and a 1 mm thickness, demonstrated an exceptionally high shielding effectiveness per unit weight (SEE/t) of 8818.2 dB cm^2 g^{-1}, surpassing that of alternative materials by several magnitudes. Bian et al. [51] pioneered the development of an ultralight MXene aerogel through freeze-drying, attaining a minimal density of 6.26 mg cm^{-3} devoid of external reinforcement (Fig. 9.6a). Throughout the freeze-drying process, ice crystals growth repelled the MXene flakes, confining them at the interfaces to establish interconnected thin MXene nanosheets forming continuous walls within the cavities (Fig. 9.6b). This aerogel showcased formidable EMI shielding capabilities of approximately 60 dB and remarkably low EM wave reflection (<1 dB) (Fig. 9.6c), primarily due to the porous conductive network's efficient electromagnetic wave absorption. In a separate investigation, carbon nanotubes (CNTs) were incorporated into the dual-directional freeze-dried $Ti_3C_2T_x$ aerogel, yielding hybrid aerogels with enhanced electrical conductivity (9.43 S cm^{-1}), EMI shielding effectiveness (103.9 dB) at a 3 mm thickness, and a more than ninefold surge in compressive modulus compared to the pristine MXene aerogel. This enhancement stemmed from the synergistic interplay between MXene and CNTs, as well as the stable lamellar porous structure they collectively constructed.

To enhance the dispersion and connection of MXene nanosheets, researchers have incorporated various additives such as cellulose nanofibers (CNFs), graphene oxide (GO), calcium alginate (CA), and sodium alginate (SA). These additives contain oxygen-containing functional groups that form strong hydrogen bonds with MXene nanosheets, constructing a robust 3D network. This network can withstand external pressure without affecting the characteristics of the composite aerogel and sponge, resulting in lightweight MXene shielding materials [52]. For example, a CNF/ Ti_3C_2Tx aerogel was fabricated using an ice-templated freeze-casting approach. The integration of CNF "mortars" with Ti_3C_2Tx MXene "bricks" in a nacre-like

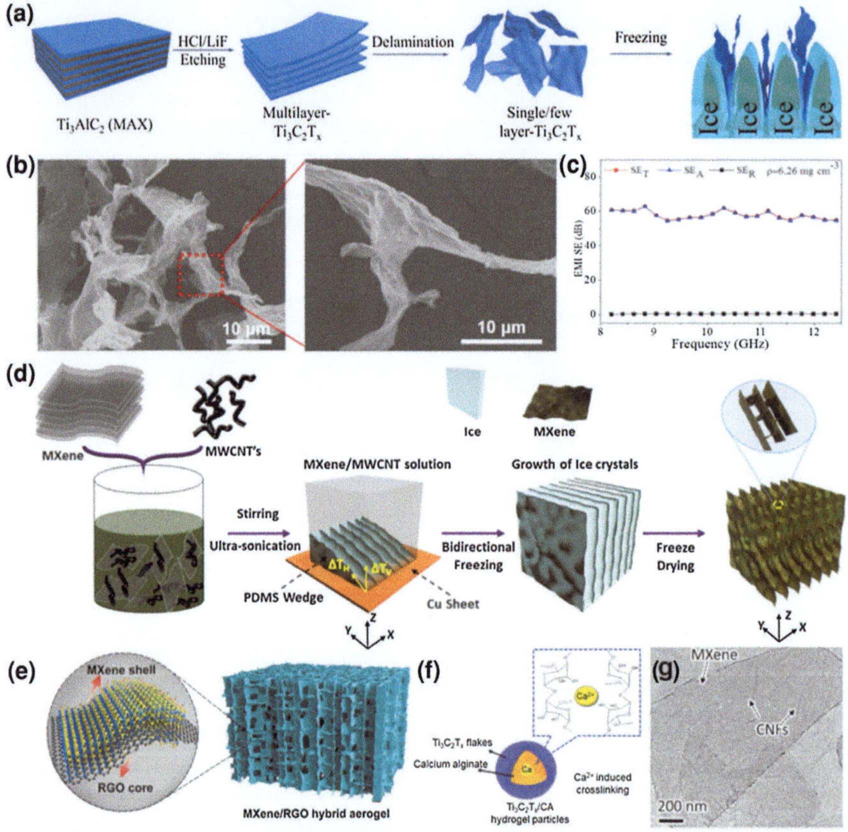

Fig. 9.6 The scheme for fabricating the ultralight MXene aerogel is presented in (**a**). SEM images capturing the MXene aerogel at various magnifications are displayed in (**b**). The EMI shielding effectiveness of the MXene aerogel, including SE_T, SE_R, and SE_A, is evaluated in (**c**). Reprinted with permission from Ref. [51]. Copyright 2019, The Royal Society of Chemistry. Co. KGaA, Weinheim

cell wall architecture led to high electrical conductivity and excellent EMI shielding performance, facilitated by interfacial polarization.

9.3.4.2 MXene-Based Foam/Sponge

Yu's team [53] initially explored the EMI shielding capabilities of pure MXene foam. They employed an efficient and straightforward approach to produce standalone, pliable, and hydrophobic $Ti_3C_2T_x$ MXene foam. In stark contrast to well-known hydrophilic MXene materials, the $Ti_3C_2T_x$ foams exhibited unexpectedly hydrophobic surfaces, demonstrating exceptional water resistance and resilience. As a result of the remarkably effective wave attenuation within the advantageous porous structure, the lightweight $Ti_3C_2T_x$ foam displayed an improved EMI shielding performance of 70 dB when compared to its unfoamed film counterpart (53 dB).

To achieve thinness, broad absorption bandwidth, and lightweight nature in EMA materials, researchers have endeavored to overcome numerous obstacles [54, 55]. Li et al. [56] integrated the wet chemical method, self-assembly technique, and sacrificial template method to fabricate a distinctive reduced graphene oxide (RGO)/$Ti_3C_2T_x$ syntactic foam and manipulate EMA properties. Briefly, as depicted in Fig. 9.7a, polymethyl methacrylate (PMMA) spheres and $Ti_3C_2T_x$ flakes were dispersed in water, blended, and centrifuged to form $Ti_3C_2T_x$/PMMA spheres through self-assembly facilitated by hydrogen bonding and van der Waals forces. GO/$Ti_3C_2T_x$/PMMA spheres were synthesized using the same process. Scanning electron microscopy (SEM) images and corresponding schematic diagrams are presented in Fig. 9.7a1–a2, where the RGO nanoflakes serve as connectors for the stability of the foam structure. Relative to all documented foam-based counterparts, the RGO/$Ti_3C_2T_x$ foam demonstrates superior EMA performance, with EAB covering nearly the entire X-band. The EMA mechanism of RGO/$Ti_3C_2T_x$ syntactic foam is illustrated in Fig. 9.7a3. Primarily leveraging the increased EMW scattering of the microporous core–shell structure, the heightened polarization loss induced by abundant heterointerfaces, and the conduction loss resulting from various imperfections, all these factors collectively play a crucial role in enhancing EMA performance.

Not only is the structural design applicable to high-performance EMA materials, but it also extends to high-performance EMI shielding materials. Ma et al. [57] effectively engineered a lightweight, self-repairing, and adjustable EMI shielding sponge using shielding capsules through a brief dip-coating process. In summary, the melamine sponge's uniform pores are enveloped by the formed MXene film, establishing a connection between the porous framework and the shielding capsule structure (see Fig. 9.7b–h). This distinct structural arrangement leads to significantly improved interaction between EMW and the MXene film, resulting in the composite sponge achieving an EMI shielding effect of 90.49 dB in the X-band. Additionally, the introduction of a PU interlayer enhances the self-repairing capability of the hybrid sponge. As a result, even after repetitive cutting and repair, the EMI SE value of this hybrid sponge can still reach 72.89 dB, demonstrating the superior potential application of the MXene/PU@MS composite sponge in wearable devices.

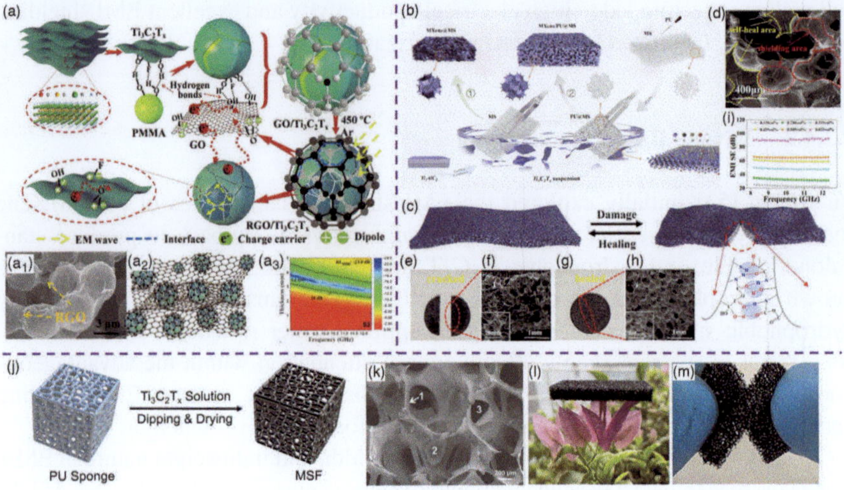

Fig. 9.7 The preparation process of MXene-supported MS (1) and MXene/PU@MS (2) is outlined in the schematic diagram in (**a**). An illustrative depiction of the dynamic fracture and self-healing mechanism of MXene/PU@MS driven by interfacial hydrogen bonding is presented in (**b**). A cross-sectional SEM image of MXene/PU@MS (16 mg mL^{-1} MXene suspension) is displayed in (**c**). Images showing the damaged and healed MXene/PU@MS samples, along with corresponding SEM images, are provided for visual reference. Reprinted with permission from Ref. [57]. Copyright 2021, Elsevier

Interestingly, the sponge foam material can also integrate with terahertz (THz) absorption technology, applicable in EMI shielding, radar stealth, and the upcoming 6G communication. Addressing the challenges of complex manufacturing processes and the limited bandwidth of THz-absorbing materials, Shui et al. [58] proposed a method to produce $Ti_3C_2T_x$ MXene sponge foam (MSF) using a dip-coating technique, resulting in a lightweight, wideband, and hydrophobic THz absorber. The preparation process of MSF is illustrated in Fig. 9.7j. The PU sponge, featuring pores larger than 300 μm, was immersed in the $Ti_3C_2T_x$ colloidal solution. Subsequent extrusion for 5 min led to rapid coverage of the sponge skeleton by $Ti_3C_2T_x$. Upon drying, the $Ti_3C_2T_x$ flakes permeated the entire sponge, as depicted in Fig. 9.7k, showcasing three distinct filling states: filling, adhering to the framework, and forming a thin film on the pores. Notably, the THz absorption properties varied across these states. Furthermore, the lightweight and flexibility of MSF are evident in Fig. 9.7l, m. Thus, the innovative approach of combining large-aperture porous structures with 2D nanosheets lays the foundation for meeting the requirements of high-performance multifunctional terahertz absorbers. Wang et al. [59] synthesized a porous 3D $Ti_3C_2T_x$ MXene/C hybrid foam (MCF) through a sol–gel process followed by a thermal reaction. The MCF/epoxy composite, achieved via vacuum-assisted impregnation and subsequent curing, exhibited an EMI SE of 46 dB and electrical conductivity of 184 S m^{-1} with 4.25 wt% MCF, surpassing that of MCF-0/epoxy nanocomposites (lacking $Ti_3C_2T_x$ MXene). Additionally, a novel ultralight carbon foam modified by $Ti_3C_2T_x$ nanosheets ($CF/Ti_3C_2T_x$), featuring a three-dimensional network structure, was prepared through vacuum impregnation and freeze-drying [60]. The $CF/Ti_3C_2T_x$ foam, with an ultralow density of only 5–7 mg cm^{-3}, exhibited exceptional flexibility and stable compression-resilience properties. Moreover, the foam demonstrated superior EM absorption compared to most foam-based EM absorbers, achieving a minimum RL of −45 dB at 8.8 GHz.

The addition of alternative conductive additives, like graphene and silver nanowires, can boost the electrical conductivity and EMI shielding capabilities of the composite foam. Through freeze-drying and heat treatment for reduction, a hybrid foam of lightweight $Ti_3C_2T_x$ MXene/graphene ($Ti_3C_2T_x$–GO) was produced [61]. Leveraging enhanced electrical conductivity within the foam and effective wave attenuation in interconnected porous structures, the $Ti_3C_2T_x$–GO hybrid foam exhibited remarkable EMI shielding performance, reaching 50.7 dB and a specific EMI shielding effectiveness of 6217 dB cm^3 g^{-1}, surpassing that of many existing EMI shielding materials. The fabrication of Silver nanowires (AgNWs)/$Ti_3C_2T_x$ foam involved integrating AgNWs as the framework and $Ti_3C_2T_x$ as the surficial adornment for the foam structure [62]. Notably, the AgNWs/$Ti_3C_2T_x$ foam demonstrated an EMI shielding effectiveness of 41.3 dB at a thickness of 1.2 μm in the X-band. The inherent free space generated during foaming facilitated effective EM wave scattering within the skin depth.

9.4 Multifunctional MXene-Based EMI Shielding Materials

The rapid development of 5G electronic communication technology has created a pressing need for the advancement and utilization of portable and wearable electronic devices [63, 64]. Key requirements for these devices include flexibility, lightweight design, and safety. In order to construct effective electromagnetic devices, it is crucial to enhance their electromagnetic performance and anti-interference capabilities [65]. Among various options, flexible textiles have emerged as the ideal choice for portable wearable electronic devices. The utilization of advanced manufacturing techniques can maximize the potential applications of flexible textiles, particularly in the fields of EMI shielding, energy storage, human thermal management, and human health monitoring [66, 67]. To address these needs, Liu et al. [68] employed a vacuum-assisted layer-by-layer assembly strategy to develop multifunctional flexible conductive textiles with exceptional EMI shielding, high susceptibility to humidity response, and super-hydrophobicity. The structure of these textiles consisted of leaf-like nanostructures composed of silver nanowires (AgNWs) and MXene nanoflakes, which provided highly conductive frameworks. The presence of MXene prevented the oxidation of AgNWs and enhanced the binding strength between AgNWs and the fabric matrix, resulting in the self-derived hydrophobicity of the flexible textiles. The conductivity of the original silk was significantly increased by the highly conductive MXene, reducing the interfacial resistance. Comparing the EMI shielding properties of silk textiles with both MXene and AgNWs to those with only MXene or

AgNWs, it becomes evident that both components are essential for achieving optimal shielding properties. The textiles demonstrated an EMI shielding effectiveness of 90 dB at 12.4 GHz with a thickness of 480 μm. Moreover, the integration of MXene also endowed the materials with humidity response and self-derived hydrophobicity, ensuring stable application in portable wearable electronic devices. In another study, Zhai et al. [69] used a combination of spraying technology and electrospinning to prepare a multifunctional material called TPU/PDA/AgNP/MXene (TAMF). This material exhibited multiple capabilities, including electric heating, EMI shielding, and biodetection. The optimized TAMF material demonstrated ultrahigh conductivity, remarkable EMI shielding effectiveness of 108.8 dB, excellent Joule heating capability (reaching 80 °C at 1 V), and highly sensitive resistive response (with an ultralow detection limit of 0.1%, ultrahigh gauge factor of 7853, and large detection range of 200%) (Fig. 9.8a–d). The TAMF material holds great promise in the fields of portable wearables and high-performance electromagnetic wave protection, particularly in applications such as protecting pregnant women. Overall, the development of these innovative materials, such as flexible conductive textiles and TAMF, showcases their potential for advancing portable wearable electronic devices and providing effective EMW protection (Fig. 9.8e–k).

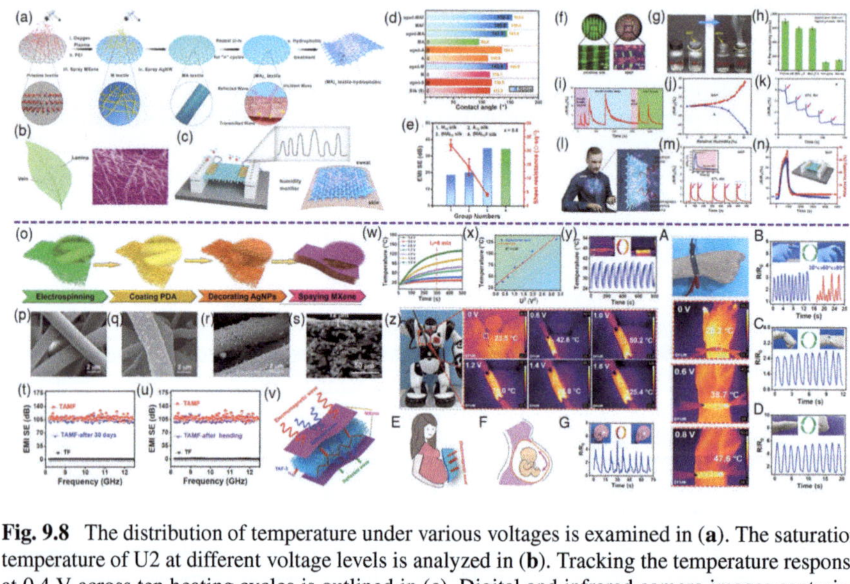

Fig. 9.8 The distribution of temperature under various voltages is examined in (**a**). The saturation temperature of U2 at different voltage levels is analyzed in (**b**). Tracking the temperature response at 0.4 V across ten heating cycles is outlined in (**c**). Digital and infrared camera images capturing electric heaters at voltages ranging from 0 to 1.8 V are presented in (**d**). The temperature of TAMF heater on the skin at 0, 0.6, and 0.8 V, along with its application in micro- and large human motion, is discussed in (**e**). Various movements including fingers flexed, wrist flexed, and elbow flexed are illustrated in (**f–h**). The concept of fetal electromagnetic protection is introduced in (**i**), supported by a schematic diagram of fetal movement in (**j**) and simulated fetal movement using a balloon in (**k**). Reprinted with permission from Ref. [69]. Copyright 2021, The Royal Society of Chemistry

The utilization of MXene conductive hydrogels is becoming increasingly prevalent in the realm of flexible smart sensor devices, attributed to their structured alignment and anisotropic properties, rendering them suitable for medical monitoring and compatibility with human–machine interfaces [70]. In their work, Feng and colleagues [71] innovatively crafted an anisotropic MXene conductive hydrogel through a directional freezing technique, addressing the shortcomings of lacking orientation and limited temperature sensitivity observed in conventional homogeneous conductive hydrogels and flexible sensors. By amalgamating MXene nanosheets with a biocompatible polymer and $ZnSO_4$ precursor, they synthesized a well-organized and adaptable intelligent hydrogel material termed PMZn. To broaden the scope of applications and enhance sensing capabilities, the PMZn material underwent a solvent replacement process, yielding PMZn-GL hydrogels (Fig. 9.9a–f). These PMZn-GL conductive hydrogels showcased versatility and were employed as pliable, wearable sensors for monitoring various human movements including hand gestures and facial expressions. Integration with a mobile phone revealed distinct images corresponding to different movements, indicating successful signal conversion (Fig. 9.9g–h). By integrating multiple flexible sensors into an array, stress variations and spatial distribution could be effectively monitored. The emergence of MXene-based smart hydrogels driven by muscle activity opens up exciting possibilities for exploring biofunctional hydrogels in the domain of flexible electronic devices.

Conventional sensors relying on physical contact suffer from issues such as generating noise, experiencing wear, compromising the reliability of triboelectric nanogenerators (TENGs), and escalating costs [72, 73]. In response to these challenges, non-contact TENGs have emerged as a viable alternative. Sohel Rana and collaborators [72, 74] introduced a non-contact dual-layer TENG (CDL-TENG) and an extremely flexible self-sustaining sensor incorporating MXene/Ecoflex and cobalt nanoporous carbon (Co-NPC)/Ecoflex nanocomposites. Each element fulfilled a unique function, collectively influencing the humidity and acceleration sensitivity of CDL-TENG through the regulation of output voltage, current density, and charge density. Within CDL-TENG, the MXene/Ecoflex composite assumed the role of the dielectric layer for charge capture, while the porous structure of Co-NPC offered a substantial surface area for charge retention. This novel configuration tackled the challenge of inadequate charge trapping in single-layer non-contact TENGs, resulting in enhanced output performance. Analysis of surface potential revealed an eightfold increase in CDL-TENG's surface potential, with the fabricated device exhibiting more than four times the output power of TENGs based solely on Ecoflex. Moreover, effective non-contact monitoring within a 20 cm radius was demonstrated. Application of CDL-TENG in vehicles and robots facilitated real-time obstacle detection and alarm activation. Additionally, CDL-TENG found utility in non-contact door locking systems. In conclusion, this material exhibits significant potential across various domains including manufacturing, self-powered sensors, wearable technology, artificial intelligence, and beyond.

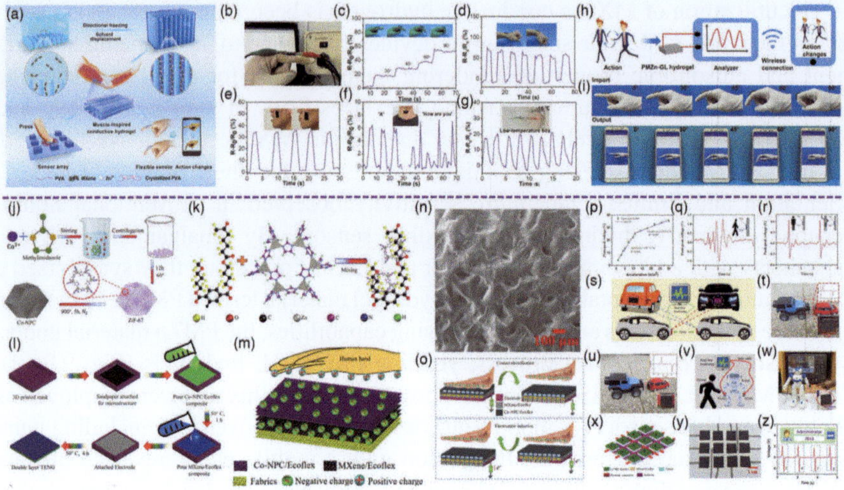

Fig. 9.9 The conductive hydrogel and the electrochemical workstation are integrated to form the sensing device in (**a**). Variations in resistance of the flexible sensor are linked to different bending angles (0, 30, 45, 60, 90) as detailed in (**b**). Actions involving bending the wrist downwards and upwards are observed in (**c**), while movements like opening and closing of the mouth are recorded in (**d**). Communication phrases "A" and "How are you" are repeated thrice by the individual in (**e**). Performing rapid finger bends approximately 30 times and repeating this process nine times within a cryostat setting is outlined in (**f**). Human activities are monitored using wearable flexible sensors, with data transmitted wirelessly to mobile devices in (**g**). Diverse finger-bending actions are captured and wirelessly relayed to mobile phones for display in (**h**). Reprinted with permission from Ref. [71]. Copyright 2021, Wiley–VCH

9.5 Conclusion and Outlook

Due to their inherent characteristics, such as exceptional metallic conductivity, 2D sheet structure, lightweight nature, adjustable surface chemistry, and effortless solution processability, MXenes display considerable promise for EMI shielding purposes. Among the various synthesized MXenes, $Ti_3C_2T_x$ MXene films ranging in thickness from nanometers to submicron demonstrate superior shielding effectiveness owing to their remarkably high intrinsic electrical conductivities. Furthermore, by incorporating extrinsic alterations such as meta-structure formation, Ti_3CNT_x MXene films exhibit cutting-edge shielding performance, surpassing the efficacy of highly conductive $Ti_3C_2T_x$ MXene and metal foils of similar thickness. Moreover, the advancement of tailored chemical composition magnetic MXenes can broaden the scope of MXene applications. The introduction of magnetic MXenes introduces an additional magnetic loss mechanism, thereby enhancing EMW absorption. The exceptional potential of 2D MXenes for EMI shielding underscores the need for the scientific community to overcome these challenges to facilitate their practical implementation in this field.

The current hydrogel-based EMI shields do not take into account the potential benefits of controllable micrometer-sized pore morphology, such as aligned channels and anisotropic pore structure, which have been demonstrated to effectively enhance EMI shielding performance in porous materials. Moreover, the utilization of hydrogels for multifunctional purposes beyond EMI shielding remains limited, failing to fully exploit their superior mechanical properties and biocompatibility. Therefore, it is essential to prioritize the exploration and comprehension of the interrelationships among controllable internal porous structures, compositions, and EMI shielding efficacy in hydrogel-based EMI shields. By integrating controllable pore structures into hydrogel-based shielding materials, it is anticipated that they will exhibit greater potential as high-performance EMI shielding materials. Concurrently, emphasis should be placed on exploring additional functionalities such as photothermal therapy, antimicrobial coatings, and adhesion to broaden the application scope of hydrogel-based EMI shields.

In the future, porous aerogels and sponges with distinctive microstructures and excellent intrinsic properties for scattering and capturing electromagnetic waves are poised to become key contenders in the realm of EMI shielding. Typically, the molded MXene aerogel and sponge frameworks are predominantly infused with polymers (e.g., epoxy, PDMS) to create nanocomposites, offering three distinct advantages: the incorporation of a non-conductive polymer can (i) enhance the impedance match, facilitating greater EM wave penetration into the shield and augmenting absorption loss; (ii) bolster the durability and mechanical robustness of the material during practical usage; (iii) reduce the quantity of MXene fillers utilized.

For researchers, it is essential to enhance their comprehension of the fundamental physical and chemical characteristics of MXene and employ various physical or chemical approaches to leverage these attributes effectively. Particularly, investigations into the impact of surface modifications on MXene's electrical conductivity and electromagnetic properties are crucial, alongside the development of diverse composite strategies to enhance both effective absorption bandwidth and electromagnetic wave performance. Notably, while MXene's electromagnetic parameters are typically assessed using vector network analyzers in laboratory settings, bridging the gap between fundamental research and practical implementation remains a significant challenge. Addressing the scalability of the manufacturing process and evaluating potential fluctuations in electromagnetic performance post-mass production necessitate further meticulous scrutiny. Specifically, in the electromagnetic attenuation (EMA) domain, MXene-based products must exhibit exceptional dielectric loss characteristics and impedance-matching capabilities. Introducing varied dimensions and structural materials holds promise for enhancing microwave attenuation in MXene-based products. Typically, MXene films are the primary form in which MXene-based materials are crafted for evaluating electromagnetic interference (EMI) shielding effectiveness, with EMI performance intricately linked to electromagnetic absorption (EMA) capabilities. Nevertheless, the heightened electrical conductivity of MXene materials may lead to increased surface reflection of incident electromagnetic waves, thereby diminishing the absorption rate. Consequently, materials featuring high conductivity derived from MXene are better suited for shielding applications rather than absorption purposes.

References

1. H. Lv, Z. Yang, H. Pan, R. Wu, Electromagnetic absorption materials: Current progress and new frontiers. Progr. Mater. Sci. **127**, 100946 (2022)
2. J.C. Shu, W.Q. Cao, M.S. Cao, Diverse metal–organic framework architectures for electromagnetic absorbers and shielding. Adv. Funct. Mater. **31**(23), 2100470 (2021)
3. J. Chang, H. Zhai, Z. Hu, J. Li, Ultra-thin metal composites for electromagnetic interference shielding. Compos. Part B: Eng. **246**, 110269 (2022)
4. B. Zhou, Z. Li, Y. Li, X. Liu, J. Ma, Y. Feng, D. Zhang, C. He, C. Liu, C. Shen, Flexible hydrophobic 2D $Ti_3C_2T_X$-based transparent conductive film with multifunctional self-cleaning, electromagnetic interference shielding and joule heating capacities. Compos. Sci. Technol. **201**, 108531(2021)
5. Y. Gogotsi, Q. Huang, MXenes: two-dimensional building blocks for future materials and devices. ACS Nano **15**(4), 5775–5780 (2021)
6. Y. Hou, Z. Sheng, C. Fu, J. Kong, X. Zhang, Hygroscopic holey graphene aerogel fibers enable highly efficient moisture capture, heat allocation and microwave absorption. Nat. Commun. **13**(1), 1227 (2022)
7. H. Ren, L. Zheng, G. Wang, X. Gao, Z. Tan, J. Shan, L. Cui, K. Li, M. Jian, L. Zhu, Y. Zhang, H. Peng, D. Wei, Z. Liu, Transfer-medium-free nanofiber-reinforced graphene film and applications in wearable transparent pressure sensors. ACS Nano **13**(5), 5541–5548 (2019)
8. Z. Qin, Y. Yin, W. Zhang, C. Li, K. Pan, Wearable and stretchable triboelectric nanogenerator based on crumpled nanofibrous membranes. ACS Appl. Mater. Interfaces **11**(13), 12452–12459 (2019)
9. A. Iqbal, P. Sambyal, C.M. Koo, 2D MXenes for electromagnetic shielding: a review. Adv. Funct. Mater. **30**(47), 2000883 (2020)
10. Z. Zeng, J. Qiao, R. Zhang, J. Liu, G. Nyström, Nanocellulose-assisted preparation of electromagnetic interference shielding materials with diversified microstructure. SmartMat **3**(4), 582–607 (2022)
11. H. Guo, F. Wang, H. Luo, Y. Li, Z. Lou, Y. Ji, X. Liu, B. Shen, Y. Peng, K. Liu, S. Jiang, Flexible TaC/C electrospun non–woven fabrics with multiple spatial-scale conductive frameworks for efficient electromagnetic interference shielding. Compos. Part A: Appl. Sci. Manuf. **151**, 106662 (2021)
12. H. Guo, Y. Chen, Y. Li, W. Zhou, W. Xu, L. Pang, X. Fan, S. Jiang, Electrospun fibrous materials and their applications for electromagnetic interference shielding: a review. Compos. Part A: Appl. Sci. Manuf. **143**, 106309 (2021)
13. X.-Y. Wang, S.-Y. Liao, Y.-J. Wan, P.-L. Zhu, Y.-G. Hu, T. Zhao, R. Sun, C.-P. Wong, Electromagnetic interference shielding materials: recent progress, structure design, and future perspective. J. Mater. Chem. C **10**(1), 44–72 (2022)
14. D. Ye, Z. Peng, J. Liu, Y. Huang, Self-limited ultraviolet laser sintering of liquid metal particles for μm-thick flexible electronics devices. Mater. Des. **223**, 111189 (2022)
15. Y. Liu, N. Wu, S. Zheng, Y. Yang, B. Li, W. Liu, J. Liu, Z. Zeng, From MXene trash to ultraflexible composites for multifunctional electromagnetic interference shielding. ACS Appl. Mater. Interfaces, 50120–50128 (2022)
16. X. Li, X. Yin, S. Liang, M. Li, L. Cheng, L. Zhang, 2D carbide MXene Ti_2CT_X as a novel high-performance electromagnetic interference shielding material. Carbon **146**, 210–217 (2019)

17. X. Guan, Z. Yang, M. Zhou, L. Yang, R. Peymanfar, B. Aslibeiki, G. Ji, 2D MXene nanomaterials: synthesis, mechanism, and multifunctional applications in microwave absorption. Small Struct. 2200102 (2022)
18. P. Song, B. Liu, H. Qiu, X. Shi, D. Cao, J. Gu, MXenes for polymer matrix electromagnetic interference shielding composites: a review. Compos. Commun. **24**, 100653 (2021)
19. X. Liu, J. Wu, J. He, L. Zhang, Electromagnetic interference shielding effectiveness of titanium carbide sheets. Mater. Lett. **205**, 261–263 (2017)
20. S. Hu, S. Li, W. Xu, J. Zhang, Y. Zhou, Z. Cheng, Rapid preparation, thermal stability and electromagnetic interference shielding properties of two-dimensional Ti_3C_2 MXene. Ceram. Int. **45**(16), 19902–19909 (2019)
21. P. He, X.X. Wang, Y.Z. Cai, J.C. Shu, Q.L. Zhao, J. Yuan, M.S. Cao, Tailoring $Ti_3C_2T_x$ nanosheets to tune local conductive network as an environmentally friendly material for highly efficient electromagnetic interference shielding. Nanoscale **11**(13), 6080–6088 (2019)
22. M. Han, X. Yin, H. Wu, Z. Hou, C. Song, X. Li, L. Zhang, L. Cheng, Ti_3C_2 MXenes with modified surface for high-performance electromagnetic absorption and shielding in the X-band. ACS Appl. Mater. Interfaces **8**(32), 21011–21019 (2016)
23. L. Wang, L. Chen, P. Song, C. Liang, Y. Lu, H. Qiu, Y. Zhang, J. Kong, J. Gu, Fabrication on the annealed $Ti_3C_2T_x$ MXene/epoxy nanocomposites for electromagnetic interference shielding application. Compos. B Eng. **171**, 111–118 (2019)
24. B. Ji, S. Fan, X. Ma, K. Hu, L. Wang, C. Luan, J. Deng, L. Cheng, L. Zhang, Electromagnetic shielding behavior of heat-treated $Ti_3C_2T_x$ MXene accompanied by structural and phase changes. Carbon **165**, 150–162 (2020)
25. Y. Yang, M. Han, W. Liu, N. Wu, J. Liu, Hydrogel-based composites beyond the porous architectures for electromagnetic interference shielding. Nano Res. **15**, 9614–9630 (2022)
26. Z. Zhan, Q. Song, Z. Zhou, C. Lu, Ultrastrong and conductive MXene/cellulose nanofiber films enhanced by hierarchical nano-architecture and interfacial interaction for flexible electromagnetic interference shielding. J. Mater. Chem. C **7**(32), 9820–9829 (2019)
27. Z. Zeng, C. Wang, T. Wu, D. Han, M. Luković, F. Pan, G. Siqueira, G. Nyström, Nanocellulose assisted preparation of ambient dried, large-scale and mechanically robust carbon nanotube foams for electromagnetic interference shielding. J. Mater. Chem. A **8**(35), 17969–17979 (2020)
28. A. Timofejeva, M. D'Este, D. Loca, Calcium phosphate/polyvinyl alcohol composite hydrogels: a review on the freeze-thawing synthesis approach and applications in regenerative medicine. Eur. Polym. J. **95**, 547–565 (2017)
29. T. Distler, A.R. Boccaccini, 3D printing of electrically conductive hydrogels for tissue engineering and biosensors: a review. Acta Biomater. **101**, 1–13 (2020)
30. Y. Yu, P. Yi, W. Xu, X. Sun, G. Deng, X. Liu, J. Shui, R. Yu, Environmentally tough and stretchable MXene organohydrogel with exceptionally enhanced electromagnetic interference shielding performances. Nanomicro Lett. **14**(1), 77 (2022)
31. B. Sarkar, X. Li, E. Quenneville, L.-P. Carignan, K. Wu, F. Cicoira, Lightweight and flexible conducting polymer sponges and hydrogels for electromagnetic interference shielding. J. Mater. Chem. C **9**(46), 16558–16565 (2021)
32. V.R. Feig, H. Tran, M. Lee, Z. Bao, Mechanically tunable conductive interpenetrating network hydrogels that mimic the elastic moduli of biological tissue. Nat. Commun. **9**(1), 2740 (2018)
33. P. Tang, Z. Deng, Y. Zhang, L.X. Liu, Z. Wang, Z.Z. Yu, H.B. Zhang, Tough, strong, and conductive graphene fibers by optimizing surface chemistry of graphene oxide precursor. Adv. Funct. Mater. **32**(28), 2112156 (2022)
34. J. Liu, L. McKeon, J. Garcia, S. Pinilla, S. Barwich, M. Mobius, P. Stamenov, J.N. Coleman, V. Nicolosi, Additive manufacturing of Ti_3C_2-MXene-functionalized conductive polymer hydrogels for electromagnetic-interference shielding. Adv. Mater. **34**(5), e2106253 (2022)
35. Y. Yang, N. Wu, B. Li, W. Liu, F. Pan, Z. Zeng, J. Liu, Biomimetic porous MXene sediment-based hydrogel for high-performance and multifunctional electromagnetic interference shielding. ACS Nano **16**(9), 15042–15052 (2022)

36. J. Liu, L. McKeon, J. Garcia, S. Pinilla, S. Barwich, M. Möbius, P. Stamenov, J.N. Coleman, V. Nicolosi, Additive manufacturing of Ti_3C_2-MXene-functionalized conductive polymer hydrogels for electromagnetic-interference shielding. Adv. Mater. **34**(5), 2106253 (2022)
37. Y. Zhang, Y. Yan, H. Qiu, Z. Ma, K. Ruan, J. Gu, A mini-review of MXene porous films: preparation, mechanism and application. J. Mater. Sci. Technol. **103**, 42–49 (2022)
38. F. Shahzad, M. Alhabeb, C.B. Hatter, B. Anasori, S. Man Hong, C.M. Koo, Y. Gogotsi, Electromagnetic interference shielding with 2D transition metal carbides (MXenes). Science **353**(6304), 1137–1140 (2016)
39. T. Yun, H. Kim, A. Iqbal, Y.S. Cho, G.S. Lee, M.K. Kim, S.J. Kim, D. Kim, Y. Gogotsi, S.O. Kim, C.M. Koo, Electromagnetic shielding of monolayer MXene assemblies. Adv. Mater. **32**(9), e1906769 (2020)
40. A. Iqbal, F. Shahzad, K. Hantanasirisakul, M.-K. Kim, J. Kwon, J. Hong, H. Kim, D. Kim, Y. Gogotsi, C.M. Koo, Anomalous absorption of electromagnetic waves by 2D transition metal carbonitride Ti_3CNT_x (MXene). Science **369**(6502), 446–450 (2020)
41. M. Han, C.E. Shuck, R. Rakhmanov, D. Parchment, B. Anasori, C.M. Koo, G. Friedman, Y. Gogotsi, Beyond $Ti_3C_2T_x$: MXenes for electromagnetic interference shielding. ACS Nano **14**(4), 5008–5016 (2020)
42. H. Xu, X. Yin, X. Li, M. Li, S. Liang, L. Zhang, L. Cheng, Lightweight Ti_2CT_x MXene/poly(vinyl alcohol) composite foams for electromagnetic wave shielding with absorption-dominated feature. ACS Appl. Mater. Interfaces **11**(10), 10198–10207 (2019)
43. Y. Zhang, W. Cheng, W. Tian, J. Lu, L. Song, K.M. Liew, B. Wang, Y. Hu, Nacre-inspired tunable electromagnetic interference shielding sandwich films with superior mechanical and fire-resistant protective performance. ACS Appl. Mater. Interfaces **12**(5), 6371–6382 (2020)
44. H. Zhao, Y. Yue, Y. Zhang, L. Li, L. Guo, Ternary artificial nacre reinforced by ultrathin amorphous alumina with exceptional mechanical properties. Adv. Mater. **28**(10), 2037–2042 (2016)
45. L. Liang, C. Yao, X. Yan, Y. Feng, X. Hao, B. Zhou, Y. Wang, J. Ma, C. Liu, C. Shen, High-efficiency electromagnetic interference shielding capability of magnetic $Ti_3C_2T_x$ MXene/CNT composite film. J. Mater. Chem. A **9**(43), 24560–24570 (2021)
46. X. Zhao, L.Y. Wang, C.Y. Tang, X.J. Zha, Y. Liu, B.H. Su, K. Ke, R.Y. Bao, M.B. Yang, W. Yang, Smart $Ti_3C_2T_x$ MXene fabric with fast humidity response and joule heating for healthcare and medical therapy applications. ACS Nano **14**(7), 8793–8805 (2020)
47. J.-Q. Luo, S. Zhao, H.-B. Zhang, Z. Deng, L. Li, Z.-Z. Yu, Flexible, stretchable and electrically conductive MXene/natural rubber nanocomposite films for efficient electromagnetic interference shielding. Compos. Sci. Technol. **182**, 107754 (2019)
48. B. Zhou, M. Su, D. Yang, G. Han, Y. Feng, B. Wang, J. Ma, J. Ma, C. Liu, C. Shen, Flexible MXene/silver nanowire-based transparent conductive film with electromagnetic interference shielding and electro-photo-thermal performance. ACS Appl. Mater. Interfaces **12**(36), 40859–40869 (2020)
49. M. Jin, W. Chen, L.-X. Liu, H.-B. Zhang, L. Ye, P. Min, Z.-Z. Yu, Transparent, conductive and flexible MXene grid/silver nanowire hierarchical films for high-performance electromagnetic interference shielding. J. Mater. Chem. A **10**(27), 14364–14373 (2022)
50. S. Han, N.U.H. Alvi, L. Granlof, H. Granberg, M. Berggren, S. Fabiano, X. Crispin, A multi-parameter pressure-temperature-humidity sensor based on mixed ionic-electronic cellulose aerogels. Adv. Sci. (Weinh) **6**(8), 1802128 (2019)
51. R. Bian, G. He, W. Zhi, S. Xiang, T. Wang, D. Cai, Ultralight MXene-based aerogels with high electromagnetic interference shielding performance. J. Mater. Chem. C **7**(3), 474–478 (2019)
52. Z. Zeng, C. Wang, G. Siqueira, D. Han, A. Huch, S. Abdolhosseinzadeh, J. Heier, F. Nuesch, C.J. Zhang, G. Nystrom, Nanocellulose-MXene biomimetic aerogels with orientation-tunable electromagnetic interference shielding performance. Adv. Sci. (Weinh) **7**(15), 2000979 (2020)
53. J. Liu, H.B. Zhang, R. Sun, Y. Liu, Z. Liu, A. Zhou, Z.Z. Yu, Hydrophobic, flexible, and lightweight MXene foams for high-performance electromagnetic-interference shielding. Adv. Mater. **29**(38), 1702367 (2017)

54. J. Li, Z. Xu, T. Li, D. Zhi, Y. Chen, Q. Lu, J. Wang, Q. Liu, F. Meng, Multifunctional antimony tin oxide/reduced graphene oxide aerogels with wideband microwave absorption and low infrared emissivity. Compos. Part B: Eng. **231**, 109565 (2022)
55. H. Wang, R. Lu, J. Yan, J. Peng, A.P. Tomsia, R. Liang, G. Sun, M. Liu, L. Jiang, Q. Cheng, Tough and conductive nacre-inspired MXene/epoxy layered bulk nanocomposites. Angew. Chem. Int. Ed. Engl. **62**(9), e202216874 (2022)
56. X. Li, X. Yin, C. Song, M. Han, H. Xu, W. Duan, L. Cheng, L. Zhang, Self-assembly core-shell graphene-bridged hollow MXenes spheres 3D foam with ultrahigh specific EM absorption performance. Adv. Funct. Mater. **28**(41), 1803938 (2018)
57. W. Ma, W. Cai, W. Chen, P. Liu, J. Wang, Z. Liu, A novel structural design of shielding capsule to prepare high-performance and self-healing MXene-based sponge for ultra-efficient electromagnetic interference shielding. Chem. Eng. J. **426**, 130729 (2021)
58. W. Shui, J. Li, H. Wang, Y. Xing, Y. Li, Q. Yang, X. Xiao, Q. Wen, H. Zhang, $Ti_3C_2T_x$ MXene sponge composite as broadband terahertz absorber. Adv. Opt. Mater. **8**(21), 2001120 (2020)
59. L. Wang, H. Qiu, P. Song, Y. Zhang, Y. Lu, C. Liang, J. Kong, L. Chen, J. Gu, 3D $Ti_3C_2T_x$ MXene/C hybrid foam/epoxy nanocomposites with superior electromagnetic interference shielding performances and robust mechanical properties. Compos. A Appl. Sci. Manuf. **123**, 293–300 (2019)
60. Y. Wang, J. Yang, Z. Chen, Y. Hu, A new flexible and ultralight carbon foam/$Ti_3C_2T_X$ MXene hybrid for high-performance electromagnetic wave absorption. RSC Adv. **9**(70), 41038–41049 (2019)
61. Z. Fan, D. Wang, Y. Yuan, Y. Wang, Z. Cheng, Y. Liu, Z. Xie, A lightweight and conductive MXene/graphene hybrid foam for superior electromagnetic interference shielding. Chem. Eng. J. **381**, 122696 (2020)
62. H. Yin, L. Bi, Z. Wu, G. Wang, M. Li, X. Zhou, S. Ji, W. Zhang, Y. Peng, J. Pan, C. Ye, Q. Huang, 2D foaming of ultrathin MXene sheets with highly conductive silver nanowires for wearable electromagnetic interference shielding applications owing to multiple reflections within created free space. Nano Futures **4**(3), 035002 (2020)
63. F. Pan, L. Cai, Y. Shi, Y. Dong, X. Zhu, J. Cheng, H. Jiang, X. Wang, S. Zhong, W. Lu, Phase engineering reinforced multiple loss network in apple tree-like liquid metal/Ni-Ni_3P/N-doped carbon fiber composites for high-performance microwave absorption. Chem. Eng. J. **435** (2022)
64. C.J. Zhang, B. Anasori, A. Seral-Ascaso, S.H. Park, N. McEvoy, A. Shmeliov, G.S. Duesberg, J.N. Coleman, Y. Gogotsi, V. Nicolosi, Transparent, flexible, and conductive 2D titanium carbide (MXene) films with high volumetric capacitance. Adv. Mater. **29**(36), 1702678 (2017)
65. H. Yang, X. Xiao, Z. Li, K. Li, N. Cheng, S. Li, J.H. Low, L. Jing, X. Fu, S. Achavananthadith, F. Low, Q. Wang, P.L. Yeh, H. Ren, J.S. Ho, C.H. Yeow, P.Y. Chen, Wireless $Ti_3C_2T_x$ MXene strain sensor with ultrahigh sensitivity and designated working windows for soft exoskeletons. ACS Nano **14**(9), 11860–11875 (2020)
66. H. Shin, W. Eom, K.H. Lee, W. Jeong, D.J. Kang, T.H. Han, Highly electroconductive and mechanically strong $Ti_3C_2T_x$ MXene fibers using a deformable MXene gel. ACS Nano **15**(2), 3320–3329 (2021)
67. Y. Wan, P. Xiong, J. Liu, F. Feng, X. Xun, F.M. Gama, Q. Zhang, F. Yao, Z. Yang, H. Luo, Y. Xu, Ultrathin, strong, and highly flexible $Ti_3C_2T_x$ MXene/bacterial cellulose composite films for high-performance electromagnetic interference shielding. ACS Nano **15**(5), 8439–8449 (2021)
68. L.X. Liu, W. Chen, H.B. Zhang, Q.W. Wang, F. Guan, Z.Z. Yu, Flexible and multifunctional silk textiles with biomimetic leaf-like MXene/silver nanowire nanostructures for electromagnetic interference shielding, humidity monitoring, and self-derived hydrophobicity. Adv. Funct. Mater. **29**(44), 1905197 (2019)
69. W. Zhai, C. Wang, S. Wang, J. Li, Y. Zhao, P. Zhan, K. Dai, G. Zheng, C. Liu, C. Shen, Ultrastretchable and multifunctional wearable electronics for superior electromagnetic interference shielding, electrical therapy and biomotion monitoring. J. Mater. Chem. A **9**(11), 7238–7247 (2021)

70. F. Pan, L. Cai, Y. Dong, X. Zhu, Y. Shi, W. Lu, Mixed-dimensional hierarchical configuration of 2D Ni2P nanosheets anchored on 1D silk-derived carbon fiber for extraordinary electromagnetic wave absorption. J. Mater. Sci. Technol. **101**, 85–94 (2022)
71. Y. Feng, H. Liu, W. Zhu, L. Guan, X. Yang, A.V. Zvyagin, Y. Zhao, C. Shen, B. Yang, Q. Lin, Muscle-inspired MXene conductive hydrogels with anisotropy and low-temperature tolerance for wearable flexible sensors and arrays. Adv. Funct. Mater. **31**(46), 2105264 (2021)
72. R. Bhardwaj, A. Hazra, MXene-based gas sensors. J. Mater. Chem. C **9**(44), 15735–15754 (2021)
73. L. Zhou, D. Liu, S. Li, Z. Zhao, C. Zhang, X. Yin, L. Liu, S. Cui, Z.L. Wang, J. Wang, Rationally designed dual-mode triboelectric nanogenerator for harvesting mechanical energy by both electrostatic induction and dielectric breakdown effects. Adv. Energy Mater. **10**(24), 2000965 (2020)
74. S.M.S. Rana, M.A. Zahed, M.T. Rahman, M. Salauddin, S.H. Lee, C. Park, P. Maharjan, T. Bhatta, K. Shrestha, J.Y. Park, Cobalt-nanoporous carbon functionalized nanocomposite-based triboelectric nanogenerator for contactless and sustainable self-powered sensor systems. Adv. Funct. Mater. **31**(52), 2105110 (2021)

Chapter 10
Role of MXenes in Biotechnology

Davinder Singh, Manpreet Singh, and Zaved Ahmed Khan

10.1 Introduction

MXenes have recently developed as a two-dimensional inorganic material and earned considerable interest due to their excellent electrical, mechanical, and optical properties and modifiable chemical structure [1, 2]. MXenes are basically represented by $M_{n+1}X_nT_x$, where M consists of thin layers of early transition metals, X is either carbon or nitrogen and T_x is a functional group (T_x), such as O, OH, F, and/or Cl and $n = 1–3$ [3, 4]. The diverse composition and structure of MXenes have resulted in the generation of a fast-growing family of 2D materials [5]. The novel MXenes and their unique properties open the door for their applications in biotechnology and biomedical sciences [6–9]. The advances in structural properties of MXenes, such as its 2D layered structure, hydrophilic nature, and surface functionalization, have opened its promising application in biomedical applications [10–13]. MXenes are discovered to be good candidates for uses in tissue engineering, photothermal therapy, antibacterial, anticancer and drug delivery, and other antimicrobial applications [7, 8, 14–16].

The use of 2D materials offers greater potential to create new technologies in the area of food packaging [9]. Since the layered structures are present in 2D materials, so their thickness is often only up to a few atomic distances. Their lateral diameters, however, can exceed 100 nm or few micrometers. Due to their intriguing physiochemical properties, 2D materials have recently attracted increasing attention in active and intelligent packing technologies [17]. The existence of heavy metals and organic pollutants in the water is a worldwide problem of grave concern. Numerous physical and chemical methods, such as adsorption, have been used in the past to eliminate these hazardous pollutants from the water [18–20]. The use of conventional adsorbents for practical applications was constrained by poor removal efficiency and

D. Singh (✉) · M. Singh · Z. A. Khan
Faculty of Sciences, Baba Farid Group of Institutions Deon Bathinda, Baba Farid College, Bathinda, India
e-mail: davindersingh110@gmail.com

© The Author(s), under exclusive license to Springer Nature Singapore Pte Ltd. 2024

O. P. Pandey and P. Sharma (eds.), *MXenes: Emerging 2D Materials*,
https://doi.org/10.1007/978-981-97-4064-2_10

amelioration [21]. In recent years, the exploding growth of nanomaterials, including MXenes and their composites, has demonstrated their large potential application in wastewater treatment due to their environment friendly approach and exceptional biocompatibility [22–24]. The present chapter is focused on the important biotechnological applications of the synthesized MXenes, and MXenes-based composites, such as therapeutics, food packaging, plant tissue culture, bioremediation, wastewater treatment, and antimicrobial applications. The current challenges and future prospects of MXenes are discussed in the conclusions section.

10.2 Therapeutics

Till date, the use of MXenes has been reported in drug administration, antimicrobial applications, tissue engineering, sensor probe development, photothermal therapy, and in many other physiological conditions. The use of MXenes is preferred in biomedical applications to hydrophobic nanoparticles as MXenes possessed a hydrophilic nature because of surface functionalization, while other materials require complex surface modifications.

In a few reports, it has been revealed that MXene-metal oxide composites have a high drug-loading capacity [25, 26]. Due to their effective treatment of malignant cells, cobalt-based nanomaterials have recently gained popularity. Doxorubicin (DOX) is widely used as a strong chemotherapeutic agent for drug administration purposes. $Ti_3C_2T_x$-CoNWs nanocarrier heterojunctions have been linked to an intriguing dual-responsive DOX release [25]. A dual responsive (pH/NIR) drug delivery system composed of layer-by-layer deposited hollow hydroxyapatite, chitosan/hyaluronic acid, AuNRs, and MXene was also synthesized by Wu and his co-workers for drug delivery applications [26]. Using this method, the initial burst release of the drug was controlled, while the AuNRs and MXene on the surface of the synthesized hybrid composite increased its photothermal conversion efficiency. This hybrid material reported excellent biocompatibility for the controlled delivery of the drug. Yang et al. [11] also used a similar method to produce a MXenes-based hydrogel for controlled drug release. This system possessed exceptional photo and magnetically sensitive characteristics and was found to be effective in treating chronic wounds. This is typically made up of poly (N-isopropyl acrylamide)-alginate hydrogel and MXene-wrapped magnetic colloids to accomplish regulated drug administration with limited hazardous side effects.

The traditional cancer treatment methods have certain drawbacks, including higher doses, hazardous side effects, limited availability of the drug, random targeting, and multidrug resistance [27, 28]. This has prompted intensive research for innovative nanosystems and drug formulations for effective cancer treatment with limited side effects [29, 30]. $Ti_3C_2T_x$-type MXenes have been intensively investigated for anticancer purposes [10], while Ti_2C [31], Mo_2C [32], Ta_4C_3 [33], Nb_2C [34], TiCN [35], and V_2C [36] have also been described as antineoplastic agents.

The exceptional photothermal properties of MXenes have shown promise for anticancer therapy [10, 37]. In addition, local hyperthermia can be induced by NIR irradiation, and localized tumor elimination could be accomplished by the synergetic effect of chemo- and photo-thermal therapy. Sonodynamic therapy is another alternative method for localized tumor elimination. An in situ production of tumor-microenvironment-specific nanosonosensitizers on a $Ti_3C_2T_x/CuO_2$@BSA catalyst was reported for sonodynamic tumor therapy [12]. The efficacy of this method was enhanced due to the production of TiO_2 nanosensitizer due to the oxidation of Ti_3C_2 by CuO_2 nanoparticles. Overall, the results demonstrated the role of 2D MXenes in developing nano-sensitizers for antitumor applications. It is important to investigate iron metabolism for anticancer therapy as iron is reported to be involved in cell proliferation and tumor growth. In this context, Xu et al. [13] fabricated a novel construct Ti_3C_2-PVP@DOXjade for antitumor applications by the inhibition of iron depletion-triggered iron transferrin receptor. They loaded the DOXjade to ultrathin MXene nanosheets, which induced an apoptotic cell death with a photothermal efficiency by 40%. Li et al. [38] demonstrated the use of nitride MXene nanocarrier (Ti_2N) for cancer therapy. This type of nanocarrier showed excellent biocompatibility because of the presence of coating of organosilica shell (Ti_2N@oSi). This nanocarrier showed high drug-loading capacity of approx. 800% and with photothermal conversion efficacy of 42% (approx.), nitride MXenes can also serve as an important nanocarrier for biomedical applications.

10.3 Food Packaging

Food packaging is essential for preserving freshness and food safety by preventing the growth of numerous food-borne pathogens. The primary roles of food packaging are preservation, containment, antimicrobial protection, and communication [39]. The ever-increasing expectations of consumers drive the packaging industry's unceasing desire for high-tech, multifunctional packaging materials. The principal objectives of packaging rely heavily on functional materials that serve as either preservatives or indicators, or both. In the packaging industry, the combination of either micro or nanofillers with the polymeric matrix is considered an essential and widespread choice for the creation of these functional materials. Numerous 2D materials, including graphene and MXene as fillers, have attracted a lot of attention owing to their enormous surface areas, physical, mechanical properties, and antibacterial activities, which are crucial and essential in the packaging industry, as shown in Fig. 10.1 [6].

MXenes are considered ideal materials for food packaging due to their barrier property offered by their multi-layered structure, large surface areas, and aspect ratios [6]. Due to the presence of different surface terminating groups, the formation of open narrow nanochannels between adjacent nanolayers has been reported in stacked MXenes laminates, indicating a distinct gas transport mechanism [40]. Therefore

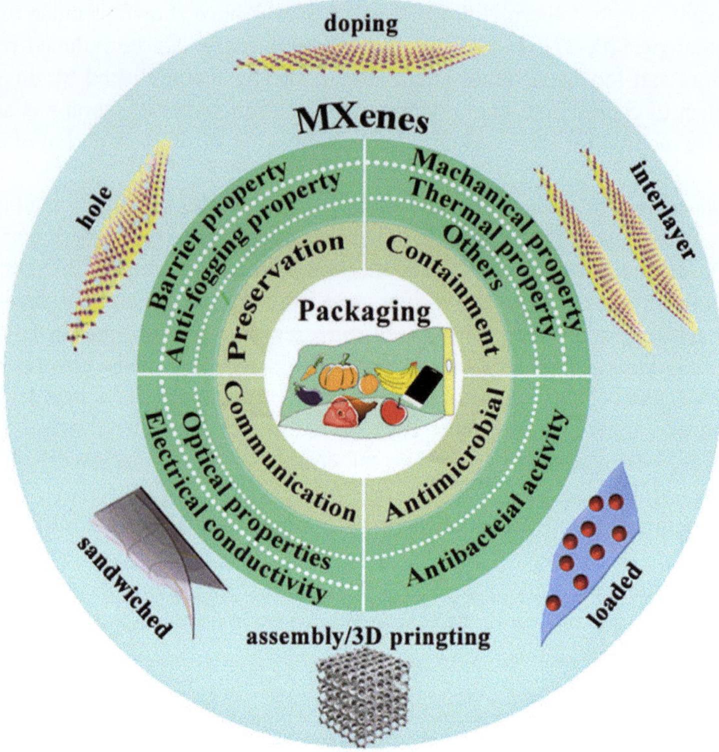

Fig. 10.1 The platform of MXenes for packaging materials of preservation, containment, communication, and antimicrobial (the out ring displays the nanostructures of MXenes-based nanomaterials). Reproduced with permission [6], copyright 2022 Wiley

the rate of diffusion of each molecule can be screened architecturally using these interlayer nanochannels between adjacent nanosheets [41, 42].

The use of MXenes in polymer films can be significant new materials for molecular sieving due to their efficient penetration and selectivity [43, 44], indicating a huge prospective in gas barrier applications [42, 45]. Owing to the unique characteristic features of MXenes, it has been proposed that the incorporation of MXenes nanofillers into polymer matrices could result in the formation of additional gas transport channels [6]. Thus, MXenes have been reported to be highly promising food packaging materials with respect to gas permeability and selectivity. The selection of gases for MXenes-based materials depends on three primary factors, such as the size of the gas molecule, the number of channels, and the stacking structures of pure MXenes [46].

Ding et al. [42] fabricated highly ordered lamellar stacked MXenes films with aligned and regular subnanometer channel structures. The experimental results of molecular dynamics simulations suggest that gas molecules (such as He and H_2) diffuse rapidly through the MXenes nanofilm due to their smaller sizes than the free gap between the surrounding nanolayers. However, gas molecules (such as O_2, N_2, and CH_4) diffuse 100 times more slowly than the free spacing due to their greater sizes. The diffusion of CO_2 is different from the other gas molecules due to the unique adsorptive characteristic and the interaction between MXene and CO_2. The authors also reported that the subnanometer interlayer gap between neighboring MXenes nanosheets provides molecular sieving channels for the separation of various sized gases. It has been reported that activated diffusion and Knudsen diffusion are the predominant mechanisms for gas diffusion in the nano-gallery [47]. Shen et al. [45] reported the fabrication of 20 nm thick nanofilms from atomically thin MXenes for gas separation. These MXenes decorated with borate and polyethyleneimine molecules resulted in the modification of nanosheet stacking characteristics and interlayer spacing. The nanofilms of MXenes exhibited channels, which led to the formation of major H_2 or CO_2 selective transport channels with superior performance. It has been reported that the solubility and diffusivity of pure MXenes nanofilms vary with the molecular size of different gases. Thus, the MXenes membrane possessed a molecular filtering action similar to the diffusion-controlled membrane. When MXenes were cross-linked by borate and polyethyleneimine, the transport channels were compressed to reduce the H_2 diffusivity [45]. The functionalized MXenes using sulfanilic acid and boric acid (BA) as the crosslinking agent, Woo et al. [48] synthesized an MXenes/polyvinyl alcohol (PVA) nanocomposite. This composite film possessed a significant decrease in gas permeability by 69%, increased tensile strength (67%), and modulus (49%) with low loadings of MXenes and BA (both 0.5 wt%) as compared to pure PVA. Thus, MXene composite indicated a potential barrier film for food packaging applications [48].

A rejection rate of greater than 90% has been reported by MXenes nanosheets for molecules with particle sizes of approximately 2.5 nm or more [49]. Due to the charged MXenes nanosheets, water molecules can bind to the surface of the film and diffuse through several channels. The shape and chemical characteristics of MXene films have been reported to be modified with various agents, such as borate, polyelectrolytes, and polyethyleneimine in order to charge its surface characteristics [50, 51]. In addition to the diameter of the channel, the diffusion of water molecules can be controlled by electrostatic attraction and the surface structure of the packaging composite films. Wang et al. [52] revealed a double-layered $Ti_3C_2T_x$ as a stiff building block in a lamellar membrane, with 2 nm interlayer ordered channels that are regular and straight. The membrane exhibited good water to organics permeation and structural stability. Liu et al. [51] employed positively charged polyelectrolytes to generate electrostatic attraction with negatively charged Ti_2CT_x nanosheets. The ionization groups increased the water affinity of the MXenes membrane, allowing the formation of ordered laminar Ti_2CT_x nanosheets. This modified MXenes membrane has the capacity to transport water rapidly and selectively. However, the presence of various salt solutions (NaCl, $MgCl_4$, Na_2SO_4, and $MgSO_4$) in water altered the rate of water transport due to the combined result of electrostatic repulsion and size sieving [50]. MXene films are also reported to possess high deformability and electrical conductivity [53]. They also possessed a maximum bending angle of 155°

under conditions of 65% relative humidity difference. It provides a potential alternative for packaging films with variable moisture inside packing, particularly for fruits and vegetables. MXene-based composites also possessed excellent moisture barrier properties, but they have been reported to be affected by various intrinsic and extrinsic factors such as humidity, temperature, respiration rate, and various other structural properties also played a vital role [6]. In addition, the differential pressure between the interior and exterior of the packaging also played a crucial role in determining the water barrier qualities. Like MXenes-based composites, the out-of-order packing characteristics of naturally stacked MXenes nanosheets can result in a low water permeation rate [54, 55]. Carey et al. [56] suggested that a low amount of MXenes added to polyamide polymers of $Ti_3C_2T_x$ and nylon-6 can reduce water vapor permeability. Wang and co-workers [57] used hydrophilic MXenes as adhering fillers for polyester textile substrates, which allowed a high level of adhesion. However, it also facilitated the oxidation and decomposition of MXenes. A silicone coating can turn the MXenes-coated textile to a hydrophobic surface with a water contact angle of around 126°. This modification method gives a strategy for achieving air-permeable, water-resistant polyester properties [57]. The barrier property of MXenes nanosheets can be changed by positively charged polyelectrolytes. Thus, barrier characteristics of MXenes-based composites can therefore be regulated by both polymers and MXenes sheets. The barrier properties of MXenes-based composite films not only extend the shelf life of the packed items but also determine their freshness and aroma.

10.4 Wastewater Treatment

The release of numerous organic pollutants into the environment is gaining more and more public attention because of the rapid growth of different industrial sectors [58]. Heavy metal ions in drinking water are difficult to decompose or metabolize once they enter the body. Consequently, they tend to accumulate until they surpass the permissible limit and pose a serious threat to human health [59]. Due to the presence of functional OH and O groups, large specific surface area, and variable surface chemistry, MXenes have a removal capacity comparable to or even superior to that of other materials (Fig. 10.2) [60–62]. MXenes have been demonstrated to be an effective heavy metal ion remover for copper (Cu), lead (Pb), and chromium (Cr), as they can adsorb the desired contaminant through the interaction of surface groups with contaminants [20, 23]. MXene-based composites have also been extensively studied for their ability to remove various organics, including dyes [63–68], phenolics [69–72], and antibiotics [73–79].

In the molecular or dispersed state, dyes impart a bright and stable color to other substances [80]. Mashtalir et al. [64] reported that the adsorption of methylene blue (MB) on $Ti_3C_2T_x$ MXene increased dramatically in the dark within the first 8 h due to the electrostatic attraction of negatively charged MXenes and cationic dye of MB, whereas there were no changes in acid blue (AB80) adsorption even after

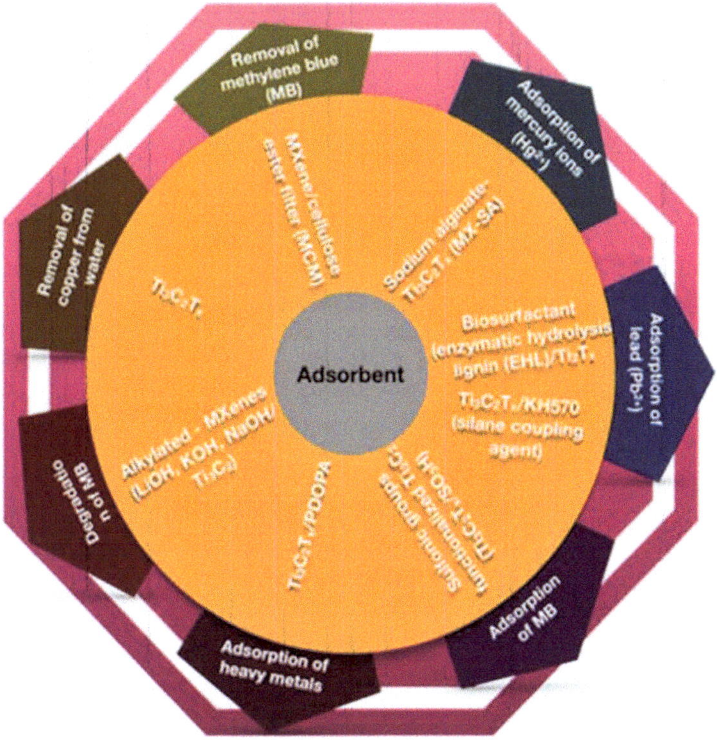

Fig. 10.2 Various applications of surface functionalized MXenes used as adsorbent materials for wastewater treatment. Reproduced with permission [62], copyright 2022 Wiley

20 h due to the electrostatic repulsion of negatively charged MXenes and anionic dye AB80. 62% of AB80 and 81% of MB were photodegraded by $Ti_3C_2T_x$ during 5 h of UV irradiation, respectively. The authors also reported that the interaction mechanism between MXene and MB was active during the first 2 h, with an increase in stacking disorder in the following 18 h due to wedging of the layered structure and/or chemical MXene transformation and oxidation resulting in the formation of titania and dissolved oxygen over long periods of time.

Antibiotics have been reported to be used in poultry metabolites, animal husbandry, and medicine to treat bacterial infections [81]. Their misuse has resulted in severe environmental degradation. As a characteristic sulfonamide antibiotic, sulfamethazine (SMZ) is durable, structurally complex, water-soluble, and chemically stable [82, 83]. Wu et al. demonstrated the high photocatalytic degradation efficiency (82.8% in 1 h) of tetracycline hydrochloride by MXene-modified MOF heterojunction [84]. It has been demonstrated that H and OH were the primary reactive species in the photocatalytic degradation of tetracycline hydrochloride. Cao et al. reported that the photocatalytic degradation rate of SMZ on $CuFe_2O_4$/MXene (CFO/Ti_3C_2) heterostructures was much greater than Ti_3C_2 MXene [76]. The degradation of SMZ on CFO/Ti_3C_2 resulted from the breaking of SeN bonds, the oxidation of aniline, and deamination, which were dominated by the attack of OH radicals.

Hexavalent chromium (Cr(VI)) is another hazardous heavy metal that has been widely utilized in leather tanning, electroplating, and metal polishing [85]. The discharge of Cr(VI) into the environment poses a risk to the health of living beings [86]. WHO stipulated that Cr(VI) concentrations in drinking water must be less than 0.05 mg/L [87]. It is therefore desired to discover novel nanomaterials that can efficiently and rapidly remove Cr(VI) from the aqueous solutions. Several studies on the removal of Cr(VI) by novel MXene [88–90] and MXene-based composites have been published so far [79, 91–93]. It has been reported that the removal mechanism of Cr(VI) on $Ti_3C_2T_x$ was based on the ion exchange of Cr(VI) by terminating OH groups, followed by reduction of adsorbed Cr(VI) into Cr(III), and ultimately surface co-precipitation of $Cr(OH)_3(s)$ [94]. MXenes also demonstrated removal efficiency at low pH. For example, He et al. [89] reported that MXene/nZVI composites had a high absorption capacity at low pH (194.87 mg/g at pH 2.0) due to the synergistic effect of nZVI and alk-Ti_3C_2 nanosheets. Another study also supported the fact that MXene exhibited significant Cr(VI) adsorption performance (99.35%) at low pH, i.e., at 2.0 [92]. Based on DFT simulations, Xie et al. proved that OH-terminated Ti_3C_2 is more suitable for $HCrO_4$ adsorption than O-terminated Ti_3C_2 [95].

Lead (Pb) is another most prevalent and hazardous heavy metal. Numerous articles reported the removal of Pb(II) by novel MXene [96–101] and MXene-based composites [102, 103]. Peng et al. [96] analyzed the Pb(II) adsorption on 2D layered alk-MXene ($Ti_3C_2(OH/ONa)_xF_{2-x}$) produced by chemical exfoliation followed by alkalization intercalation. The batch studies revealed that MXene displayed a rapid adsorption equilibrium (120 s), a substantial sorption capacity (140 mg/g), and an exceedingly minimal Pb(II) effluent, whereas Guo et al. [102] confirmed the adsorption capacity of Pb on Ti_2C_1 to be as high as 2560 mg/g.

The Cd(II) is recognized as an essential environmental concern due to the severe damage caused to human health, including kidney disease, neurological disturbance, and dyspnea [104–106]. Consequently, the efficient removal of Cd(II) from wastewater has garnered considerable attention in recent decades, including flocculation [68, 107, 108], chemical precipitation [109–111], and resin exchange [112–115]. Adsorption has been widely used to remove Cd(II) due to its environmentally friendly approach, low cost, and ease of regeneration [116]. Recently, the removal of Cd(II) from MXenes and MXene-based composites has been studied [98, 117–121]. The remarkable adsorption capacity of Alk Ti_2Csheet toward Cd(II) (325.89 mg/g) has been reported due to various properties like high porosity, large specific surface area, oxygenated functional groups, and distinctive structure [118]. Due to the increased diffusion rate of Cd(II) on the microgrid structure, Zhu et al. discovered that $Bi@Ti_3C_2T_x$ nanocomposites had extremely low detection limits for Cd(II) (0.4 mg/L) [117]. According to many cycles of adsorption–desorption, MXenes displayed excellent Cd(II) reusability [119]. The MXene nanostructures created give a new method for eliminating ppb-level Cd(II) from the environment. Due to the large negative surface charge of MXenes, the electrostatic attraction was the predominant adsorption mechanism for Cd(II) on MXenes. FTIR and XPS studies, however, revealed that the Pb(II) adsorption mechanism was ion exchange and inner-sphere complexation [119].

Numerous other studies have examined the removal of Cu(II) [122–124], Hg(II) [125, 126], and Ag(I) from MXenes [127, 128]. For example, Shahzad et al. [125] discovered that the highest adsorption capacity of Hg(II) on magnetic $Ti_3C_2T_x$ was as high as 1128 mg/g, which was much higher than MXene-based composites such as MX-SA2:20 (364.52 mg/g) and MX-SA4:20 (932.84 mg/g). The high adsorption capacity of MXene was reported due to its high porosities, large specific surface area, and abundance of oxygen-containing functional groups [126].

10.5 Antimicrobial Properties

The presence of microorganisms poses a grave health risk. MXenes (especially $Ti_3C_2T_x$) have emerged as a potential choice among several 2D materials, exhibiting antibacterial activity even greater than graphene oxide [14]. The loss of cell membrane integrity and DNA damage because of the sharp edges of MXenes and its lipopolysaccharide molecules of cell membranes have resulted in increased antibacterial activity in MXenes [15]. MXenes and their composites have reported antibacterial activity against numerous bacterial pathogens such as *Escherichia coli, Bacillus subtilis*, and many more [129]. It has also been reported that functional groups of MXenes hinder bacterial growth by blocking the uptake of nutrients, thereby inactivating cells [8]. It has been reported that the atomic structures of MXenes play a vital role in determining their antibacterial activity [129]. Several MXenes, such as $Ti_3C_2T_x$ and $TiVCT_x$, possessed inherent antibacterial characteristics (Fig. 10.3) [14, 15]. Additionally, the transport of reactive electrons from the bacterial cell to the external environment is reported to be enabled by the creation of a conductive bridge over the lipid bilayer [8]. Environmental conditions and the shape of bacterial cell walls play a significant role among several other factors, in limiting the antibacterial efficacy of MXenes. Environmental conditions contribute to membrane averaging surface oxidation of $Ti_3C_2T_x$ in air resulted in the formation of nanocrystals of anatase TiO_2 [130]. The TiO_2 catalyzed free radical formation is reported to enhance the antibacterial property of $Ti_3C_2T_x$ by inducing oxidative stress on bacterial cell walls [130]. Since the thickness of peptidoglycan differs between gram-negative and gram-positive bacteria [131, 132], there is a corresponding difference in MXene resistance. It has been reported that the stoichiometry of MXenes also exerts a substantial effect on their antibacterial activity [129]. Xu et al. [133] reported an MXene-based multimodal antibacterial platform. A nanofibrous antibacterial membrane was prepared from amoxicillin, MXene, and polyvinyl alcohol (PVA) using electrospinning. In this study, the PVA matrix restricted the release of amoxicillin, whereas the MXene converted the NIR laser into heat, resulting in local hyperthermia that stimulated the release of amoxicillin. Ultimately, local hypothermia and amoxicillin had a synergistic effect that rendered the microorganisms inactive. In addition to functioning as a physical barrier to co-load amoxicillin and MXene, the membrane reported not only displayed a higher antibacterial property but also a rapid wound-healing capacity. Liu et al. [134] described a novel NIR-activated MXene/cobalt nanowire 2D/1D heterojunction for antibacterial applications. Due to NIR-induced hyperthermia and ROS production, the heterojunction applied to an orthopedic implant exhibited 90% antibacterial efficacy within 20 min.

MXenes coupled with metal oxides, polymers, nanoparticles, and bacteriophages drew considerable study interest owing to their increasing antibacterial properties. Mansoorianfar et al. [135] studied the bacterial-targeting ability of bacteriophages combined with the physical interaction between MXene nanofragments and the bacterial cell membrane causing the cell wall to rupture, and ultimately death of the bacterial cell. The results indicated that $Ti_3C_2T_x$ MXene considerably improved the bacteriophage adsorption rate and stability during long-term cultivation in aquatic conditions, resulting in significantly higher antibacterial activity. The

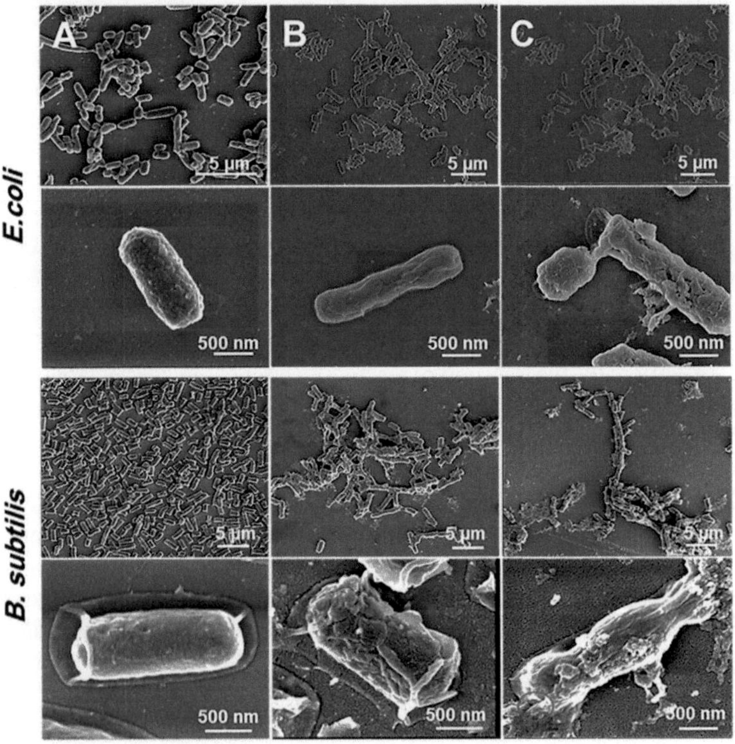

Fig. 10.3 SEM images of the *E. coli* (top panel) and *B. subtilis* (bottom panel) treated with 0 μg/mL [control] (**a**), 50 μg/mL (**b**), and 100 μg/ mL (**c**) of $Ti_3C_2T_x$, at low and high magnification, respectively. Control bacterial cells were viable with no observed membrane damage or cell death and the higher magnification shows that the bacterium is protected by intact cytoplasmic membrane (panel A). At 50 and 100 μg/mL of $Ti_3C_2T_x$, both bacteria suffered from prevalent cell lysis indicated by a severe membrane disruption and cytoplasm leakage (see the red circles at high magnification in panel C). Reproduced with permission [14], copyright 2016 American Chemical Society

bacteriophage containing MXene decreased the artificial contamination of water samples by 99.99%.

A hybrid of TiO_2 exposing facets and monolayer MXene $Ti_3C_2T_x$ nanosheet was synthesized and utilized as a photocatalyst for inactivating airborne bacteria [136]. By injecting MXene, the recombination of photo-induced electrons and holes was minimized, resulting in a 30% improvement in photocatalytic efficiency [136]. Moreover, photothermal ablation with Ti_3C_2 MXenes and near-infrared (NIR) light has been found to be a quick and broad-spectrum antibacterial technique [137]. In around 20 min, MXenes with 808 nm light reported potent antibacterial properties. This approach can be used on various bacterial species, including methicillin-resistant *S. aureus* and vancomycin-resistant *Enterococci*, by physically destroying their structures through contact and photothermal effect. After evaluating the antibacterial processes of these MXenes with light, it was determined that insertion/attachment and photothermal effects accounted for the majority of bacterial death [137].

MXene was also used to create a novel reactive catalyst for photocatalytic and bacteriostatic applications [138]. The high catalytic activity of the designed hybrid structure can be attributed to the formation of multiple interfacial junctions by graphitic carbon, TiO_2 nanoparticles, and $Ti_3C_2T_x$, which created distinct intermediate energy levels for electron–hole pair separation and boosted the catalytic activity. Due to the free radical-induced oxidative stress on bacterial cells, these nanostructures exhibited significant antibacterial performance (97.4%) against *E. coli* when exposed to light [138]. In addition, it was reported that the antibacterial performance of $Ti_3C_2T_x$ and Ag/ $Ti_3C_2T_x$ could be significantly enhanced by exposure to NIR light [139]. When exposed to NIR light (808 nm), these nanoplatforms inhibited bacteria (*E. coli* and *S. aureus*) with significant efficiency and could be used for wound healing [139].

In colloidal solution, the antimicrobial properties of single- and few-layer $Ti_3C_2T_x$ MXene flakes were also examined against *E. coli* and *B. subtilis* on the basis of the bacterial growth curve [14]. Consequently, $Ti_3C_2T_x$ displayed greater antibacterial efficacy against these pathogens than graphene oxide. The response was found to be concentration-dependent and at 200 g/mL of MXene, both bacterial cells lost more than 98% of their viability within 4 h. It has been reported that MXenes disrupted the cell membrane, resulting in the release of cytoplasmic materials and ultimately induced secondary oxidative stress (Fig. 10.3) [14]. Additionally, $Ti_3C_2T_x$ MXenes of different lateral sizes were designed to examine the antibacterial properties against pathogenic bacteria [15]. The results demonstrated that the smaller nanosheets exhibited higher antibacterial activities against both *E. coli* and *B. subtilis.* The growth kinetics measurements clearly demonstrated that direct physical contact between the sharp edges of nanosheets and bacterial membrane surface played a significant role in determining the efficacy of the nanosheet. These nanosheets could significantly injure bacterial cells in less than three hours, resulting in the loss of cellular content and the dispersion of bacterial cells [15].

Electrospinning was used to produce $Ti_3C_2T_x$ MXene flakes encased in chitosan nanofibers for antibacterial applications [140]. Consequently, in vitro tests indicated higher antibacterial efficacy against *E. coli* (95%) than against *S. aureus* (62%) after four hours of treatment with nanofibers loaded with 0.75 wt% $Ti_3C_2T_z$ [140].

Moreover, $Ti_3C_2T_x$ MXene membranes produced by filtration on a polyvinylidene fluoride substrate also demonstrated antibacterial activity [16]. Consequently, the antibacterial rate of fresh MXenes membranes was greater than 73% against *B. subtilis* and 67% against *E. coli* compared to that of control polyvinylidene fluoride, whereas aged $Ti_3C_2T_x$ membrane exhibited over 99% growth inhibition of both bacteria under the same conditions [16].

Using model pathogens, Pandey and his co-workers [141] examined the antibacterial activity of two niobium carbide MXenes (Nb_2CT_x and $Nb_4C_3T_x$). The bactericidal properties of Nb_2CT_x and $Nb_4C_3T_x$ against *E. coli* and *Staphylococcus aureus* have shown to be dependent on the sheet size and atomic structure of both MXenes. Within three hours of incubation, 120 g/mL delaminated-Nb_2CT_x and delaminated-$Nb_4C_3T_x$ inhibited the growth of 94.2 and 96.1% of *E. coli* cells and 91.6 and 93.7% of *S. aureus* cells, respectively. The cell viability was decreased by decreasing the lateral sheet size of delaminated-Nb_2CT_x and delaminated-$Nb_4C_3T_x$. Notably, MXenes' antibacterial properties originate from their 2D shape, which allows their sharp edges to penetrate and destroy the bacterial cell wall. However, these nano knife effects of MXene nanosheets may be less effective due to the loss of 2D shape at this extremely small particle size, which can no longer harm the surface of bacterial membranes [141, 142].

Cuprous oxide (Cu_2O)-anchored MXene nanosheets reported higher antibacterial activity against *S. aureus* (97.04%) than *Pseudomonas aeruginosa* (95.59%) [143]. Moreover, 2D $Ti_3C_2T_x$ MXene combined with silver nanoparticles was also developed [144]. This composite membrane inhibited *E. coli* growth by more than 99%, whereas the MXene membrane inhibited by approximately 60%.

In addition, the antibacterial properties of MnO_2 and MoS_2 produced on 2D nanosheets of graphene oxide, reduced graphene oxide, and Ti_3C_2 MXene were also examined [145]. Analyses by scanning electron microscopy (SEM) revealed that the 2D nanomaterials inhibited growth by targeting the peptidoglycan layer of the bacterial cells. The vertically oriented nanosheets of MnO_2 demonstrated the greatest antibacterial activity, suggesting that the nanosheet edges weakened the cell wall upon contact [145]. Additionally, a synergistic antibacterial agent was created by conjugating ultra-small gold nanoclusters with MXene nanosheets. The delivery of synthesized nanoclusters into bacteria resulted in ROS production, which causes secondary oxidative stress to bacterial DNA and cell membrane, and eventually leads to the death of gram-positive and gram-negative bacteria [146]. Notably, the nanosystem's crumpled shape has the synergistic capability of antibacterial action, a hydrophobic surface to prevent the attachment of bacteria, and a large surface area containing a larger concentration of bactericides, all of which can inhibit biofilm formation [146].

10.6 Plant Biotechnology

Numerous biotic and abiotic stresses are hampering plant growth and ultimately reducing their productivity [147, 148]. Plant nano bionics, an interdisciplinary approach of nanobiotechnology and plant biology, has enabled researchers to manage the existing germplasm to achieve tolerance against these stresses [149–151]. The recent developments in nanotechnology have revolutionized the use of nanomaterials in agriculture for improved plant productivity and protection from various environmental stresses [152, 153]. The potential of 1D and 2D nano fertilizers has also been exploited in recent years to meet the global challenge of food production [154, 155]. Nanoparticles (NPs), due to their small size, have a large surface area and multiple binding sites, which make them ideal for the delivery of poorly soluble micronutrients to promote plant growth even under adverse environmental conditions [156]. Various techniques such as foliar application, hydroponics, or direct soil route were used for the delivery of NPs, but their environmental risk needs to be predicted before their safe use under natural conditions [157]. The implementation of an optimum dose of NPs seems to be an important factor while analyzing their fate and transfer to the environment. Previous studies have reported that most of the NPs at low concentrations increased plant growth by several folds, whereas high concentrations were reported to have toxic or minimal effects [158]. The extensive use of NPs for ameliorating abiotic stresses has also been studied in various plant species [159]. They have been reported to impart salt stress tolerance via enhancing chlorophyll content, osmolytes, antioxidant enzyme activities, and improving the K^+/Na^+ ratio [151, 160]. The direct testing of NPs under natural conditions is a long-lasting process, which has been reported to be easily facilitated by the use of a tissue culture approach [151, 161]. NPs were easily suspended in a culture medium and tested for different explants and species simultaneously under control and sterile conditions, minimizing their leakage into the environment [159]. Numerous publications reported the effectual use of synthesized NPs, from the establishment of aseptic cultures [162] to imparting stress tolerance against various biotic/abiotic stress tolerance [151, 154]. In the recent findings, it has been reported that $Ti_3C_2T_x$ nanosheets at very low amounts can increase the tolerance of *Nicotiana benthamiana* and *Arabidopsis thaliana* to tobacco mosaic virus and *Pseudomonas syringae* infection, respectively (Fig. 10.4) [150].

10.7 Conclusion and Future Challenges

The increasing number of studies have proven the role of MXenes and their composites in therapeutics, food packaging, wastewater treatment, and plant biotechnology. This has aroused great scientific interest owing to the unique functional and structural properties of MXenes and their derivates. The various combinations of characteristics for this enormous, underexplored family of MXenes could pave the way for numerous

Fig. 10.4 Assessment of biocompatibility of aqueous $Ti_3C_2T_x$ MXene colloids after eight days of treatment of A, *Nicotiana benthamiana* and B, *Arabidopsis thaliana* seedlings. Reproduced with permission [150], copyright 2022

biomedical applications. Although the cytotoxicity of MXenes is reported, a particularly worrisome scenario is that the long-term effects of MXenes on the human body still need to be studied. The unexplored MXenes and their composites have the potential to accumulate in the body and cause harm. Because the cytotoxicity of MXenes is determined by its synthesis technique, surface functional groups, and shape, it would be incorrect to make a universal claim of biocompatibility. This is especially essential as MXenes are increasingly included in biomedical applications. The route of synthesis and materials used for synthesizing MXenes must be judged safe for clinical usage. The use of 2D MXenes in improving the stress tolerance of plants has been also reported. There is an urgent need to explore its mechanism of action before implementing its use in natural environments.

MXenes-based food packaging materials emerged as eco-friendly alternatives due to their superior performance and other multiple properties. However, the cost, law, consumer acceptance, as well as the migratory behaviors and toxicity of MXenes and their composites in packaging materials, must be further investigated so that MXenes-based food packaging methods can become economically feasible.

Although MXene-based materials have been extensively used in the realm of environmental remediation, the majority of these studies have focused on their characteristics and laboratory applications. In other words, the study on large-scale practical uses of MXenes is unfinished, and there are still numerous obstacles to overcome before MXenes may be utilized in the field. Activated carbon is nowadays the cost-effective material for water treatment. In comparison to activated carbon, the raw materials for MXenes are costly and the production method is intricate. All existing methods for synthesizing MXenes have a high cost and result in a low yield and this has severely hampered the manufacture of MXenes in industrial quantities. The use of MXenes for treating wastewater and other applications is crucial to create a safe, environmentally friendly, low-cost preparation method with a high yield.

References

1. U.U. Rahman et al., MXenes as emerging materials: synthesis, properties, and applications. Molecules **27**(15), 4909 (2022)
2. X. Zhan et al., MXene and MXene-based composites: synthesis, properties and environment-related applications. Nanoscale Horiz. **5**(2), 235–258 (2020)
3. M. Naguib et al., MXene: a promising transition metal carbide anode for lithium-ion batteries. Electrochem. Commun. **16**(1), 61–64 (2012)
4. Y. Dall'Agnese et al., High capacitance of surface-modified 2D titanium carbide in acidic electrolyte. Electrochem. Commun. **48**, 118–122 (2014)
5. K. Deshmukh, T. Kovářík, S.K. Pasha, State of the art recent progress in two dimensional MXenes based gas sensors and biosensors: a comprehensive review. Coord. Chem. Rev. **424**, 213514 (2020)
6. X. Zhou et al., MXenes: an emergent materials for packaging platforms and looking beyond. Nano Select (2022)
7. A. Koyappayil et al., Advances of MXenes; perspectives on biomedical research. Biosensors **12**(7), 454 (2022)
8. M. Khatami et al., MXenes for antimicrobial and antiviral applications: recent advances. Mater. Technol. 1–16 (2021)
9. Y. Yu et al., Applications of two-dimensional materials in food packaging. Trends Food Sci. Technol. **110**, 443–457 (2021)
10. S. Iravani, R.S. Varma, MXenes for cancer therapy and diagnosis: recent advances and current challenges. ACS Biomater. Sci. Eng. **7**(6), 1900–1913 (2021)
11. X. Yang et al., Multiple stimuli-responsive MXene-based hydrogel as intelligent drug delivery carriers for deep chronic wound healing. Small **18**(5), 2104368 (2022)
12. M. Zhang et al., Two-dimensional MXene-originated in situ nanosonosensitizer generation for augmented and synergistic sonodynamic tumor nanotherapy. ACS Nano **16**(6), 9938–9952 (2022)
13. Y. Xu et al., 2D-ultrathin MXene/DOXjade platform for iron chelation chemo-photothermal therapy. Bioactive Mater. **14**, 76–85 (2022)
14. K. Rasool et al., Antibacterial activity of $Ti_3C_2T_x$ MXene. ACS Nano **10**(3), 3674–3684 (2016)
15. A. Arabi Shamsabadi et al., Antimicrobial mode-of-action of colloidal $Ti_3C_2T_x$ MXene nanosheets. ACS Sustain. Chem. Eng. **6**(12), 16586–16596 (2018)
16. K. Rasool et al., Efficient antibacterial membrane based on two-dimensional $Ti_3C_2T_x$ (MXene) nanosheets. Sci. Rep. **7**(1), 1–11 (2017)
17. Y. Chen et al., Two-dimensional metal nanomaterials: synthesis, properties, and applications. Chem. Rev. **118**(13), 6409–6455 (2018)
18. C.F. Carolin et al., Efficient techniques for the removal of toxic heavy metals from aquatic environment: a review. J. Environ. Chem. Eng. **5**(3), 2782–2799 (2017)
19. M. Bagheri, S.A. Mirbagheri, Critical review of fouling mitigation strategies in membrane bioreactors treating water and wastewater. Biores. Technol. **258**, 318–334 (2018)
20. M.N. Chong et al., Recent developments in photocatalytic water treatment technology: a review. Water Res. **44**(10), 2997–3027 (2010)
21. S. Lu et al., Spectroscopic and modeling investigation of efficient removal of U(VI) on a novel magnesium silicate/diatomite. Sep. Purif. Technol. **174**, 425–431 (2017)
22. Y. Sun, Y. Li, Potential environmental applications of MXenes: a critical review. Chemosphere **271**, 129578 (2021)
23. S. Yu et al., MXenes as emerging nanomaterials in water purification and environmental remediation. Sci. Total. Environ. **811**, 152280 (2022)
24. F. Dixit et al., Application of MXenes for water treatment and energy-efficient desalination: a review. J. Hazard. Mater. **423**, 127050 (2022)
25. Y. Liu et al., Two-dimensional MXene/cobalt nanowire heterojunction for controlled drug delivery and chemo-photothermal therapy. Mater. Sci. Eng. C **116**, 111212 (2020)

26. Z. Wu et al., Chitosan/hyaluronic acid based hollow microcapsules equipped with MXene/gold nanorods for synergistically enhanced near infrared responsive drug delivery. Int. J. Biol. Macromol. **183**, 870–879 (2021)
27. V. Jain, S. Jain, S. Mahajan, Nanomedicines based drug delivery systems for anti-cancer targeting and treatment. Curr. Drug Deliv. **12**(2), 177–191 (2015)
28. P. Kumar, R. Srivastava, Nanomedicine for cancer therapy: from chemotherapeutic to hyperthermia-based therapy (2016)
29. H.A. Havel, Where are the nanodrugs? An industry perspective on development of drug products containing nanomaterials. AAPS J. **18**(6), 1351–1353 (2016)
30. J.M. Caster et al., Investigational nanomedicines in 2016: a review of nanotherapeutics currently undergoing clinical trials. Wiley Interdiscipl. Rev. Nanomed. Nanobiotechnol. **9**(1), e1416 (2017)
31. A. Szuplewska et al., 2D Ti_2C (MXene) as a novel highly efficient and selective agent for photothermal therapy. Mater. Sci. Eng. C **98**, 874–886 (2019)
32. W. Feng et al., Ultrathin molybdenum carbide MXene with fast biodegradability for highly efficient theory-oriented photonic tumor hyperthermia. Adv. Func. Mater. **29**(22), 1901942 (2019)
33. C. Dai et al., Two-dimensional tantalum carbide (MXenes) composite nanosheets for multiple imaging-guided photothermal tumor ablation. ACS Nano **11**(12), 12696–12712 (2017)
34. H. Yin et al., Nanomedicine-enabled photonic thermogaseous cancer therapy. Adv. Sci. **7**(2), 1901954 (2020)
35. W. Kong et al., One-step hydrothermal synthesis of fluorescent MXene-like titanium carbonitride quantum dots. Inorg. Chem. Commun. **105**, 151–157 (2019)
36. Y. Cao et al., Engineered exosome-mediated near-infrared-II region V2C quantum dot delivery for nucleus-target low-temperature photothermal therapy. ACS Nano **13**(2), 1499–1510 (2019)
37. C. Korupalli et al., Engineering the surface of Ti_3C_2 MXene nanosheets for high stability and multimodal anticancer therapy. Pharmaceutics **14**(2), 304 (2022)
38. L. Li et al., A Ti_2N MXene-based nanosystem with ultrahigh drug loading for dual-strategy synergistic oncotherapy. Nanoscale **13**(44), 18546–18557 (2021)
39. R. Auras, B. Harte, S. Selke, An overview of polylactides as packaging materials. Macromol. Biosci. **4**(9), 835–864 (2004)
40. D.S. Sholl, Understanding macroscopic diffusion of adsorbed molecules in crystalline nanoporous materials via atomistic simulations. Acc. Chem. Res. **39**(6), 403–411 (2006)
41. W.L. Xu et al., Self-assembly: a facile way of forming ultrathin, high-performance graphene oxide membranes for water purification. Nano Lett. **17**(5), 2928–2933 (2017)
42. L. Ding et al., MXene molecular sieving membranes for highly efficient gas separation. Nat. Commun. **9**(1), 1–7 (2018)
43. Y. Su et al., Impermeable barrier films and protective coatings based on reduced graphene oxide. Nat. Commun. **5**(1), 1–5 (2014)
44. N. Yousefi et al., Highly aligned, ultralarge-size reduced graphene oxide/polyurethane nanocomposites: mechanical properties and moisture permeability. Compos. A Appl. Sci. Manuf. **49**, 42–50 (2013)
45. J. Shen et al., 2D MXene nanofilms with tunable gas transport channels. Adv. Func. Mater. **28**(31), 1801511 (2018)
46. W. Lei et al., Eco-friendly waterborne polyurethane reinforced with cellulose nanocrystal from office waste paper by two different methods. Carbohyd. Polym. **209**, 299–309 (2019)
47. L. Li et al., Selective gas diffusion in two-dimensional MXene lamellar membranes: insights from molecular dynamics simulations. J. Mater. Chem. A **6**(25), 11734–11742 (2018)
48. J.H. Woo et al., Effects of the addition of boric acid on the physical properties of MXene/polyvinyl alcohol (PVA) nanocomposite. Compos. B Eng. **199**, 108205 (2020)
49. L. Ding et al., A two-dimensional lamellar membrane: MXene nanosheet stacks. Angew. Chem. Int. Ed. **56**(7), 1825–1829 (2017)
50. B. Meng et al., Fabrication of surface-charged MXene membrane and its application for water desalination. J. Membr. Sci. **623**, 119076 (2021)

51. G. Liu et al., Polyelectrolyte functionalized Ti_2CT_x mxene membranes for pervaporation dehydration of isopropanol/water mixtures. Ind. Eng. Chem. Res. **59**(10), 4732–4741 (2020)
52. J. Wang et al., A regularly channeled lamellar membrane for unparalleled water and organics permeation. Angew. Chem. Int. Ed. **57**(23), 6814–6818 (2018)
53. J. Wang et al., Highly conductive MXene film actuator based on moisture gradients. Angew. Chem. Int. Ed. **59**(33), 14029–14033 (2020)
54. C. Chen et al., Self-assembled free-standing graphite oxide membrane. Adv. Mater. **21**(29), 3007–3011 (2009)
55. J. Shen et al., Subnanometer two-dimensional graphene oxide channels for ultrafast gas sieving. ACS Nano **10**(3), 3398–3409 (2016)
56. M. Carey et al., Nylon-6/$Ti_3C_2T_z$ MXene nanocomposites synthesized by in situ ring opening polymerization of ε-caprolactam and their water transport properties. ACS Appl. Mater. Interfaces **11**(22), 20425–20436 (2019)
57. Q.W. Wang et al., Multifunctional and water-resistant MXene-decorated polyester textiles with outstanding electromagnetic interference shielding and joule heating performances. Adv. Func. Mater. **29**(7), 1806819 (2019)
58. A. Alsbaiee et al., Rapid removal of organic micropollutants from water by a porous β-cyclodextrin polymer. Nature **529**(7585), 190–194 (2016)
59. R. Chen et al., Facile synthesis of a sandwiched $Ti_3C_2T_x$ MXene/nZVI/fungal hypha nanofiber hybrid membrane for enhanced removal of Be (II) from Be $(NH_2)_2$ complexing solutions. Chem. Eng. J. **421**, 129682 (2021)
60. A.K. Fard et al., Two-dimensional MXene for efficient arsenic removal from aqueous solutions: experimental and molecular dynamics simulation. Desalin. Water Treat. **211**, 280–295 (2021)
61. S. Wang et al., Removal of heavy metal cations and co-existing anions in simulated wastewater by two separated hydroxylated MXene membranes under an external voltage. J. Membr. Sci. **638**, 119697 (2021)
62. L. Damptey et al., Surface functionalized MXenes for wastewater treatment—a comprehensive review. Glob. Chall. **6**(6), 2100120 (2022)
63. K.M. Kang et al., Selective molecular separation on $Ti_3C_2T_x$–graphene oxide membranes during pressure-driven filtration: comparison with graphene oxide and MXenes. ACS Appl. Mater. Interfaces **9**(51), 44687–44694 (2017)
64. O. Mashtalir et al., Dye adsorption and decomposition on two-dimensional titanium carbide in aqueous media. J. Mater. Chem. A **2**(35), 14334–14338 (2014)
65. Y. Lu et al., Preparation and photocatalytic performance of $Ti_3C_2/TiO_2/CuO$ ternary nanocomposites. J. Nanomater. **2017** (2017)
66. Z. Wei et al., Alkali treated $Ti_3C_2T_x$ MXenes and their dye adsorption performance. Mater. Chem. Phys. **206**, 270–276 (2018)
67. S. Kim, M. Yu, Y. Yoon, Fouling and retention mechanisms of selected cationic and anionic dyes in a $Ti_3C_2T_x$ MXene-ultrafiltration hybrid system. ACS Appl. Mater. Interfaces **12**(14), 16557–16565 (2020)
68. H. Lei et al., Insight into adsorption performance and mechanism on efficient removal of methylene blue by accordion-like V_2CT_x MXene. J. Phys. Chem. Lett. **11**(11), 4253–4260 (2020)
69. M. Ming et al., Promoted effect of alkalization on the catalytic performance of Rh/alk-$Ti_3C_2X_2$ (XO, F) for the hydrodechlorination of chlorophenols in base-free aqueous medium. Appl. Catal. B **210**, 462–469 (2017)
70. Y. Liu et al., Sandwich-like Co_3O_4/MXene composite with enhanced catalytic performance for Bisphenol A degradation. Chem. Eng. J. **347**, 731–740 (2018)
71. L. Wu et al., 2D transition metal carbide MXene as a robust biosensing platform for enzyme immobilization and ultrasensitive detection of phenol. Biosens. Bioelectron. **107**, 69–75 (2018)
72. H. Huang et al., One-step in-situ preparation of N-doped TiO_2@C derived from Ti_3C_2 MXene for enhanced visible-light driven photodegradation. Appl. Catal. B **251**, 154–161 (2019)

73. Q. Liu et al., MXene as a non-metal charge mediator in 2D layered CdS@ Ti_3C_2@TiO_2 composites with superior Z-scheme visible light-driven photocatalytic activity. Environ. Sci. Nano **6**(10), 3158–3169 (2019)
74. J. Shen et al., Built-in electric field induced CeO_2/Ti_3C_2-MXene Schottky-junction for coupled photocatalytic tetracycline degradation and CO_2 reduction. Ceram. Int. **45**(18), 24146–24153 (2019)
75. H. Wang et al., Electrical promotion of spatially photoinduced charge separation via interfacial-built-in quasi-alloying effect in hierarchical $Zn_2In_2S_5/Ti_3C_2(O, OH)_x$ hybrids toward efficient photocatalytic hydrogen evolution and environmental remediation. Appl. Catal. B **245**, 290–301 (2019)
76. Y. Cao et al., Fabrication of novel $CuFe_2O_4$/MXene hierarchical heterostructures for enhanced photocatalytic degradation of sulfonamides under visible light. J. Hazard. Mater. **387**, 122021 (2020)
77. C. Cui et al., Bi_2WO_6/Nb_2CT_x MXene hybrid nanosheets with enhanced visible-light-driven photocatalytic activity for organic pollutants degradation. Appl. Surf. Sci. **505**, 144595 (2020)
78. B. Shao et al., $Ti_3C_2T_x$ MXene decorated black phosphorus nanosheets with improved visible-light photocatalytic activity: experimental and theoretical studies. J. Mater. Chem. A **8**(10), 5171–5185 (2020)
79. D. Zhao, C. Cai, Preparation of Bi_2MoO_6/Ti_3C_2 MXene heterojunction photocatalysts for fast tetracycline degradation and Cr (vi) reduction. Inorg. Chem. Front. **7**(15), 2799–2808 (2020)
80. N. Zhang et al., Recent investigations and progress in environmental remediation by using covalent organic framework-based adsorption method: a review. J. Clean. Prod. **277**, 123360 (2020)
81. J. Tolls, Sorption of veterinary pharmaceuticals in soils: a review. Environ. Sci. Technol. **35**(17), 3397–3406 (2001)
82. Z. Chen et al., PH-dependent sorption of sulfonamide antibiotics onto biochars: sorption mechanisms and modeling. Environ. Pollut. **248**, 48–56 (2019)
83. P. Li et al., Controllable synthesis of cerium zirconium oxide nanocomposites and their application for photocatalytic degradation of sulfonamides. Appl. Catal. B **259**, 118107 (2019)
84. Y. Wu et al., Mxene-modulated dual-heterojunction generation on a metal-organic framework (MOF) via surface constitution reconstruction for enhanced photocatalytic activity. Chem. Eng. J. **390**, 124519 (2020)
85. M. Owlad et al., Removal of hexavalent chromium-contaminated water and wastewater: a review. Water Air Soil Pollut. **200**(1), 59–77 (2009)
86. D.E. Kimbrough et al., A critical assessment of chromium in the environment. Crit. Rev. Environ. Sci. Technol. **29**(1), 1–46 (1999)
87. G. WHO, *Guidelines for Drinking-Water Quality*, vol. **216** (World Health Organization, 2011), pp. 303–304
88. X. Xie et al., Microstructure and surface control of MXene films for water purification. Nat. Sustain. **2**(9), 856–862 (2019)
89. L. He et al., Nanoscale zero-valent iron intercalated 2D titanium carbides for removal of Cr(VI) in aqueous solution and the mechanistic aspect. J. Hazard. Mater. **388**, 121761 (2020)
90. L. Jin et al., Two-dimensional titanium carbides ($Ti_3C_2T_x$) functionalized by poly (m-phenylenediamine) for efficient adsorption and reduction of hexavalent chromium. Int. J. Environ. Res. Public Health **17**(1), 167 (2020)
91. G. Zou et al., Synthesis of urchin-like rutile titania carbon nanocomposites by iron-facilitated phase transformation of MXene for environmental remediation. J. Mater. Chem. A **4**(2), 489–499 (2016)
92. H. Wang et al., Facile synthesis of heterojunction of MXenes/TiO_2 nanoparticles towards enhanced hexavalent chromium removal. J. Colloid Interface Sci. **561**, 46–57 (2020)
93. H. Zhang et al., Preparation of magnetic α-$Fe_2O_3/ZnFe_2O_4$@Ti_3C_2 MXene with excellent photocatalytic performance. Ceram. Int. **46**(1), 81–88 (2020)

94. Y. Ying et al., Two-dimensional titanium carbide for efficiently reductive removal of highly toxic chromium (VI) from water. ACS Appl. Mater. Interfaces **7**(3), 1795–1803 (2015)
95. G. Zou et al., Synthesis of MXene/Ag composites for extraordinary long cycle lifetime lithium storage at high rates. ACS Appl. Mater. Interfaces **8**(34), 22280–22286 (2016)
96. Q. Peng et al., Unique lead adsorption behavior of activated hydroxyl group in two-dimensional titanium carbide. J. Am. Chem. Soc. **136**(11), 4113–4116 (2014)
97. J. Guo et al., Theoretical interpretation on lead adsorption behavior of new two-dimensional transition metal carbides and nitrides. J. Alloy. Compd. **684**, 504–509 (2016)
98. X. Zhu et al., Alkaline intercalation of Ti_3C_2 MXene for simultaneous electrochemical detection of Cd(II), Pb(II), Cu(II) and Hg(II). Electrochim. Acta **248**, 46–57 (2017)
99. P. Gu et al., Experimental and theoretical calculation investigation on efficient Pb(II) adsorption on etched Ti_3AlC_2 nanofibers and nanosheets. Environ. Sci. Nano **5**(4), 946–955 (2018)
100. P. Gu et al., Two-dimensional MAX-derived titanate nanostructures for efficient removal of Pb(II). Dalton Trans. **48**(6), 2100–2107 (2019)
101. Y. Du et al., Efficient removal of Pb(II) by $Ti_3C_2T_x$ powder modified with a silane coupling agent. J. Mater. Sci. **54**(20), 13283–13297 (2019)
102. X. Guo et al., High adsorption capacity of heavy metals on two-dimensional MXenes: an ab initio study with molecular dynamics simulation. Phys. Chem. Chem. Phys. **18**(1), 228–233 (2016)
103. X. Dong et al., Sustainable and scalable in-situ synthesis of hydrochar-wrapped Ti_3AlC_2-derived nanofibers as adsorbents to remove heavy metals. Biores. Technol. **282**, 222–227 (2019)
104. X. Liu et al., Human health risk assessment of heavy metals in soil–vegetable system: a multi-medium analysis. Sci. Total. Environ. **463**, 530–540 (2013)
105. J. Huang et al., Surface complexation modeling of adsorption of Cd(II) on graphene oxides. J. Mol. Liq. **209**, 753–758 (2015)
106. Q. Yang et al., A review of soil heavy metal pollution from industrial and agricultural regions in China: pollution and risk assessment. Sci. Total. Environ. **642**, 690–700 (2018)
107. W. Wei et al., Synergetic effects and flocculation behavior of anionic polyacrylamide and extracellular polymeric substrates extracted from Klebsiella sp. J1 on improving soluble cadmium removal. Bioresource Technol. **175**, 34–41 (2015)
108. L. Pan et al., Removal of copper and cadmium from simulated wastewater using coupled internal electrolysis with flocculation: a comparison of effects and investigation of mechanisms. Desalin. Water Treat. **189**, 329–337 (2020)
109. P. Grimshaw, J.M. Calo, G. Hradil, Cyclic electrowinning/precipitation (CEP) system for the removal of heavy metal mixtures from aqueous solutions. Chem. Eng. J. **175**, 103–109 (2011)
110. N. Colantonio, Y. Kim, Cadmium (II) removal mechanisms in microbial electrolysis cells. J. Hazard. Mater. **311**, 134–141 (2016)
111. J. Zieliński et al., Kinetic modelling of cadmium removal from wet phosphoric acid by precipitation method. Hydrometallurgy **190**, 105157 (2019)
112. W. Wang, V. Fthenakis, Kinetics study on separation of cadmium from tellurium in acidic solution media using ion-exchange resins. J. Hazard. Mater. **125**(1–3), 80–88 (2005)
113. M.F. Hamza et al., Synthesis and adsorption characteristics of grafted hydrazinyl amine magnetite-chitosan for Ni(II) and Pb(II) recovery. Chem. Eng. J. **362**, 310–324 (2019)
114. S. Bao et al., One-pot synthesis of magnetic graphene oxide composites as an efficient and recoverable adsorbent for Cd(II) and Pb(II) removal from aqueous solution. J. Hazard. Mater. **381**, 120914 (2020)
115. Z. Wu et al., A facile foaming-polymerization strategy to prepare 3D MnO_2 modified biochar-based porous hydrogels for efficient removal of Cd(II) and Pb(II). Chemosphere **239**, 124745 (2020)
116. G. Zhao et al., Few-layered graphene oxide nanosheets as superior sorbents for heavy metal ion pollution management. Environ. Sci. Technol. **45**(24), 10454–10462 (2011)

117. X. Zhu et al., A micromilled microgrid sensor with delaminated MXene-bismuth nanocomposite assembly for simultaneous electrochemical detection of lead (II), cadmium (II) and zinc (II). Microchim. Acta **186**(12), 1–7 (2019)
118. A. Shahzad et al., Exfoliation of titanium aluminum carbide (211 MAX phase) to form nanofibers and two-dimensional nanosheets and their application in aqueous-phase cadmium sequestration. ACS Appl. Mater. Interfaces **11**(21), 19156–19166 (2019)
119. B.-M. Jun et al., Effective removal of Pb (ii) from synthetic wastewater using $Ti_3C_2T_x$ MXene. Environ. Sci. Water Res. Technol. **6**(1), 173–180 (2020)
120. X. Lv et al., Facile synthesis of protonated carbon nitride/$T_{i3}C_2T_x$ nanocomposite for simultaneous detection of Pb^{2+} and Cd^{2+}. J. Electrochem. Soc. **167**(6), 067509 (2020)
121. P. Pandey et al., $CsPbBr_3$–$Ti_3C_2T_x$ MXene QD/QD heterojunction: photoluminescence quenching, charge transfer, and Cd ion sensing application. ACS Appl. Nano Mater. **3**(4), 3305–3314 (2020)
122. Y. Su et al., Self-templated microwave-assisted hydrothermal synthesis of two-dimensional holey hydroxyapatite nanosheets for efficient heavy metal removal. Environ. Sci. Pollut. Res. **26**(29), 30076–30086 (2019)
123. S. Elumalai, M. Yoshimura, M. Ogawa, Simultaneous delamination and rutile formation on the surface of $Ti_3C_2T_x$ MXene for copper adsorption. Chem. Asian J. **15**(7), 1044–1051 (2020)
124. D. Gan et al., Bioinspired functionalization of MXenes ($Ti_3C_2T_X$) with amino acids for efficient removal of heavy metal ions. Appl. Surf. Sci. **504**, 144603 (2020)
125. A. Shahzad et al., Mercuric ion capturing by recoverable titanium carbide magnetic nanocomposite. J. Hazard. Mater. **344**, 811–818 (2018)
126. A. Shahzad et al., $Ti_3C_2T_x$ MXene core-shell spheres for ultrahigh removal of mercuric ions. Chem. Eng. J. **368**, 400–408 (2019)
127. M.L. Desai et al., Ultra-small two dimensional MXene nanosheets for selective and sensitive fluorescence detection of Ag^+ and Mn^{2+} ions. Colloids Surfaces A: Physicochem. Eng. Asp. **565**, 70–77 (2019)
128. Y. Wang et al., Titanium carbide MXenes mediated in situ reduction allows label-free and visualized nanoplasmonic sensing of silver ions. Anal. Chem. **92**(6), 4623–4629 (2020)
129. A.M. Jastrzębska, et al., The atomic structure of Ti_2C and Ti_3C_2 MXenes is responsible for their antibacterial activity toward E. coli bacteria. J. Mater. Eng. Perform. **28**(3), 1272–1277 (2019)
130. C. Peng et al., (111) TiO_{2-x}/Ti_3C_2: synergy of active facets, interfacial charge transfer and Ti^{3+} doping for enhance photocatalytic activity. Mater. Res. Bull. **89**, 16–25 (2017)
131. K.C. Huang et al., Cell shape and cell-wall organization in Gram-negative bacteria. Proc. Natl. Acad. Sci. **105**(49), 19282–19287 (2008)
132. E.I. Tocheva et al., Peptidoglycan transformations during Bacillus subtilis sporulation. Mol. Microbiol. **88**(4), 673–686 (2013)
133. X. Xu et al., A multimodal antimicrobial platform based on MXene for treatment of wound infection. Colloids Surfaces B: Biointerfaces **207**, 111979 (2021)
134. Y. Liu et al., Synergism of 2D/1D MXene/cobalt nanowire heterojunctions for boosted photo-activated antibacterial application. Chem. Eng. J. **410**, 128209 (2021)
135. M. Mansoorianfar et al., MXene–laden bacteriophage: a new antibacterial candidate to control bacterial contamination in water. Chemosphere **290**, 133383 (2022)
136. S. Lu et al., Photocatalytic inactivation of airborne bacteria in a polyurethane foam reactor loaded with a hybrid of MXene and anatase TiO_2 exposing 0 0 1 facets. Chem. Eng. J. **404**, 126526 (2021)
137. F. Wu et al., Rapid eradication of antibiotic-resistant bacteria and biofilms by MXene and near-infrared light through photothermal ablation. Sci. China Mater. **64**(3), 748–758 (2021)
138. K. Rajavel et al., Photocatalytic and bactericidal properties of MXene-derived graphitic carbon-supported TiO_2 nanoparticles. Appl. Surf. Sci. **538**, 148083 (2021)
139. X. Zhu et al., A near-infrared light-mediated antimicrobial based on $Ag/Ti_3C_2T_x$ for effective synergetic antibacterial applications. Nanoscale **12**(37), 19129–19141 (2020)

140. E.A. Mayerberger et al., Antibacterial properties of electrospun $Ti_3C_2T_z$ (MXene)/chitosan nanofibers. RSC Adv. **8**(62), 35386–35394 (2018)
141. R.P. Pandey et al., Effect of sheet size and atomic structure on the antibacterial activity of Nb-MXene nanosheets. ACS Appl. Nano Mater. **3**(11), 11372–11382 (2020)
142. F. Perreault et al., Antimicrobial properties of graphene oxide nanosheets: why size matters. ACS Nano **9**(7), 7226–7236 (2015)
143. W. Wang et al., A photo catalyst of cuprous oxide anchored MXene nanosheet for dramatic enhancement of synergistic antibacterial ability. Chem. Eng. J. **386**, 124116 (2020)
144. R.P. Pandey et al., Ultrahigh-flux and fouling-resistant membranes based on layered silver/MXene ($Ti_3C_2T_x$) nanosheets. J. Mater. Chem. A **6**(8), 3522–3533 (2018)
145. F. Alimohammadi et al., Antimicrobial properties of 2D MnO_2 and MoS_2 nanomaterials vertically aligned on graphene materials and Ti_3C_2 MXene. Langmuir **34**(24), 7192–7200 (2018)
146. K. Zheng et al., Synergistic antimicrobial titanium carbide (MXene) conjugated with gold nanoclusters. Adv. Healthc. Mater. **9**(19), 2001007 (2020)
147. D. Singh, A. Kumar, In vitro screening and characterization of selected elite clones of eucalyptus tereticornis Sm. for salt stress. J. Plant Growth Regul. **40**, 694–706 (2021)
148. D. Singh, A. Kumar, Multivariate screening approach indicated adaptive tolerance to salt stress in the seedlings of an agroforestry tree, Eucalyptus tereticornis Sm. Plant Cell Tissue Organ Cult. **145**, 545–560 (2021)
149. H. Wu, N. Tito, J.P. Giraldo, Anionic cerium oxide nanoparticles protect plant photosynthesis from abiotic stress by scavenging reactive oxygen species. ACS Nano **11**(11), 11283–11297 (2017)
150. S. Khanahmadi, et al., *$Ti_3C_2T_x$ Mxene Nanosheets Enhance Systemic Plant Disease Resistance* (2022)
151. D. Singh et al., Dual nanozyme characteristics of iron oxide nanoparticles alleviate salinity stress and promote the growth of an agroforestry tree, *Eucalyptus tereticornis Sm.* Environ. Sci. Nano **8**(5), 1308–1325 (2021)
152. P. Wang et al., Nanotechnology: a new opportunity in plant sciences. Trends Plant Sci. **21**(8), 699–712 (2016)
153. G. Gohari et al., Titanium dioxide nanoparticles (TiO_2 NPs) promote growth and ameliorate salinity stress effects on essential oil profile and biochemical attributes of Dracocephalum moldavica. Sci. Rep. **10**(1), 1–14 (2020)
154. J. Singhal, S. Verma, S. Kumar, The physio-chemical properties and applications of 2D nanomaterials in agricultural and environmental sustainability. Sci. Total Environ. 155669 (2022)
155. M. Usman, et al., Nanotechnology in agriculture: current status, challenges and future opportunities. Sci. Total Environ. 137778 (2020)
156. Y. Shang et al., Applications of nanotechnology in plant growth and crop protection: a review. Molecules **24**(14), 2558 (2019)
157. T. Guha et al., Nanocomposites for delivering agrochemicals: a comprehensive review. J. Agric. Food Chem. **68**(12), 3691–3702 (2020)
158. P. Landa, Positive effects of metallic nanoparticles on plants: overview of involved mechanisms. Plant Physiol. Biochem. **161**, 12–24 (2021)
159. D.H. Kim, J. Gopal, I. Sivanesan, Nanomaterials in plant tissue culture: the disclosed and undisclosed. RSC Adv. **7**(58), 36492–36505 (2017)
160. F. Zulfiqar, M. Ashraf, Nanoparticles potentially mediate salt stress tolerance in plants. Plant Physiol. Biochem. **160**, 257–268 (2021)
161. D. Singh et al., In vitro screening of Indian potato cultivars for the salt stress and associated physio-biochemical changes. Biologia (2021)
162. M. Sarmast, H. Salehi, M. Khosh-Khui, Nano silver treatment is effective in reducing bacterial contaminations of Araucaria excelsa R. Br. var. glauca explants. Acta Biol. Hung. **62**(4), 477–484 (2011)

Chapter 11
Application of MXenes on Separation Processes

Shanli Nezami, Farzad Moazami, Ahad Ghaemi, and Alireza Hemmati

11.1 Introduction

Today, with increasing the world's population, the need for energy is also increasing. As well as the rapid expansion of industrial activities, agriculture, urbanization, and the use of nuclear energy (in the last few decades) has caused the greatest human challenge, which is the increase of environmental pollution. Environmental pollution (e.g., air pollution, organic substances, heavy metals, and radionuclides) due to chemical toxicity and carcinogenic and harmful effects on human health (due to toxicity, indestructibility, and bioaccumulation in the food chain and causing various diseases and disorders), is a major concern [1–8]. For this purpose, several methods have been developed to clean the environment like purification methods (e.g., biological purification, membrane, advanced oxidation, and electrochemical method). However, most of the disadvantages of these methods can be mentioned as high investment values, low efficiency, production of toxic wastewater, and complex or incomplete removal of pollutants. On the other hand, the adsorption is significant due to its cost-effectiveness, non-destruction of the environment, high efficiency, phase separation, and availability, as an identified and promising method among different technologies for treating air and water [9–13]. Since adsorption is based on the interaction between adsorbents and pollutants, each of the mutual forces of adsorption, as electrostatic elements, ion exchange, van der Waals interactions, and chemical bonding that exists between these adsorbents and pollutants, helps the removal of pollutants [13, 14]. As a result, the adsorption method is used to remove a large number of organic/inorganic/polar/non-polar pollutants and radionuclides from water and toxic and dangerous gases from the air. To date, various adsorbents have been investigated to remove pollutants from wastewater and gases from the air, such as activated

S. Nezami · F. Moazami · A. Ghaemi · A. Hemmati (✉)
School of Chemical, Petroleum and Gas Engineering, Iran University of Science and Technology, Tehran, Iran
e-mail: alireza_hemmati@iust.ac.ir

© The Author(s), under exclusive license to Springer Nature Singapore Pte Ltd. 2024

O. P. Pandey and P. Sharma (eds.), *MXenes: Emerging 2D Materials*,
https://doi.org/10.1007/978-981-97-4064-2_11

carbons, natural minerals, zeolites, alumina, clay, amberlite, bio-absorbent, nanomaterials, and metal oxides. However, the relatively low adsorption capacity and slow adsorption kinetics along with the poor selectivity of some of these adsorbents limit their widespread implementation. Therefore, researches is being done to develop a cost-effective but efficient attractant that also has good selectivity [14–16]. The development of suitable materials with excellent adsorption capacity is still a key challenge, and it is necessary to formulate a suitable adsorbent with high adsorption capacity for the removal of pollutants in harsh environmental conditions. In the last few decades, the adsorption potential of two-dimensional nanomaterials, including graphene, transition metals dichalcogenides (MoS_2), boron nitride, and phosphorene, due to having multiple active sites on surfaces and edges with an extremely high specific surface area to purification the environment and adsorb pollutants such as heavy metals, Causal dyes, radionuclides, and gases have attracted the attention of researchers. As these two-dimensional materials have the quantum properties in the main direction, they show amazing properties and chemical compounds and have a very high surface-to-volume ratio. However, the application of traditional 2D materials is limited due to their single-element characteristics (use of carbon network), poor selectivity in aqueous environments, low efficiency, and lack of surface functional groups. In recent years, a new class of two-dimensional materials, called MXene, has attracted attention due to their unique physical and chemical properties. MXenes are two-dimensional layered materials that are usually synthesized from the selective etching of A elements by chemical wet etching method from three-dimensional carbides, nitrides, or carbonitrides (D3), MAX phases. The MAX phase has the general formula $M_{n+1}AX_n$ ($n = 1–3$), where "M" is a primary transition metal (e.g., Mo, Ti, Zr, Cr, etc.). Element "A" corresponds to group III A or IV A elements such as Al, Ga, Ge, and Si. Where "X" can be carbon, nitrogen, or a mixture of them. After removing the A-layers, the formula of the produced MXene becomes $M_{n+1}X_nT_x$, where T_x refers to the surface-terminating functional groups (–O, –F, or –OH) which produced in the etching steps and demonstrate with "x" indicates the end of the surface [1, 14, 17–22]. The unique structural and compositional features of MXenes make them an exceptional family of materials that exhibit excellent chemical, electrical, optical, physical, and mechanical properties, making them promising options for a wide range of applications, including energy storage, batteries, electromagnetic interference shielding, catalyst, optoelectronics, plasmonic, medicine, sensors, etc. [1, 14, 17–22]. In addition to these applications, research on MXenes for adsorption treatment of pollutants has received attention, due to their hydrophilic nature, tunable surface chemistry, and environmentally friendly properties. MXenes can remove toxins through interaction pathways such as ion exchange. MXenes show a high adsorption tendency towards various pollutants, including heavy metal ions, organic dyes, radionuclides, and gas molecules. Furthermore, the presence of many active sites formed by multiple functional groups at the termination of MXene can act as preferential sites for the adsorption of pollutant molecules through electrostatic and chemical interactions in addition to direct ion exchange. Since MXene-based adsorbents can absorb various environmental pollutants due to their unique structure, it is important to fully understand the mechanisms of adsorption and interaction between pollutants and adsorbents. Furthermore, the adsorption process of both types of pollutants in water is affected not only by the characteristics of the adsorbent, but also by the water quality conditions such as temperature, pH, and the presence of natural organic matter. Therefore, in this chapter, the potential of adsorption for heavy metals, gases, and the effect of environmental conditions on this adsorption are investigated.

11.2 MXene

Generally, MXenes are obtained via delamination and etching from MAX phases (a family of ternary nitrides and carbides) [23]. These phases have a common formula of $M_{n+1}AX_n$, where A is group 13 or 14 elements (Al, Si, Ga, etc.), M is the transition metal ion, and X is carbon and or nitrogen [24, 25]. The MAX phases are initially converted into multi-layered M-X form, by using etchants, then intercalation s done to yield single-layered MXenes with greater interlayer spaces [26]. Physical adsorption, electrostatic attraction, hydrogen-bond interaction, covalent, and non-covalent are the surface modification methods of MXene-based. Surface-modified MXene, in the pH range of 1–10, shows dominant adsorption due to electrostatic interaction. Another excellent properties and high efficiencies of MXene-based material grades are magnetic-MXene nanocomposites. The efficiency degradation of pollutants can multiply by combining MXenes with magnetic. On the other hand, the removal efficiency of heavy metals from water resources can increase considerably by magnetic properties [13, 27–31]. The various advantages are preparation, good electrical conductivity, small size, high surface-to-volume ratio, fast process, liquid/solid separation, magnetic separation, reusability, and more outstanding biocompatibility; MXenes are promising alternatives in wastewater decontamination and recovery of sewage and water resources [9, 32–35]. The main removal techniques for eliminating radionuclides, heavy metals, and dyes are adsorption, membranes, and capacitive deionization.

11.2.1 Synthesis of MXenes

The HF method and the HF in situ method are two usual methods for the exfoliation of layers from MAX phases through the etching process. The HF method, as the most widely adopted etchant, is employed for the exfoliation of A-layers and synthesis of MXenes [36]. The MXenes are formed by etching the "A" atoms from the related MAX phase and replaced by the terminating groups (F–, OH–) [37]. In 2011, Naguib et al. synthesized multi-layered $Ti_3C_2T_x$ through selective etching of the layered precursor's "Al" atoms and using the etching agent HF. Experimental conditions, including HF concentration, duration of immersion, and temperature had considerable effect on the morphology and properties of synthesized MXenes

(0.1). The density functional theory (DFT) and molecular dynamic (MD) studies showed that etching is more favorable at higher HF concentrations. Since critical corrosiveness of HF is risky, high concentrations of HF are not employed; rather HF in situ is used as an alternative [38]. The use of reagents in the HF in situ method will have similar effects because they contain HF or lead to the formation of HF during the reaction. Therefore, etchants such as FeF_3/HCl, LiF/HCl, bifluorides, and ammonium fluorides are used for exfoliation and synthesis of MXene from the MAX phase [36, 38, 39]. The main advantages of HF in situ method include the simultaneous occurrence of etching and intercalation, no need to sonication, and use of less hazardous compounds are among; all these can result in a better etching process, and hence, homogeneous product [38, 39]. In the HCl + LiF strategy, the interlayer space between the MXene nanosheets is increased by the presence of Li^+ and water. Recently, the synthesis of 2D-MXenes with high stability and on a larger scale by synthetic methods without fluoride has been more and more attention [27, 40]. Moreover, milder concentrations of etchant, smaller MAX precursor particle size, lesser time, and lower temperature can lead to better exfoliation [41].

11.2.2 Delamination and Intercalation

Etching is followed by interpolation and sonication to layer a two-dimensional structure and create space between the layers. The final manufacturing step is ultrasonication to control the size and distance between the MXene layers. Centrifuging to separate small and large layers of the colloidal solution is one of the ways to control the distance between the layers. Alhabeb et al. showed that the interlayer distance increases by using larger ions such as TBA^+ in the successful synthesis of $Ti_3C_2T_x$ regardless of the HF concentration (Fig. 11.1) [1]. The concentration of MXene fragments is increased by performing sonication before centrifugation in the colloidal solution [44]. It is worth noting, when synthesizing MXene using (LiF + HCl), centrifugation and manual shaking are used to produce smaller sizes. The properties of MXene ($Ti_3C_2T_x$) are significantly improved by ultrasonication, thus showing a high capacity in the absorption process. In addition, separating the layers and increasing the distance between the layers is created by performing ultrasound in proportion to hand movements [38, 45]. The need for sonication is reduced by using higher concentrations of LiF/HCl [46]. The space between the layers of MXenes increases with the placement of molecules or ions. As mentioned, the surface of the synthesized MXene is covered with OH^- and F^- functional groups that during intercalation, the cations in the intercalants replace some of the functional groups in the MXene. By performing intercalation, MXenes are effective in selective adsorption [45]. Some of the intercalants used include tetrabutylammonium hydroxide, tetramethylammonium hydroxide, and dimethyl sulfoxide [28, 42]. Various characterization techniques such as XRD, SEM, and TEM can be used to confirm the intercalation. During intercalation, the distance between the layers increases as new molecules/ions are placed between the layers. By increasing the distance between

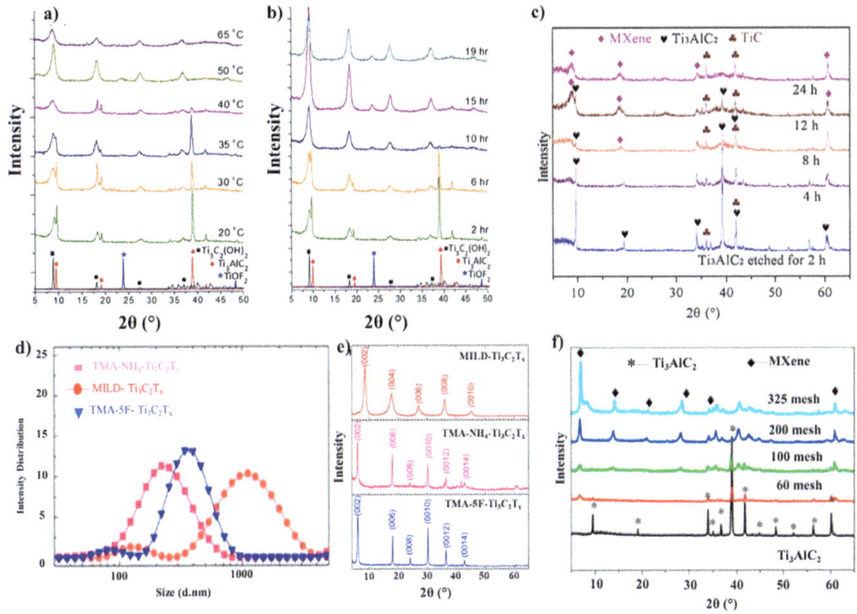

Fig. 11.1 **a** XRD analysis for exfoliation of Ti_3AlC_2 for 2 h at different temperatures for 2 h using 50% HF solution [42]; **b** XRD peak for Ti_3AlC_2 etching varying with the time of exfoliation at room temperature using 50% HF solution [42]; **c** XRD patterns of exfoliating Ti_3AlC_2 at 60 °C by 49% HF, Adapted with permission from [37] Copyright (2023) Li, Z., et al.; **d** XRD analysis for the respective particle size with TMA-5F-$Ti_3C_2T_x$, TMA-NH_4-$Ti_3C_2T_x$, and MILD-$Ti_3C_2T_x$ methods [1]; **e** XRD analysis dried films with TMA-5F-$Ti_3C_2T_x$, TMA-NH_4-$Ti_3C_2T_x$, and MILD-$Ti_3C_2T_x$ methods [1]; **f** XRD peak for Ti_3AlC_2 exfoliation using NH_4HF_2 for different sized particles, Adapted with permission from [43] Copyright (2023) Feng, A., et al.

the layers, the c-lattice parameter (c-LP) or the distance between the surfaces (d-spacing) increases and the MXene peak shifts to a smaller angle [29, 42]. It is worth noting that the increase/decrease of the interfacial distance depends on the size of the entangled molecules. For example, larger interlayer spacing results from using larger molecules. Therefore, the interfacial distance changes with the change in the size of the entangled molecules. By increasing the distance between the layers, the layers will be thinner and the functional groups (F^-, OH^-, and O^-) will be better accessible [47].

11.3 Heavy Metal Ions Adsorption

Considering that heavy metals are easily decomposed or metabolized in drinking water, they easily enter the food cycle and leave harmful effects by accumulating in body tissues [48]. The adsorption capacity of MXenes is higher and in some cases similar to other adsorbents due to adjustable surface chemistry, abundant surface functional groups, and high specific surface area [30, 31]. MXenes can absorb various heavy metal ions such as mercury (Hg), lead (Pb), chromium (Cr), and copper (Cu) as adsorbents [41]. The adsorption of target pollutants is increased by using MXenes by reacting with its surface groups (Table 11.1).

11.3.1 Lead

Lead (Pb), a hazardous pollutant, is regularly released from industry into the environment due to various industrial processes such as manufacturing batteries, paints, fertilizers, metal plating, and mining. For Pb(II) removal, carbon-based adsorbents, especially graphene oxide (GO), are used with advantages such as high adsorption capacity and fast equilibration time [64]. One of the main obstacles to using them as adsorbents in wastewater treatment is the difficulty of synthesis and high costs. Recently, $Ti_3C_2T_x$-MXene with excellent adsorption capacity through the mechanism of electrostatic interaction and ion exchange has been used as a super scavenger of heavy ions [2, 65]. Peng et al. synthesized 2D-alk MXene ($Ti_3C_2(OH/ONa)_xF_{2-x}$) via chemical exfoliation and alkalization intercalation lead adsorption equilibrium time of less than 2 min and an adsorption capacity of about 140.1 mg g^{-1} have been Obtained by 2D-alk MXene. Peng et al. (2014) have reduced Pb concentration to less than 2 μg/L in 4500 kg of water/kg of 2D-alk MXene [53]. June et al. (2020) showed that MXenes with a lower surface area (10 m^2 g^{-1}) compared to the surface area of powdered activated carbon (~470 m^2 g^{-1}) and higher negative surface charge have a higher adsorption capacity of Pb(II). Using XPS and FTIR analysis, it was shown that the main mechanism of attraction was ion exchange, electrostatic attraction, and intrasphere complex formation [66]. Wang et al. (2020) have functionalized Ti_2CT_X MXene nanosheets by EHL (a non-ionic biosurfactant). The EHL-functionalized Ti_2CT_X's adsorption capacity increased to 232.9 mg g^{-1} for Pb (II) ions. Furthermore, the main reason for the Pb(II) adsorption ability of alk-MXene-NH_2 was the interaction of ONa, NH_2, and OH complex. In addition to the increase in surface area with MXene modification, the adsorption of Pb(II)ions increases with van der Waals and electrostatic interactions. Various biosurfactants, including lignosulfonate (LS, an anionic surfactant), enzymatically hydrolyzed lignin (EHL, a non-ionic surfactant), and chitosan (CS, a cationic surfactant), were used to modify Ti_2CT_x nanosheets. The favorable elimination capacity of 232.9 mg g^{-1} for lead ions has been exhibited by EHL-functionalized Ti_2CT_x [67].

11.3.2 Copper

Having an important role in various physiological processes of the human body, copper (Cu) can cause numerous health problems such as cardiovascular diseases

Table 11.1 Summary of the adsorption of heavy metal ions onto $Ti_3C_2T_x$–MXene nanosheets [41]

MXene materials	Target	Reaction conditions					q_{max} (mg/g)	Isotherms	Kinetics	Reusability (Condition/cycles)	Mechanism	Ref.
		m/V (g/L)	C_0 (mg/L)	pH	Time (m)	Temperature (K)						
$Ti_3C_2T_x$	Ba(II)	5	55	3–9	60	298	9.3	Freundlich	Pseudo-second order	–	Chelation	[49]
Alk-$Ti_3C_2T_x$	Ba(II)	0.5	100	7	24 h	298	46.46	Langmuir	Pseudo-second order	–	Ion exchange	[50]
$Ti_3C_2T_x$	Ba(II)	–	–	9	1 h	298	175.1	Freundlich	Pseudo-second order	–	Ion exchange with H^+ or Na^+	[51]
	Sr(II)	–	–	9	1 h	298	225.2					
$Ti_3C_2T_x$-DMSO-hydrated	U(VI)	0.4	100	5	360	298	214	Freundlich	Pseudo-second order	0.2 M HNO_3	Electrostatic interaction; encapsulation	[36]
Zero valent Fe/ Alk-Ti_3C_2Tx	U(VI)	0.8	200	3.5	24 h	298	1315	–	Pseudo-second order	–	Reduction	[31]
$Ti_3C_2T_x$	Cs(I)	1	10	6	30	298	25.4	Freundlich	Pseudo-second order	0.2 M HCL/5 cycles	Electrostatic interaction	[52]
$Ti_3C_2T_x$	Cs(I)	0.1	2	7	60	293	148	Freundlich	Pseudo-second order	1 M HCL/4 cycles	Ion exchange	[51]
$Ti_3C_2T_x$-Nanosheet	Pb(II)	1	20	5	120	298	285.9	Langmuir	Pseudo-second order	–	Ion exchange; surface complexation	[53]
$Ti_3C_2T_x$-Nanofibres	Pb(II)	1	20	5	120	298	285.9	Langmuir	Pseudo-second order	–	Ion exchange; surface complexation	[53]

(continued)

Table 11.1 (continued)

MXene materials	Target	Reaction conditions					q_{max} (mg/g)	Isotherms	Kinetics	Reusability (Condition/cycles)	Mechanism	Ref.
		m/V (g/L)	C_0 (mg/L)	pH	Time (m)	Temperature (K)						
$Ti_3C_2T_x$-KH570	Pb(II)	3.2	500	–	30	303	148.0	Langmuir	Pseudo-second order	0.5 M Na_2EDTA/4 cycles	Chelation or coordination	[54]
Alk-MXene-NH_2	Pb(II)	0.1	25	6.3	20	298	384.6	Langmuir	Pseudo-second order	0.5 M Na_2EDTA/3 cycles	Complexation; van der Waals forces; elctrostatic effects	[55]
DL-$Ti_3C_2T_x$	Cu(II)	0.5	25	5.0	3	298	78.5	Freundlich	Pseudo-second order	HNO_3 and $CA(NO_3)_2$/3 cycles	Ion exchange; inner-sphere complexation	[56]
	Cu(II)	1.0	95.3	–	15	–	87.6					
$Ti_3C_2T_x$-PDOPA	Cu(II)	0.2	10	7.0	–	298	65.1	Freundlich	Pseudo-first order	–	Electrostatic interaction; chelation	[57]
SA/MXene/CFO	Cu(II)	0.225	–	5.5	–	298	234.3	Langmuir	Pseudo-second order	0.05% HCl/5 cycles	Surface complexation	[58]
$Ti_3C_2T_x$/PMPD	Cr(VI)	0.1	100	2.0	10	303	540.5	Redlich-Peterson	Pseudo-second order	M NaoH/5 cycles	Reduction; chelation; electrostatic interaction	[59]
$Ti_3C_2T_x/TiO_2$	Cr(VI)	0.05	50	2	100	–	–				Reduction	[60]

(continued)

Table 11.1 (continued)

MXene materials	Target	Reaction conditions					q_{max} (mg/g)	Isotherms	Kinetics	Reusability (Condition/cycles)	Mechanism	Ref.
		m/V (g/L)	C_0 (mg/L)	pH	Time (m)	Temperature (K)						
NH_2-$Ti_3C_2T_x$	Cr(VI)	0.67	100	2.5	120	298	107.4	Langmuir	Pseudo-second order	M NaoH and 0.5 M HCl/6 cycles	Adsorption; reduction	[61]
Ti_3C_2-Imidazole hydride	Cr(VI)	0.33	30	2	45	298	184	Langmuir	Pseudo-second order	–	Reduction	[62]
$Ti_3C_2T_x$-alginate aerogel	Cr(VI)	0.2	–	2	100	–	539	Langmuir	Pseudo-second order	–	Reduction	[62]
MnO_2-$Ti_3C_2T_x$	Cr(VI)	1	5	3	120	–		Freundlich	Pseudo-second order	0.75 M HCL/5 cycles	–	[63]

(CVDs), disorders related to oxidative stress, and cognitive decline in cases of increased concentration in the human body [68]. MXene, with its unique structure and abundant groups, is used to absorb Cu(II) ions from wastewater. In addition, the surface-activated hydroxyls impart properties such as fast adsorption rate, high adsorption capacity, reversible adsorption, and high sensitivity to titanium MXene substrates. Shahrazad et al. (2017) found that (DL)-$Ti_3C_2T_x$ showed a high adsorption capacity for Cu(II) ions (78.45 mg g^{-1}) in less than one min [56]. Accordingly, in other environmental problems, functionalized $Ti_3C_2T_x$ can be used to absorb other heavy metal ions. In addition, MXene can be functionalized with various functional groups for efficient adsorption of Cu(II) ions. For example, high adsorption (87.6 mg g^{-1}) at short equilibration time (15 min) through ion-exchange bonding mechanisms with abundant oxygen-containing groups and chemical coordination has been demonstrated for alginate-MXene composites. Amino acids react with $Ti_3C_2T_x$ to form a rutile phase of titanium dioxide (TiO_2) on MXene layers (d-$Ti_3C_2T_x$). With this modification, Elumalai et al. (2020) showed an adsorption capacity of 94.6 mg/g during 5 min. It is worth noting that the potential application of 2D $Ti_3C_2T_x$ will be effective in real pollution treatment by converting halides to a low oxidation state [32, 41].

11.3.3 Chromium

Chromium (Cr), an inorganic pollutant, is generated by industries such as refractory, metallurgy, electroplating, and leather tanning. It typically exists in oxidation states of +2, +3, and +6; however, Cr^{2+} is unstable and its hydrolysis is not well understood. The hydrolysis of Cr(III) involves a complex process resulting in the formation of polynuclear species such as $Cr_2(OH)_2$ and $Cr_3(OH)_4{}^{5+}$, as well as neutral species like $Cr(OH)_3{}^0$, and mononuclear species including $CrOH^{2+}$, $Cr(OH)_2{}^+$, and $Cr(OH)_4{}^-$. Conversely, the hydrolysis products of Cr^{6+} consist primarily of anionic and neutral species, notably $Cr_2O_7{}^{2-}$, $CrO_4{}^{2-}$, and $HCrO_4{}^{2-}$ [33]. $CrO_4{}^{2-}$ occurs when pH is <6.5, and $Cr_2O_7{}^{2-}$ is dominant when pH is low, and the concentration of Cr (VI) solution is high [34]. Compared to Cr(III), the higher mobility and solubility of Cr(VI) in water gives rise to toxic Cr(VI) compounds. In many studies, the practical method of effective conversion of Cr(VI) to Cr(III) is used to reduce Cr pollution. Recently, MXenes have been used as scavengers to remove Cr(VI). High adsorption speed, high adsorption capacity, and reversible adsorption are created as a result of the presence of active hydroxyls on the surface of MXene. Interestingly, $Ti_3C_2T_x$ adsorbs pollutants by changing them to low oxidation states. By etching by HF, and layering by ultrasonic, they were able to synthesize $Ti_3C_2T_x$ nanosheets. These nanoplates show a high potential in removing Cr(VI) ions from wastewater by reducing chromium ions to lower oxidation and with an adsorption capacity of 250 mg/g [35]. Karthikeyan et al. (2021) modified Mxenes using 0.1 M NaOH and showed the efficiency of adsorption and desorption in 5 steps with a slight decrease in adsorption. Mechanisms of surface complexation, ion exchange, and electrostatic interaction are involved in the elimination of Cr(VI) removal by Mxenes. The potential of MXenes in the adsorption of Cr(VI) ions can be increased by adjusting the distance between the layers and the surface charge [69]. $Ti_3C_2T_x$ multilayer nanosheets showed a maximum adsorption capacity of 107.4 mg/g for Cr(VI) based on the Langmuir model. $Ti_3C_2T_x$ is covalently bonded with triethoxysilane (3-aminopropyl) (APTES) and has a positive charge. Meanwhile, in the removal process of Cr(VI), $-NH_3^+$, and Ti(II) are oxidized to NO_3^- and Ti(VI), respectively. In addition, nanosheets show reusability and high selectivity [41, 61].

11.4 Radionuclide Pollutant Adsorption

As a result of using nuclear energy to supply the lack of traditional fossil fuels, a large amount of radionuclides such as uranium (235238U), strontium (90Sr), cesium (137Cs), barium (133140Ba), and thorium (232Th) enter the environment. These radionuclides with high solubility in water and having a long half-life, even in low concentrations with long-term chemical and radiological toxicity, are extremely harmful to human health and can cause irreparable damage to human health and the environment. In recent years, $Ti_3C_2T_x$ has attracted attention in nuclear waste management due to its unique advantages as an adsorbent [60, 70].

11.4.1 Uranium

Uranium (U) is widely used in the nuclear fuel cycle with high radioactivity and chemical toxicity [71]. The main uranium species found in water solutions is U(VI), which can be readily released into water during the disposal of nuclear waste, leading to widespread radioactive contamination. Consequently, there is an urgent need for the extraction and immobilization of U(VI) ions to prevent their dispersion and transformation within aquatic environments [72]. To remove U(VI) ions, various adsorbents such as nZVI, GO, CNT, COF, and MOF with high adsorption potential have been investigated [73–78]. Hydroxylated titanium carbide $Ti_3C_2(OH)_2$, with a theoretical adsorption capacity of 595.3 mg/g, is a promising adsorbent for removing uranyl ions [71]. Uranyl ions preferentially bind to deprotonated O adsorption sites on the hydroxyl surface, rather than protonated sites. The dominant mechanisms of chemical bonds and chemical interaction are effective in adsorption. The role of the ion exchange mechanism in the adsorption of U(VI) can be described by the deprotonation of the hydroxyl group after bonding with U(VI). Increasing the interlayer space of $Ti_3C_2T_x$ during synthesis is a simple solution to maintain the stability of MXenes in hydrated conditions. U(VI) ions can be extensively trapped inside $Ti_3C_2T_x$ multilayers by increasing the interlayer space. High adsorption capacity of U(VI) ions was obtained with $Ti_3C_2T_x$ MXene prepared with DMSO, (214 mg g^{-1}), compared to

$Ti_3C_2T_x$-dry (26 mg g^{-1}) [36]. In another study, an adsorption capacity of 470 mg/g was shown at pH 3.0 for the efficient removal of U(VI) through the strategy of reducing highly mobile U(VI) to less soluble U(IV) using TiC_2T_x. The researchers confirmed simultaneous adsorption and reduction of U(VI) ions over a wide pH range by Ti-based MXene. In addition, immobilization due to U(VI) reduction by Ti_2CT_x MXenes was also demonstrated. This promotes the study and investigation of removing and other oxidized pollutants using titanium-based materials [41, 79]. Furthermore, Zhang et al. (2021) reported that the surface of $Ti_3C_2T_x$ MXenes was functionalized by linking diazonium salt with amidoxime functional groups, and as a result, the selectivity for uranium adsorption should be increased with this modification. The stability of $Ti_3C_2T_x$ in aqueous and uranyl solutions and the selectivity for uranium adsorption by functionalizing the surface of $Ti_3C_2T_x$ provide an increase and efficient and rapid absorption of uranium from aqueous solutions containing competitive metal ions [80].

11.4.2 Strontium

Strontium (Sr) appears during the nuclear fuel cycle as one of the radioactive isotopes 89Sr and 90Sr [81]. The solubility of Sr ions easily enters the food chain in drinking water and thus leads to bone cancer by depositing in the bones. Sr^{2+}, unlike $SrOH^+$ which is present only in high pH environments, dominates in almost the entire pH range of the species. Many studies have been conducted to use an effective and affordable adsorbent that can be regenerated and recovered. For example, hydroxyapatite materials [82, 83], ion exchangers [84], Aerogel [85, 86], magnetic nanomaterials [87], carbon materials [88], and clay minerals [89] have been employed to remove Sr (II) ions. Jun et al. (2020) evaluated the adsorption of Sr(II) in the presence of $CaCl_2$, NaCl, and other ions at different pH values with $Ti_3C_2T_x$ MXene adsorbent. Electrostatic attraction is the main mechanism of adsorption of Sr(II) ions using $Ti_3C_2T_x$ MXene. The adsorption rate of 225 mg/g in 1-h shows that the adsorption process of Sr(II) ions with $Ti_3C_2T_x$ MXene has fast kinetics and high potential. In addition, the adsorption mechanisms of ion exchange and surface complexation of the inner sphere have been shown through XPS and FTIR results [51]. Shahzad et al. (2021) exfoliated Ti_2AlC (211) MAX phase via hydrothermal alkalization to produce different nanostructured materials for removal of Sr(II) ions. Because of its porosity and higher specific surface area, the fabricated Alk-Ti_2C nanosheet effectively removed Sr(II) ions with the maximum removal rate of 296.46 mg g^{-1}. Furthermore, the excellent adsorption of Sr(II) by Alk-Ti_2C in seawater (3543.33 μg g^{-1}) means its promising future applications in different environments [90].

11.4.3 Barium

Barium (Ba), a white and shiny metal, exists in the environment in the form of various compounds such as barium halide, barium nitrite, barium sulfide, and barium oxide. Hydrogen in water is produced from the oxidation of Ba ions. By spreading in wide areas, barium species will have a great impact on the health of humans and aquatic animals due to the solubility and compatibility of Ba ions in water [91, 92]. The EPA of the United States of America and the World Health Organization (WHO) have determined the Ba ion limit in water to be 2 mg l^{-1} and 0.7 mg l^{-1}, respectively. Fred et al. (2017) showed that 90% of Ba ions were selectively removed in the first 10 min of contact time by exfoliated MXene as adsorbent (with a removal ability of 9.3 mg g^{-1} and a removal efficiency of 100%) through the mechanism Physical adsorption was removed from the solution. In addition, by using MXene surface functional groups (F^-, O^-, and OH^-) adsorption is also done chemically. Barium ions in solution with functional groups produce barium hydroxide and barium fluoride through chemical bonding [49]. Mu et al. (2018) used NaOH to improve Ba adsorption potential by $Ti_3C_2T_x$ and increase surface activity [50]. The surface interaction affinity towards Ba ion in solution was increased by increasing the c-lattice parameter up to 2.09 nm by this modification. Through physical and chemical adsorption with O^-, OH^-, and F^- surface functional groups, a high adsorption potential of Ba ions up to 46.46 mg/g was obtained with $Ti_3C_2T_x$ modified with NaOH. The following reaction shows the chemical adsorption mechanism of Ba ions with $Ti_3C_2T_x$ surface functional groups:

$$Ba^{2+} + 2OH^- \rightarrow Ba(OH)_2 \text{ and } Ba^{2+} + 2F^- \rightarrow Ba(F)_2 \quad (1)$$

Jun et al. (2020) increased the adsorption of barium and strontium ions from nuclear wastewater using $Ti_3C_2T_x$ and through the electrostatic attraction of the negatively charged surface of nanosheets. XPS, FTIR, and kinetic and isothermal studies revealed the mechanisms of chemical ion exchange, inner sphere complexation, and chemisorption for barium uptake by $Ti_3C_2T_x$. The adsorption potential of barium with $Ti_3C_2T_x$ in this study is reported to be 180 mg/g. Four cycles of desorption and reabsorption have been performed for this adsorbent [41, 51].

11.4.4 Cesium

Cesium-137 is created as a radioactive isotope during the fission process of radioactive metals, especially uranium. In nuclear energy production, the fission process is widely used. A huge amount of nuclear waste has been released into the environment in nuclear disasters such as Fukushima and Chernobyl [93]. Recently, Cs-137, a strong beta and gamma emitter with a half-life of 30.17, has turned in to a major problem in nuclear waste management [94]. For the first time, Cs^+ removal from nuclear wastewater has been performed, using $Ti_3C_2T_x$. In this work, the researchers evaluated their findings of Cs^+ adsorption by $Ti_3C_2T_x$ with the Langmuir and Freundlich isotherm models and reported that the adsorption process is consistent with the Freundlich isotherm model. In addition, in the contact time of 1 min, the maximum adsorption capacity of 25.4 mg g^{-1} was obtained. The data show that the behavior of the multilayer adsorbent is highly influenced by the active adsorption sites with different energy (containing hydroxyl, fluorine, and oxygen groups). Also, the researchers show the high selectivity of the adsorbent in the presence of a high concentration of competing cations (Li^+, Sr^{2+}, K^+, Na^+, and Mg^{2+}) in the space between the layers and the active sites of adsorption [52]. In another study, the removal of Cs^+ ions from nuclear wastewater was evaluated with $Ti_3C_2T_x$ MXene and compared with the adsorption potential of porous activated carbon (PAC). The researchers showed that $Ti_3C_2T_x$, having a more negative surface charge, has a higher adsorption potential of 148 mg g^{-1} for Cs^+ ions compared to PAC with an adsorption potential of 80 mg g^{-1}). On the other hand, the surface area of $Ti_3C_2T_x$ (10 m^2 g^{-1}) is much less compared to the surface area of PAC (470 m^2 g^{-1}). The authors indicated that the key role in determining the adsorption capacity of Cs^+ ions is the electrostatic interactions between Cs^+ ions and the adsorbent. Also, based on their previous experiments on Cs^+ ions in the presence of various cations (Ca^{2+}, Mg^{2+}, K^+) and organic acids (sodium oleate, oxalic acid, and citric acid), they had predicted that the ion exchange mechanism plays a role in removing Cs^+ ions [51]. In another study, various mechanisms including immobilization, electrostatic interaction, ion exchange, reduction, and coordination, the main mechanisms of absorption of radionuclides with $Ti_3C_2T_x$ have been shown (Table 11.1). Since the concentration of radionuclides is much lower than other pollutants in the environment, it is necessary to consider the selectivity and stability of the adsorbent in the adsorption process with $Ti_3C_2T_x$. In order to selectively adsorb radionuclides from wastewater, $Ti_3C_2T_x$ can be synthesized using specific functional groups through surface modification [41]. Further studies are needed to demonstrate the widespread applications of titanium-based adsorbents.

11.5 Factors Effective on Adsorption

11.5.1 pH Solution

pH is an important parameter affecting the adsorption process. The concentration of H^+ in water determines the pH value and the surface charge of MXenes and is effective on the adsorption efficiency of target cations by surface functional groups. pHpzc is the specific pH at which the zeta potential value is 0. At acidic pH, the concentration of H^+ ions in the medium is high and competes with the target cations to react with negatively charged functional groups such as F^-, OH^-, and O^-. As a result of this competition, H^+ ions protonate the MXene surface by electrostatic reactions with negatively charged functional groups [14, 56, 95]. Therefore, no electrostatic reaction occurs between the surface of protonated MXene (no negative charge) and target cations [95]. As the concentration of H^+ ions decreases and the pH increases, the surface of MXenes will be negatively charged. Therefore, with the negative surface charge of MXenes, the electrostatic force of attraction between the adsorbent and the adsorbed increases, and the rate of absorption of heavy metal ions and radionuclides increases [10]. With the increase of hydroxide ions in the environment and increasing the pH value to basic conditions, the metal ions in the environment are oxidized and metal absorption with MXenes decreases. Although more negatively charged MXenes occur at higher pH values [49]. Electrostatically, co-precipitation and strong surface coordination under neutral pH conditions increase the rate of removal. Therefore, higher pH is not always preferred for absorption of heavy metals and radionuclides [45, 50, 52, 70, 79, 96, 97]. The negatively charged target pollutants could adsorbed by the positively charged MXenes. In Wang et al.'s study, in the pH range of 1–10, MXenes remained positively charged and adsorption by MXene was almost independent of the pH value [60]. A limited option is the adsorption of anions by MXenes in a highly acidic medium. The positive charge of MXenes to remove ReO_4 anions in a variable pH range of 1–10 is shown by analyzing the zeta potential value of Ti_2CT_x nanosheets. The dissolution Ti at very low pH values reduces the stability of MXenes. It has been shown that this reduction in stability can be controlled to some extent by modification. The surface area of Ti_2CT_x MXenes in the reaction with diallylmethylammonium chloride decreased with increasing solution pH. Therefore, the adsorption efficiency of anions by Ti_2CT_x-modified multilayer MXenes was lower than other MXenes. On the other hand, electrostatic repulsion is created between the inner layers of unmodified MXene with a negative charge [60].

11.5.2 Temperature

Temperature is an effective parameter on physical and chemical characteristics, as the temperature increases, the absorption efficiency of heavy metal ions increases. By evaluating the standard free energy, standard entropy, and standard enthalpy, the effect of temperature on the absorption process can be checked [14, 95]. With increasing temperature, standard free energy, entropy, and enthalpy decrease, increase, and increase respectively. A negative change in the standard free energy indicates a spontaneous reaction, a positive change in the standard enthalpy indicates that the reaction is endothermic, and a positive change in the standard entropy is a possible reaction. The effect of temperature on the adsorption process has been widely investigated [52, 53, 56, 70, 96]. The effect of temperature on adsorption behavior is shown by adsorption isotherm curves. Dubinin Radushkevich, Langmuir and Freundlich models are the main adsorption isotherms [98]. The agreement of the adsorption of heavy metal ions by MXene with the Langmuir model shows that the adsorption is carried out as a monolayer of MXene covered with homogeneous functional groups. In addition, with the Langmuir isotherm simulation model, the maximum rate of adsorption of heavy metal ions and radionuclides can be achieved [52, 53, 56, 70, 96]. By studying the effect of temperature on the adsorption process of copper and lead ions by alginate/Ti_3C_2 adsorbent, Dong et al. showed that the absorption process increases by 10% in the temperature range of 298–333 K. However, they have finally shown the main mechanism of absorption increases independent of temperature [99]. Adsorption of Pb(II) ions with MXenes at different temperatures (293, 303, and 313 K) has been investigated and shown that, unlike previous studies, the values of enthalpy and entropy changes were negative. The negative values indicate the reversible, spontaneous, and exothermic nature of the adsorption reaction with MXene [100].

11.5.3 Background Ions

Various ions such as Na^+, Mg^{2+}, Ca^{2+}, SO_4^{2-}, PO_4^{3-}, and Cl^- are often present in industrial wastewater. The presence of these ions in competition with the target cation or anion can harm the adsorption performance of MXenes. Therefore, it is necessary to check the adsorption performance of MXenes for the target cation or anion in the presence of other available ions to use MXenes in urban and industrial wastewater treatment. In general, one of the effective features of the adsorbent for the target pollutant is the selectivity and high affinity of the adsorbent to the desired pollutant in the presence of other ions [36, 50, 101]. The presence of Ca^{2+} and Mg^{2+} cations in wastewater makes it hard. Therefore, it is necessary to investigate the ionic competition of MXene in the presence of Ca^{2+} and Mg^{2+} cations. Karthikeyan et al. (2020) investigated the selectivity of chromium adsorption by $Ti_3C_2T_x$ in the presence of bicarbonate, chloride, and sulfate anions and showed that the chromium adsorption potential decreased in the presence of anions. The adsorption potential of chromium by $Ti_3C_2T_x$ is reported to be the lowest in the presence of negatively charged sulfate and the highest in the presence of negatively charged bicarbonate. The simultaneous presence of anions or cations harms the adsorption of pollutants by MXene, which is due to electrostatic adsorption, the main mechanism of adsorption, between the target pollutant and the MXene surface [69]. By investigating the adsorption of Re by DL-methylammonium chloride in the presence of ClO_4^-, Cl^-, SO_4^{2-}, and NO_3^- anions at pH 4, the effect of the presence of anions on the adsorption of the target anion with MXene was investigated. Wang et al. (2019) showed that in the presence of SO_4^{2-}, Cl^-, methyl ammonium chloride DALLD has a high selectivity for adsorbing Re ions. The ionic radius, chemical composition, and hydration energy similar to Re reduces the adsorption capacity of ClO_4^-. On the other hand, NO_3^- anion has been shown to have the highest competition with Re for adsorption by DLMD [60].

11.5.4 Contact Time

Contact time is an important parameter in deciding the possible use of adsorbent for the adsorption process and the efficiency of the process. In addition, increasing the contact time leads to an increase in the dimensions and size of the column and higher energy requirements. Therefore, increasing the contact time can lead to an increase in the costs of the recruitment process [101]. Therefore, to perform the absorption process in a shorter time, it is necessary to use faster kinetics. Elwich, Zeldwich, Lagergren, quasi-first-order, and pseudo-second-order kinetic models are used to investigate the practical potential and application of the adsorbent and the reaction speed of heavy metal absorption and contact time on the absorption process [10, 102, 103]. In general, the heavy metal ions by $Ti_3C_2T_x$ are two stages until reaching the channel equilibrium: the first stage where the highest rate of adsorption occurs and is very fast, and the second stage is slow and continues until the equilibrium is reached. The pseudo-second-order kinetic model shows the best agreement with the experimental data. This model shows the predominance of the chemisorption mechanism in the adsorption of heavy metals by MXenes.

11.5.5 Adsorbent Dosage

The adsorbent dose is a parameter that affects the absorption capacity, and its optimization is necessary to save the absorption process. To have the maximum interaction between the adsorbent and the adsorbed and to achieve the maximum adsorption efficiency, the optimization of the adsorbent dosage is required [71]. With the increase of active sites and MXenes content, the adsorption efficiency increases. By increasing the adsorbent dose, the reaction sites on the surface are involved and there is a possibility of decreasing the adsorption capacity [49] (Fig. 11.2).

11.6 Heavy Metals Ion and Radionuclide Adsorption Mechanism

Adsorption is a highly efficient technique for capturing heavy metal ions by MXenes through hydroxide, fluoride, and oxygen-containing functional groups [10]. The existence of too many functional groups and reaction sites on the surface of MXenes have made them typical removers to efficiently capture radionuclides and heavy metal ions through the mechanism of adsorption reaction. The common removal mechanisms are electrostatic adsorption, hydrogen bonding, specific surface bonding, and ion exchange. Radionuclides and heavy metals are adsorbed onto the surface of Ti_3C_2Tx MXenes through different mechanisms, especially electrostatic interaction, which are highly dependent on the oxidation state of metal ions, as well as the solution

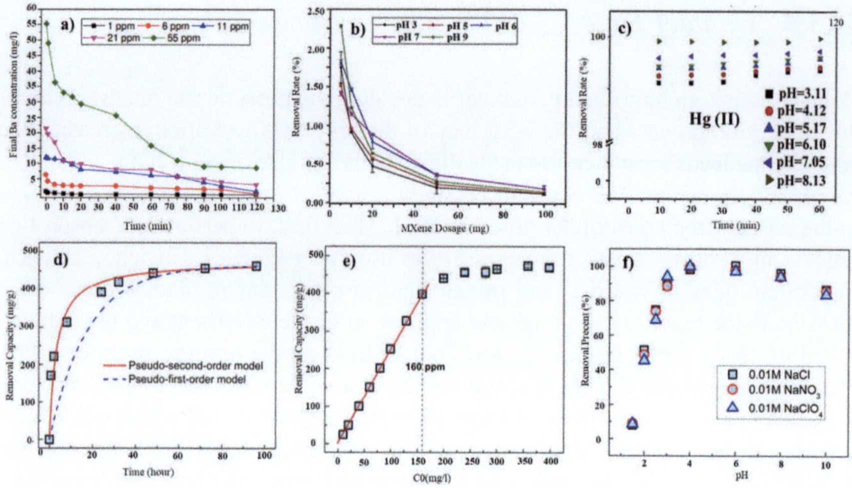

Fig. 11.2 **a** Effect of contact time on removal of Ba(II) ions by MXenes at different pH values as function of final Ba(II) concentration [49]; **b** Effect of pH on the removal efficiency of Ba(II) ions using MXene, Adapted with permission from [49] Copyright (2023 Fard, A.K., et al.; **c** Effect of pH on removing 60 mg L^{-1} of Hg(II) ions onto Ti_3C_2Tx nanosheets (Liu et al., 2014); **d** Removal of U(VI) ions from aqueous solution by multi-layered Ti_2CT_x as a function of contact time; **e** Initial U(VI) concentration; and **f** Initial U(VI) pH [104]

pH because it changes the adsorbent surface's characteristics. Titanium has a great similarity to other metal ions, and may be trapped in the layered structure of MXene nanosheets having large surface areas. Moreover, the ion-exchange between the functional groups and metal ions on MXene surface can lead to electrostatic interaction. One can use Freundlich model to better describe the isotherm data of the multilayer removal of heavy metals by MXene nanomaterials. The mechanisms of interaction of heavy metal ions with novel MXene are usually multi-factorial in the real environment [41] (Fig. 11.3).

11.7 Removal of Dye

Among many organic pollutants, one of the main components of textile industry wastewater is dye. Color, with huge annual (about 100 tons), its wide variety and release is a drastic threat to the environment. MXenes with excellent photocatalytic performance are very efficient for the removal of various dyes from wastewater [107]. Tirumal et al. (2021) degraded dye molecules with $Ti_3C_2T_x$ with 98% efficiency and showed the great potential of MXene-based materials for remove toxic organic pollutants from wastewater. The wide use of titanium dioxide (TiO_2) in the photocatalytic degradation of pollutants is considered as the most promising catalytic degradation

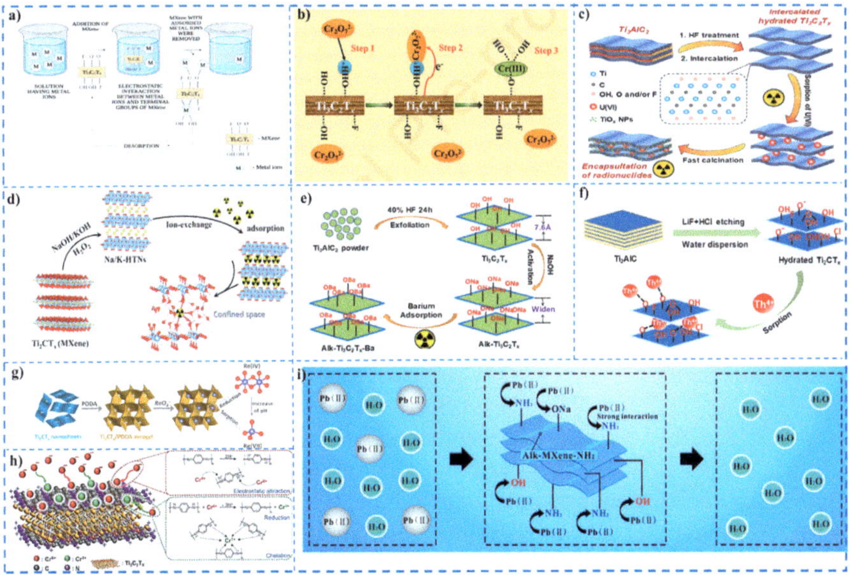

Fig. 11.3 **a** Removal of metal ions by MXenes, Adapted with permission from [105] Copyright (2023) Velusamy, K., et al.; **b** The mechanism of removing Cr(VI) by $Ti_3C_2T_x$ nanomaterial [35]; **c** Hydrated intercalation synthesis strategy of $Ti_3C_2T_x$ MXenes for efficient uptake and imprisonment of U(VI) ions [97]; **d** Synthesis of MXene-derived hierarchical titanate nanostructures (HTNs) for effective sequestration of radionuclides [106]; **e** Adsorption of radioactive Ba ions, Adapted with permission from [50] Copyright (2023) Mu, W., et al.; **f** Sorption of thorium ions onto Ti_2CTx nano-sheets, Adapted with permission from [70] Copyright (2023) Li, S., et al.; **g** Schematic of the sorption of Re(VII) ions by Ti_2CTx/PDDA aerogel [60]; **h** adsorption mechanism Cr(VI) of $Ti_3C_2T_x$/PmPD [59] and **i** Adsorption mechanism of lead ions by amino functionalized $Ti_3C_2T_x$ [55]

material [108]. In addition, Hieu et al. (2021) found that the degradation of the Ti_3C_2–TiO_2 photocatalyst for MO increased to an efficiency of about 99% after exposure to light for 40 min. In addition, TiO_2–Ti_3C_2 composite showed good photocatalytic activity even after 5 cycles. The band gap of BFO can be modified by metal ions such as gadolinium (Gd^{3+}) and tin (Sn^{4+}) [109]. Tariq et al. (2018) doped BFO in different ratios with Sn^{4+} and Gd^{3+} through sol–gel technique. The nanohybrid showed the degradation efficiency of CR dye with 100% efficiency in 120 min [110]. Furthermore, BFO nanoparticles doped with La and Mn embedded in $Ti_3C_2T_x$ MXene were obtained through a dual solvent sol–gel approach and the catalytic activity for CR degradation reached 100% within 20 min (Iqbal et al. 2019a). $BiOBr/Ti_3C_2$ (BTC) layered composites showed higher visible light photocatalytic degradation activity for RhB than pure BiOBr [111]. Chen et al. (2020) fabricated TiO_2/MXene composites after modification with various reagents such as C_2H_6O and C_3H_8O with thermal solution approach. They found that the electron transfer efficiency increases with the uniform growth of TiO_2 on the surface of $Ti_3C_2T_x$ MXene, and $Ti_3C_2T_x$ is used as a cocatalyst to capture the electrons generated by light. Moreover, the $TiO_2/Ti_3C_2T_x$

composite boosts the breakdown of organic substances and efficiently hinders the recombination of electron–hole pairs generated during photodegradation. Among the composite materials, the $TiO_2/Ti_3C_2T_x–C_3H_8O$ composite has shown the best photocatalytic degradation efficiency for methyl orange (MO, 90.5%) with an efficiency of about 1.5 times higher than the single TiO_2 material [112].

11.8 Gas Separation

During the past decade, two-dimensional nanomaterials have been the center of focus for advanced separations by their structural, thermal, and mechanical properties [113, 114]. MXene is useful in many different fields, such as the potential to adsorb gases such as carbon dioxide, and the conductivity of metals such as electricity, it is used in a large number of applications, including stored energy, optoelectronics, biomedicine, communications, and the environment such as the removal of pollutants. Using MXene for gas separation gain exceptional interest, as their properties to control and be selective over the kind of gases and also low cost and regenerable [115, 116].

11.8.1 Surface Chemistry of MXenes

The Effect of the Synthesis Method on Functional Groups

As mentioned, MXenes are synthesized by usual top-down method via wet chemical etching always end with functional groups such as –O, –F, and –OH. Functional groups are important in MXenes because their properties such as energy storage capacity, magnetism, band gap, gas adsorption, and gas storage are directly affected by them [117]. To better control material properties, it is critical to understand the effect of preparation sample on the type and surface concentration of functional groups. Even so diffraction methods have the advantage of obtaining moderate structural information from a large sample volume, XRD is not sensitive to elements with low atomic number such as hydrogen, and quantitative analysis of OH and hydrogen bonds is difficult. In addition, a comprehensive structure model should be constructed to correctly represent the structure of MXene, considering the complex and heterogeneous structures and crystal structure of the MXene phase, such as the distance between layers and the displacement between horizontal layers after etching. In multilevel structural modeling, neutron diffraction has been used to investigate MXene $Ti_3C_2T_x$ functional groups [118].

According to Wang et al. results, diffraction of neutrons has the advantage which H atoms are strong scatterers for neutrons and allow better quantification of –OH functional groups. Compared to conventional small box models multilevel structural modeling allows more desired structural details, including sequence or disorder, surface species characterization, distribution, and interactions. Analysis of structural by neutron diffraction and pair distribution function (PDF) was performed on three samples (HF48, HF10, and HF48-NH_4OH) prepared using different synthesis conditions. HF48 and HF10 were prepared by etching Ti_3AlC_2 powders in 48 wt% HF and 10 wt% HF solutions, respectively. Higher HF concentration leads to higher nanohole and atomic hole concentrations. HF concentration is expected to affect other structural parameters. The HF48-NH_4OH sample was prepared by soaking HF48 in an aqueous solution of ammonium hydroxide (NH_4OH) (28–30% based on NH_3) at room temperature. PDF files of three samples obtained by Fourier transform of the total scattering structure coefficient S(Q) to Q_{max} 25 Å-1 collected using neutron diffraction on the NOMAD beamline at the Spallation Neutron Source, Oak Ridge National Laboratory. All three PDFs exhibit similar positive and negative peaks but with different peak shapes. The first three peaks have a negative peak at 0.97 Å (O–H bond distance), a negative peak at 2.13 Å (Ti-C/O/F bond distance), and a positive peak at 3.04 Å (Ti–Ti and C-in-plane distance are). The peaks of HF10 are generally sharper and narrower than those of HF48 and HF48-NH_4OH, specifying that HF10 has a more regular crystal structure. Even for a very large bond distance ($r > \sim 45$ Å), distinct peaks can be easily detected above the noise level of the HF10 PDF, while the signal is more or less white noise at large bond distances for HF48 and HF48-NH_4OH is, therefore, the MXene structure of lower concentration HF has better long-range ordering than the larger coherent lateral dimensions of $Ti_3C_2T_x$ sheets, the lower degree of stacking error between Ti layers and smaller voids, which is consistent with the results presented using atomic resolution STEM imaging [118] (Fig. 11.4).

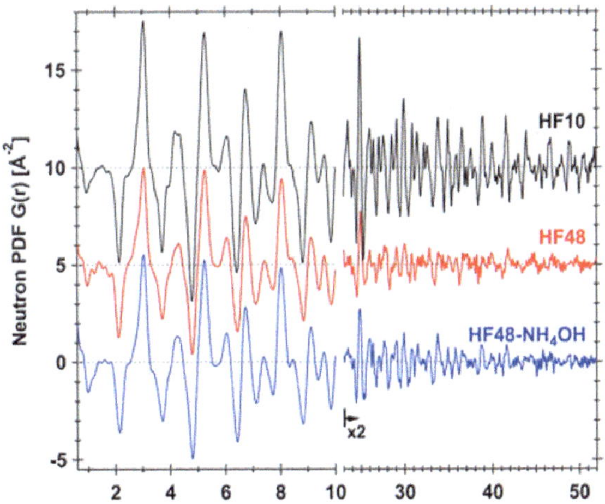

Fig. 11.4 PFD image of the three MXenes mentioned above [118]

11.8.2 Gas Adsorption

11.8.2.1 CO_2

Global warming become one of the challenges that we should consider due to using plenty of fossil fuels by mankind. To reduce the amount of CO_2, using various ways are common such as dry solid adsorbents with large surface volume such as zeolites, graphene oxide, and carbonaceous materials but the demonstration of two-dimensional transition metal carbides (goes by name of MXene) is a new ray of hope due to their high potential to separation and capture CO_2 [119]. However, despite these materials have various advantages, they have poor selectivity over N_2 in the action of CO_2 capture, [120] but they are predicted to capture 2–8 mol CO_2 kg^{-1} with high efficiency [121]. MXenes are stable on two-parameter compared with other adsorbents such as zeolite, which are chemical and thermal and therefore making them more capable of regeneration [119].

As it was mentioned, MXene has less CO_2 selectivity over N_2, therefore Ingemar Persson et al. demonstrate that using O-depleted MXene can lead to immediate complete CO_2 saturation of surface and constant 2D structure and no sight of N_2 adsorption [119]. MXene's CO_2 absorption capacity can be shown as Ti_2CT_x > V_2CT_x > Zr_2CT_x > Nb_2CT_x > Mo_2CT_x > Hr_2CT_x > Ta_2CT_x > W_2CT_x. The adsorption capacity was still high even at low CO_2 partial pressure and relatively high temperature. The higher CO_2 adsorption capacity of MXene can be attributed to the higher surface area, more adsorption sites, MXene → CO_2 charge transfer, and higher adsorption energy [122]. It was predicted that a higher CO conversion efficiency would be achieved by MXene due to its high CO_2 adsorption capacity even at high temperatures and low CO_2 partial pressure. The work assumed adsorption strength and adsorption structure is the main influence on CO_2 desorption.

Another important aspect for the industrial application of carbo absorption is the regeneration of CO_2 from the adsorbent and it is desirable if CO_2 removal occurs at mild temperature and mild pressure. Apart from adsorption strength, the number of interaction sites may be critical for CO_2 uptake and regeneration. Multiple sites contribute to higher CO_2 absorption capacity and lower energy consumption for CO_2 removal [123]. This may apply to MXenes and MXene-based composites. As a solid adsorbent, a desirable method to improve CO_2 absorption may be the addition of increasing ratio, size, and distribution of pores in MXenes. Furthermore, the absorption of CO_2 by MXene may be affected by many other factors. From the studies of Zhou et al., it can be seen that a certain higher level of prepared MXene increases the absorption of CO_2. In particular, the adsorption capacity by $Ti_3C_2T_x$ (21 m^2 g^{-1} surface area) was 5.79 mmol g^{-1}, which was comparable to conventional CO_2 adsorbent. In particular, the theoretical capacity of CO_2 adsorption by MXene with the highest surface area can reach 44.2 mmol g^{-1} [124]. This exhibited that adjusting the MXene level would be a favorable way to improve the CO_2 absorption capacity. One of the effective methods of designing and preparing ultra-thin MXene. However, using density functional theory shows the influence of MXene atomic layers on CO_2

absorption is relatively small. Experiments results on nine MXenes (i.e., Ti_2CT_x, Zr_2CT_x, Hf_2CT_x, V_2CT_x, Nb_2CT_x, Ta_2CT_x, Cr_2CT_x, Mo_2CT_x, and W_2CT_x) showed that the Gibbs free energy of CO_2 adsorption on MXene was all negative [125], which is indicating CO_2 absorption by MXene. It was thermodynamically favorable. Specifically, the values of Gibbs free energies of CO_2 adsorption by MXene were ordered as $Ti_2CT_x < Hf_2CT_x < Zr_2CT_x < V_2CT_x < Nb_2CT_x < Ta_2CT_x < Cr_2CT_x < Mo_2CT_x < W_2CT_x$.

Improving Mxene's Structure for CO_2 Capture

The inherent properties of MXenes are needed for development in various directions with wide variety of applications. their characteristics are established during the synthesis phase and are influenced by multiple factors including the precursor phase of Max, the etchant used, reaction duration, and temperature. The targeted design of MXene properties remains an important challenge as these parameters are difficult to control precisely. Here, the post-synthesis alteration of MXenes becomes crucial for conveniently and effectively adjusting and enhancing their properties. These approaches encompass techniques such as atomic doping, modification with functional groups, and integration with other materials, to tailor the electrical, mechanical, and chemical characteristics, as well as the hydrophilicity and hydrophobicity of MXenes [24, 126, 127]. In the layering stage, MXenes undergo exfoliation and triggered by mechanical vibrations or chemical agents (such as positively charged ions, polar organic molecules, and large organic alkaline molecules), they assemble into either multi-layered or single-layer [128]. However, during sonication, severe vibration reduces the size of MXene nanosheets, thereby leading to a significant loss in electrical and mechanical properties. While it is difficult to completely remove these factors in the subsequent processes, the use of chemical composition prevents the reduction of the MXene size to a certain extent. MXenes applications in various fields, including energy storage, catalysis, gas adsorption, electromagnetic shielding, flexible electronics, and chemical sensing are limited due to difference in practical properties from theoretical values. Therefore, post-processing modification methods theoretical calculations and different applications for MXenes are necessary to investigate. In Fig. 11.5, various methodologies for post-processing modification of MXenes are methodically outlined, aiming to offer a theoretical foundation and methodological assistance for enhancing the properties of MXenes. Meanwhile, doping with non-metallic elements that have weak electronegativity promotes the transition of MXenes from semiconductor to metal. The ratio of functional groups on the surface of MXenes is regulated by different post-treatment processes. The introduction of external functional groups gives new properties to MXene, and the combination of strengths of polymer materials and the connection between MXene nanosheets [129].

A series of research advances on the post-processing modification of MXenes in this section briefly states that they are divided into two categories: chemical modification (atomic doping and termination modification) and physical composites

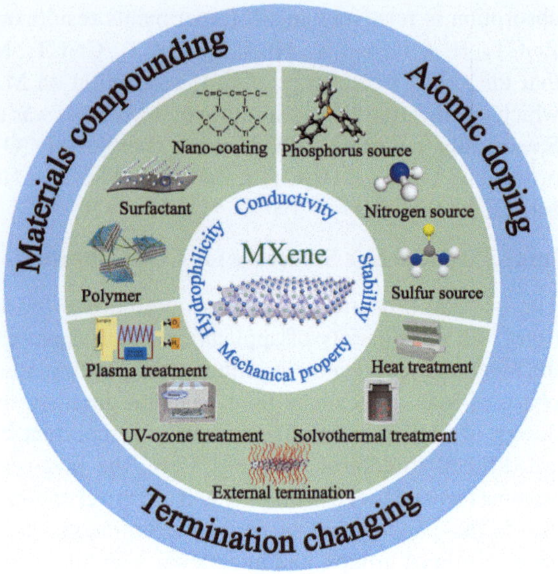

Fig. 11.5 Post-processing approaches and classification of MXenes, Adapted with permission from [129] Copyright (2023) Hao Tang, et al.

(composition of materials). In Table 11.2, methods of the specific process, conductance changes, stability, mechanical properties, and hydrophilicity of MXenes are summarized.

Modified MXenes by Atomic Doping

Based on findings examined by other researchers, incorporating additional electrons from elements like nitrogen, sulfur, phosphorus, and other dopants into MXene carbides can alter their band structure, thereby enhancing electrical conductivity [130]. Meanwhile, to promote the semiconductor-to-metal transition in power-enhancing nitrogen, carbide MXene was widely studied. In addition, the previously discovered carbides are now more commonly known in the MXene family. Mo_2CT_x MXene at 60 °C is transformed into two-dimensional nitrides, whose power enhancement ratio is large enough and can provide an unconventional route for the synthesis of nitride-based MXenes. Scientists have proposed a process based on nitrogen [130]. Yang et al., introduced a nitrogen enhancement process via a thermal solvent method, where an alcohol solution saturated with urea or monoethanolamine serves as a liquid nitrogen source for exfoliating Ti_3C_2 (d-Ti_3C_2). Deconvolution of the N1s peaks in the XPS spectra of UN-Ti_3C_2 (urea-saturated alcohol solution) and MN-Ti_3C_2 (monoethanolamine) was conducted to assess the successful incorporation of nitrogen atoms into Ti_3C_2. It is important to highlight that the introduction of N-5 significantly enhances the conductivity of Ti_3C_2 nanosheets owing to its exceptional electron-donating capability. Likewise, N-6 facilitates the connection of an electron pair and π-interlaced rings, demonstrating its electron-donating nature in UN-Ti_3C_2. Experimental findings corroborate this observation, with the resistance of MN-Ti_3C_2, UN-Ti_3C_2, and d-Ti_3C_2 measured at 3.1, 3.3, and 3.4 Ω

Table 11.2 Post-processing modification for MXenes [129]

Types	Process method	Conductivity	Stability	Mechanical property	Hydrophilicity
Atomic doping	NH_3; solvothermal	Increase	–	–	–
	$(C_6H_5)_3P$; heat treatment	Increase	–	–	–
	$(NH_2)_2CS$; heat treatment	Increase	–	–	–
Original termination changing	Vacuum; heat treatment	Increase	Increase	–	Decrease
	Ar; heat treatment	Increase	Increase	Increase	Decrease
	H_2; heat treatment	Increase	Decrease	–	–
	H_2 or O_2; plasma treatment	Increase (H_2); Decrease (O_2)	–	–	–
	Solvothermal in EtOH and DI	Increase	–	–	Increase
	UV-ozone treatment	Increase	–	–	–
External termination introducing	*p*-phenyl-SO_3H	Decrease	–	Decrease	–
	PFDTMS	–	Increase	–	Decrease
	PDMAEMA	–	–	–	Increase (low T/ CO_2 uptake); Decrease (high T/CO_2 release)
Materials compounding	PEDOT: PSS; vacuum filtration	Decrease	–	Increase	–
	PVA; vacuum filtration	Decrease	Increase	Increase	Decrease
	NR; vacuum filtration	Decrease	Increase	Increase	Decrease
	CNF; vacuum filtration	Decrease	–	Increase	–
	Glucose; hydrothermal and heat treatment	–	Increase	–	Decrease
	CTAB; regulating pH of aqueous phase	–	–	–	Increase (acidic); Decrease (alkaline)

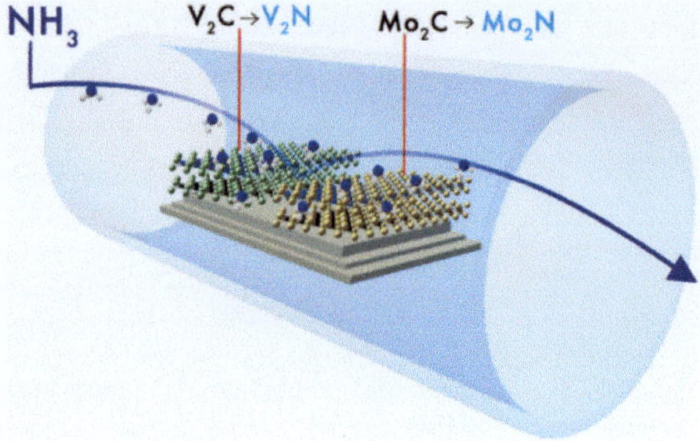

Fig. 11.6 Synthesis process from transition metal carbides (Mo_2CT_x and V_2CT_x) to MXene nitride, Adapted with permission from [129] Copyright (2023) Hao Tang, et al.

respectively. Another commonly used nitrogen enhancement process is heat treatment in an ammonia atmosphere, through which MXene carbide can be completely converted to nitride by adjusting the operating temperature [129, 130] (Fig. 11.6).

MXenes Modified by Terminal Element Rule

Changing the termination with the most common method to change the surface of MXene which is heat treatment. Heat treatment is considered a separate section. Various atmospheres and additives can cause MXene surface terminations to form in completely different directions. It was shown in previous studies that if Ti_2CT_x is tempered in air, it transforms into titanium oxide nanoparticles (TiO_2) and graphite carbon [129, 131], Therefore, to keep the atomic and chemical integrity of MXenes, heat treatment should be conducted in a vacuum or under an inert atmosphere. The impact of vacuum and inert atmosphere is analogous, both serving to inhibit the interaction between MXenes and oxygen. In Hart's study, the thermal treatment of Ti_3CT_x in a vacuum unveiled the intrinsic correlation between the sample's surface conductivity, assessed through electrical bias method, and spectroscopic analysis [132]. Figure 11.7 presents the real-time resistance profile of TiC_2T_x at various temperatures, illustrating the phases of resistance decline attributed to the detachment of adsorbed species, elimination of water molecules, and reduction in fluorine content. DFT calculations indicate that –OH is the first option that should be eliminated during heat treatment in the mixed terminals. Because of the coverage of water molecules by thermogram mass spectrometer (TGMS), we are not able to detect its signal.

In addition, it can be deactivated with the help of high electronegativity –F, causing changes in the structure and improving electrical conductivity in MXenes. The work of Sardich [129] investigated the effect of MXenes were processed at a temperature of 1500 and in a Helium, atmosphere using TG-MS analysis. The $Ti_3C_2T_x$ layers were etched and layered by HF, $Ti_3C_2T_x$, and it was shown by TMAOH that the water signal was changed by HF at high concentrations at 200 and 320 °C for several Ti_3C_2T layers synthesized. HF method is used to release water molecules inside the material between nanoplates (weak water-water interaction) if water molecules (TMAOH) are exposed to the –OH end of $Ti_3C_2T_x$ (strong water-surface interaction). The good

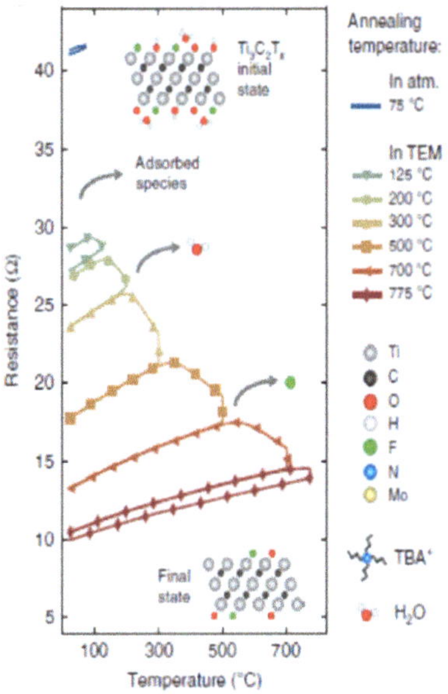

Fig. 11.7 The effect of in situ vacuum annealing temperature on the resistance of $Ti_3C_2T_x$, Adapted with permission from [129] Copyright (2023) Hao Tang, et al.

and pristine hydrophilicity of MXene is affected when the processing involves water molecules and OH termination. In the sample presented by TMAOH, it can be seen a significant change in the signal of $NH_2/NH_3/NH_4$ caused by the change of the cationic group of TMA^+. Also, with high heat treatment and removal of two groups from the terminal and MXene phase transition which causes in material properties and crystal structure. When the phase transition temperature (800 °C) is reached, the chemical structure of $Ti_3C_2T_\cdot$ is destroyed, which is transformed into cubic carbide (TiC) by the release of carbon monoxide (CO), and in this condition, the conductivity of other basic properties of the material is lost [129].

11.8.2.2 Other Gases

MXenes and O-MXene

As results of examinations of four MXenes using the O function for gas absorption. Figure 11.8 illustrated that the calculated E_{ad} of ten gases is plotted and compared on four O-MXenes. As you can see, the outcomes derived from PBE and PBE-D3 are contrasted and depicted in Fig. 11.8a, b. Ten gas molecules on Ti_2CO_2 [35, 131, 133]. The tendency of gas absorption power is from top to bottom as shown in Table 11.3:

Therefore, Ti_2CO_2 with an E_{ad} of 0.37 eV has the strongest interaction with NH_3, where the N of NH_3 is bonded above the Ti site by a distance of 2.43 Å. The rest of the gaseous types have a weak binding interaction with Ti_2CO_2. The adsorption energy (E_{ad}) of SO_2/Ti_2CO_2 is marginally less than the E_{ad} of −0.20 eV reported in the literature using DFT-D2 calculation under identical absorption configuration.

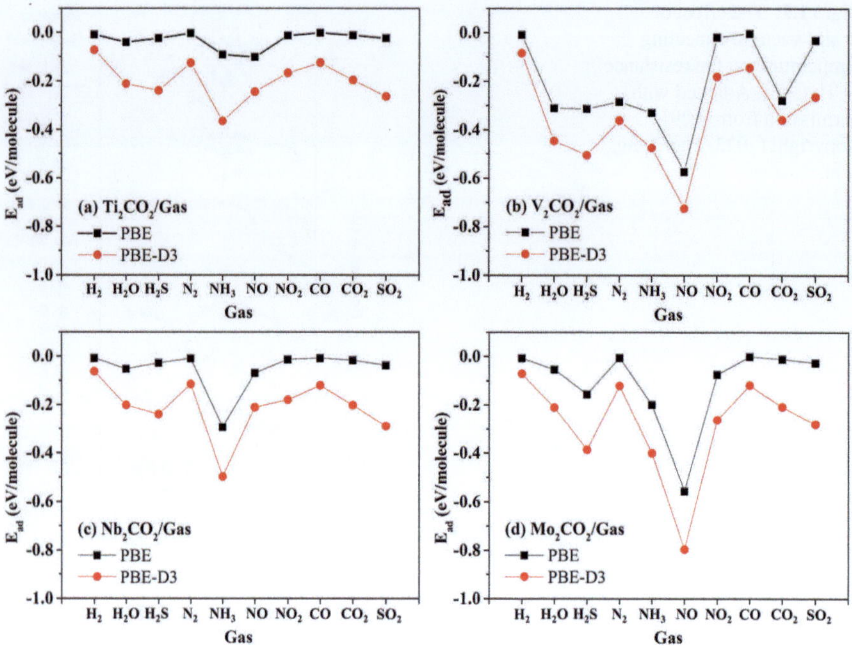

Fig. 11.8 E_{ad} comparison for gas absorption [133]

Table 11.3 The tendency of gas absorption power [133]

1	NH_3/Ti_2CO_2 ($E_{ad} = -0.37$ eV)	6	CO_2/Ti_2CO_2 ($E_{ad} = -0.20$ eV)
2	SO_2/Ti_2CO_2 ($E_{ad} = -0.26$ eV)	7	NO_2/Ti_2CO_2 ($E_{ad} = -0.17$ eV)
3	NO/Ti_2CO_2 ($E_{ad} = -0.25$ eV)	8	N_2/Ti_2CO_2 ($E_{ad} = -0.13$ eV)
4	H_2S/Ti_2CO_2 ($E_{ad} = -0.24$ eV)	9	CO/Ti_2CO_2 ($E_{ad} = -0.13$ eV)
5	H_2O/ Ti_2CO_2 ($E_{ad} = -0.21$ eV)	10	H_2/Ti_2CO_2 ($E_{ad} = -0.07$ eV)

H_2S and H_2O bond with their hydrogen atoms to O sites in Ti_2CO_2. CO_2, NO_2, CO, N_2, and H_2 exhibit adsorption energies (E_{ad}) of −0.21 eV and −0.24 eV, respectively, both exceeding −0.2 eV. H_2 has the weakest interaction among the ten gas species with E_{ad} −0.7 eV. Appropriately, Ti_2CO_2 has good selectivity towards NH_3, which is in good agreement with the previous work. According to the calculated E_{ad} values, the absorption power is as follows (Table 11.4)

The most notable absorption capability is for the NO/V_2CO_2 system, characterized by an E_{ad} value of −0.73 eV. The nitrogen of NO forms a connection with the surface oxygen site at a distance of 2.27 Å, while the O–N–O angle measures 109°. Similarly, H_2S and H_2O, akin to Ti_2CO_2, engage in an H–O interaction with surface oxygen sites at distances ranging from approximately 2.3–2.4 Å. As for NO_2, CO, and SO_2, they interact weakly with the surface. Compared to the Ti_2CO_2 surface, the binding

Table 11.4 E_{ad} values [133]

1	NO/V_2CO_2 ($E_{ad} = -0.73$ eV)	6	CO_2/V_2CO_2 ($E_{ad} = -0.36$ eV)
2	H_2S/V_2CO_2 ($E_{ad} = -0.51$ eV)	7	SO_2/V_2CO_2 ($E_{ad} = -0.27$ eV)
3	NH_3/V_2CO_2 ($E_{ad} = -0.48$ eV)	8	NO_2/V_2CO_2 ($E_{ad} = -0.18$ eV)
4	H_2O/V_2CO_2 ($E_{ad} = -0.45$ eV)	9	CO/V_2CO_2 ($E_{ad} = -0.14$ eV)
5	N_2/ V_2CO_2 ($E_{ad} = -0.36$ eV)	10	H_2/V_2CO_2 ($E_{ad} = -0.09$ eV)

strength of each gas in V_2CO_2 is increased. Likewise, H_2 demonstrates the least robust bond affinity with the surface in comparison to other gas molecules. Consequently, NO can be selective by V_2CO_2. Next, the adsorption of gases on Nb_2CO_2 is examined, and the findings are in Fig. 8c. The absorption power is as follows (Table 11.5).

The strongest interaction can be seen in NH_3 by Nb_2CO_2 (E_{ad} of -0.50 eV) in comparison to other nine molecules. The nitrogen atom of NH_3 is bonded to the Nb site at a distance of 2.41 Å. The arrangement of H_2O and H_2S on the Nb_2CO_2 surface, along with their Ead values, mirrors those observed on the Ti_2CO_2 surface. Due to the non-distortion of the structure of gas molecules physical absorption can be detected and as a result, Nb_2CO_2 provides a good choice over NH_3. For the last system, in Fig. 8d, a plot of E_{ad} values of gas adsorption on Mo_2CO_2 is shown. By notice of the calculation of E_{ad} values, the tendency for the strength of the connection between the gas and the surface is as follows (Table 11.6)

Mo_2CO_2 demonstrates a reaction reminiscent of V_2CO_2, wherein it establishes stronger bonds with NO in contrast to the other nine molecules. Furthermore, the arrangement of NH_3 on Mo_2CO_2 differs from that of other –O MXenes. Three H atoms of NH_3 interact with the O sites of the substrate. NO/Mo_2CO_2, NH_3/Mo_2CO_2,

Table 11.5 Gas adsorption on Nb_2CO_2 [133]

1	NH_3/Nb_2CO_2 ($E_{ad} = -0.50$ eV)	6	H_2O/Nb_2CO_2 ($E_{ad} = -0.20$ eV)
2	SO_2/Nb_2CO_2 ($E_{ad} = -0.29$ eV)	7	NO_2/Nb_2CO_2 ($E_{ad} = -0.18$ eV)
3	H_2S/Nb_2CO_2 ($E_{ad} = -0.24$ eV)	8	CO/Nb_2CO_2 ($E_{ad} = -0.12$ eV)
4	NO/Nb_2CO_2 ($E_{ad} = -0.21$ eV)	9	N_2/Nb_2CO_2 ($E_{ad} = -0.12$ eV)
5	CO_2/Nb_2CO_2 ($E_{ad} = -0.20$ eV)	10	H_2/Nb_2CO_2 ($E_{ad} = -0.06$ eV)

Table 11.6 The tendency for the strength of the connection between the gas and the surface [133]

1	NO/Mo_2CO_2 ($E_{ad} = -0.80$ eV)	6	H_2O/Mo_2CO_2 ($E_{ad} = -0.21$ eV)
2	NH_3/Mo_2CO_2 ($E_{ad} = 0.40$ eV)	7	CO_2/Mo_2CO_2 ($E_{ad} = -0.21$ eV)
3	H_2S/Mo_2CO_2 ($E_{ad} = -0.39$ eV)	8	CO/Mo_2CO_2 ($E_{ad} = -0.12$ eV)
4	SO_2/ Mo_2CO_2 ($E_{ad} = -0.28$ eV)	9	N_2/Mo_2CO_2 ($E_{ad} = -0.12$ eV)
5	NO_2/ Mo_2CO_2 ($E_{ad} = -0.26$ eV)	10	H_2/Mo_2CO_2 ($E_{ad} = -0.07$ eV)

and H_2S/Mo_2CO_2 have E_{ad} values below −0.4 eV, whereas the remaining combinations have E_{ad} values exceeding −0.3 eV. Mo_2CO_2 shows selectivity towards NO and the interaction between them is slightly stronger than the Nb_2CO_2 system.

While both PBE and PBE-D3 methods yield comparable bond configurations and lengths for pristine surfaces (i.e., M_2C and bare M_2CO_2), the inclusion of scattering effects in the PBE-D3 calculation enhances the adsorption strength relative to the PBE calculation. In absorption complexes, E_{ad} can be reduced from approximately 0.1–0.3 eV, when the D3 modification is incorporated, the variety of configurations is different from system to system to absorb NH_3 on these surfaces. In case of NO/V_2CO_2, when the D3 modification is incorporated, the distance between NO and the substrate decreases by about 0.06 Å. In NO/Mo_2CO_2, there are variations in the computed adsorbed configurations between the PBE and PBE-D3 methods, with the PBE calculation exhibiting a more pronounced deviation of the Mo_2CO_2 surface. Specifically, the distance between O' and Mo_2 in the Mo_2CO_2 substrate is longer than the other O'-Mo bonds, leading to changes in the values of r(M-O) and r(M-C) due to the interaction between the substrates and gas molecules. Within the group of ten molecules, dispersion plays a larger role in the adsorption energy of H_2O, H_2S, NH_3, CO_2, and SO_2 across most –O MXenes. The alteration in adsorption strength due to the incorporation of van der Waals forces is also influenced by the adsorption configuration, specifically the number of bonds between two adsorbents in each gas-substrate complex. The atomic distances exhibit slight variations when incorporating D3 in all cases except for H_2S/Ti_2CO_2 and H_2S/V_2CO_2. As depicted in Fig. 11.8, Mo_2CO_2 and V_2CO_2 exhibit significant selectivity towards NO, while Nb_2CO_2 and Ti_2CO_2 show strong preference for NH_3. NO and NH_3 show a tendency to preferentially bind with oxygen and metal sites, respectively. Conversely, H_2, N_2, CO, CO_2, NO_2, and SO_2 typically form weak bonds in the majority of –O MXenes, with H_2 displaying the weakest adsorption strength ($E_{ad} < -0.1$ eV) across all instances. To summarize, the interaction between gases and O-MXenes in this study is primarily governed by physical adsorption. This characteristic of physisorption is evident from both the Ead values and the structural resemblance of the absorbed gases to their gas phase counterparts. The structure of O-tetra MXenes is not twisted due to gas absorption. This is a beneficial characteristic of reusable absorbents. Furthermore, the strong binding energy of oxygen on MXenes can describe the high stability of O-MXenes, which can result in prevention of oxygen loss during use. They tend to remain stable even at elevated temperatures. With its low Ead value, SO_2 shows weak adsorption on O-MXenes. This indicates that –O MXenes display high selectivity towards SO_2, a corrosive gas commonly found in gas streams. These distinctive features are broadening the potential applications of these functionalized –O MXenes.

In the result of this investigation, it can be expressed that the wavy plane-based DFT calculations were used to examine the absorption of gas molecules in four MXenes and their O-functionalized surfaces. Not only the energy and structure properties but also the electronic charge properties were determined and used to explain their nature. In absorption process of original MXenes, most of the molecules are separated. The prevalent chemical adsorption process, characterized by low adsorption energy, indicates the high reactivity of MXenes but limited selectivity towards these gas molecules. Among the eleven gas molecules, only NH_3 exhibits molecular adsorption on these four MXenes. In the case of CO and N_2, while they are not fully dissociated, their bonds become activated upon adsorption on MXenes. These observations imply that empty MXenes could potentially serve as catalysts for the conversion of CO_2 or the fixation of nitrogen into ammonia. O-containing MXenes show weaker gas adsorption capacity than pristine MXenes in contrast to their bare surfaces, but they are more selective towards certain gas species. Among the adsorption energies, Ti_2CO_2 and Nb_2CO_2 exhibit stronger affinity for NH_3 compared to others, whereas Mo_2CO_2 and V_2CO_2 show a preference for NO. Additionally, the analysis of PDOS and ELF provides a clear depiction of the characteristics of MXenes and O-MXenes. These results suggest the potential utilization of O-MXenes in various applications such as gas separation, sensing, recording, and catalysis. Moreover, O-MXenes demonstrate promising stability and durability [133].

11.8.3 Gas Separation Mechanism

To achieving high selectivity and efficiency in the CO_2 reduction reaction (CO_2RR), including metals (copper, gold, platinum, silver, etc.), alloys, and oxides, extensive work has been done to discover suitable catalysts. However, challenges such as energy efficiency and product selection need to be addressed. Therefore, in CO_2 conversion and utilization, exploring new CO_2RR catalysts with high activity and selectivity is still a significant mission. In general, CO_2RR processes can be briefly described as follows: First, CO_2 becomes CO_2^- (activation of CO_2) by a reaction that produces one electron and generates various compounds through a multi-electron reaction. Suitable catalysts can perform fast multiple transitions.

Electrons play a crucial role in efficiently reducing electricity through effective physicochemical interaction with gas molecules and reaction intermediates. Hence, it's imperative to fully acknowledge the following two significant challenges in designing innovative CO_2RR catalysts. First, as a prerequisite for CO_2RR, the adsorption and activation of stable intrinsic CO_2 molecules are challenging, which always function as a limiting factor for the total electrochemical CO_2RR. Therefore, a fundamental necessity for an appropriate CO_2RR catalyst is the capability to adsorb and activate CO_2 on its surface. Second, CO_2 can be changed into various products such as CO, HCOOH, HCHO, CH_3OH, and CH_4, depending on the total number of electron-proton pairs transferred (from two to eight). Since a specific energy-rich product is preferable over mixed products, the CO_2RR catalyst is expected to show high catalytic selectivity for the specific product of interest, which is closely associated with the ability of the substrate to selectively adsorb specific intermediates and/or products. Recently, a new class of graphene-like 2D carbon/metal nitride MXenes has been produced by selective etching of the A layer from max phases. MXene materials have demonstrated potential applications in numerous domains, including catalysis. For instance, several investigations have indicated that certain

MXenes exhibit catalytic prowess in oxygen reduction reactions, oxygen evolution reactions, and nitrogen chemical fixation with notable efficiency and stability. However, the understanding of MXene materials as catalysts for CO_2 reduction reactions remains relatively limited. Until recently, Zhang et al. predicted that the O-functionalized MXene Ti_2C (Ti_2CO_2) surface exhibits promising photocatalytic performance for the reduction of CO_2 to HCOOH. Soon after, Lee et al. showed that Mo_3C_2 and Cr_3C_2 from M_3C_2 type MXene have suitable adsorption characteristics for CO_2 molecules and their reduction intermediates, theoretically. Nevertheless, higher-energy hydrocarbon fuels are highly desired and the experimental synthesis of Mo_3C_2 and MXene Cr_3C_2 remains a great challenge. In contrast, at least four M_2C-type MXenes, including Ti_2C, V_2C, Nb_2C, and Mo_2C, have been experimentally produced, while in the near future, others such as Zr_2C, Hf_2C, Ta_2C, and Cr_2C are expected to be synthesized as their major Max phases are fabricated. In this regard, communities will be more interested in whether M_2C-type MXene can be utilized as CO_2RR catalysts for higher energy hydrocarbons [125, 128, 134–137].

11.9 Conclusion

MXenes, a novel category of 2D nanomaterials, present numerous advantages. Despite a decade of research on these nanomaterials, MXenes are still in the early phases of their development. By standardization and optimization of all factors via controlling the reaction parameters, $Ti_3C_2T_x$ MXenes can be developed as commercial nanomaterials with stable features like long durability, homogeneous dispersion, better efficiency, and increased biocompatibility comparing to other available 2D nanomaterials for removal of water pollutants, especially radionuclides and heavy metal ions.

Due to the unique optical and electronic structure of MXene nanosheets as well as excellent biocompatibility and water solubility, MXenes have attracted more attention in terms of surface modification, application, and synthesis. At present, various strategies such as top-down and bottom-up techniques are used to synthesize MXene-based nanomaterials with diverse properties. Small molecule/polymer modified MXene composites outperform the original MXenes due to having more functional groups, better light adsorption, lower band gap, and larger specific surface area. Hence, MXene-based nanomaterials can remove radionuclides and heavy metal ions rapidly and efficiently. Accordingly, they are promising candidates to replace traditional catalysts and adsorbents, and thus, find a wide range of applications in wastewater remediation. While numerous studies have explored MXene-based nanomaterials for environmental remediation, the majority have concentrated on their laboratory applications and properties. Consequently, there remains a notable gap in research regarding the large-scale practical utilization of MXenes. Addressing this gap is crucial, as there are numerous challenges to overcome before MXenes can be swiftly deployed in field applications. One major challenge is the economics of MXene-based nanomaterials. At present, in the field of water treatment, activated carbon is the most economical material. On the other hand, MXenes raw materials are expensive and the production process is very complicated compared to activated carbon. All existing methods employed thus far for synthesizing MXene-based nanomaterials are costly and result in low yields, thus constraining mass production. Consequently, the imperative need arises for the development of a cost-effective and high-yield synthesis method to facilitate the broader application of MXene-based nanomaterials in purification processes. Despite these challenges, MXene-based nanomaterials hold significant promise for widespread utilization in environmental remediation, particularly in the removal of heavy metal ions and other pollutants from wastewater, provided that their limitations can be addressed.

References

1. M. Alhabeb et al., Guidelines for synthesis and processing of two-dimensional titanium carbide ($Ti_3C_2T_x$ MXene). Chem. Mater. **29**(18), 7633–7644 (2017)
2. Y. Ibrahim et al., Unveiling fabrication and environmental remediation of MXene-based nanoarchitectures in toxic metals removal from wastewater: strategy and mechanism. Nanomaterials **10**(5), 885 (2020)
3. M. Vakili et al., Regeneration of chitosan-based adsorbents used in heavy metal adsorption: a review. Sep. Purif. Technol. **224**, 373–387 (2019)
4. P. Haseena et al., Adsorption of ammonium nitrogen from aqueous systems using chitosan-bentonite film composite. Procedia Technol. **24**, 733–740 (2016)
5. M. Wu et al., Facile synthesis of chitosan/gelatin filled with graphene bead adsorbent for orange II removal. Chem. Eng. Res. Des. **144**, 35–46 (2019)
6. S. Pap et al., Low-cost chitosan-calcite adsorbent development for potential phosphate removal and recovery from wastewater effluent. Water Res. **173**, 115573 (2020)
7. S.N. Malik et al., Hybrid ozonation process for industrial wastewater treatment: principles and applications: a review. J. Water Process Eng. **35**, 101193 (2020)
8. J.A. Kumar et al., Methods of synthesis, characteristics, and environmental applications of Mxene: a comprehensive review. Chemosphere **286**, 131607 (2022)
9. P. Zhao et al., A new paradigm of ultrathin 2D nanomaterial adsorbents in aqueous media: graphene and GO, MoS_2, MXenes, and 2D MOFs. J. Mater. Chem. A **7**(28), 16598–16621 (2019)
10. Y. Wu et al., Environmental remediation of heavy metal ions by novel-nanomaterials: a review. Environ. Pollut. **246**, 608–620 (2019)
11. R. Ahmad, A. Mirza, Facile one pot green synthesis of Chitosan-Iron oxide (CS-Fe_2O_3) nanocomposite: removal of Pb (II) and Cd (II) from synthetic and industrial wastewater. J. Clean. Prod. **186**, 342–352 (2018)
12. J. Mittal et al., Expeditious and enhanced sequestration of heavy metal ions from aqueous environment by papaya peel carbon: a green and low-cost adsorbent. Desalin. Water Treat. **210**, 365–376 (2021)
13. W.S. Chai et al., A review on conventional and novel materials towards heavy metal adsorption in wastewater treatment application. J. Clean. Prod. **296**, 126589 (2021)
14. Y. Sheth et al., Prospects of titanium carbide-based MXene in heavy metal ion and radionuclide adsorption for wastewater remediation: a review. Chemosphere 133563 (2022)
15. K. Li, B. Lee, Y. Kim, High performance reverse osmosis membrane with carbon nanotube support layer. J. Membr. Sci. **592**, 117358 (2019)
16. Y.A. Al-Hamadani et al., Applications of MXene-based membranes in water purification: a review. Chemosphere **254**, 126821 (2020)

17. G. Yury, B. Anasori, The rise of MXenes. ACS Nano **13**, 8491–8494 (2019)
18. B. Anasori, Û.G. Gogotsi, *2D Metal Carbides and Nitrides (MXenes)*, vol. 416. (Springer, 2019)
19. Y. Zhang et al., Adsorptive environmental applications of MXene nanomaterials: a review. RSC Adv. **8**(36), 19895–19905 (2018)
20. A. Rafieerad et al., Fabrication of smart tantalum carbide MXene quantum dots with intrinsic immunomodulatory properties for treatment of allograft vasculopathy. Adv. Func. Mater. **31**(46), 2106786 (2021)
21. M.W. Barsoum, P. Eklund, The for MXenes Mn+ 1AXn phases: the precursors, in *2D Metal Carbides and Nitrides (MXenes): Structure, Properties and Applications* (2019), p. 15
22. L. Verger et al., MXenes: an introduction of their synthesis, select properties, and applications. Trends Chem. **1**(7), 656–669 (2019)
23. J.-C. Lei, X. Zhang, Z. Zhou, Recent advances in MXene: preparation, properties, and applications. Front. Phys. **10**(3), 276–286 (2015)
24. M. Naguib et al., Two-dimensional transition metal carbides. ACS Nano **6**(2), 1322–1331 (2012)
25. C.E. Shuck et al., Scalable synthesis of $Ti_3C_2T_x$ mxene. Adv. Eng. Mater. **22**(3), 1901241 (2020)
26. G.G. Chan, C.M. Koch, L.H. Connors, Blood proteomic profiling in inherited (ATTRm) and acquired (ATTRwt) forms of transthyretin-associated cardiac amyloidosis. J. Proteome Res. **16**(4), 1659–1668 (2017)
27. X. Yang et al., Surface functional groups of carbon-based adsorbents and their roles in the removal of heavy metals from aqueous solutions: a critical review. Chem. Eng. J. **366**, 608–621 (2019)
28. M.R. Lukatskaya et al., Cation intercalation and high volumetric capacitance of two-dimensional titanium carbide. Science **341**(6153), 1502–1505 (2013)
29. T. Zhang et al., Synthesis of two-dimensional $Ti_3C_2T_x$ MXene using HCl^+ LiF etchant: enhanced exfoliation and delamination. J. Alloy. Compd. **695**, 818–826 (2017)
30. A.K. Fard et al., Two-dimensional MXene for efficient arsenic removal from aqueous solutions: experimental and molecular dynamics simulation. Desalin. Water Treat. **211**, 280–295 (2021)
31. S. Wang et al., Removal of heavy metal cations and co-existing anions in simulated wastewater by two separated hydroxylated MXene membranes under an external voltage. J. Membr. Sci. **638**, 119697 (2021)
32. S. Elumalai, M. Yoshimura, M. Ogawa, Simultaneous delamination and rutile formation on the surface of $Ti_3C_2T_x$ MXene for copper adsorption. Chem. Asian J. **15**(7), 1044–1051 (2020)
33. D. Mohan, C.U. Pittman Jr., Activated carbons and low cost adsorbents for remediation of tri-and hexavalent chromium from water. J. Hazard. Mater. **137**(2), 762–811 (2006)
34. D. Mohan, K.P. Singh, V.K. Singh, Removal of hexavalent chromium from aqueous solution using low-cost activated carbons derived from agricultural waste materials and activated carbon fabric cloth. Ind. Eng. Chem. Res. **44**(4), 1027–1042 (2005)
35. Y. Ying et al., Two-dimensional titanium carbide for efficiently reductive removal of highly toxic chromium (VI) from water. ACS Appl. Mater. Interfaces **7**(3), 1795–1803 (2015)
36. L. Wang et al., Rational control of the interlayer space inside two-dimensional titanium carbides for highly efficient uranium removal and imprisonment. Chem. Commun. **53**(89), 12084–12087 (2017)
37. Z. Li et al., Synthesis and thermal stability of two-dimensional carbide MXene Ti_3C_2. Mater. Sci. Eng. B **191**, 33–40 (2015)
38. J. Halim et al., Transparent conductive two-dimensional titanium carbide epitaxial thin films. Chem. Mater. **26**(7), 2374–2381 (2014)
39. M. Naguib et al., Large-scale delamination of multi-layers transition metal carbides and carbonitrides "MXenes." Dalton Trans. **44**(20), 9353–9358 (2015)

40. F. Han et al., Boosting the yield of MXene 2D sheets via a facile hydrothermal-assisted intercalation. ACS Appl. Mater. Interfaces **11**(8), 8443–8452 (2019)
41. S. Nezami, A. Ghaemi, T. Yousefi, Application of titanium carbide/nitride (MXene)-based NPs in adsorption of radionuclides and heavy metal ions for wastewater remediation: a review, in *Case Studies in Chemical and Environmental Engineering* (2023), p. 100326
42. O. Mashtalir et al., Intercalation and delamination of layered carbides and carbonitrides. Nat. Commun. **4**(1), 1–7 (2013)
43. A. Feng et al., Fabrication and thermal stability of NH_4HF_2-etched Ti_3C_2 MXene. Ceram. Int. **43**(8), 6322–6328 (2017)
44. C. Backes et al., Guidelines for exfoliation, characterization and processing of layered materials produced by liquid exfoliation. Chem. Mater. **29**(1), 243–255 (2017)
45. P. Zhang et al., Sorption of Eu(III) on MXene-derived titanate structures: the effect of nano-confined space. Chem. Eng. J. **370**, 1200–1209 (2019)
46. A. Lipatov et al., Effect of synthesis on quality, electronic properties and environmental stability of individual monolayer Ti_3C_2 MXene flakes. Adv. Electron. Mater. **2**(12), 1600255 (2016)
47. A. Feng et al., Two-dimensional MXene Ti_3C_2 produced by exfoliation of Ti_3AlC_2. Mater. Des. **114**, 161–166 (2017)
48. R. Chen et al., Facile synthesis of a sandwiched $Ti_3C_2T_x$ MXene/nZVI/fungal hypha nanofiber hybrid membrane for enhanced removal of Be(II) from Be $(NH_2)_2$ complexing solutions. Chem. Eng. J. **421**, 129682 (2021)
49. A.K. Fard et al., Barium removal from synthetic natural and produced water using MXene as two dimensional (2-D) nanosheet adsorbent. Chem. Eng. J. **317**, 331–342 (2017)
50. W. Mu et al., Improving barium ion adsorption on two-dimensional titanium carbide by surface modification. Dalton Trans. **47**(25), 8375–8381 (2018)
51. B.-M. Jun et al., Adsorption of Ba^{2+} and Sr^{2+} on $Ti_3C_2T_x$ MXene in model fracking wastewater. J. Environ. Manag. **256**, 109940 (2020)
52. A.R. Khan et al., Two-dimensional transition metal carbide ($Ti_3C_2T_x$) as an efficient adsorbent to remove cesium (Cs^+). Dalton Trans. **48**(31), 11803–11812 (2019)
53. P. Gu et al., Experimental and theoretical calculation investigation on efficient Pb (II) adsorption on etched Ti_3AlC_2 nanofibers and nanosheets. Environ. Sci. Nano **5**(4), 946–955 (2018)
54. Y. Du et al., Efficient removal of Pb (II) by $Ti_3C_2T_x$ powder modified with a silane coupling agent. J. Mater. Sci. **54**(20), 13283–13297 (2019)
55. G. Zhang et al., Synthesis of amino-functionalized $Ti_3C_2T_x$ MXene by alkalization-grafting modification for efficient lead adsorption. Chem. Commun. **56**(76), 11283–11286 (2020)
56. A. Shahzad et al., Two-dimensional $Ti_3C_2T_x$ MXene nanosheets for efficient copper removal from water. ACS Sustain. Chem. Eng. **5**(12), 11481–11488 (2017)
57. D. Gan et al., Bioinspired functionalization of MXenes ($Ti_3C_2T_X$) with amino acids for efficient removal of heavy metal ions. Appl. Surf. Sci. **504**, 144603 (2020)
58. J. Ren et al., Enhanced adsorption performance of alginate/MXene/$CoFe_2O_4$ for antibiotic and heavy metal under rotating magnetic field. Chemosphere **284**, 131284 (2021)
59. L. Jin et al., Two-dimensional titanium carbides ($Ti_3C_2T_x$) functionalized by poly (m-phenylenediamine) for efficient adsorption and reduction of hexavalent chromium. Int. J. Environ. Res. Public Health **17**(1), 167 (2020)
60. L. Wang et al., Effective removal of anionic Re (VII) by surface-modified Ti_2CT_x MXene nanocomposites: implications for Tc (VII) sequestration. Environ. Sci. Technol. **53**(7), 3739–3747 (2019)
61. A. Kong et al., Amino-functionalized MXenes for efficient removal of Cr (VI). Colloids Surf. A **617**, 126388 (2021)
62. Y. Feng et al., Fabrication of MXene/PEI functionalized sodium alginate aerogel and its excellent adsorption behavior for Cr (VI) and Congo Red from aqueous solution. J. Hazard. Mater. **416**, 125777 (2021)

63. A.R. Khan et al., 3D flower like δ-MnO_2/MXene nano-hybrids for the removal of hexavalent Cr from wastewater. Ceram. Int. **47**(18), 25951–25958 (2021)
64. M. Ghorbani, O. Seyedin, M. Aghamohammadhassan, Adsorptive removal of lead (II) ion from water and wastewater media using carbon-based nanomaterials as unique sorbents: a review. J. Environ. Manag. **254**, 109814 (2020)
65. Q. Shafiq et al., Rational design of MXene coated polyurethane foam for the removal of Pb^{2+}. Mater. Lett. **304**, 130600 (2021)
66. B.-M. Jun et al., Effective removal of Pb (ii) from synthetic wastewater using $Ti_3C_2T_x$ MXene. Environ. Sci.: Water Res. Technol. **6**(1), 173–180 (2020)
67. S. Wang et al., Facile preparation of biosurfactant-functionalized Ti_2CT_X MXene nanosheets with an enhanced adsorption performance for Pb (II) ions. J. Mol. Liq. **297**, 111810 (2020)
68. S. Lee et al., The effects of picolinic acid and pH on the adsorption of Cu (II) by activated carbon fibers. Carbon **40**(3), 329–334 (2002)
69. P. Karthikeyan et al., Effective removal of Cr (VI) and methyl orange from the aqueous environment using two-dimensional (2D) $Ti_3C_2T_x$ MXene nanosheets. Ceram. Int. **47**(3), 3692–3698 (2021)
70. S. Li et al., Efficient thorium (IV) removal by two-dimensional Ti_2CT_x MXene from aqueous solution. Chem. Eng. J. **366**, 192–199 (2019)
71. Y.-J. Zhang et al., Adsorption of uranyl species on hydroxylated titanium carbide nanosheet: a first-principles study. J. Hazard. Mater. **308**, 402–410 (2016)
72. E.A. Gendy et al., Removal of heavy metals by covalent organic frameworks (COFs): a review on its mechanism and adsorption properties. J. Environ. Chem. Eng. **9**(4), 105687 (2021)
73. G. Cheng et al., Extremely stable amidoxime functionalized covalent organic frameworks for uranium extraction from seawater with high efficiency and selectivity. Sci. Bull. **66**(19), 1994–2001 (2021)
74. S. Li et al., Rapid and selective uranium extraction from aqueous solution under visible light in the absence of solid photocatalyst. Sci. China Chem. **64**(8), 1323–1331 (2021)
75. Y. Liu et al., Zeolitic imidazolate framework-based nanomaterials for the capture of heavy metal ions and radionuclides: a review. Chem. Eng. J. **406**, 127139 (2021)
76. H. Tang et al., Recent advances in nanoscale zero-valent iron-based materials: characteristics, environmental remediation and challenges. J. Clean. Prod. **319**, 128641 (2021)
77. H. Yang et al., Functionalized iron–nitrogen–carbon electrocatalyst provides a reversible electron transfer platform for efficient uranium extraction from seawater. Adv. Mater. **33**(51), 2106621 (2021)
78. S. Zhang et al., Applications of water-stable metal-organic frameworks in the removal of water pollutants: a review. Environ. Pollut. **291**, 118076 (2021)
79. L. Wang et al., Efficient U (VI) reduction and sequestration by Ti_2CT_x MXene. Environ. Sci. Technol. **52**(18), 10748–10756 (2018)
80. P. Zhang et al., Aryl diazonium-assisted amidoximation of MXene for boosting water stability and uranyl sequestration via electrochemical sorption. ACS Appl. Mater. Interfaces **12**(13), 15579–15587 (2020)
81. R. Cheng et al., Adsorption of Sr (II) from water by mercerized bacterial cellulose membrane modified with EDTA. J. Hazard. Mater. **364**, 645–653 (2019)
82. S. Feng et al., Casein-hydroxyapatite composite microspheres for strontium-containing wastewater treatment. ACS ES&T Water **1**(4), 900–909 (2021)
83. B. Park et al., Dual-functional micro-adsorbents: application for simultaneous adsorption of cesium and strontium. Chemosphere **263**, 128266 (2021)
84. P. Amesh et al., Tuning the ion exchange behavior of cesium and strontium on sodium iron titanate. Sep. Purif. Technol. **267**, 118678 (2021)
85. S. Eun et al., Simultaneous removal of radioactive cesium and strontium from seawater using a highly efficient Prussian blue-embedded alginate aerogel. J. Environ. Manage. **297**, 113389 (2021)
86. J. Yin et al., Biogene-derived aerogels for simultaneously selective adsorption of uranium (VI) and strontium (II) by co-imprinting method. Sep. Purif. Technol. **271**, 118849 (2021)

87. J.-W. Choi, S. Cho, S.-J. Choi, Ecofriendly, selective removal of radioactive strontium ions in aqueous solutions using magnetic banana peels. Sci. Total. Environ. **778**, 146327 (2021)
88. J. Shin et al., Changes in adsorption mechanisms of radioactive barium, cobalt, and strontium ions using spent coffee waste biochars via alkaline chemical activation: Enrichment effects of O-containing functional groups. Environ. Res. **199**, 111346 (2021)
89. H. Wu et al., Comparative study of strontium adsorption on muscovite, biotite and phlogopite. J. Environ. Radioact. **225**, 106446 (2020)
90. A. Shahzad et al., Strontium ions capturing in aqueous media using exfoliated titanium aluminum carbide (Ti_2AlC MAX phase). J. Nucl. Mater. **549**, 152916 (2021)
91. A. Ghaemi, M. Torab-Mostaedi, M. Ghannadi-Maragheh, Characterizations of strontium (II) and barium (II) adsorption from aqueous solutions using dolomite powder. J. Hazard. Mater. **190**(1–3), 916–921 (2011)
92. M. Torab-Mostaedi et al., Removal of strontium and barium from aqueous solutions by adsorption onto expanded perlite. Can. J. Chem. Eng. **89**(5), 1247–1254 (2011)
93. S.Y. Lee et al. Innovative biological method to remove Cesium-137, in *Proceedings of the Korean Radioactive Waste Society Conference* (Korean Radioactive Waste Society, 2017)
94. Y.Z. Tan et al., Photothermal-enhanced and fouling-resistant membrane for solar-assisted membrane distillation. J. Membr. Sci. **565**, 254–265 (2018)
95. M. Jeon et al., A review on MXene-based nanomaterials as adsorbents in aqueous solution. Chemosphere **261**, 127781 (2020)
96. A. Shahzad et al., Mercuric ion capturing by recoverable titanium carbide magnetic nanocomposite. J. Hazard. Mater. **344**, 811–818 (2018)
97. X. Wang et al., A new etching environment (FeF_3/HCl) for the synthesis of two-dimensional titanium carbide MXenes: a route towards selective reactivity vs. water. J. Mater. Chem. A **5**(41), 22012–22023 (2017)
98. V.V. Dev et al., Zwitterion-Chitosan bed for the simultaneous immobilization of Zn (II), Cd (II), Pb (II) and Cu (II) from multi-metal aqueous systems. J. Clean. Prod. **255**, 120309 (2020)
99. Y. Dong et al., Mxene/alginate composites for lead and copper ion removal from aqueous solutions. RSC Adv. **9**(50), 29015–29022 (2019)
100. W. Mu et al., Removal of radioactive palladium based on novel 2D titanium carbides. Chem. Eng. J. **358**, 283–290 (2019)
101. Y. Sheth et al., An environment friendly approach for heavy metal removal from industrial wastewater using chitosan based biosorbent: a review. Sustain. Energy Technol. Assess. **43**, 100951 (2021)
102. R. Sivaranjanee, A. Saravanan, Carbon sphere: synthesis, characterization and elimination of toxic Cr (VI) ions from aquatic system. J. Ind. Eng. Chem. **60**, 307–320 (2018)
103. P.S. Kumar, S.J. Varjani, S. Suganya, Treatment of dye wastewater using an ultrasonic aided nanoparticle stacked activated carbon: kinetic and isotherm modelling. Biores. Technol. **250**, 716–722 (2018)
104. S.-W. Li, R.-M. Gao, J.-S. Zhao, Deep oxidative desulfurization of fuel catalyzed by modified heteropolyacid: the comparison performance of three kinds of ionic liquids. ACS Sustain. Chem. Eng. **6**(11), 15858–15866 (2018)
105. K. Velusamy et al., Functionalization of MXene-based nanomaterials for the treatment of micropollutants in aquatic system: a review. Environ. Pollut. 119034 (2022)
106. S.K. Hwang et al., MXene: an emerging two-dimensional layered material for removal of radioactive pollutants. Chem. Eng. J. **397**, 125428 (2020)
107. I. Ihsanullah, MXenes (two-dimensional metal carbides) as emerging nanomaterials for water purification: progress, challenges and prospects. Chem. Eng. J. **388**, 124340 (2020)
108. V. Thirumal et al., Efficient photocatalytic degradation of hazardous pollutants by home-made kitchen blender novel technique via 2D-material of few-layer MXene nanosheets. Chemosphere **281**, 130984 (2021)
109. V.Q. Hieu et al., Photocatalytic degradation of methyl orange dye by Ti_3C_2–TiO_2 heterojunction under solar light. Chemosphere **276**, 130154 (2021)

110. A. Tariq et al., Efficient visible-light photocatalysis of 2D-MXene nanohybrids with Gd^{3+}-and Sn^{4+}-codoped bismuth ferrite. ACS Omega **3**(10), 13828–13836 (2018)
111. C. Liu et al., Layered BiOBr/Ti_3C_2 MXene composite with improved visible-light photocatalytic activity. J. Mater. Sci. **54**(3), 2458–2471 (2019)
112. J. Chen et al., Morphology and photocatalytic activity of TiO_2/MXene composites by in-situ solvothermal method. Ceram. Int. **46**(12), 20088–20096 (2020)
113. M. Najafi et al., Polysulfone membranes incorporated with reduced graphene oxide nanoparticles for enhanced Olefin/Paraffin separation. ChemistrySelect **5**(12), 3675–3681 (2020)
114. B.M. Yoo et al., Graphene and graphene oxide membranes for gas separation applications. Curr. Opin. Chem. Eng. **16**, 39–47 (2017)
115. L. Chen et al., Ion sieving in graphene oxide membranes via cationic control of interlayer spacing. Nature **550**(7676), 380–383 (2017)
116. K.K. Sadasivuni, et al., *MXenes and Their Composites: Synthesis, Properties and Potential Applications* (Elsevier, 2021)
117. X. Zha, et al., *2D Metal Carbides and Nitrides (MXenes)* (Springer International Publishing, Cham, 2019)
118. H.-W. Wang et al., Resolving the structure of $Ti_3C_2T_x$ MXenes through multilevel structural modeling of the atomic pair distribution function. Chem. Mater. **28**(1), 349–359 (2016)
119. I. Persson et al., 2D transition metal carbides (MXenes) for carbon capture. Adv. Mater. **31**(2), 1805472 (2019)
120. R. Ben-Mansour et al., Carbon capture by physical adsorption: materials, experimental investigations and numerical modeling and simulations–a review. Appl. Energy **161**, 225–255 (2016)
121. N. Li et al., Understanding of electrochemical mechanisms for CO_2 capture and conversion into hydrocarbon fuels in transition-metal carbides (MXenes). ACS Nano **11**(11), 10825–10833 (2017)
122. C. Giordano, M. Antonietti, Synthesis of crystalline metal nitride and metal carbide nanostructures by sol–gel chemistry. Nano Today **6**(4), 366–380 (2011)
123. X. Luo et al., Significant improvements in CO_2 capture by pyridine-containing anion-functionalized ionic liquids through multiple-site cooperative interactions. Angew. Chem. **126**(27), 7173–7177 (2014)
124. B. Wang et al., Carbon dioxide adsorption of two-dimensional carbide MXenes. J. Adv. Ceram. **7**, 237–245 (2018)
125. Z. Guo et al., M2C-type MXenes: promising catalysts for CO_2 capture and reduction. Appl. Surf. Sci. **521**, 146436 (2020)
126. M.W. Barsoum, T. El-Raghy, The MAX phases: unique new carbide and nitride materials: ternary ceramics turn out to be surprisingly soft and machinable, yet also heat-tolerant, strong and lightweight. Am. Sci. **89**(4), 334–343 (2001)
127. Y. Gogotsi, Q. Huang, *MXenes: Two-Dimensional Building Blocks for Future Materials and Devices* (ACS Publications, 2021), pp. 5775–5780
128. W.Y. Chen et al., Surface functionalization of $Ti_3C_2T_x$ MXene with highly reliable superhydrophobic protection for volatile organic compounds sensing. ACS Nano **14**(9), 11490–11501 (2020)
129. H. Tang et al., Post-processing strategies for improving the electrical and mechanical properties of MXenes. Chem. Eng. J. **425**, 131472 (2021)
130. C. Yang et al., Flexible nitrogen-doped 2D titanium carbides (MXene) films constructed by an ex situ solvothermal method with extraordinary volumetric capacitance. Adv. Energy Mater. **8**(31), 1802087 (2018)
131. F. Xie et al., Ultrathin MXene/aramid nanofiber composite paper with excellent mechanical properties for efficient electromagnetic interference shielding. Nanoscale **11**(48), 23382–23391 (2019)
132. J.L. Hart et al., Control of MXenes' electronic properties through termination and intercalation. Nat. Commun. **10**(1), 522 (2019)

133. A. Junkaew, R. Arroyave, Enhancement of the selectivity of MXenes (M_2C, M= Ti, V, Nb, Mo) via oxygen-functionalization: promising materials for gas-sensing and-separation. Phys. Chem. Chem. Phys. **20**(9), 6073–6082 (2018)
134. R. Khaledialidusti, A.K. Mishra, A. Barnoush, Atomic defects in monolayer ordered double transition metal carbide ($Mo_2TiC_2T_x$) MXene and CO_2 adsorption. J. Mater. Chem. C **8**(14), 4771–4779 (2020)
135. Y. Lee et al., Considerably improved water and oil washability of highly conductive stretchable fibers by chemical functionalization with fluorinated silane. J. Mater. Chem. C **7**(39), 12297–12305 (2019)
136. A.A. Shamsabadi et al., Pushing rubbery polymer membranes to be economic for CO_2 separation: embedment with $Ti_3C_2T_x$ MXene nanosheets. ACS Appl. Mater. Interfaces **12**(3), 3984–3992 (2019)
137. D. Yoo et al., Highly reliable superhydrophobic protection for organic field-effect transistors by fluoroalkylsilane-coated TiO_2 nanoparticles. ACS Nano **12**(11), 11062–11069 (2018)

Chapter 12
Application of MXenes in Solution-Processed Optoelectronic Devices

Ping Cai, Ling Ding, Kefan Chen, Can Song, and Baiquan Liu

12.1 Introduction

Transition metal carbides and/or nitrides as the emerging family of two-dimensional (2D) nanomaterials, called MXenes, possess the general chemical formula of $M_{n+1}X_nT_x$ (n = 1–4). MXenes are synthesized by selectively etching the A-site atomic layers of the precursors $M_{n+1}AX_n$ (MAX) phase, where M represents the early transition metals (e.g. Ti, V, Nb, Mo, Zr, and Hf), A usually represents IIIA and IVA elements (e.g. Al, Si, and Ga), X represents carbon and/or nitrogen, and T_x represents surface functional groups (e.g. =O, –F, –OH, and –Cl) [1]. At present, more than 30 kinds of MXenes have been synthesized from various MAX phases via different etching methods, such as fluorine-containing acidic solution etching method, Lewis acid molten salt etching method, and electrochemical etching method. As shown in Fig. 12.1, MXenes with 2D layered structure and abundant surface functional groups usually show metal or semiconductor-like characteristics, such as excellent conductivity, high optical transmittance, and tunable work function, as well as solution processability; moreover, the optical and electrical properties of MXenes are related to their structures, compositions and surface functional groups and can be regulated by selecting different MAX precursors, etching methods, and surface modifier, which facilitate to the versatile applications of MXenes in solution-processed optoelectronic devices, mainly including organic/perovskite light-emitting diodes and solar cells.

Solution-processed organic/perovskite light-emitting diodes and solar cells, as a new generation of semiconductor optoelectronic devices, have attracted broad

P. Cai · L. Ding · K. Chen · C. Song
School of Materials Science and Engineering & Guangxi Key Laboratory of Information Materials, Guilin University of Electronic Technology, Guilin 541004, China

B. Liu (✉)
School of Electronics and Information Technology, Sun Yat-sen University, Guangzhou 510275, China
e-mail: liubq33@mail.sysu.edu.cn

© The Author(s), under exclusive license to Springer Nature Singapore Pte Ltd. 2024
O. P. Pandey and P. Sharma (eds.), *MXenes: Emerging 2D Materials*,
https://doi.org/10.1007/978-981-97-4064-2_12

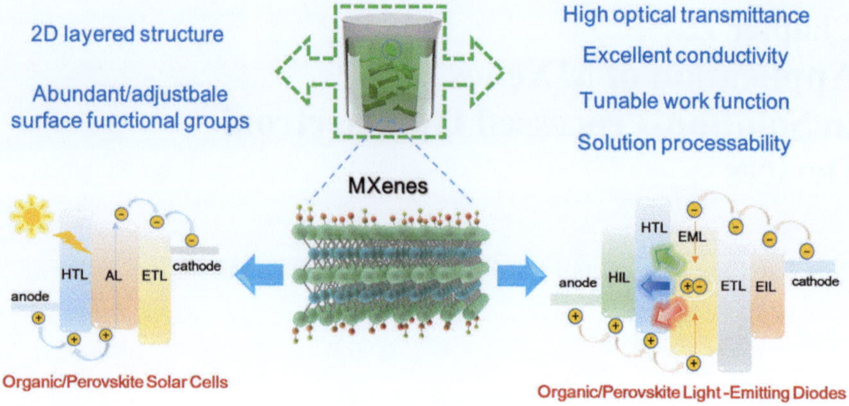

Fig. 12.1 Schematic illustration of MXenes with excellent properties and their applications in solution-processed optoelectronic devices, including organic/perovskite light-emitting diodes and solar cells

attention and research due to the distinct advantages such as low cost, light weight, flexibility, and large-area printing fabrications [2]. As shown in Fig. 12.1, solution-processed organic/perovskite light-emitting diodes and solar cells, usually possess the similar sandwiched device structures, where organic solar cells (OSCs) and perovskite solar cells (PeSCs) are typically composed of anode, hole transport layer (HTL), organic/perovskite active layer, electron transport layer (ETL) and cathode, and organic light-emitting diodes (OLEDs) and perovskite light-emitting diodes (PeLEDs) are typically composed of anode, hole injection layer (HIL), hole transport layer (HTL), organic/perovskite active layer, electron transport layer (ETL), electron injection layer (EIL) and cathode. The excellent electrical conductivity and tunable work function of MXenes (especially $Ti_3C_2T_x$) facilitate MXene application in electrode, and the easily tunable optoelectronic properties and abundant surface functional groups of MXenes facilitate MXene applications in interface layer and emitting layer/active layer.

Herein, this review comprehensively summarizes the versatile applications of MXenes in different functional layers, including electrode, interface layer, and emitting layer/active layer, for high-performance solution-processed optoelectronic devices (organic/perovskite light-emitting diodes and solar cells). Some major challenges and future opportunities of MXenes applications in solution-processed optoelectronic devices are discussed and prospected.

12.2 Applications of MXenes in Solution-Processed Optoelectronic Devices

12.2.1 Electrode

i. Pristine MXene electrode

As shown in Fig. 12.2a, Ahn et al. [3] reported a transparent conductive $Ti_3C_2T_x$ MXene film with low sheet resistance of 108 Ω sq^{-1}, high transmittance of 85%, high work function (WF) of 5.1 eV and excellent flexibility, where the $Ti_3C_2T_x$ MXene film was prepared by spin-coating its solution and then vacuum annealing treatment at 200 °C. Utilizing the $Ti_3C_2T_x$ MXene electrode, flexible green OLED with polymeric HIL (n-GraHIL) and emitting layer (EML) of TCTA:Ir(ppy)$_2$acac/CBP:Ir(ppy)$_2$acac showed high current efficiency (CE) of 102 cd/A and external quantum efficiency (EQE) of 28.5%, similar to the performance of OLED with flexible ITO electrode. As shown in Fig. 12.2b, Lee et al. [4] prepared a $Ti_3C_2T_x$ MXene electrode via one-step spin-coating of MXene flakes dispersed in water on a glass substrate. Notably, the glass substrate was sequentially treated with O_2 plasma and (3-aminopropyl) triethoxysilane (APTES) before spin-coating MXene, and the APTES provided a strong electrostatic interaction with the terminal groups of MXene, leading to a homogeneous close-packed MXene layer with a low surface roughness. Based on the $Ti_3C_2T_x$ MXene transparent electrode, OLEDs with the EML of Super Yellow (PDY-132) exhibited turn-on voltage (V_{on}) of 2.1 V, CE of 7 cd/A, and brightness of 12,547 cd/m^2 at 1 kHz alternating current, showing better performance compared to the devices with other flexible transparent electrodes including carbon nanotubes, reduced graphene oxide, and silver nanowires.

As shown in Fig. 12.2c, Cao et al. [5] used the combined processes of spray-coating, vacuum-assisted filtration, and hot-pressing to prepare Ti_3C_2 MXene as the back electrode in PeSCs, and the PeSCs with Ti_3C_2 electrode and $CH_3NH_3PbI_3$ ($MAPbI_3$) active layer showed higher PCE compared to the control devices with carbon electrode. Qin et al. [6] utilized small-size $Ti_3C_2T_x$ nanosheets (average size of below 1 micron) with ultrasonic treatment and large-size $Ti_3C_2T_x$ nanosheets without ultrasonic treatment as the electrodes of OSCs and supercapacitors (SCs), respectively, and prepared semitransparent and flexible solid-state photovoltaic supercapacitors, where the configuration and working principle of the photovoltaic supercapacitors are shown in Fig. 12.2d. Interestingly, $Ti_3C_2T_x$ MXene electrodes in OSCs can be separately utilized as anode and cathode by interface modification of PEI CIL and PEDOT:PSS AIL, respectively, and the photovoltaic supercapacitors with $Ti_3C_2T_x$ MXene electrodes and P3HT:ICBA active layer showed good PCE of 2.2%.

ii. MXene-based composite electrode

In order to further improve optoelectronic properties and solution processability of MXene-based electrodes, MXene was also usually blended with other conductive materials, such as PEDOT:PSS and Ag NWs. As shown in Fig. 12.2e, Bai et al. [7]

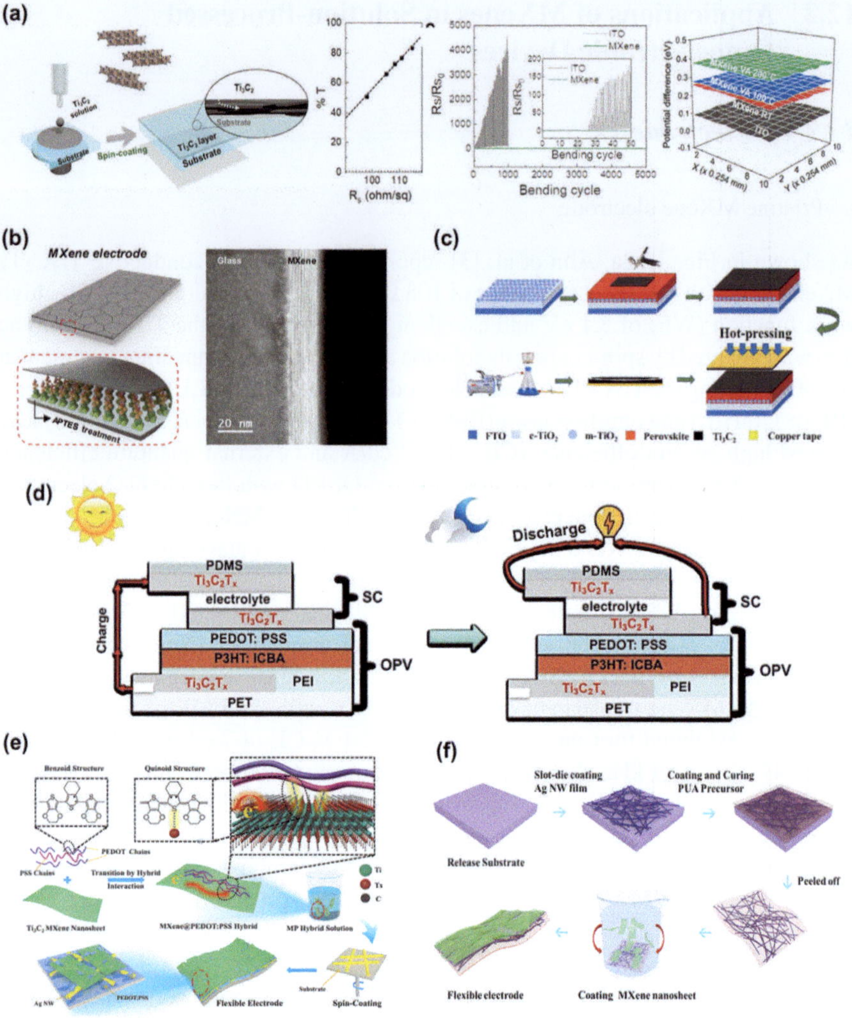

Fig. 12.2 **a** Schematic illustration and optoelectronic properties and flexibility of $Ti_3C_2T_x$ film spin-coated from its colloid solution. Reproduced with permission [3]. Copyright 2019, Wiley-VCH. **b** Schematic illustration (left) and cross-sectional scanning TEM image (right) of the thin-film electrode composed of MXene flakes. Reproduced with permission [4]. Copyright 2020, Wiley-VCH. **c** Schematic diagram of the fabrication process of Ti_3C_2 electrode by hot-pressing method. Reproduced with permission [5]. Copyright 2019, The Royal Society of Chemistry. **d** The configuration and working principle of the photovoltaic supercapacitors, including the charging process (left) and discharging process (right). Reproduced with permission [6]. Copyright 2019, The Royal Society of Chemistry. **e** Schematic illustration of AgNW-MXene@PEDOT:PSS flexible transparent electrode. Reproduced with permission [7]. Copyright 2020, Elsevier B.V. **f** Fabrication process of $Ti_3C_2T_x$/Ag NWs–PUA flexible electrode. Reproduced with permission [9]. Copyright 2020, American Chemical Society

developed a novel MXene-based composite electrode composed of Ag NWs, Ti_3C_2 MXene, and PEDOT:PSS, namely AgNW-MXene@PEDOT:PSS (AgNW-MP) electrode. The AgNW-MP electrode showed ultrahigh transmittance of 97.6%, ultralow sheet resistance of 17 Ω/m^2, and excellent flexibility and stability, and flexible OLED with the AgNW-MP electrode and (2,3-dpqx-F2)2Ir(pzrl) EML exhibited high CE of 21 cd/A and EQE of 25.9%. Yang et al. [8] developed a mixed-dimensional (0D–1D–2D–3D) composite electrode consisting of 0D Ag nanoparticles (AgNPs)/1D AgNWs/2D MXene/3D PEDOT:PSS, which combined the versatile advantages of metal, inorganic, and organic semiconductors and thus offered the electrode with excellent optoelectronic properties, flexibility, and a smooth surface. Based on the composite electrode, the flexible PeLEDs with a large emitting area of 8 cm^2 exhibited high CE and EQE of 54.1 cd/A and 16.5%, respectively, with excellent mechanical stability.

Tang et al. [9] prepared $Ti_3C_2T_x$/Ag NWs composite electrode by using a slot-die coating process (Fig. 12.2f) and the composite electrode possessed a low sheet resistance of 26 Ω/m^2 and maximum transmittance of 83.3%. Flexible OSCs with the composite electrode achieved high PCE of 7.16%, 7.70%, and 8.30% for the active layer of PTB7-Th:PC_{71}BM, PBDB-T:ITIC, and PBDB-T:ITIC:PC_{71}BM, respectively. Mi et al. [10] introduced 1D carbon nanotubes (CNTs) and 2D Ti_3C_2-MXene nanosheets into a commercial carbon paste to prepare a mixed carbon electrode (carbon/CNT/MXene) for inorganic PeSCs. The carbon/CNT/MXene mixed electrode can provide a network structure and multi-dimensional charge transfer pathways and thus effectively increase the conductivity of the carbon electrode, so the CsPbBr3-based PeSCs with the mixed electrode showed a good PCE of 7.09%.

12.2.2 Interface Layer

i. Pristine MXene as interface layer

Miao et al. [11] synthesized $Ti_3C_2T_x$-MXene nanosheets with a lateral dimension of 1–2 μm and appropriately adjusted the work function (5.14 eV) of $Ti_3C_2T_x$-MXene with UV-ozone treatment to well match with ITO. As shown in Fig. 11.3a, Ir(ppy)3-, FIrpic-, and Ir(pq)$_2$acac-based green, blue, and red OLEDs with $Ti_3C_2T_x$-MXene as hole injection layer (HIL) have all showed significantly improved device performance with EQE of 20.34%, 22.07%, and 21.90%, respectively, compared to the corresponding PEDOT:PSS-based devices.

Yu et al. [12] utilized the $Ti_3C_2T_x$ MXene nanosheets with good light transmission and high conductivity as hole transport layer (HTL) in OSCs. As shown in Fig. 11.3b, compared with the bare ITO-based OSCs (4.21%) with PBDB-T:ITIC active layer, the OSCs with the $Ti_3C_2T_x$ HTL possessed superior charge extraction capability and exhibited a significantly higher PCE of 10.53%, which also outperformed that of the PEDOT:PSS-based devices (10.11%). In $CsPbBr_3$-based PeSCs with carbon electrode, Chen et al. [13] introduced Ti_3C_2-MXene nanosheets as interlayer between

$CsPbBr_3$ layer and carbon electrode, where the Ti_3C_2-MXene with the work function of 4.5 eV can improve energy-level match and holes extraction from $CsPbBr_3$ layer to carbon electrode, as shown in Fig. 12.3c. In addition, the surface functional groups (=O, –F, etc.) of Ti_3C_2-MXene can passivate the under-coordinated Pb atom of $CsPbBr_3$ and thus reduce the defects of the $CsPbBr_3$ layer. As shown in Fig. 12.3d, the PeSCs with Ti_3C_2-MXene interlayer showed higher PCE and reduced hysteresis effect and more excellent environment stability (T_{80} under the environment stress of 45% humidity for 1900 h and 80 °C for 600 h) compared to the control device with bare carbon electrode.

ii. Functionalized MXene as interface layer

Hole Transport Layer (HTL)

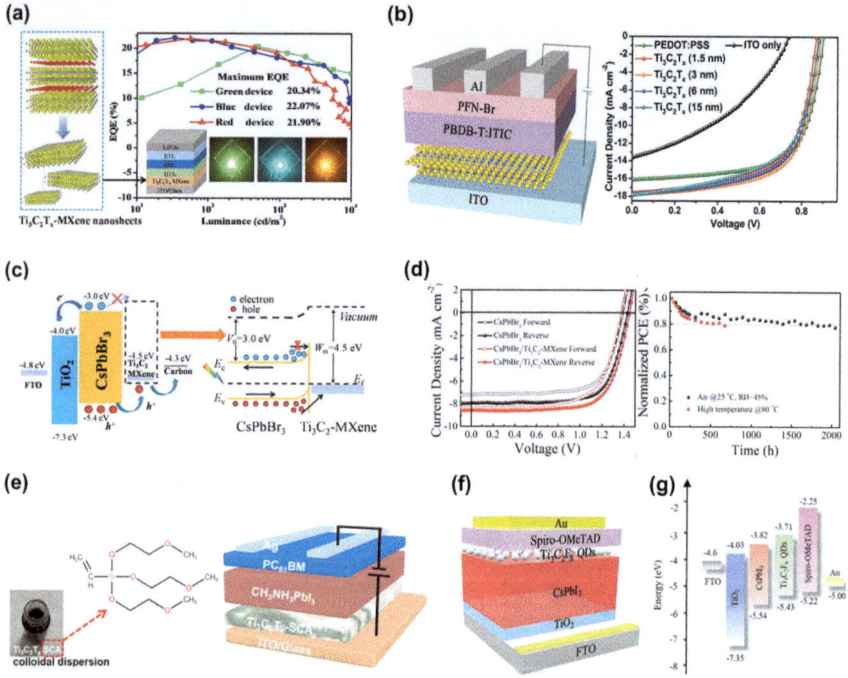

Fig. 12.3 **a** Green, blue, and red OLEDs with $Ti_3C_2T_x$-MXene as hole injection layer (HIL). Reproduced with permission [11]. Copyright 2022, Elsevier B.V. **b** Device structure and photovoltaic performance of the OSCs with the $Ti_3C_2T_x$ HTL. Reproduced with permission [12]. Copyright 2019, The Royal Society of Chemistry. **c** Schematic illustration of energy level and carrier transport of the PeSCs with Ti_3C_2-MXene interlayer, and **d** *J–V* curves and environmental stability of the PeSCs with Ti_3C_2-MXene interlayer. Reproduced with permission [13]. Copyright 2019, The Royal Society of Chemistry. **e** Structural formula of vinyl tris (2-methoxyethoxy) silane (SCA) and schematic diagram of the PeSCs with $Ti_3C_2T_x$-SCA HTL. Reproduced with permission [14]. Copyright 2021, Elsevier B.V. **f** Schematic image and **g** energy-level diagram of the $CsPbI_3$-based PeSCs with $Ti_3C_2F_x$ QDs interface passivators. Reproduced with permission [16]. Copyright 2022, Wiley-VCH

Wang et al. [14] utilized silane coupling agent vinyltris (2-methoxyethoxylsilane) (SCA) to modify $Ti_3C_2T_x$ MXene nanosheets for preparing $Ti_3C_2T_x$-SCA colloidal dispersion, and the $Ti_3C_2T_x$-SCA colloidal dispersion was spray-coated as HTL in $MAPbI_3$-based PeSCs (Fig. 12.3e). The $Ti_3C_2T_x$-SCA (0.3 V/V%) HTL with the advantage of SCA possessed flat, pin-hole free film, and enhanced hole extraction at the perovskite/HTL interface, thus leading to the PCE increase from 11.12 to 13.65% for the PeSCs. Yu et al. [15] synthesized Nb_2CT_x MXene with excellent optoelectronic properties and utilized oxygen plasma treatment to increase the work function of Nb_2CT_x as HTL in $MAPbI_3$-based PeSCs. The oxygen plasma-treated Nb_2CT_x HTL possessed enhanced surface dipoles due to the increased –O terminated groups and improved energy band alignment at the perovskite/HTL interface, facilitating increased hole extraction and reduced charge recombination. The $MAPbI_3$-based PeSCs with the oxygen plasma-treated Nb_2CT_x HTL achieved a high PCE of 20.74% and good stability, and the large area (0.99 cm^2) and flexible devices also achieved excellent PCEs of 17.94% and 17.26%, respectively. He et al. [16] synthesized fluorine functionalized $Ti_3C_2F_x$ MXene quantum dots (QDs) and used $Ti_3C_2F_x$ QDs as interface passivators in $CsPbI_3$-based PeSCs (Fig. 12.3f). The p-type $Ti_3C_2F_x$ QDs can tune the energy level of perovskite layer and improve hole transfer (Fig. 12.3g), reduce interface recombination by passivating defects, and improve device stability by forming a barrier layer to prevent water invasion. The $CsPbI_3$-based PeSCs with $Ti_3C_2F_x$ QDs interface passivators exhibited a high PCE of 20.44% with a high V_{oc} of 1.22 V and excellent air-stability.

Electron Transport Layer (ETL)

Wang et al. [17] first introduced $Ti_3C_2T_x$ MXene nanosheets as ETL in PeSCs and utilized simple UV-ozone treatment to improve $Ti_3C_2T_x$ MXene suitability as ETL, where UV-ozone treatment can increase the surface Ti–O bonds of $Ti_3C_2T_x$ MXene without change in its bulk properties. In $MAPbI_3$-based PeSCs, the $Ti_3C_2T_x$ MXene with 30 min of UV-ozone treatment can efficiently improve electron transfer and suppressed recombination at the ETL/perovskite interface and thus enable much higher PCE of 17.17% than that (5%) of the control device with $Ti_3C_2T_x$ without UV-ozone treatment. As shown in Fig. 12.4a, Wang et al. [18] further carried out different oxidation of $Ti_3C_2T_x$ hydrocolloid in air to tune its properties for an excellent ETL in PeSCs, where oxidized $Ti_3C_2T_x$ ETL, including lightly oxidized $Ti_3C_2T_x$ (LO-$Ti_3C_2T_x$) and heavily oxidized $Ti_3C_2T_x$ (HO-$Ti_3C_2T_x$), as well as a composite of HO-$Ti_3C_2T_x$ with pristine $Ti_3C_2T_x$ (HO-$Ti_3C_2T_x$@$Ti_3C_2T_x$), were spin-coated from its hydrocolloid. Oxidation of $Ti_3C_2T_x$ can generate more Ti–O bonds and reduce the macroscopic defects of the spin-coated film, while obvious energy-level changes and transition from metallic material to semiconductor occurred with the degree increase of oxidation (Fig. 12.4b), contributing to better energy-levels alignment between $MAPbI_3$ perovskite layer and ETL. As shown in Fig. 12.4c, the $MAPbI_3$-based PeSCs with HO-$Ti_3C_2T_x$@$Ti_3C_2T_x$ ETL exhibited the best photovoltaic performance with PCE of 18.29%, which was significantly better than that (16.50%) of the PeSCs with pristine $Ti_3C_2T_x$ ETL.

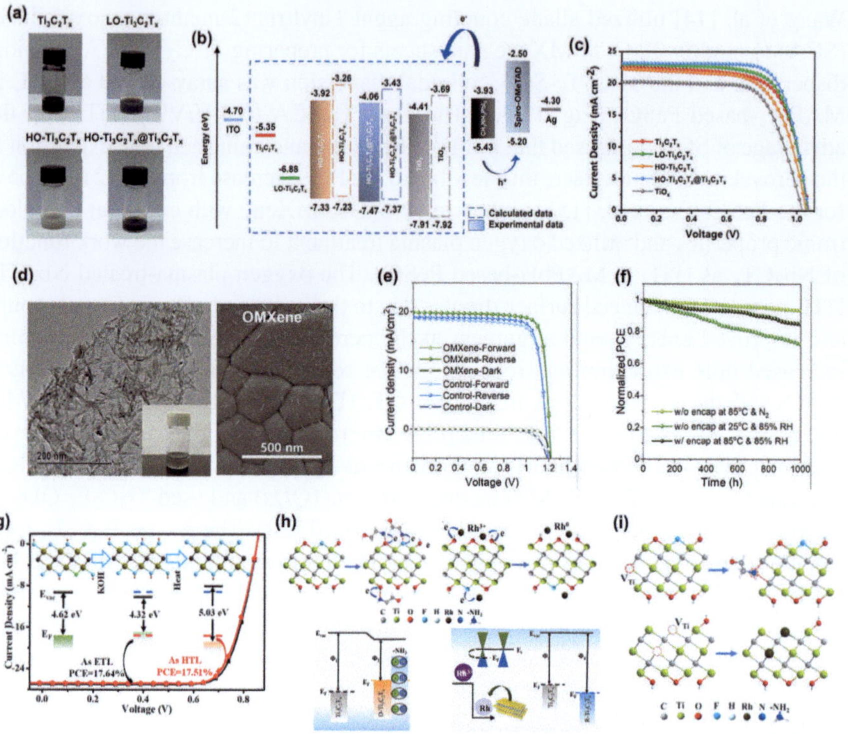

Fig. 12.4 **a** Digital photos of $Ti_3C_2T_x$, LO-$Ti_3C_2T_x$, HO-$Ti_3C_2T_x$ and HO-$Ti_3C_2T_x$@$Ti_3C_2T_x$ hydrocolloid, **b** schematic diagram of energy levels of PeSCs based on different ETLs determined by DFT calculation and UPS data, and **c** *J–V* curves of PeSCs based on different ETLs. Reproduced with permission [18]. Copyright 2021, The Royal Society of Chemistry. **d** TEM images of OMXene and SEM surface images of the $CsPbI_3$ perovskite film with OMXene passivation layer, **e** *J–V* curves of the PeSCs without and with OMXene passivation layer, and **f** long-term stability test of the encapsulated minimodule with OMXene passivation layer under simultaneous damp heat (85 °C/85% RH) and 1-sun illumination. Reproduced with permission [20]. Copyright 2022, Elsevier Inc. **g** Work function changes of Nb_2CT_x with alkali and annealing treatments and the corresponding device performance. Reproduced with permission [22]. Copyright 2021, American Chemical Society. **h** Schematic of formation mechanism and work function changes mechanism in D-$Ti_3C_2T_x$ and R-$Ti_3C_2T_x$, and **i** schematic illustration of the $Ti_3C_2T_x$ surface passivated by ethanolamine (top) and Rh^{3+} (bottom) Reproduced with permission [23]. Copyright 2022, Wiley-VCH

Chen et al. [19] used the $Ti_3C_2T_x$ MXene with oxygen plasma treatment as ETL in PeSCs. The oxygen plasma treatment can generate abundant Ti–O bonds on MXene surface by breaking parts of Ti–C bonds, which resulted in tunable work functions of MXene and reduced trap states and improved electron transport at the perovskite/MXene interface. In addition, the Pb–O interactions at the perovskite/MXene interface improved the device stability. As shown in Fig. 12.4d, Im et al. [20] synthesized oxidized $Ti_3C_2T_x$ MXene (OMXene) nanoplates dispersion in air and spray-coated the O-MXene nanoplates dispersion on the surface of $CsPbI_3$ layer in PeSCs. The

O-MXene can enhance the interface electric field and improve charge separation at the perovskite/ETL interface and form a physical barrier against moisture. Compared with the control PeSCs without OMXene, the PeSCs with OMXene passivation layer showed enhanced photovoltaic performance with PCEs of 19.69% for 0.096 cm^2 devices (Fig. 12.4e) and 14.64% for 25 cm^2 minimodules, respectively, and superior device stability (Fig. 12.4f).

Hole Transport Layer (HTL) and Electron Transport Layer (ETL)

Ouyang et al. [21] utilized the treatments of UV-ozone and N_2H_4 to separately increase and reduce the work function of $Ti_3C_2T_x$ MXene. Compared with the work function (4.68 eV) of pristine $Ti_3C_2T_x$ MXene, the work function of UV-ozone-treated MXene (U-MXene) was increased to 5.01 eV and the work function of UV-ozone-then-N_2H_4-treated MXene (UH-MXene) was reduced to 4.30 eV. Based on PBDB-T:ITIC active layer, the conventional OSCs with U-MXene as HTL showed PCE of 9.02%, and the inverted OSCs with UH-MXene as ETL showed PCE of 9.06%, which were comparable with that (9.67%) of the control device. Yu et al. [22] utilized alkali treatment and annealing treatment to separately reduce and increase the work function of Nb_2CT_x. As shown in Fig. 12.4g, the work function of Nb_2CT_x with alkali treatment was reduced to 4.32 eV, and the work function of Nb_2CT_x with annealing treatment was increased to 5.03 eV. Based on the PM6:BTP-eC9:PC_{71}BM active layer, the OSCs with the alkali treated Nb_2CT_x as ETL and the OSCs with the annealing treated Nb_2CT_x as HTL showed PCE of 17.64% and 17.51%, respectively. Yu et al. [23] also used ethanolamine and rhodium chloride ($RhCl_3$) treatments to develop surface-engineered $Ti_3C_2T_x$ with tunable work functions. As shown in Fig. 12.4h, the NH_2 of ethanolamine can adsorb on $Ti_3C_2T_x$ surface via hydrogen bonding, thus decreasing the work function of $Ti_3C_2T_x$ (D-$Ti_3C_2T_x$), while the $RhCl_3$ doping can downshift the Femi level of $Ti_3C_2T_x$, thus increasing the work function of $Ti_3C_2T_x$ (R-$Ti_3C_2T_x$). In addition, the ethanolamine and $RhCl_3$ can effectively passivate the Ti vacancies of MXene (Fig. 12.4i) and enhance the interfacing characteristics between the MXene and active layer. The PM6:Y6-based inverted OSCs with D-$Ti_3C_2T_x$ as ETL and the PM6:Y6-based conventional OSCs with R-$Ti_3C_2T_x$ as HTL exhibited PCEs of up to 15.88% and 15.54%, respectively. Xu et al. [24] utilized LiOH and annealed treatments to manipulate surface terminal groups of Nb_2CT_x nanosheets, and the Nb_2CT_x nanosheets were treated with LiOH solution (10 wt%) for 6 h to prepare $Nb_2CT_{OH\text{-}6h}$ and then further annealed at 150 °C for 4 h to prepare $Nb_2CT_{O\text{-}4h}$. As shown in Fig. 12.5a, the $Nb_2CT_{OH\text{-}6h}$ and $Nb_2CT_{O\text{-}4h}$ were used as interface layers at ETL/perovskite interface and perovskite/HTL interface, respectively, in PeSCs. The $Nb_2CT_{OH\text{-}6h}$ and $Nb_2CT_{O\text{-}4h}$ interface layers improved carrier mobility and energy-levels alignment at perovskite/CTLs interfaces (Fig. 12.5b), and moreover, $Nb_2CT_{O\text{-}4h}$ located at grain boundaries of perovskite layer efficiently accelerated hole current transport to HTL, enabling charge transport balance in devices. Based on the synergy effect, the optimized PeSCs achieved a progressively enhanced PCE of 24.11% with negligible hysteresis (Fig. 12.5c), as well as excellent thermal and light stability.

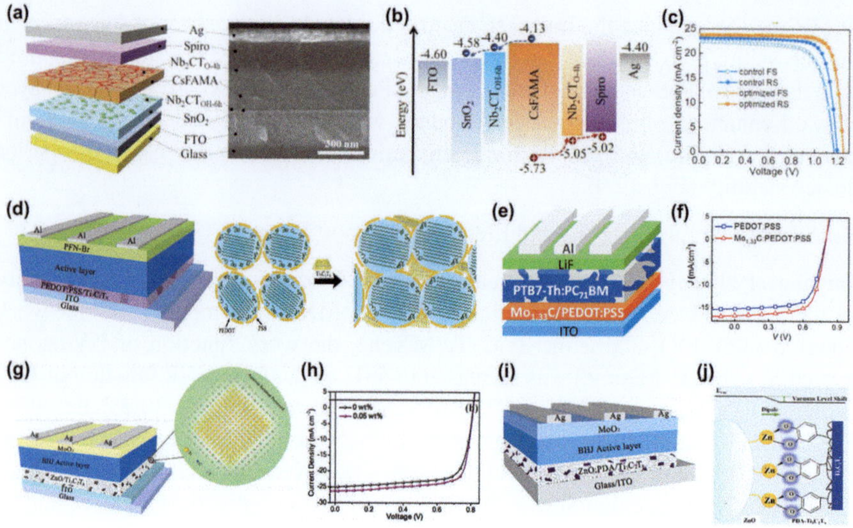

Fig. 12.5 **a** Schematic diagram of structure and cross-section SEM, **b** energy-levels diagram of the optimized device, and **c** *J*–*V* curves of the best-performance control and optimized PeSCs in FS and RS directions. Reproduced with permission [24]. Copyright 2022, Wiley-VCH. **d** Schematic illustration of OSCs structure and morphology changes of PEDOT:PSS with $Ti_3C_2T_x$. Reproduced with permission [25]. Copyright 2020, The Royal Society of Chemistry. **e** Schematic diagram and **f** $J-V$ curves of the OSCs with the $Mo_{1.33}C$:PEDOT:PSS HTL. Reproduced with permission [26]. Copyright 2019, American Chemical Society. **g** Schematic structures and **h** $J-V$ curves of the OSCs with ZnO/$Ti_3C_2T_x$ (0.05 wt%) ETL. Reproduced with permission [28]. Copyright 2020, Elsevier B.V. (i) Device configuration of the OSCs and (j) molecular dipole-induced work function reduction on ZnO nanocrystals surface. Reproduced with permission [29]. Copyright 2021, Elsevier B.V.

iii (Functionalized-)MXene-based composite interface layer

HTL

As shown in Fig. 12.5d, Hou et al. [25] introduced $Ti_3C_2T_x$ nanosheets into PEDOT:PSS to prepare PEDOT:PSS/$Ti_3C_2T_x$ composite HTL, where the surface functional groups of $Ti_3C_2T_x$ can form electrostatic interaction with PSS chains and induce the conformational transition from a coil to a linear/expanded-coil structure for PEDOT chains and thus improve hole transport/extraction in OSCs. Utilizing the PEDOT:PSS/$Ti_3C_2T_x$ HTL, the OSCs with PBDB-T:ITIC and PM6:Y6 active layers achieved high PCE of 11.02% and 14.55%, respectively. Zhang et al. [26] mixed $Mo_{1.33}C$ MXene and PEDOT:PSS as composite HTL in OSCs (Fig. 12.5e), and the PTB7-Th:$PC_{71}BM$-based OSCs with $Mo_{1.33}C$:PEDOT:PSS HTL showed higher PCE of 9.24% compared to the control devices with PEDOT:PSS HTL (Fig. 12.5f), which may be ascribed to the improved charge transport and extraction of $Mo_{1.33}C$:PEDOT:PSS HTL, along with the morphology improvement of the active layer.

ETL

In addition, MXene-based mixture/composite was also usually used as interface layers for improving device performance. Lu et al. [27] reported a ZnO–Ti_3C_2 (ZTC) mixture as ETL in all-inorganic perovskite nanocrystal LEDs and explored the influence of different Ti_3C_2 MXene contents on device performance. The energy level of ZTC ETL can be optimized by regulating the Ti_3C_2 MXene contents and thus offer more balanced charge carrier transport in LEDs. Therefore, the LEDs with ZTC ETL (10% Ti_3C_2) achieved a higher EQE of 17.4% than that (14.2%) of the devices without Ti_3C_2.

Yu et al. [28] introduced $Ti_3C_2T_x$ MXene as additive in ZnO nanocrystals to prepare ZnO/$Ti_3C_2T_x$ composite ETL in OSCs (Fig. 12.5g), and the $Ti_3C_2T_x$ additive can increase electron transport pathways in ETL and passivate the ZnO surface defects by forming Zn–O–Ti bonds, facilitating to increase charge extraction in the devices. The OSCs with PBDB-T:ITIC active layer and ZnO/$Ti_3C_2T_x$ ETL showed higher PCE of 12.20% than that (10.56%) of the control device with pristine ZnO ETL, and moreover the OSCs with PM6:Y6 active layer and ZnO/$Ti_3C_2T_x$ ETL also exhibited higher PCE of 16.51% than that (14.99%) of the control device with pristine ZnO ETL (Fig. 12.5h). Yu et al. [29] synthesized polydopamine (PDA) modified $Ti_3C_2T_x$ (PDA-$Ti_3C_2T_x$) and introduced PDA-$Ti_3C_2T_x$ into ZnO nanocrystals to prepare ZnO:PDA-$Ti_3C_2T_x$ composite ETL in OSCs (Fig. 12.5i). As shown in Fig. 12.5j, ZnO and PDA molecules can form strong and stable chelate interactions between Zn^{2+} and –OH and thus passivate the surface trap states of ZnO, and moreover, the interface dipoles formation between PDA-$Ti_3C_2T_x$ and ZnO can downshift the vacuum level of ETL and improve electron extraction. Based on the ZnO:PDA-$Ti_3C_2T_x$ ETL, the inverted OSCs with PBDB-T:ITIC and PM6:Y6 active layer exhibited PCE of 12.07% and 16.69%, respectively, outperforming the ZnO-based devices.

Wang et al. [30] prepared SnO_2–Ti_3C_2 MXene nanocomposites with different contents of Ti_3C_2 (0, 0.5, 1.0, 2.0, and 2.5 wt‰) and utilized SnO_2–Ti_3C_2 MXene nanocomposites as ETL in $MAPbI_3$-based PeSCs (Fig. 12.6a). The Ti_3C_2 MXene with excellent conductivity can improve charge transfer pathways of the SnO_2–Ti_3C_2 composite ETL with appropriate energy level, and the $MAPbI_3$-based PeSCs with SnO_2–Ti_3C_2 (1.0 wt‰) ETL exhibited higher PCE of 18.34% than that (17.23%) of the control devices with SnO_2 ETL (Fig. 12.6b). As shown in Fig. 12.6c, Ren et al. [31] developed 2D $Ti_3C_2T_x$ MXene-modified SnO_2 as composite ETL by depositing $Ti_3C_2T_x$ MXene on FTO substrate and then depositing SnO_2 on MXene in PeSCs and the $Ti_3C_2T_x$ MXene can enhance the electron mobility and charge transfer ability of SnO_2 ETL with strong interface interaction. The MXene-modified SnO_2 ETL also improved perovskite film morphology with reduced trap density and offered better energy-level alignment, facilitating the reduction of non-radiative recombination and charge transport losses in devices (Fig. 12.6d). The PeSCs with the MXene-modified SnO_2 ETL achieved an enhanced PCE of 20.65% with negligible hysteresis. Yang et al. [32] introduced $Ti_3C_2T_x$ MXene quantum dots (MQDs) into SnO_2 to prepare MQDs-SnO_2 composite ETL in PeSCs (Fig. 12.6e). According to in situ synchrotron-based 2D grazing-incidence X-ray diffraction measurements, the MQDs-SnO_2 ETL rapidly induced perovskite nucleation from the precursor solution and then formed an intermediate perovskite phase after anti-solvent treatment (Fig. 12.6f), which obviously improved the crystal quality and phase stability of the perovskite film (Fig. 12.6g). Based on the MQDs-SnO_2 ETL, the PeSCs with $FA_{0.9}MA_{0.05}Cs_{0.05}PbI_{0.98}Br_{0.02}$ perovskite layer achieved an enhanced PCE of up to 23.3% (Fig. 12.6h), with excellent stability against humidity and light soaking.

Li et al. [33] doped monolayer $Ti_3C_2T_x$ MXene nanosheets into TiO_2 to prepare $Ti_3C_2T_x$@TiO_2 composite ETL in PeSCs with $Cs_2AgBiBr_6$ perovskite layer (Fig. 12.6i). The MXene nanosheets can significantly increase the electrical conductivity of TiO_2 and improve the surface wettability of ETL, promoting perovskite crystallization (Fig. 12.6j). As shown in Fig. 12.6k, the $Ti_3C_2T_x$@TiO_2-based PeSCs

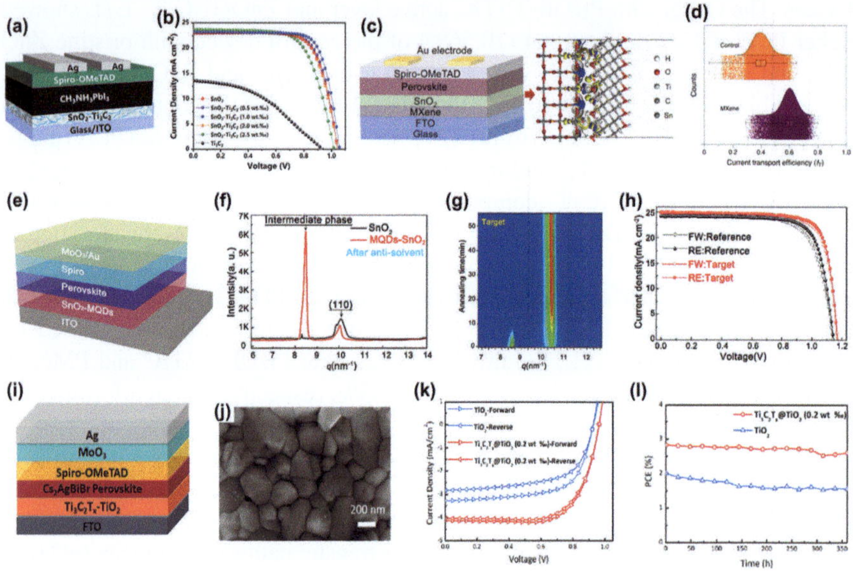

Fig. 12.6 **a** Device structure of the PeSCs with SnO_2–Ti_3C_2 ETL, and **b** *J–V* curves of the PeSCs with different ETLs. Reproduced with permission [30]. Copyright 2019, The Royal Society of Chemistry. **c** Device configuration of the MXene-modified PeSCs and interface interaction between SnO_2 and MXene based on DFT calculation, and **d** the corresponding statistical distributions of local current transport efficiencies. Reproduced with permission [31]. Copyright 2020, American Chemical Society. **e** Schematic structure of the PeSCs with the MQDs-SnO_2 ETL, **f** integrated 1D-XRD spectra of the perovskite films after antisolvent treatment, **g** integrated 1D-GIXRD spectra of target perovskite film (on the MQDs-SnO_2 ETL) at different annealing time points, and **h** *J–V* curves of the best SnO_2-based and MQDs-SnO_2-based PeSCs. Reproduced with permission [32]. Copyright 2021, The Royal Society of Chemistry. **i** Schematic diagram of the $Cs_2AgBiBr_6$ solar cell structure, **j** top-view SEM images of $Cs_2AgBiBr_6$ film on $Ti_3C_2T_x$@TiO_2 ETL, **k** *J–V* curves displaying the hysteretic responses of the TiO_2 and $Ti_3C_2T_x$@TiO_2 based PeSCs, and **l** Ambient stability characteristics of the PeSCs based on TiO_2 and $Ti_3C_2T_x$@TiO_2. Reproduced with permission [33]. Copyright 2021, American Chemical Society

achieved a greatly enhanced PCE of 2.81% with suppressed hysteresis compared to the TiO_2-based devices. Furthermore, the $Ti_3C_2T_x$@TiO_2-based device also showed better long-term stability with PCE maintaining 93% of the initial value after storing for 15 days in ambient air (Fig. 12.6l). Huang et al. [34] developed a multi-dimensional conductive network (MDCN) structured ETL composed of TiO_2, SnO_2, and $Ti_3C_2T_x$ MXene in PeSCs (Fig. 12.7a). As shown in Fig. 12.7b, the MDCN structured ETL was fabricated in a controlled low-temperature anneal method in air and then in N_2 atmospheres, where 0D anatase TiO_2 quantum dots were surrounded on 2D $Ti_3C_2T_x$ MXene nanosheets and moreover in situ rooted on 3D SnO_2 nanoparticles, constructing the MDCN structure. The MDCN structured ETL with the optimal MXene content can efficiently improve the optical property and crystallinity of perovskite layer and internal interfaces. Therefore, the PeSCs with the MDCN structured ETL and $(FAPbI_3)_{0.97}(MAPbBr_3)_{0.03}$ active layer exhibited higher PCE of 19.14% than that (16.83%) of the SnO_2-based device (Fig. 12.7c).

Xu et al. [35] first introduced Nb_2C MXene as an additive into SnO_2 ETL for PeSCs. The incorporation of Nb_2C MXene accelerated the SnO_2 grains growth and increased the lattice spacing facets of SnO_2, contributing to the improved roughness, surface energy, and defects of the SnO_2–Nb_2C ETL, and thus the subsequent perovskite film on the SnO_2–Nb_2C ETL achieved a higher quality with superior crystalline. The PeSCs with the SnO_2–Nb_2C ETL showed a higher PCE of 22.86% and better device stability than that (18.96%) of the control devices with pristine SnO_2 ETL. Carlo et al. [36] introduced $Ti_3C_2T_x$ MXene into bathocuproine (BCP) to prepare BCP:MXene composite interlayer between ETL and cathode in PeSCs with $Cs_{0.2}FA_{0.8}PbI_3$ perovskite layer. The PeSCs with the BCP:MXene interlayer achieved an enhanced PCE of 17.5% more than that (16.5%) of the reference devices with pristine BCP interlayer. Notably, the BCP:MXene interlayer dramatically improved the out of Glove-Box stability for the PeSCs: the T80 period increased from 460 to >2300 h under ISOS-L-2 light soaking stress at 63 ± 1.5 °C (Fig. 12.7d).

Wang et al. [37] introduced in-situ-oxidized $Ti_3C_2T_x$ MXene (O-$Ti_3C_2T_x$) into SnO_2 to form O-$Ti_3C_2T_x$/SnO_2 ETL with nanoscale heterojunction in $MAPbI_3$-based PeSCs. The O-$Ti_3C_2T_x$ showed the transition from metallicity to semiconductivity and achieved energy levels well matched with the perovskite layer, and the O-$Ti_3C_2T_x$ improved the electron mobility of ETL, which facilitated better electron extraction and reduced recombination between the O-$Ti_3C_2T_x$/SnO_2 ETL and the perovskite layer. The PeSCs with the O-$Ti_3C_2T_x$/SnO_2 ETL achieved an enhanced PCE of 20.09% and better stability compared to the control devices with pristine SnO_2 ETL (17.68%). Zhao et al. [38] developed two functionalized $Ti_3C_2T_x$ MXenes (MXene-F and MXene-H) as nanosheet dopants in SnO_2 ETL of PeSCs (Fig. 12.7e), where MXene-H and MXene-F were prepared by utilizing $Ti_3C_2T_x$ MXenes modified by dodecyltrimethoxysilane and fluoroalkylsilane molecules (Fig. 12.7f), respectively, and SnO_2 ETL doped with MXene-H and MXene-F were denoted as SnO_2-MH and SnO_2-MF ETLs, respectively. The functionalized $Ti_3C_2T_x$ MXenes improved the energy level of ETL and the SnO_2-MH ETL offered the better band alignment in devices (Fig. 12.7g). Meanwhile, the functionalized MXenes with high electrical conductivity can form zero Schottky barrier heterojunction with SnO_2 and thus

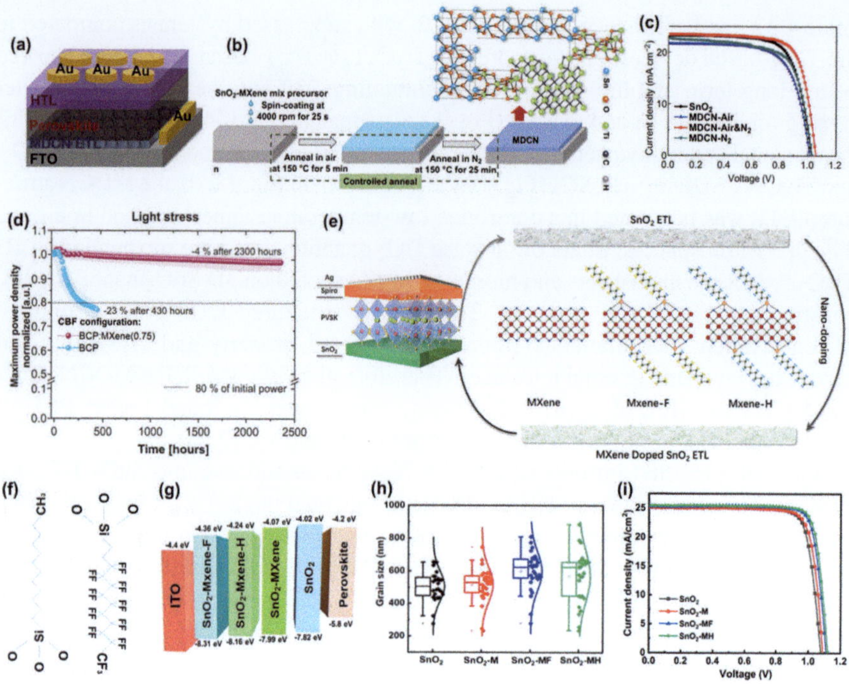

Fig. 12.7 **a** Device structure of the PeSC with MDCN ETL, **b** fabrication process and schematic structure of MDCN ETL, and **c** *J–V* curves of the PeSC with MDCN ETL. Reproduced with permission [34]. Copyright 2020, Springer Nature. **d** Light soaking stability of the BCP:MXene (0.75)-based PeSCs in comparison with the reference devices. Reproduced with permission [36]. Copyright 2022, Wiley-VCH. **e** Device structure and crystal structures of MXene, MXene-F, and MXene-H, **f** the molecule structures of dodecyltrimethoxysilane and fluoroalkylsilane, **g** energy-level diagram of the ITO/ETL/perovskite structure based on different ETLs, **h** grain-size box chart of the perovskite film on different ETLs, and **i** *J–V* curves of the champion devices based on different ETLs under reverse scans for $FAPbI_3$ system. Reproduced with permission [38]. Copyright 2022, Elsevier

enhanced charge transfer, and the SnO_2-MH and SnO_2-MF ETLs achieved suitable surface energy, facilitating to enlarge the grain size of perovskite film (Fig. 12.7h). The SnO_2-MH based PeSCs with $FAPbI_3$ perovskite layer achieved a significantly improved PCE from 22.36 to 24.12% (Fig. 12.7i), as well as improved moisture resistance and operational stability.

12.2.3 Active Layer

i. Pristine MXene additive in active layer

Ma et al. [39] first introduced $Ti_3C_2T_x$ MXene into the perovskite layer for PeSCs. The termination groups (e.g. –OH, –F) of $Ti_3C_2T_x$ MXene can retard the crystallization rate, thus increasing the crystal size of $MAPbI_3$ perovskite layer, and $Ti_3C_2T_x$ MXene with the high electrical conductivity and mobility can accelerate the charge transfer in perovskite layer. Utilizing 0.03 wt% amount of $Ti_3C_2T_x$ MXene additive, the $MAPbI_3$-based PeSCs achieved 12% enhancement in device performance. As shown in Fig. 12.8a, the compact PbI_2 cannot be completely converted to perovskite in the two-step process, while Qi et al. [40] innovatively introduced 2D monolayer $Ti_3C_2T_x$ (T_x = O, OH, and F) MXene nanosheets into the PbI_2 layer to form porous channels and enhance the PbI_2 reactivity, which can facilitate to the subsequent reaction with MAI to completely convert to $MAPbI_3$, and thus effectively reduce the residual PbI_2 amount and enlarge the perovskite grain size in the perovskite film. In addition, the surface functional groups of the $Ti_3C_2T_x$ can passivate the defects by interacting with the under-coordinated Pb^{2+} in perovskite, effectively mitigating hysteresis and suppressing non-radiative recombination. Moreover, the $Ti_3C_2T_x$ can regulate the work function of the perovskite, thus enabling better energy-level alignment (Fig. 12.8b) and improving carrier extraction and injection. The $MAPbI_3$-based PeSCs with 0.03 wt% of the $Ti_3C_2T_x$ additive achieved higher PCE of 19.27% than that (16.45%) of the control devices without $Ti_3C_2T_x$ additive (Fig. 12.8c). Wang et al. [41] utilized 2D $Ti_3C_2T_x$ MXene nanosheets as a nanosized additive to prepare 2D Ruddlesden–Popper $BA_2MA_4Pb_5I_{16}$ (BA, namely $CH_3(CH_2)_3NH_3$) perovskite films in PeSCs. The $Ti_3C_2T_x$ nanosheets with an optimized concentration can improve the crystallinity and passivate trap states in the 3D phase, accelerating the charge transfer process in the vertical direction, so the PeSCs with 0.3 mM $Ti_3C_2T_x$ additive achieved higher PCE of 15.71% than that (13.69%) of the control devices without additive. Yang et al. [42] separately introduced 2D $Ti_3C_2T_x$ and V_2CT_x MXene as additives in PbI_2 precursor solution to fabricate $MAPbI_3$ perovskite film via a one-step deposition method. The $Ti_3C_2T_x$ and V_2CT_x additives can improve the morphology and increase the grain size of perovskite film, and especially the perovskite film with V_2CT_x additive possessed unique morphology with uniform grain size and ordered layered crystal particles. The $MAPbI_3$-based PeSCs with V_2CT_x additive (0.0013 wt%) showed the highest PCE of 17.61% and the best stability compared to the control devices without additive and with $Ti_3C_2T_x$ additive. Li et al. [43] introduced $Ti_3C_2T_x$ MXene nanosheets as an additive in the $CsPbI_2Br$ perovskite layer, and the MXene additive facilitated to enable the improved crystallinity/reduced defect density in the perovskite layer and form the Schottky junction between the perovskite and MXene, enhancing charge separation and transfer. Based on the device configuration of ITO/SnO_2/$CsPbI_2Br$/P3HT/Ag, the PeSCs with MXene additive achieved a significantly boosted PCE of 15.10% and enhanced humidity stability compared to that (12.94%) of the control devices without MXene additive.

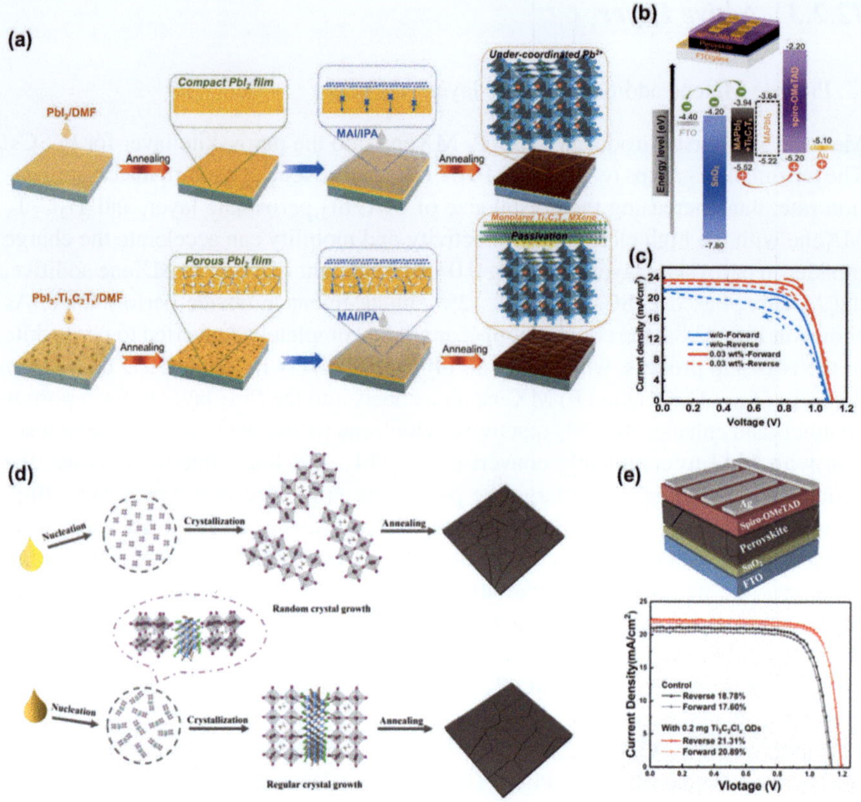

Fig. 12.8 **a** Mechanism diagram of preparing high-quality two-step-processed perovskite films assisted by $Ti_3C_2T_x$ additive, **b** Device configuration and the energy-level diagram of the PeSCs with and without $Ti_3C_2T_x$ doping, and **c** *J–V* curves of the best performed PeSCs with 0.03 wt% $Ti_3C_2T_x$ additive and without $Ti_3C_2T_x$ additive. Reproduced with permission [40]. Copyright 2020, Elsevier B.V. **d** Schematic illustration of the nucleation and crystallization process of the perovskite films without and with 0.2 mg/mL $Ti_3C_2T_x$ QDs additives, **e** schematic illustration of the PeSCs structure and *JV* curves of the PeSCs without and with 0.2 mg/mL $Ti_3C_2T_x$ QDs additives. Reproduced with permission [46]. Copyright 2021, Elsevier B.V.

In OSCs, Sun et al. [44] added 2D $Ti_3C_2T_x$ nanosheets into the organic active layer, where the 2D $Ti_3C_2T_x$ nanosheets can increase the short-circuit current density (J_{SC}) by enhancing the light scattering and provide additional charge transport pathway, thus improving the carrier dissociation and transfer and reducing bimolecular recombination in devices. After adding the 2D $Ti_3C_2T_x$ nanosheets, the PCEs of OSCs are increased from 9.34% to 10.72% and from 14.64% to 16.25% for PBDB-T:ITIC active layer and PM6:Y6 active layer, respectively.

ii. Functionalized MXene additive in active layer

Shapter et al. [45] synthesized cesium (Cs) functionalized $Ti_3C_2T_x$ MXene nanosheets ($Cs\text{-}Ti_3C_2T_x$) and introduced $Cs\text{-}Ti_3C_2T_x$ with the advantages of both additives into PbI_2 precursor solution for PeSCs via two-step deposition method. The $Cs\text{-}Ti_3C_2T_x$ additives can improve perovskite crystallization, and thus enlarge crystal grains and reduce charge recombination, so the PeSCs with $Cs\text{-}Ti_3C_2T_x$ additives achieved a high PCE of 21.57% with excellent thermal stability, demonstrating the feasibility of MXene doped with some monovalent cations (e.g. Cs^+) in perovskite.

Chen et al. [46] synthesized Cl-terminated Ti_3C_2 quantum dots ($Ti_3C_2Cl_x$ QDs), and introduced the $Ti_3C_2Cl_x$ QDs as additives in perovskite precursor solution via one-step deposition method. The Cl terminations of the $Ti_3C_2Cl_x$ QDs can strongly interact with Pb^{2+} of perovskite, which can retard the perovskite crystallization and deliver a high-quality perovskite film with high crystallinity and few trap states (Fig. 12.8d). Moreover, the $Ti_3C_2Cl_x$ QDs additives with top–bottom increasing gradient distribution can also facilitate the charge extraction and improve energy band alignment between perovskite layer and SnO_2 ETL. As shown in Fig. 12.8e, the PeSCs with the $Ti_3C_2Cl_x$ QDs additives achieved a significantly enhanced efficiency of 21.31% with negligible hysteresis. In addition, the $Ti_3C_2Cl_x$ QDs with Cl substituting the OH terminations of $Ti_3C_2T_x$ can avoid possible deprotonation of protonated organic amine in perovskite, so the unencapsulated device with the $Ti_3C_2Cl_x$ QDs additives exhibited better long-term humidity stability compared to the devices with $Ti_3C_2T_x$ QDs additives and $Ti_3C_2OH_x$ QDs additives.

12.2.4 Active Layer and Interface Layer

i. Pristine MXene in both active layer and interface layer

Carlo et al. [47] utilized 2D $Ti_3C_2T_x$ MXene with various termination groups (T_x) as additive doping into the perovskite layer and TiO_2 ETL, and as interface layer between perovskite layer and TiO_2 ETL in PeSCs. The $Ti_3C_2T_x$ additive can tune the work function of perovskite layer and TiO_2 ETL without affecting other electronic properties and moreover the dipole induced by the $Ti_3C_2T_x$ at the perovskite/ETL interface can improve the energy band alignment between these layers, which contributed to the 26% increase of PCE and hysteresis reduction in MXene-modified PeSCs compared to the reference devices without MXene. As shown in Fig. 12.9a, Xu et al. [48] introduced $Ti_3C_2T_x$ quantum dots (TQDs) into perovskite layer and ETL and doped $Cu_{1.8}S$ nanoparticles into HTL in PeSCs. The TQDs additive in the perovskite can increase the crystallization size and conductivity and reduce the grain boundaries and the intrinsic defect density in the perovskite layer, thereby improving charge separation and transport. The TQDs additive in the mesoporous TiO_2 can reduce the interfacial defects and improve the electron extraction from the perovskite layer. In addition, $Cu_{1.8}S$ nanoparticles in the HTL layer can improve hole mobility and conductivity, suppress the Spiro-OMeTAD aggregation, and slow down the perovskite decomposition. The PeSCs with TQDs and $Cu_{1.8}S$ nanoparticles doping achieved a remarkable hysteresis-free PCE of 21.64% (Fig. 12.9b). As shown in Fig. 12.9c, Li et al. [49] introduced TQDs into the perovskite layer and inserted Ti_3C_2 nanosheets into the TiO_2 ETL in $MAPbI_3$-based PeSCs. The TQDs can effectively passivate the defects of perovskite layer and the Ti_3C_2 nanosheets can improve the charge transport of the ETL, so the PeSCs with the functions of 0D and 2D Ti_3C_2 achieved an enhanced PCE of 17.1% with reduced hysteresis effect (Fig. 12.9d). Agresti et al. [50] separately doped $Ti_3C_2T_x$ MXene into perovskite layer and PCBM ETL in $MAPbI_3$-based PeSCs. The MXene can tune the energy-level alignment and passivate the trap states between perovskite layer and charge transporting layer, thereby improving charge extraction and collection at the electrodes. The MXene-based engineered PeSCs achieved superior performance with PCE exceeding 19% and improved stabilized power output compared to the control devices.

ii. Functionalized MXene in both active layer and interface layer

Duan et al. [51] synthesized 2D Cl-terminated Ti_3C_2 ($Ti_3C_2Cl_x$) MXene via molten salt-based etching method, and introduced $Ti_3C_2Cl_x$ into the bulk of $CsPbBr_3$ film and as the interface layer on perovskite layer in carbon-based PeSCs (Fig. 12.9e). As shown in Fig. 12.9f, the Cl terminations in $Ti_3C_2Cl_x$ can strongly interact with the under-coordinated Pb^{2+} in $CsPbBr_3$ perovskite lattice; the Pb–Cl bond can play the role of "glue" and the 2D Ti_3C_2 can immobilize the perovskite lattice, so that the expanded perovskite lattice can be compressed and confined to act as a lattice "tape" and the defective surface can also be healed by $Ti_3C_2Cl_x$. The carbon-based PeSCs with the $CsPbBr_3$ active layer and $Ti_3C_2Cl_x$ additives achieved a high PCE of 11.08% with an ultrahigh V_{oc} of 1.702 V and excellent stability. Yu et al. [52] prepared 2D Nb_2CT_x MXene nanosheets, and the work function of Nb_2CT_x was reduced from 4.65 to 4.32 eV by replacing the surface –F groups with $-NH_2$ groups via hydrazine (N_2H_4) treatment. The N_2H_4-treated Nb_2CT_x (T-Nb_2CT_x) can be used as ETL, improving energy band alignment with the $FA_{0.85}Cs_{0.15}PbI_3$ perovskite layer. Moreover, T-Nb_2CT_x with abundant $-NH_2$ groups were incorporated into the perovskite precursor to retard the perovskite crystallization by forming hydrogen bonds with I^-, and thus induced to form high-quality perovskite film. The PeSCs with T-Nb_2CT_x ETL and T-Nb_2CT_x nanosheets additive exhibited the highest PCE of 21.79% with excellent storage stability, and the corresponding flexible and large-area devices achieved the highest PCE of 19.15% and 18.31%, respectively.

12.3 Conclusion and Prospect

MXenes with unique layered structures and abundant and modifiable surface functional groups show excellent and controllable optical/electrical properties and solution processability, which facilitate the versatile applications of MXenes in

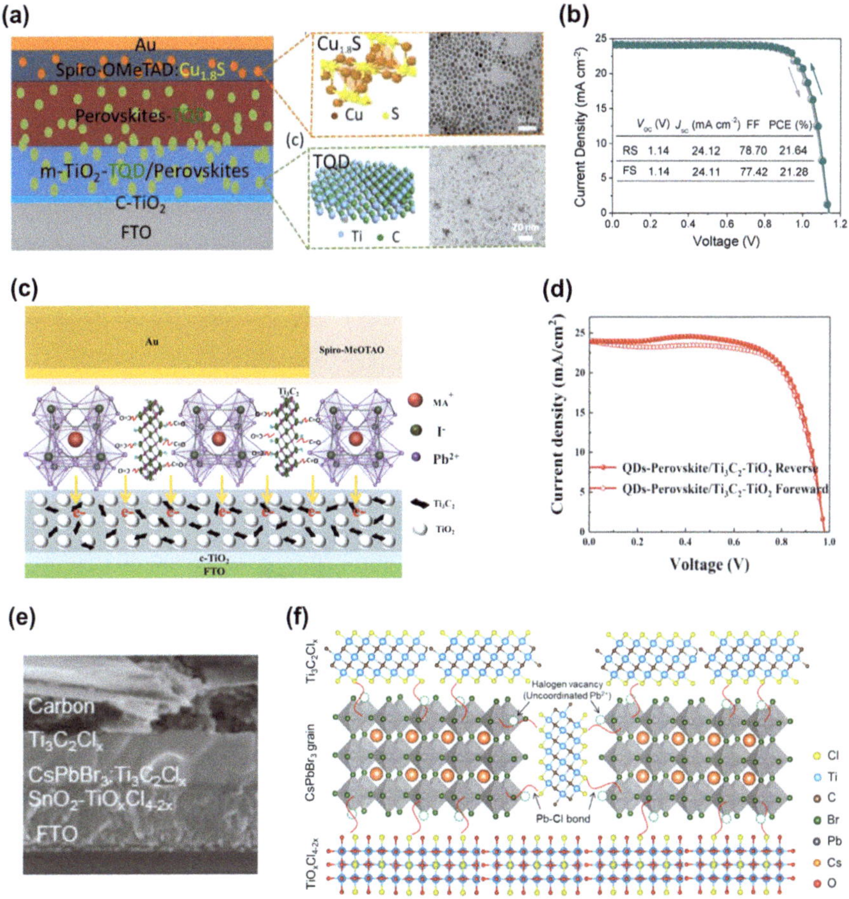

Fig. 12.9 **a** Architecture of the PeSCs and schematic structures and TEM images of $Cu_{1.8}S$ nanoparticles and $Ti_3C_2T_x$ quantum dots (TQD), and **b** *J–V* curves of the PeSCs with TQD and $Cu_{1.8}S$ nanoparticles doping. Reproduced with permission [48]. Copyright 2020, Wiley-VCH. **c** Schematic diagram and **d** *J–V* curves of the TQDs-modified PeSCs with Ti_3C_2–TiO_2 ETL. Reproduced with permission. [49] Copyright 2020, Elsevier B.V. **e** Cross-sectional SEM image of carbon-based PeSCs with the $CsPbBr_3$ perovskite layer and $Ti_3C_2Cl_x$ additives, and **f** schematic diagram of full defect passivation in $CsPbBr_3$ film by $Ti_3C_2Cl_x$. Reproduced with permission [51]. Copyright 2021, Wiley-VCH

solution-processed organic/perovskite light-emitting diodes and solar cells. Especially, MXenes show broad and great application potential in electrode, interface layer, and active layer for high-performance solution-processed organic/perovskite solar cells, as summarized in Table 12.1. Here, we conclude the main properties/functions of MXenes application in electrode, interface layer, and active layer, and simply discuss the future exploration of MXenes for further enhancing device performance.

Table 12.1 Photovoltaic parameters of organic/perovskite solar cells with applications of MXenes in electrode, interface layer (HTL and ETL), and active layer (AL)

Device structure	J_{sc} /(mA·cm^{-2})	V_{oc} /V	FF /%	PCE /%	Features	Ref.
Glass/$Ti_3C_2T_x$/PEDOT:PSS/PM6:Y6/PFN-Br/Al	24.97	0.84	64.9	13.62	Electrode	[6]
PET/$Ti_3C_2T_x$/PEDOT:PSS/PM6:Y6/PFN-Br/Ag	24.78	0.83	64	13.15	Electrode	[6]
PET/$Ti_3C_2T_x$/PEI/PTB7-Th:$PC_{71}BM$/PEDOT:PSS/$Ti_3C_2T_x$	14.43	0.75	48.73	5.26	Electrode	[6]
PUA-Ag NWs/$Ti_3C_2T_x$/PEDOT:PSS/PBDB-T:ITIC:$PC_{71}BM$/PDINO/Al	14.85	0.88	63	8.3	Electrode	[9]
FTO/c-TiO_2/$CsPbBr_3$/Carbon/CNT/Ti_3C_2	7.16	1.357	72.97	7.09	Electrode	[10]
ITO/$Ti_3C_2T_x$/PBDB-T:ITIC/PFN-Br/Al	17.85	0.88	67.06	10.53	HTL	[12]
FTO/TiO_2/$CsPbBr_3$/$Ti_3C_2T_x$/carbon	8.54	1.44	73.08	9.01	HTL	[13]
ITO/$Ti_3C_2T_x$-SCA/$CH_3NH_3PbI_3$–$Ti_3C_2T_x$/$PC_{61}BM$/Ag	21.30	0.87	73.47	13.65	HTL	[14]
ITO/T-$NbCT_x$/$FA_{0.95}Cs_{0.05}PbI_3$/PCBM/Ag	23.06	1.126	79.75	20.74	HTL	[15]
FTO/TiO_2/$CsPbI_3$/$Ti_3C_2F_x$ QDs/Spiro-OMeTAD/Au	20.59	1.22	81.55	20.44	ETL	[16]
ITO/UV–$Ti_3C_2T_x$/$CH_3NH_3PbI_3$/Spiro-OMeTAD/Ag	22.63	1.08	70	17.17	ETL	[17]
ITO/HO–$Ti_3C_2T_x$@$Ti_3C_2T_x$/$CH_3NH_3PbI_3$/spiro-OMeTAD/Ag	23.11	1.07	74	18.29	ETL	[18]
ITO/$Ti_3C_2T_x$/$(FA_{0.95-x}MA_xCs_{0.05})Pb(I_{1-x}Br_x)_3$/Spiro-OMeTAD/Au	24.17	1.03	80	19.92	ETL	[19]
ITO/PTAA/$CsPbI_3$/OMXene/CTBA/Ag	19.86	1.21	81.96	19.79	ETL	[20]
ITO/U–$Ti_3C_2T_x$/PBDB-T:ITIC/Ca/Al	15.98	0.89	64	8.91	HTL	[21]
ITO/UH–$Ti_3C_2T_x$/PBDBT:ITIC/MoO_3/Al	17.36	0.87	60	9.00	ETL	[21]
ITO/$Nb_2CO_{1.2}OH_{0.8}$/PM6:Y6/MoO_3/Ag	25.36	0.84	71.45	15.22	ETL	[22]
ITO/$Nb_2CO_{1.36}OH_{0.2}$/PM6:Y6/PFN-Br/Al	25.08	0.84	71.33	15.03	HTL	[22]
ITO/R-$Ti_3C_2T_x$-15/PM6:Y6/MoO_3/Ag	25.46	0.85	72.07	15.60	ETL	[23]
ITO/D-$Ti_3C_2T_x$-50/PM6:Y6/PFN-Br/Al	25.46	0.85	72.63	15.88	HTL	[23]
FTO/SnO_2/Nb_2CT_{OH}/CsFAMA/Nb_2CT_O/Spiro-OMeTAD/Ag	23.73	1.253	81.07	24.11	ETL + HTL	[24]

(continued)

Table 12.1 (continued)

Device structure	J_{sc} /(mA·cm^{-2})	V_{oc} /V	FF /%	PCE /%	Features	Ref.
ITO/PEDOT:PSS–$Ti_3C_2T_x$/PBDB-T:ITIC/PFN-Br/Al	17.08	0.91	70.93	11.02	HTL	[25]
ITO/PEDOT:PSS–$Ti_3C_2T_x$/PM6:Y6/PFN-Br/Al	25.63	0.83	68.40	14.55	HTL	[25]
ITO/$Mo_{1.33}C$:PEDOT:PSS/PTB7-Th:$PC_{71}BM$/LiF/Al	16.86	0.8	68.7	9.24	HTL	[26]
ITO/ZnO–$Ti_3C_2T_x$/PBDB-T:ITIC/MoO_3/Ag	18.63	0.93	70.39	12.20	ETL	[28]
ITO/ZnO–$Ti_3C_2T_x$/PM6:Y6/MoO_3/Ag	26.38	0.83	75.40	16.51	ETL	[28]
ITO/ZnO:PDA-$Ti_3C_2T_x$/PBDB-T:ITIC/MoO_3/Ag	18.14	0.937	71.01	12.19	ETL	[29]
ITO/ZnO:PDA-$Ti_3C_2T_x$/PM6:Y6/MoO_3/Ag	26.41	0.837	75.52	16.80	ETL	[29]
ITO/ZnO:PDA-$Ti_3C_2T_x$/PTB7:$PC_{71}BM$/MoO_3/Ag	17.60	0.776	68.90	9.51	ETL	[29]
ITO/SnO_2–$Ti_3C_2T_x$/$CH_3NH_3PbI_3$/Spiro-OMeTAD/Ag	23.14	1.06	75	18.34	ETL	[30]
ITO/$Ti_3C_2T_x$/SnO_2/$Cs_{0.05}FA_{0.76}MA_{0.19}PbI_{2.715}Br0_{.285}$/Spiro-OMeTAD/Ag	24.34	1.11	74	20.65	ETL	[31]
ITO/MQDs-SnO_2/$FA_{0.9}MA_{0.05}Cs_{0.05}PbI_{0.98}Br_{0.02}$/Spiro-OMeTAD/$MoO_3$/Ag	24.96	1.172	79.8	23.34	ETL	[32]
FTO/$Ti_3C_2T_x$–TiO_2/$Cs_2AgBiBr_6$/PCBM/Ag	4.14	0.96	70	2.81	ETL	[33]
FTO/TiO_2–SnO_2–$Ti_3C_2T_x$/$(FAPbI_3)_{0.97}(MAPbBr_3)_{0.03}$/Spiro-OMeTAD/Au	23.13	1.07	74.62	18.44	ETL	[34]
ITO/SnO_2–Nb_2C/$Cs_{0.05}FA_{0.88}MA_{0.07}PbI_3$/Spiro-OMeTAD/$MoO_3$/Ag	25.29	1.138	79.5	22.86	ETL	[35]
ITO/NiO_x/$CsFAPbI_3$/C60/BCP:$Ti_3C_2T_x$/Cu	22.31	1.02	76	17.46	ETL	[36]
FTO/O-$Ti_3C_2T_x$/SnO_2/$CH_3NH_3PbI_3$/Spiro-OMeTAD/Ag	23.88	1.073	78.4	20.09	ETL	[37]
ITO/SnO_2-MXene-H/$FAPbI_3$/Spiro-OMeTAD/MoO_3/Ag	23.5	1.114	77.40	20.26	ETL	[38]
ITO/SnO_2-MXene-H/CsFAMAPbIBr/Spiro-OMeTAD/MoO_3/Ag	25.4	1.121	84.42	24.12	ETL	[38]
ITO/SnO_2/$CH_3NH_3PbI_3$–$Ti_3C_2T_x$/Spiro-OMeTAD/Au	22.26	1.03	76	17.41	AL	[39]
FTO/SnO_2/$CH_3NH_3PbI_3$–$Ti_3C_2T_x$/Spiro-OMeTAD/Au	23.4	1.12	73.66	19.27	AL	[40]
ITO/SnO_2/$(BA)_2(MA)_4Pb_5I_{16}$–$Ti_3C_2T_x$/Spiro-OMeTAD/Ag	20.87	1.11	67.84	15.71	AL	[41]

(continued)

Table 12.1 (continued)

Device structure	J_{sc} /(mA·cm^{-2})	V_{oc} /V	FF /%	PCE /%	Features	Ref.
ITO/PTAA/$CH_3NH_3PbI_3$–$Ti_3C_2T_x$/PCBM:BCP/Ag	23.34	1.00	73.60	17.20	AL	[42]
ITO/PTAA/$CH_3NH_3PbI_3$–V_2CT_x/PCBM:BCP/Ag	23.12	1.03	73.90	17.61	AL	[42]
ITO/SnO_2/$CsPbI_2Br$-$Ti_3C_2T_x$/P3HT/Ag	15.92	1.22	78	15.10	AL	[43]
ITO/PEDOT:PSS/PBDB-T:ITIC–$Ti_3C_2T_x$/PDINO/Al	16.28	0.915	72.0	10.72	AL	[44]
ITO/PEDOT:PSS/PBDB-T:IT-M–$Ti_3C_2T_x$/PDINO/Al	17.65	0.936	60.3	9.96	AL	[44]
ITO/PEDOT:PSS/PM6:Y6–$Ti_3C_2T_x$/PDINO/Al	27.02	0.839	71.7	16.25	AL	[44]
ITO/SnO_2/Cs-$Ti_3C_2T_x$/Spiro-OMeTAD/Au	26.00	1.1	76	21.57	AL	[45]
ITO/SnO_2/$Rb_{0.05}Cs_{0.05}(FA_{0.83}MA_{0.17})_{0.90}Pb(I_{0.83}Br_{0.17})_3$-$Ti_3C_2Cl_x$/Spiro-OMeTAD/Ag	22.27	1.19	80.42	21.31	AL	[46]
FTO/c-TiO_2/m-TiO_2–TQD/perovskites–TQD/Spiro-OMeTAD:$Cu_{1.8}S$/Au	24.12	1.14	78.70	21.64	ETL + AL	[48]
FTO/TiO_2/$CH_3NH_3PbI_3$–TQD/Spiro-OMeTAD/Au	23.9	1.00	67	16.0	AL + ETL	[49]
ITO/NiO/$MAPbI_3$–$Ti_3C_2T_x$/PCBM–$Ti_3C_2T_x$/BCP/Ag	22.88	1.09	77	19.20	AL + ETL	[50]
FTO/SnO_2–$TiOxCl_{4-2x}$/$CsPbBr_3$:$Ti_3C_2Cl_x$/$Ti_3C_2Cl_x$/Carbon	7.87	1.702	82.7	11.08	AL + HTL	[51]
ITO/NH-$NbCT_x$/$FA_{0.85}Cs_{0.15}PbI_3$:NH-$NbCT_x$/Spiro-OMeTAD/Ag	25.07	1.124	77.36	21.79	ETL + AL	[52]

$Ti_3C_2T_x$ MXene with excellent conductivity and appropriate work function (WF) can be directly used as an efficient electrode and also mixed with other conductive materials, including AgNWs and PEDOT:PSS, as excellent composite electrodes. Ohmic contacts, excellent optical transmittance, and reasonable WFs are required for excellent transparent electrodes, and extra flexibility is essential for flexible transparent electrodes. However, it is still challenging for pristine MXene electrodes to have excellent electrical conductivity, light transmission, and flexibility at the same time. Therefore, MXene-based composite electrodes may be more promising for excellent transparent/flexible electrodes via appropriately selecting other conductive materials.

Compared with the pristine MXenes interface layer, functionalized MXenes and MXene-based composite interface layers usually show greater potential for enhancing device performance. The surface functional groups of functionalized MXenes are reasonably designed and used to tune WF/energy level and passivate defects/trap states, and can thus further enhance charge transfer and reduce charge recombination in devices. In addition, (functionalized) MXenes mixed with other interfacial materials are used as composite interface layers and can further offer synergistic effects for enhancing device performance. Therefore, more functionalized MXenes with targeted surface functional groups need to be designed and can be appropriately mixed with other interfacial materials for excellent interface layers.

MXenes additives can increase the charge transfer pathway and conductivity in the active layer. Moreover, functionalized MXenes with tunable work function/energy level can be used to improve energy-level alignment between perovskite layer and interface layer, and the surface functional groups of functionalized MXenes can passivate the uncoordinated Pb^{2+} in perovskite layer and thus improve perovskite crystallinity and reduce trap states in perovskite layer, which together contribute to the higher device performance. Therefore, more functionalized MXenes with targeted surface functional groups need to be designed for excellent additives in the active layer.

Acknowledgements The authors acknowledge the financial support of the National Natural Science Foundation of China (nos. 52063010), and the Natural Science Foundation of Guangxi Province (nos. 2021GXNSFAA220050).

Conflict of Interest Ping Cai, Ling Ding, Kefan Chen, Can Song, and Baiquan Liu declare that they have no conflict of interest.

References

1. A.V. Mohammadi, J. Rosen, Y. Gogotsi, The world of two-dimensional carbides and nitrides (MXenes). Science **372**(6547), 1165 (2021). https://doi.org/10.1126/science.abf1581
2. K.F. Chen, P. Cai, H.L. Peng, X.G. Xue, Z.M. Wang, L.X. Sun, $Ti_3C_2T_x$ MXene for organic/perovskite optoelectronic devices. J. Cent. South Univ. **28**(12), 3935–3958 (2021). https://doi.org/10.1007/s11771-021-4846-z

3. S. Ahn, T.H. Han, K. Maleski, J.N. Song, Y.H. Kim, M.H. Park, H.Y. Zhou, S. Yoo, Y. Gogotsi, T.W. Lee, A 2D titanium carbide MXene flexible electrode for high-efficiency light-emitting diodes. Adv. Mater. **32**(23) (2020). https://doi.org/10.1002/adma.202000919
4. S. Lee, E.H. Kim, S. Yu, H. Kim, C. Park, T.H. Park, H. Han, S.W. Lee, S. Baek, W. Jin, C.M. Koo, C. Park, Alternating-current MXene polymer light-emitting diodes. Adv. Funct. Mater. **30**(32) (2020). https://doi.org/10.1002/adfm.202001224
5. J.M. Cao, F.N. Meng, L.G. Gao, S.Z. Yang, Y.L. Yan, N. Wang, A.M. Liu, Y.Q. Li, T.L. Ma, Alternative electrodes for htms and noble-metal-free perovskite solar cells: 2D MXenes electrodes. RSC Adv. **9**(59), 34152–34157 (2019). https://doi.org/10.1039/c9ra06091j
6. L.Q. Qin, J.X. Jiang, Q.Z. Tao, C.F. Wang, I. Persson, M. Fahlman, P.O.A. Persson, L.T. Hou, J. Rosen, F. Zhang, A flexible semitransparent photovoltaic supercapacitor based on water-processed MXene electrodes. J. Mater. Chem. A. **8**(11), 5467–5475 (2020). https://doi.org/10.1039/d0ta00687d
7. S.C. Bai, X.Z. Guo, T.R. Chen, Y. Zhang, X.Y. Zhang, H. Yang, X.Y. Zhao, Solution processed fabrication of silver nanowire-MXene@pedot: Pss flexible transparent electrodes for flexible organic light-emitting diodes. Compos. Part A-Appl. S **139** (2020). https://doi.org/10.1016/j.compositesa.2020.106088
8. F. Cao, M.Q. You, L.M. Kong, Y.J. Dou, Q.Q. Wu, L. Wang, B. Wei, X.Y. Zhang, W.Y. Wong, X.Y. Yang, Mixed-dimensional MXene-based composite electrodes enable mechanically stable and efficient flexible perovskite light-emitting diodes. Nano Lett. **22**(10), 4246–4252 (2022). https://doi.org/10.1021/acs.nanolett.2c01517
9. H.H. Tang, H.R. Feng, H.K. Wang, X.J. Wang, J.J. Liang, Y. Chen, Highly conducting MXene-silver nanowire transparent electrodes for flexible organic solar cells. ACS Appl. Mater. Interfaces **11**(28), 25330–25337 (2019). https://doi.org/10.1021/acsami.9b04113
10. L.F. Mi, Y. Zhang, T.T. Chen, E.Z. Xu, Y. Jiang, Carbon electrode engineering for high efficiency all-inorganic perovskite solar cells. RSC Adv. **10**(21), 12298–12303 (2020). https://doi.org/10.1039/d0ra00288g
11. Q.Q. Lu, J.K. Wang, Y.Q. Miao, Y.Y. Guo, G.L. Wang, J.H. Dong, M. Zhao, H. Wang. UV-ozone treated $Ti_3C_2T_X$-MXene nanosheets as hole injection layer for organic light-emitting diodes. Chem. Eng. J. **450** (2022). https://doi.org/10.1016/j.cej.2022.138439
12. C.L. Hou, H.Z. Yu, C.W. Huang, Solution-processable $Ti_3C_2T_X$ nanosheets as an efficient hole transport layer for high-performance and stable polymer solar cells. J. Mater. Chem. C **7**(37), 11549–11558 (2019). https://doi.org/10.1039/c9tc03415c
13. T.T. Chen, G.Q. Tong, E.Z. Xu, H. Li, P.C. Li, Z.F. Zhu, J.X. Tang, Y.B. Qi, Y. Jiang, Accelerating hole extraction by inserting 2D Ti_3C_2-MXene interlayer to all inorganic perovskite solar cells with long-term stability. J. Mater. Chem. A **7**(36), 20597–20603 (2019). https://doi.org/10.1039/c9ta06035a
14. J.B. Du, L. Yang, X. Jin, C.L. Liu, H.H. Wang, X.F. Wang, Spray deposition of vinyl tris(2-methoxyethoxy) silane-doped $Ti_3C_2T_X$ MXene hole transporting layer for planar perovskite solar cells. J. Alloys Compd. **900** (2022). https://doi.org/10.1016/j.jallcom.2021.163372
15. J. Zhang, C. Huang, H. Yu, Modulate the work function of Nb_2CT_X MXene as the hole transport layer for perovskite solar cells. Appl. Phys. Lett. **119**(3) (2021). https://doi.org/10.1063/5.0057978
16. D. Xu, T. Li, Y. Han, X. He, S. Yang, Y. Che, J. Xu, H. Zou, X. Guo, J. Wang, X. Lei, Z. Liu, Fluorine functionalized MXene QDs for near-record-efficiency $CsPbI_3$ solar cell with high open-circuit voltage. Adv. Funct. Mater. **32**(33) (2022). https://doi.org/10.1002/adfm.202203704
17. L. Yang, C. Dall'Agnese, Y. Dall'Agnese, G. Chen, Y. Gao, Y. Sanehira, A.K. Jena, X.F. Wang, Y. Gogotsi, T. Miyasaka, Surface-modified metallic $Ti_3C_2T_X$ MXene as electron transport layer for planar heterojunction perovskite solar cells. Adv. Funct. Mater. **29**(46), (2019). https://doi.org/10.1002/adfm.201905694
18. L. Yang, D.X. Kan, C. Dall'Agnese, Y. Dall'Agnese, B.N. Wang, A.K. Jena, Y.J. Wei, G. Chen, X.F. Wang, Y. Gogotsi, T. Miyasaka, Performance improvement of MXene-based perovskite solar cells upon property transition from metallic to semiconductive by oxidation of $Ti_3C_2T_X$ in air. J. Mater. Chem. A **9**(8), 5016–5025 (2021). https://doi.org/10.1039/d0ta11397b

19. J. Wang, Z. Cai, D. Lin, K. Chen, L. Zhao, F. Xie, R. Su, W. Xie, P. Liu, R. Zhu, Plasma oxidized $Ti_3C_2T_X$ MXene as electron transport layer for efficient perovskite solar cells. ACS Appl. Mater. Interfaces **13**(27), 32495–32502 (2021). https://doi.org/10.1021/acsami.1c07146
20. J.H. Heo, F. Zhang, J.K. Park, H. Joon Lee, D.S. Lee, S.J. Heo, J.M. Luther, J.J. Berry, K. Zhu, S.H. Im, Surface engineering with oxidized $Ti_3C_2T_X$ MXene enables efficient and stable p-i-n-structured $CsPbI_3$ perovskite solar cells. Joule **6**(7), 1672–1688 (2022). https://doi.org/10.1016/j.joule.2022.05.013
21. Z.M. Yu, W. Feng, W.H. Lu, B.C. Li, H.Y. Yao, K.Y. Zeng, J.Y. Ouyang, MXenes with tunable work functions and their application as electron- and hole-transport materials in non-fullerene organic solar cells. J. Mater. Chem. A **7**(18), 11160–11169 (2019). https://doi.org/10.1039/c9ta01195a
22. C. Huang, S. Shi, H. Yu, Work function adjustment of Nb_2CT_X nanoflakes as hole and electron transport layers in organic solar cells by controlling surface functional groups. ACS Energy Lett. **6**(10), 3464–3472 (2021). https://doi.org/10.1021/acsenergylett.1c01656
23. C. Hou, C. Huang, H. Yu, S. Shi, Surface-engineered $Ti_3C_2T_X$ with tunable work functions for highly efficient polymer solar cells. Small **18**(21), e2201046 (2022). https://doi.org/10.1002/smll.202201046
24. Y. Zhang, L. Xu, J. Sun, Y. Wu, Z. Kan, H. Zhang, L. Yang, B. Liu, B. Dong, X. Bai, H. Song, 24.11% high performance perovskite solar cells by dual interfacial carrier mobility enhancement and charge-carrier transport balance. Adv. Energy Mater. **12**(37) (2022). https://doi.org/10.1002/aenm.202201269
25. C.L. Hou, H.Z. Yu, Modifying the nanostructures of PEDOT:PSS/$Ti_3C_2T_X$ composite hole transport layers for highly efficient polymer solar cells. J. Mater. Chem. C **8**(12), 4169–4180 (2020). https://doi.org/10.1039/d0tc00075b
26. Y. Liu, Q. Tao, Y. Jin, X. Liu, H. Sun, A. E. Ghazaly, S. Fabiano, Z. Li, J. Luo, J. Rosen, F. Zhang, $Mo_{1.33}c$ MXene-assisted PEDOT:PSS hole transport layer for high-performance bulk-heterojunction polymer solar cells. ACS Appl. Electron. Mater. **2**(1), 163–169 (2019). https://doi.org/10.1021/acsaelm.9b00668
27. P. Lu, J. Wu, X. Shen, X. Gao, Z. Shi, M. Lu, W.W. Yu, Y. Zhang, ZnO–Ti_3C_2 MXene electron transport layer for high external quantum efficiency perovskite nanocrystal light-emitting diodes. Adv. Sci. **7**(19) (2020). https://doi.org/10.1002/advs.202001562
28. C.L. Hou, H.Z. Yu, ZnO/$Ti_3C_2T_X$ monolayer electron transport layers with enhanced conductivity for highly efficient inverted polymer solar cells. Chem. Eng. J. **407** (2021). https://doi.org/10.1016/j.cej.2020.127192
29. C. Hou, H. Yu, C. Huang, ZnO:Bio-inspired polydopamine functionalized $Ti_3C_2T_X$ composite electron transport layers for highly efficient polymer solar cells. J. Alloys Compd. **900** (2022). https://doi.org/10.1016/j.jallcom.2021.163381
30. L. Yanga, Y. Dall'Agnesea, K. Hantanasirisakulb, C.E. Shuckb, K. Maleskib, M. Alhabebb, G. Chena, Y. Gaoa, Y. Sanehirac, A.K. Jenac, L. Shend, C. Dall'Agnesea, X.-F. Wanga, Y. Gogotsi, T. Miyasakac, SnO_2–Ti_3C_2 mxene electron transport layers for perovskite solar cells. J. Mater. Chem. A **7**(10), 5635–5642 (2019). https://doi.org/10.1039/C8TA12140K
31. Y. Wang, P. Xiang, A. Ren, H. Lai, Z. Zhang, Z. Xuan, Z. Wan, J. Zhang, X. Hao, L. Wu, M. Sugiyama, U. Schwingenschlögl, C. Liu, Z. Tang, Z. W. Jiang Wu, D. Zhao, Mxene-modulated electrode-SnO_2 interface boosting charge transport in perovskite solar cells. ACS Appl. Mater Interfaces **12**(48), 53973–53983 (2020). https://doi.org/10.1021/acsami.0c17338
32. Y. Yang, H. Lu, S. Feng, L. Yang, H. Dong, J. Wang, C. Tian, L. Li, H. Lu, J. Jeong, S.M. Zakeeruddin, Y. Liu, M. Grätzel, A. Hagfeldt, Modulation of perovskite crystallization processes towards highly efficient and stable perovskite solar cells with MXene quantum dot-modified SnO_2. Energy Environ. Sci. **14**(6), 3447–3454 (2021). https://doi.org/10.1039/d1ee00056j
33. Z.X. Li, P. Wang, C. Ma, F. Igbari, Y.K. Kang, K.L. Wang, W.Y. Song, C. Dong, Y.J. Li, J.S. Yao, D. Meng, Z.K. Wang, Y. Yang, Single-layered MXene nanosheets doping TiO_2 for efficient and stable double perovskite solar cells. J. Am. Chem. Soc. **143**(6), 2593–2600 (2021). https://doi.org/10.1021/jacs.0c12739

34. L.S. Huang, X.W. Zhou, R. Xue, P.F. Xu, S.L. Wang, C. Xu, W. Zeng, Y. Xiong, H. Q. Sang, D. Liang, Low-temperature growing anatase TiO_2/SnO_2 multi-dimensional heterojunctions at MXene conductive network for high-efficient perovskite solar cells. Nano-Micro Lett. **12**(1) (2020). https://doi.org/10.1007/s40820-020-0379-5
35. Y. Niu, C. Tian, J. Gao, F. Fan, Y. Zhang, Y. Mi, X. Ouyang, L. Li, J. Li, S. Chen, Y. Liu, H.-L. Lu, X. Zhao, L. Yang, H. Ju, Y. Yang, C.-F. Ding, M. Xu, Q. Xu, Nb_2C MXenes modified SnO_2 as high quality electron transfer layer for efficient and stability perovskite solar cells. Nano Energy **89** (2021). https://doi.org/10.1016/j.nanoen.2021.106455
36. A. Yakusheva, D. Saranin, D. Muratov, P. Gostishchev, H. Pazniak, A. Di Vito, T.S. Le, L. Luchnikov, A. Vasiliev, D. Podgorny, D. Kuznetsov, S. Didenko, A. Di Carlo, Photo stabilization of p-i-n perovskite solar cells with bathocuproine: MXene. Small **18**(37), e2201730 (2022). https://doi.org/10.1002/smll.202201730
37. L. Yang, B. Wang, C. Dall'Agnese, Y. Dall'Agnese, G. Chen, A.K. Jena, X.-F. Wang, T. Miyasaka, Hybridization of SnO_2 and an in-situ-oxidized $Ti_3C_2T_X$ MXene electron transport bilayer for high-performance planar perovskite solar cells. ACS Sustain. Chem. Eng. **9**(40), 13672–13680 (2021). https://doi.org/10.1021/acssuschemeng.1c05245
38. L. Yin, C. Liu, C. Ding, C. Zhao, I.Z. Mitrovic, E.G. Lim, H. Wang, Y. Sun, Y. Han, Z. Li, L. Yang, C.-Q. Ma, C. Zhao, Functionalized-MXene-nanosheet-doped tin oxide enhances the electrical properties in perovskite solar cells. Cell Rep. Phys. Sci. **3**(6) (2022). https://doi.org/10.1016/j.xcrp.2022.100905
39. Z.L. Guo, L.G. Gao, Z.H. Xu, S. Teo, C. Zhang, Y. Kamata, S. Hayase, T.L. Ma, High electrical conductivity 2D MXene serves as additive of perovskite for efficient solar cells. Small **14**(47) (2018). https://doi.org/10.1002/smll.201802738
40. Y. Zhao, X. Zhang, X.F. Han, C.Y. Hou, H.Z. Wang, J.B. Qi, Y.G. Li, Q.H. Zhang, Tuning the reactivity of PbI_2 film via monolayer $Ti_3C_2T_X$ MXene for two-step-processed $CH_3NH_3PbI_3$ solar cells. Chem. Eng. J. **417** (2021). https://doi.org/10.1016/j.cej.2020.127912
41. X. Jin, L. Yang, X.F. Wang, Efficient two-dimensional perovskite solar cells realized by incorporation of $Ti_3C_2T_X$ MXene as nano-dopants. Nano-Micro Lett. **13**(1) (2021). https://doi.org/10.1007/s40820-021-00602-w
42. Y. Li, D. Wang, L. Yang, S. Yin, Preparation and performance of perovskite solar cells with two dimensional MXene as active layer additive. J. Alloys Compd. **904** (2022). https://doi.org/10.1016/j.jallcom.2022.163742
43. Y. Wang, J. Li, X. Yao, C. Xie, Q. Chen, W. Liu, Z. Gao, Y. Fu, Q. Liu, D. He, Y. Li, Improved comprehensive photovoltaic performance and mechanisms by additive engineering of $Ti_3C_2T_X$ MXene into $CsPbI_2Br$. ACS Appl. Mater. Interfaces **14**(36), 40930–40938 (2022). https://doi.org/10.1021/acsami.2c10417
44. Y. Zhao, X.J. Liu, X. Jing, X.K. Wang, C.Y. Gao, S.X. Dai, L. M. Yu, M.L. Sun, Addition of 2D $Ti_3C_2T_X$ to enhance photocurrent in diodes for high-efficiency organic solar cells. Sol. Rrl. **5**(4) (2021). https://doi.org/10.1002/solr.202100127
45. A.S.R. Bati, A.A. Sutanto, M. Hao, M. Batmunkh, Y. Yamauchi, L. Wang, Y. Wang, M.K. Nazeeruddin, J.G. Shapter, Cesium-doped $Ti_3C_2T_X$ MXene for efficient and thermally stable perovskite solar cells. Cell Rep. Phys. Sci. **2**(10) (2021). https://doi.org/10.1016/j.xcrp.2021.100598
46. X. Liu, Z. Zhang, J. Jiang, C. Tian, X. Wang, L. Wang, Z. Zhang, X. Wu, Y. Zheng, J. Liang, C.-C. Chen, Chlorine-terminated MXene quantum dots for improving crystallinity and moisture stability in high-performance perovskite solar cells. Chem. Eng. J. **432** (2022). https://doi.org/10.1016/j.cej.2021.134382
47. A. Agresti, A. Pazniak, S. Pescetelli, A. Di Vito, D. Rossi, A. Pecchia, M.A.D. Maur, A. Liedl, R. Larciprete, D.V. Kuznetsov, D. Saranin, A. Di Carlo, Titanium-carbide MXenes for work function and interface engineering in perovskite solar cells. Nat. Mater. **18**(11), 1264–1264 (2019). https://doi.org/10.1038/s41563-019-0527-9
48. X. Chen, W. Xu, N. Ding, Y.N. Ji, G.C. Pan, J.Y. Zhu, D.L. Zhou, Y.J. Wu, C. Chen, H.W. Song, Dual interfacial modification engineering with 2D MXene quantum dots and copper sulphide nanocrystals enabled high-performance perovskite solar cells. Adv. Funct. Mater. **30**(30) (2020). https://doi.org/10.1002/adfm.202003295

49. J. Ge, W. Li, X. He, H. Chen, W. Fang, X. Du, Y. Li, L. Zhao, Charge behavior modulation by titanium-carbide quantum dots and nanosheets for efficient perovskite solar cells. Mater. Today Energy **18** (2020). https://doi.org/10.1016/j.mtener.2020.100562
50. D. Saranin, S. Pescetelli, A. Pazniak, D. Rossi, A. Liedl, A. Yakusheva, L. Luchnikov, D. Podgorny, P. Gostischev, S. Didenko, A. Tameev, D. Lizzit, M. Angelucci, R. Cimino, R. Larciprete, A. Agresti, A. Di Carlo, Transition metal carbides (MXenes) for efficient NiO-based inverted perovskite solar cells. Nano Energy **82** (2021). https://doi.org/10.1016/j.nanoen.2021.105771
51. Q. Zhou, J. Duan, J. Du, Q. Guo, Q. Zhang, X. Yang, Y. Duan, Q. Tang, Tailored lattice "tape" to confine tensile interface for 11.08%-efficiency all-inorganic $CsPbBr_3$ perovskite solar cell with an ultrahigh voltage of 1.702 v. Adv. Sci. (Weinh) **8**(19), e2101418 (2021). https://doi.org/10.1002/advs.202101418
52. J.K. Zhang, C.W. Huang, Y.P. Sun, H.Z. Yu, Amino-functionalized niobium-carbide MXene serving as electron transport layer and perovskite additive for the preparation of high-performance and stable methylammonium-free perovskite solar cells. Adv. Funct. Mater. **32**(24) (2022). https://doi.org/10.1002/adfm.202113367

Chapter 13
Future Prospective and Research Avenues

Ajit K. Katiyar, Ravi P. Srivastava, and Mayank Gupta

13.1 Introduction

Miracle two-dimensional (2D) material Graphene, discovered in 2004 by Prof. Geim and Prof. Novoselov, was recognized for the Nobel prize in 2010. The unique properties of the material provided an excellent platform for its use in various fields and opened a new research direction for the discovery of other 2D materials. The 2D materials show interesting properties such as a high percentage of exposed atoms, large specific surfaces, superior quantum confinement effect, and resistance to mechanical strain, etc. [1, 2]. Due to their superior properties, these materials perform much better as compared to 3D counterparts and have a wide range of applications. To promote the design and development of 2D materials, researchers have combined the experimental and theoretical aspects, as the study at only one front provides limited insight into the properties of materials. These efforts resulted in significant development in the 2D materials and their properties. Various 2D materials such as transition metal dichalcogenides (TMDs), metal oxides and hydroxides, graphitic carbon nitride, hexagonal boron nitride, covalent organic frameworks (COFs), metal-organic frameworks (MOFs), silicene, germanene, monoelemental 2D Xenes (graphdiyne, arsenene, borophene, etc.) have been developed and investigated [3, 4]. In addition to the aforementioned materials, the family of 2D materials is blessed with another

A. K. Katiyar
Department of Electrical and Electronic Engineering, Yonsei University, Seoul 03722, South Korea

R. P. Srivastava (✉)
Department of Materials Science and Engineering, Yonsei University, Seoul 03722, South Korea
e-mail: rpsrivastava@yonsei.ac.kr

Department of Applied Sciences, Institute of Engineering and Technology, Sitapur 261001, India

M. Gupta
Department of Chemical Engineering, Indian Institute of Technology Kanpur, Kanpur 208016, India

© The Author(s), under exclusive license to Springer Nature Singapore Pte Ltd. 2024
O. P. Pandey and P. Sharma (eds.), *MXenes: Emerging 2D Materials*,
https://doi.org/10.1007/978-981-97-4064-2_13

wonderful class of materials, known as MXenes. MXenes with a unique combination of properties, such as high electrical conductivity, high surface area, biocompatibility, excellent hydrophilicity, etc., opened a new and fascinating research field of 2D materials [5]. In 2012, the research team of Prof. Gogotsi and Prof. Barsoum at Drexel University discovered the first MXene ($Ti_3C_2T_x$) via hydrofluoric (HF) etching of titanium aluminum carbide (Ti_3AlC_2) [6]. The materials were named MXene, as these materials have been developed with MAX phase and showed properties similar to that of Graphene [7]. These 2D materials are layered transition metal carbides, carbonitrides, or nitrides and can usually be produced by removing the *A* layers from their parent MAX phase. The MAX phases are a large family of ternary layered carbide, nitride, or ternary layer carbonitride with general formula $M_{n+1}AX_n$ (n = 1 to 3), where M stands for an early transition metal (Ti, V, Cr, etc.), *A* belongs to an element of group 13–15 (Al, Ga, In, etc.), and *X* stands for carbon and/or nitrogen [8]. In MAX phases, *M* layers are intercalated with the *A* layers forming a hexagonal layered structure, in which octahedral sites are occupied by the *X* atoms [9]. The bonding between M and X (M-X) has a mixed (metallic/covalent/ionic) nature, while M-A bonds are metallic in nature and slightly weaker as compared to M-X, which provides an opportunity to produce MXenes by selective etching of *A* layers.

MXenes produced after selective etching of *A* layers have the general formula $M_{n+1}X_nT_x$, where T represents the surface termination group (–F, –Cl, –O, –OH, etc.) and x stands for a number of terminations [10]. Surface termination groups have a prominent effect on the properties and can significantly tune the optical, electronic, and transport properties of MXenes. Combining M, A, and X elements, more than 100 MAX phases have been developed, but only a few MXenes have been experimentally produced and investigated. A timeline of the journey of MXenes from 2011 to 2022 is shown in Fig. 13.1. From the figure, it can be seen that family of MXenes is growing with time, and fields of application are increasing continuously. MXenes have found applications in various fields such as energy storage, biomedical, sensing devices, optoelectronics, etc. Apart from interesting properties and versatile applications, there are various challenges associated with these materials that led the researches to discover easy, eco-friendly methods along with new MXenes toward their commercialization. In this chapter, we will start with a brief introduction to various synthesis methods, followed by different research avenues, associated challenges, and future prospects suggesting possible strategies to overcome those issues.

13.2 Synthesis of MXenes

The synthesis approach to produce MXenes has effective control on the morphology, thickness, lateral size, and surface termination leading to different properties of MXenes obtained by different methods. In the past, a variety of methods have been employed to produce MXenes, which can be broadly classified into top-down and bottom-up approaches and is shown in Fig. 13.2. A timeline for the typical synthesis routes for MXenes in the past decade is represented in Fig. 13.2g.

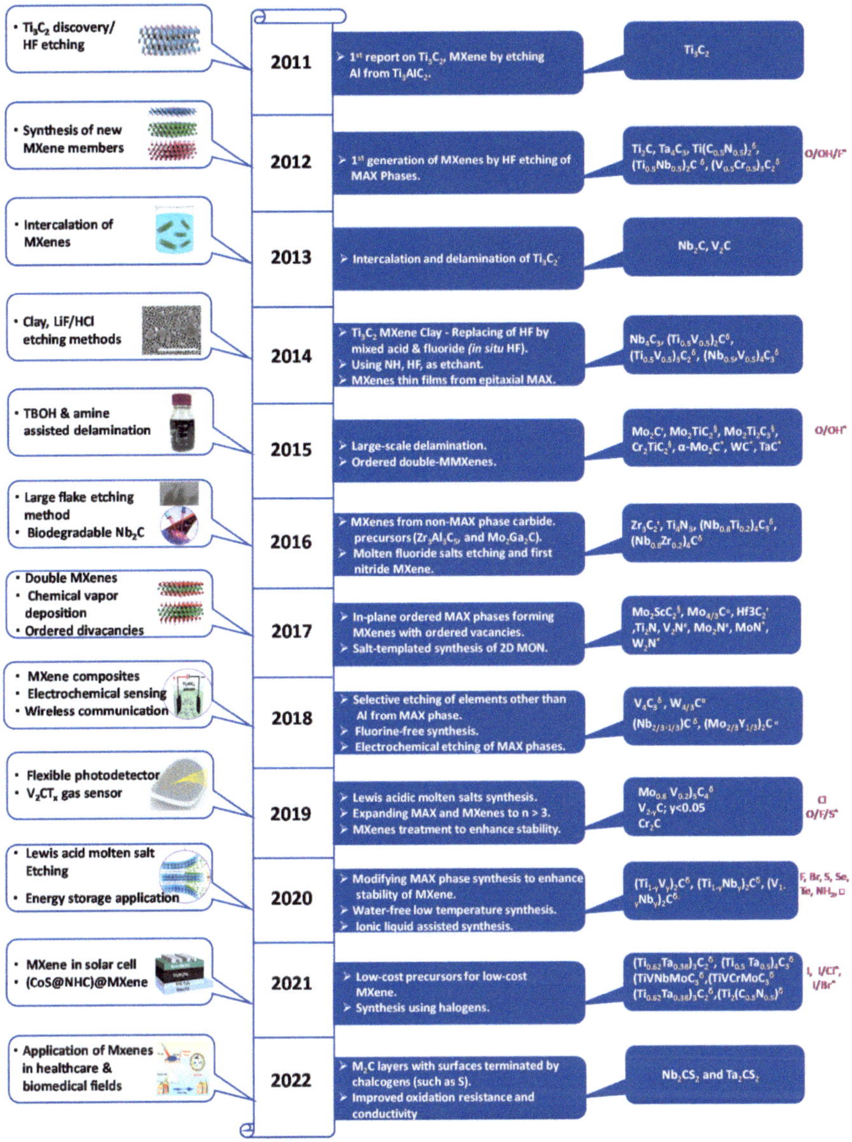

Fig. 13.1 Timeframe, development, and discoveries of MXenes in the past

13.2.1 Top-Down Methods

In top-down synthesis, ***A*** layers of parent MAX/non-MAX phase are selectively etched/removed to produce 2D layered MXenes. This is due to metallic bonding and the relatively weak nature of M-A bonds, which gives the feasibility to produce

MXenes by breaking these bonds. These bonds can be broken at high temperatures also, but high temperatures can distort the layered structure and can produce rock salt structure [11]. In order to tackle this problem, selective chemical etching of ***A*** layers is commonly used. Among the various etching methods to remove the ***A*** layer, HF etching, alkali etching, acid-fluoride salt etching, molten salt etching, and electrochemical etching are a few popular methods. Besides these, some other etching approaches such as electromagnetic, and algae extraction, etc., have been explored in the past [12, 13]. The etching process converts the MAX phase into multilayered MXenes, which on further exfoliation can produce few-layered MXenes. It was observed that the etching of Al-C layers is energetically more favorable as compared to Al layers, suggesting the conversion of non-MAX phase materials into corresponding MXenes [14]. Mo_2CT_x was the first MXene obtained from non-MAX material Mo_xGa_2C using HF etching of Ga layers [15].

i. **HF etching**

Among the various etching, HF etching with easy operation and low-temperature requirement is a common approach to produce MXenes. In this method, a MAX phase material is immersed into an aqueous HF solution to selectively remove ***A*** layers (Fig. 13.2a). Etching results in the formation of surface termination groups like –F, –O, –OH, etc. Ti_3C_2 was the first HF etched MXenes, produced by the etching of Ti_3AlC_2 (MAX phase) with 50% HF at room temperature for 2 h [6]. In the synthesized MXenes, 2D layers have accordion-like structures bonded with hydrogen and van der Waals bonds. In the etching process, HF concentration, etching time, and/or temperature play a very important role and significantly affect the properties of synthesized MXenes. A longer etching time and higher temperature can result in faster synthesis but compromised with structural stabilities and sheet resistance. The lateral size of MXenes and defects can also be controlled with the concentration of etchant [16]. Although, HF etching is the most common synthesis approach to obtain MXenes, still it has its own limitations. It is also important to note that, the method is only suitable for carbides and carbonitrides-based MXenes and cannot be applied to nitrides-based MXenes. Moreover, HF etching usually produces MXenes with –F terminated surfaces, which have limitations for certain applications [17]. Furthermore, HF is a corrosive and hazardous chemical. It can penetrate through human skin, if not properly handled, and can be extremely harmful to bone and body tissues. These concerns motivated the researchers to develop other HF-free methods for the synthesis of MXenes.

ii. **In-situ formed HF etching**

As mentioned earlier, HF is not eco-friendly and has adverse effects on the human body. Therefore, the hunt to avoid HF suggested the use of in-situ formed HF, produced by the dissolution of fluoride salts (LiF, $NaHF_2$, KF, etc.) in acids (like HCl, H_2SO_4, etc.), in the synthesis of MXenes [18, 19]. In-situ formed HFs were found to be less aggressive and more effective. Among the various combinations of acids and fluoride salts, HCl/LiF etching, with smaller intercalation time and simultaneous etching and intercalating properties, is the first reported and most studied [20]. In comparison to HF, in-situ formed HF etching is simple, fast, and produces high-purity MXenes with fewer defects. In addition, the surface termination of MXenes can also be tailored with in-situ formed HF etching. An environment friendly O-terminated MXenes can be produced with the HCl/LiF etching as compared to F-terminated MXenes obtained with HF etching [17]. A general map of the wet chemical synthesis of $Ti_3C_2T_x$ using HF and in-situ formed HF etching is presented in Fig. 13.2b.

iii. **Alkali etching**

In the MAX phase, Al layers have an amphoteric nature, and therefore, can be etched with acid as well as alkali treatments. MXenes with lower F terminations and greater interlayer spacing can be obtained with the alkali etching. The interaction of Ti_3AlC_2 with an organic base (tetramethylammonium hydroxide) provides delaminated MXene sheets of $Ti_3C_2T_x$ [23]. In another study, high-purity multilayered, totally fluorine-free MXenes were prepared by NaOH-assisted hydrothermal method [24].

iv **Molten salt etching**

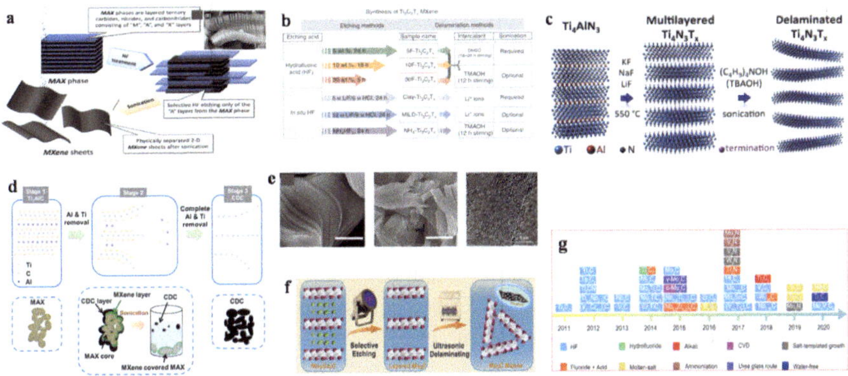

Fig. 13.2 **a** Schematic representation of carbide-based 2D-Mxene synthesis using fluorine method (Right panel shows SEM image of obtained sample after HF etching). Adapted with permission from Ref. [6]; Copyright (2012) American Chemical Society. **b** General map of the liquid-phase synthesis process of $Ti_3C_2T_x$ from Ti_3AlC_2 MAX phase. Adapted with permission from Ref. [21]; Copyright (2017) American Chemical Society. **c** Schematic process showing synthesis of nitride-based $Ti_4N_3T_x$ Mxenes by molten salt approach. Adapted with permission from Ref. [25]; Copyright (2016) Royal Society of Chemistry. **d** Proposed mechanism of electrochemical etching of Ti_2AlC in HCl aqueous electrolyte. **e** SEM images of MXenes from different etching approaches; left electrochemically etched and center chemically etched sample. Scale bars 2 mm. Adapted with permission from Ref. [27]; Copyright (2017) Royal Society of Chemistry. **f** Schematic illustration of UV-induced selective etching synthesis of two-dimensional (2D) mesoporous Mo_2C, HRTEM image of Mo_2C MXene showing the layered structure (right panel of **e**). Adapted with permission from Ref. [13]; Copyright (2020) Elsevier **g** timeline of typical synthesis routes for Mxenes in the past decade. Adapted with permission from Ref. [22]; Copyright (2020) Elsevier

Nitride-based MXenes have higher formation energy and lower cohesive energy, therefore, the method developed for carbide-based MXenes could not be applied to nitride-based MXenes [9]. Besides this, the HF etching is not good for the nitride-based MXenes due to their lower stability, leading to their dissolution in HF solutions. In order to overcome the issues involved in the synthesis of nitride-based MXenes, the molten salt etching method was explored. In this method, molten fluoride salts in an argon (Ar) environment at 550 °C were used to etch the Al layers from Ti_4AlN_3 MAX phase to produce $Ti_4N_3T_x$ [25]. The schematic of the process is shown in Fig. 13.2c. The obtained MXenes contained –O, –OH, and –Cl termination instead of –F termination.

v. **Electrochemical etching**

In a typical electrochemical etching method, MXenes are synthesized by selective removal of *A* layers by application of voltage, using MAX phase material as an electrode in electrolytes such as HCl, NaCl, NH_4Cl, etc. In a binary aqueous electrolyte, fluoride-free Ti_2C MXenes were synthesized from titanium aluminum carbide Ti_3AlC_2 [26]. By tailoring the composition of electrolytes, single or bilayer sheets with high yield (>90%) and large average dimensions were produced. In another study, Sun et al. [27] reported electrochemical etching of Al from the MAX phase (Ti_2AlC) and obtained Ti_2CT_x MXenes in a low-concentration hydrochloric acid solution (Fig. 13.2d and e (left and middle)).

vi. **Green synthesis routes**

The HF etching is a commonly used method to produce MXenes but it involves handling hazardous HF acid. In order to avoid it, making the process scalable and low cost, green synthesis approaches were explored. In this direction, Limbu et al. [28] reported a facile, room temperature, and environmentally benign synthesis of reduced $Ti_3C_2T_x$ MXene by simple treatment with L-ascorbic acid. This simple method was highly reproducible, and produced MXenes with improved properties, such as better oxidation stability and enhanced electrical conductivity, in comparison to conventional methods. In another green synthesis approach, thermal reduction followed by ultrasound-assisted delamination was investigated to produce the Ti_2C nanosheets from Ti_2SC [29]. A novel green synthesis method to produce MXenes from V_2AlC MAX phase utilizing algae extraction was reported by Zada et al. [12].

vii. **Other methods**

In addition to HF etching, acid/base etching few other etching methods utilizing electromagnetic/mechanical waves were also investigated to remove *A* layers in the MAX phases. These methods are relatively fast compared to conventional methods and can produce fluoride-free MXenes. Mei et al. [13] synthesized mesoporous Mo_2C from Mo_2Ga_2C in mild H_3PO_4 solution via UV-induced selective etching (Fig. 13.2f). The method resulted in a high yield of fluorine-free MXenes with high quality. The process was completed in only a few hours, whereas much longer etching time was required in conventional processes. In another investigation, Ghazaly et al. [30]

converted the parent MAX phase into MXenes with the utilization of surface acoustic waves (SAWs). In this ultrafast method, an aqueous mixture of LiF solution and MAX phase was exposed to SAWs. The fluoride ions produced from the LiF, combined with protons produced from the SAW, lead to selective etching of *A* layers.

13.2.2 Bottom-Up Methods

As discussed in the previous section, a variety of top-down methods have been investigated to produce MXenes. Although each method has its own advantages, these methods are limited to irregular thickness, small crystal sizes, and defects [31]. To overcome these issues, bottom-up methods have been explored for the synthesis of MXenes. In bottom-up methods, MXenes can be synthesized by combining various constituent elements. MXenes with higher crystallinity can be obtained with bottom-up methods, particularly CVD, as compared to selective etching methods. Moreover, the bottom-up approach can produce MXenes with different stoichiometries and heterostructures, which are hard to produce with the top-down methods. In addition, MXenes free from –F terminations can also be obtained with this method. Chemical vapor deposition (CVD), plasma enhanced pulsed laser deposition (PE-PLD), and template methods are a few common bottom-up approaches employed in the past.

i. **Chemical Vapor Deposition (CVD)**

In 2015, Xu et al. [32] produced high-quality transition metal carbides (TMCs) on a bilayer (copper/transition metal) substrate using the CVD method. The methane gas with a low flow rate was introduced on the Cu/Mo foils, which was previously heated to a high (1085 °C) temperature in hydrogen. It results in the formation of Mo_2C crystals with a few nanometers of thickness on the liquid Cu surface. Geng et al. [33] investigated structural control in MXenes with carbon supersaturation. On increasing carbon supersaturation, the shape of crystals transformed from triangular to polyhedron (Fig. 13.3a–e). In another study, Geng et al. [34] reported one-step direct synthesis of Mo_2C on the in-situ grown graphene under different methane flow rates. At a low flow rate, thick crystals with random shapes were formed, whereas a high flow rate resulted predominantly in the formation of thin hexagonal-shaped crystals.

ii. **Template Method**

The template method is another useful bottom-up approach to synthesizing MXenes. In comparison to the CVD method, the template methods have much higher yields. In this method, transition metal oxides (TMOs) nanosheets are used as templates, which are carbonized or N-doped to form the 2D TMCs/TMNs. The TMOs used for the synthesis play an important role in the structure of synthesized TMCs/TMNs. Jia et al. [35] obtained single-crystal and ultrathin (thickness of ~1.0 nm) N-Mo_2C NSs via a two-step approach by using ultrathin MoO_2 NSs as a template (Fig. 13.3f–h).

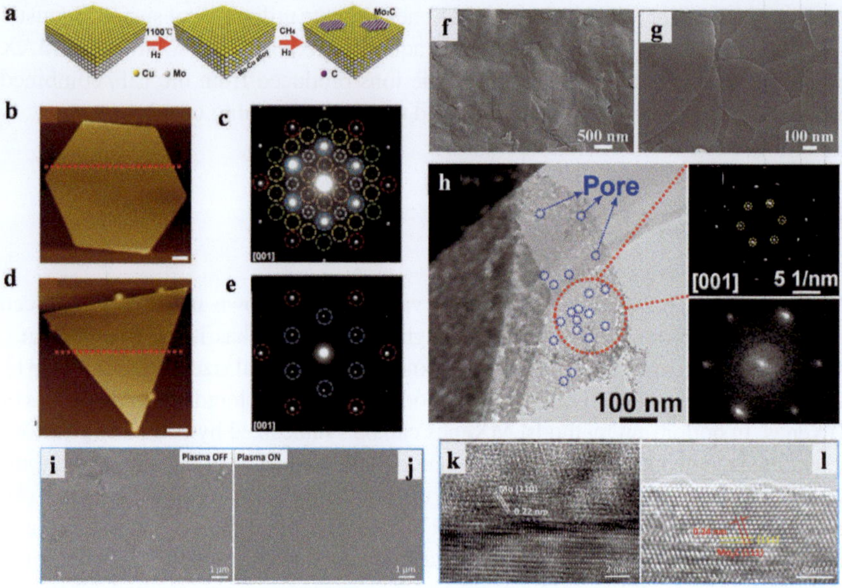

Fig. 13.3 **a** Schematic representation of the growth process for MO_2C crystals. **b**, **d** AFM images of hexagonal and triangular α-MO_2C crystals with a uniform thickness of ~250 nm. **c**, **e** SAED patterns of hexagonal and triangular α-MO_2C crystals with a uniform thickness of ~250 nm. Adapted with permission from Ref. [33]; Copyright (2017) IOP Publishing. SEM images of **f** MoO_2 NSs and **g** N-Mo_2C NSs, **h** TEM image of N-Mo_2C NSs with corresponding electron diffraction patterns, the single layer (right, top) and fast Fourier transform (right, bottom) of the hexagonal NSs. Adapted with permission from Ref. [35]; Copyright (2017) American Chemical Society. **i** and **j** Top view SEM images of thin film grown without and with plasma, respectively. **k** and **l** HRTEM images of thin films grown without and with plasma. Adapted with permission from Ref. [36]; Copyright (2017) Royal Society of Chemistry

As-synthesized N-Mo_2C NSs showed excellent hydrogen evolution reaction (HER) performance, which was mainly attributed to the 2D structure of Mo_2C NSs with large electrochemical area. In addition, particular crystalline phase, high-proportioned C planes of Mo_2C NSs, and synergistic manipulation of N doping produce more active sites for hydrogen evolution.

iii. **Plasma-Enhanced Pulse Laser Deposition**

In addition to the abovementioned bottom-up approaches, the PE-PLD method was explored by Zhang et al. [36] for synthesizing the superconducting Mo_2C thin films on the sapphire substrates. The methane plasma used in the synthesis played a crucial role in the low-temperature deposition of Mo_2C films (Fig. 13.3i–l). The study suggested a relationship between the growth temperature and crystal structure of the deposited films. The films obtained at the low temperature showed nanocrystals with orthorhombic structure, whereas films synthesized at high temperatures showed face-centered cubic structure with <111> as the preferred orientation, which was governed by a coherent orientation relationship and small in-plane lattice mismatch between grown films and substrates.

13.3 Research Avenues for MXenes

MXenes, a wonderful class of 2D materials, with interesting electronic, optical, and mechanical properties are rising stars in the present time. The material has been explored in various fields such as energy conversion and storage, optoelectronic devices, sensors, biomedical, water harvesting, catalysis, etc., which are continuously growing day by day. In this section, we will discuss the various research avenues for MXenes and the progress made in those fields. The associated challenges, future prospects, and possible strategies to take care of concerned issues will be discussed below in the next section.

13.3.1 Lithium-Ion Batteries

Global energy consumption, which is increasing daily, strongly suggests exploring inexpensive renewable energy sources for energy generation and storage. Currently, most electronic equipment such as mobile phones, laptops, cameras, etc., use lithium-ion batteries as the main source of energy, due to their improved energy density. The excellent electronic properties of MXenes led to their introduction into lithium-ion batteries. In 2011, the first time Gogosti et al. [6] implemented the MXenes in lithium-ion batteries. Further in 2013, Naguib et al. [37] explored the possibility of carbide electrodes in lithium-ion batteries by removing Al layers from the vanadium and niobium MAX phase (Fig. 13.4a, b). Both these electrode materials showed reversible capacities of 170 and 260 mAh/g at 1 C. Initial studies concerning to molecular weight of the electrode materials suggest that low molecular weight MXenes, due to their higher gravimetric capacities, are the most suitable candidates for the electrode material in energy storage. However, progressive research revealed that the molecular weight alone is not a deciding factor for the capacity of MXenes-based electrodes, as high molecular weight Nb_2CT_x showed higher capacity compared to low molecular weight Ti_2CT_x [38]. Furthermore, the performance of these materials in energy storage devices was found to be affected by the surface terminal groups. The Ti-MXenes with –OH and –F surface terminations showed a higher diffusion rate, while the bare Ti_3C_2 showed easy migration of Li-ions owing to small path lengths and low energy barriers [39]. In addition, the –O terminated MXenes are theoretically predicted as superior candidates for energy storage applications, which was also experimentally observed by Mashtalir et al. [40]. They produced –O terminated Nb_2CT_x MXenes with amine-assisted delamination of MAX phase and applied in lithium-ion storage devices. These devices showed excellent capacity and stability. Besides, surface termination groups the surface cations also affect the performance of energy storage devices. Luo et al. [41] studied the effect of surface cation on the device performance by incorporating Sn^{4+} cation into the MXenes and observed a reversible volumetric capacity of 1375 mA/g along with excellent cyclic stability (Fig. 13.4c–f). The hybridized MXenes were also explored for lithium-ion batteries. Lin et al. [42] utilized the carbon nanofibers (CNFs) to produce Ti_3C_2/CNF and observed an improved reversible capacity as compared to Ti_3C_2 (Fig. 13.4g–i). Other TMO/MXenes and metal nanoparticles decorated with MXenes have also shown an improved performance of the devices [43, 44]. In short, it can be said that MXenes, explored for lithium-ion batteries, showed significant improvement in the properties and device performance by changing the type of MXenes, surface termination groups, cations as well as combinations of MXenes with other materials. These studies are still limited and only a few combinations have been explored. Various other MXenes with different stoichiometry as well as combinations with other materials should be utilized for improved performance of the lithium-ion batteries.

13.3.2 Supercapacitors

MXenes also emerge as very promising candidates for supercapacitor electrodes because of their distinctive layered structures which facilitate enhanced electrolyte ion transport and provide transition metal active redox sites on the surface. The electrochemical double layer capacitors (EDLC), hybrid, and pseudocapacitors are the three main categories of supercapacitors developed with MXenes. The electrochemical performance of most supercapacitors depends mainly on the characteristics of electrodes. A highly conducting electrode having a high specific area with rich redox sites is needed to realize high-performance supercapacitors. Several candidates of the MXene family possess these beneficial characteristics with the additional uniqueness of having interlayer spacing which provides a natural channel for ion transport [45–49]. The charge storage mechanism in EDLC-based MXene supercapacitors relies on the larger surface area of electrodes, whereas in pseudocapacitors, charge storage is based on ion intercalation [50]. The existence of a wide range of M, X, and T elements in MXenes is another merit that enables design flexibility to achieve improved performance [51]. The performance of MXene-based supercapacitors depends on several aspects such as; the MXene synthesis approaches, structural morphology of MXene, electrolyte used, and electrode layouts. Out of these factors, the architectural design of the electrodes is most crucial. In a study, Zhu et al. [52] reported the preparation of expanded MXene and MXene foam using MgO nanoparticles and tried to address the re-stacking issue of exfoliated layers. The electrode material based on expanded MXene exhibited a capacitance of 180 F g^{-1} at 1 A g^{-1}. The expanded Ti_3C_2Tx MXene layers also displayed excellent cyclic stability by maintaining 99% of its initial capacitance at 5 A g^{-1} over the 5000 charge–discharge cycles. The pseudocapacitive performance of MXene-based electrodes can be further enhanced by improving their specific surface area which enables more access of electrolytes to electrodes. In recent work, Huang et al. [53] developed a facile strategy to fabricate high-performance MXene hydrogel supercapacitor electrodes. They utilized the combination of unidirectional freezing and thawing of slurry in H_2SO_4 to produce the MXene electrode (Fig. 13.4s). These specifically designed three-dimensional macrostructures delivered an ultrahigh areal capacitance of 2.0 F cm^{-2} at 1.2 mA cm^{-2} with retaining 1.2 F cm^{-2} at 60 mA cm^{-2} together with a record-high energy density of 0.1 mWh cm^{-2} at 0.38 mW cm^{-2} (Fig. 13.4t).

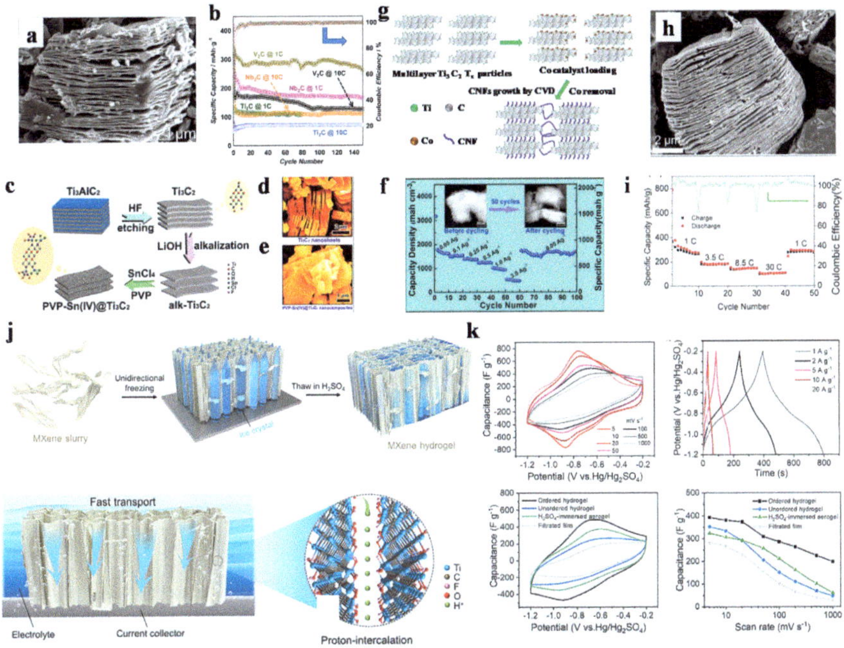

Fig. 13.4 **a** SEM images of Nb_2CT_X. **b** Specific lithiation (circles) and delithiation (squares) capacities (per mass of active material) versus cycle number at different rates for Nb_2CT_X and V_2CT_X-based electrodes (produced by HF treatment of attrition milled V_2AlC at RT for 8 h). Adapted with permission from Ref. [37]; Copyright (2013) American Chemical Society. **c** Schematic illustration of the fabrication process of PVP-Sn(IV)@Ti_3C_2 nanocomposites. **d** and **e** SEM image of Ti_3C_2 nanosheets and PVP-Sn(IV)@Ti_3C_2 nanocomposites. **f** Rate performance of nanocomposite electrode adapted with permission from Ref. [41]; Copyright (2016) American Chemical Society. **g** Schematic showing the preparation of Mxene/CNF hybrid particles. **h** SEM image of as-etched $Ti_3C_2T_X$ particles. **i** Rate performance of the Ti_3C_2-CNF-1-10 sample. Adapted with permission from Ref. [42]; Copyright (2015) American Chemical Society. **j** Schematic illustration of the preparation of MXene hydrogel supercapacitor electrode via MSES. **k** CV curves and GCD profiles of ordered MXene hydrogel at different scan rates adapted with permission from Ref. [53]; Copyright (2021) Wiley-VCH

13.3.3 Optoelectronics

Due to their distinctive electrical and optical properties, the candidates of the MXenes family exhibit significant potential for a broad spectrum of optoelectronic applications. Most of the MXenes exhibit a high density of states (DOS) at the Fermi level with tunable work functions [54]. Also, the MXenes exhibit all three electronic phases such as; metallic phase (non-terminated MXenes) and semi-metallic or semiconducting phase (surface group terminated MXenes). These properties make MXenes very suitable candidates to be explored in optoelectronic applications. The most explored optoelectronic application of MXenes is the transparent electrical contacts (TECs) used in various photonic devices. A typical TEC requires a suitable work function that leads to the Ohmic contacts and a sufficient ratio of direct-current to optical-conductivity. The use of MXenes provides an opportunity to tune the work functions to a larger span than metals which ultimately enables the utilization of the Ohmic contacts for efficient devices. High conductivity, tunable bandgaps, and hydrophilic 2D surfaces make MXenes very suitable candidates for photovoltaics. So far, MXenes have been utilized in several roles such as additives in light-absorber, electron/hole transport layers (ETL/HTL), and electrodes in perovskite-, silicon-, organic/polymer-, and QD sensitized-solar cells (QDSSC). Perovskite solar cells (PSCs) are one of the most recent classes of third-generation photovoltaics which are extensively studied [55–58]. A power conversion efficiency (PCE) close to single-junction silicon and thin film solar cells has already been achieved in typical PSCs, however, their commercial deployment is still in progress due to their limitation in long-term stability. The use of 2D MXenes as additives in perovskite helps to slow down the crystallization which ultimately helps to increase stability. The vigorous interactions between perovskites and MXene additionally provide reduced defects and a good interface with the constituent layers. The use of MXene in PSCs was introduced by Guo and co-workers in 2018 [59] as a light-absorbing layer. Recently, in June 2022, Xu et al. [60] reported the utilization of fluorine-functionalized MXene QDs as a surface passivator to increase the efficiency and stability of a $CsPbI_3$ perovskite. They have observed that the fluorine-doped MXene sheets provide a barrier against moisture sensitivity, improve charge transfer, and reduce recombination which results in a power conversion efficiency as high as 20.44%. Wang et al. [61] demonstrated an interface engineering strategy by using MXene-modified SnO_2 as an ETL and achieved a power conversion efficiency of 20.65% in perovskite solar cells. The enhanced power conversion efficiency was attributed to the improved charge transfer ability along with the reduced non-radiative recombination in MXene-modified SnO_2-based PSCs (Fig. 13.5a, b). The existence of MXene in the hole or electron transporting layer can significantly change the band diagram which provides enhanced charge transport and better interfaces. In a study conducted by Yang et al. [62], the UV-ozone treated Ti_3C_2Tx nanosheets were utilized as the ETL in perovskite solar cells. In another study, the use of Ti_3C_2Tx provided accelerated hole transport with reduced carrier recombination [63]. Few attempts have also been made to utilize the high conductivity and transparency of MXenes for transparent electrodes in silicon-based solar cells [64, 65]. He and co-workers reported the fabrication of Ti_3C_2Tx MXene/n^+np^+-Si solar cells [10]. In another study, Yu et al. [65] fabricated Ti_3C_2Tx MXene/n-silicon heterojunction solar cells based on the $Ti_3C_2T_x$ MXene. The use of MXene was not only limited to the Perovskite and Si solar cells, but it has also been explored in organic solar cells. For example, Yu et al. [66] used chemically modified T_3C_2Tx MXene as ETL in organic solar cells. In another report, Huang and co-workers explored the utility of Ti_3C_2Tx as HTL in organic solar cells with PBDB-T: ITIC as active material [67]. Chen and co-workers used a combination of silver nanowires and Ti_3C_2Tx nanosheets as a flexible transparent electrode in organic solar cells [68]. Along with the photovoltaic applications, MXenes have also been well explored in high-performance photodetectors and phototransistors. Kang et al. [69] fabricated a MXene-silicon vdW heterostructure which worked as a self-driven Schottky photodiode. The Ti_3C_2Tx/n-Si Schottky junction was prepared by drop-casting colloidal solution of Ti_3C_2Tx on the pieces of n-type silicon substrates. The prepared device exhibited an on/off ratio of 10^5 with a responsivity value as high as 26.95 mA W^{-1} for 405 nm light. Guo et al. [70] reported the fabrication of nano-heterojunction of crossed Zn_2GeO_4 nanowire networks with Ti_3C_2Tx and used it for deep-UV photodetection. Montazeri et al. [71] utilized Ti_3C_2Tx as electrodes in GaAs-based photodetectors as an alternative to the commonly used conventional gold electrodes. Yang et al. [72] used patterned plasmonic Ti_2CT_x nanoribbons as electrodes for InSe-based avalanche phototransistors. To fabricate this device, a Ti_2CTx flake was mechanically transferred onto an InSe nano sheet followed by patterning and etching of the unwanted material. The fabricated P-InSe/Ti_2CTx nano-heterojunction devices can successfully detect light with intensities as low as 17 W cm^{-2} (Fig. 13.5c). P-InSe/Ti_2CTx devices show better responsivity and detectivity values in comparison to that of the InSe/Ti_2CTx devices under the incident light of different wavelengths. As mentioned above, owing to their good transmittance and high conductivity, MXenes are also very suitable candidates to be used as transparent conducting electrodes in light-emitting devices. Ahn et al. [73] explored the spin-coated Ti_3C_2Tx film as a transparent electrode in flexible organic LEDs. A spin-coated film of 7.9 nm exhibited a transmittance of 85% at 550 nm and a sheet resistance of 108 Ω sq^{-1}. Cao et al. [74] explored the MXene-based electrodes to fabricate flexible PeLEDs which demonstrated a record external quantum efficiency of 16.5% and luminance of as high as 50,000 cd/m^2 (Fig. 13.5d). Although the new 2D materials of the MXenes family exhibited a wonderful potential in optoelectronics, several challenges are there that need to be addressed with delicate observation and intense efforts.

13.3.4 Biomedical Applications

The diverse chemistry and structure of MXenes, leading to interesting chemical and physical properties made them potential candidates for biomedical and clinical applications. These materials can not only diagnose, but also treat various diseases.

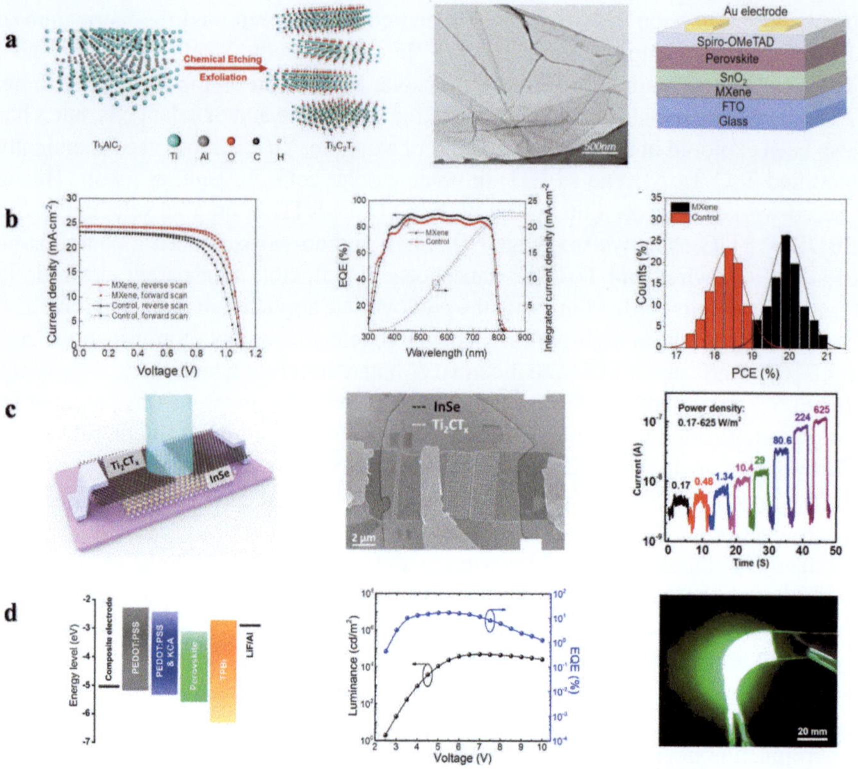

Fig. 13.5 Optoelectronic applications of MXenes; **a** Schematic process of Ti_3C_2Tx synthesis (left). TEM image of Ti_3C_2Tx nanoflakes on a copper grid (middle). Device configuration of the photovoltaic device fabricated with Ti_3C_2Tx (Right). **b** Standard J–V characteristics of control and MXene-modified PSCs under AM1.5G 100 mW cm^{-2} illumination (left) and the spectral response with the integrated current densities for both control and MXene-modified PSCs (middle). Histogram for the control and MXene-modified PSCs measured from 50 devices (right). Adapted with permission from Ref. [61]; Copyright (2020) American Chemical Society. **c** Schematic of InSe/Ti_2CTx photodetector under light illumination. SEM image and photoresponse curves of the fabricated device adapted with permission from Ref. [72]; Copyright (2019) American Chemical Society. **d** Energy level diagram and the L-V-EQE characteristic of PeLEDs based on KCA-modified PEDOT:PSS HTL. Photograph of a large-area flexible PeLED with a size of 40 mm × 20 mm operating at 5 V (right). Adapted with permission from Ref. [74]; Copyright (2022) American Chemical Society

This section will focus on various MXenes, and their composites for application in drug delivery, cancer treatment, and antimicrobial activity. 2D layered MXenes provide an excellent platform for drug delivery but lack of controllability is limiting its progress. To overcome this problem, Liu et al. [75] developed titanium carbide-cobalt heterostructured (Ti_3C_2-CoNWs) nanocarriers for synergetic anticancer therapy with dual stimuli-responsive drug release, and chemo-photothermal therapy (Fig. 13.6a, b). The nanocarriers can be controlled under the external magnetic field owing to

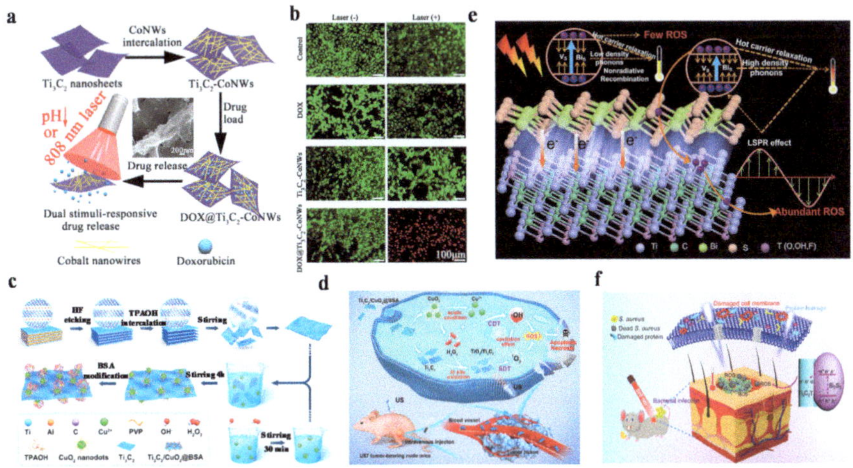

Fig. 13.6 **a** Schematic illustration of a combination of Ti_3C_2 nanosheets and CoNWs and their dual stimuli-responsive drug release (inset SEM image of Ti_3C_2-CoNWs). **b** Fluorescence images of 4T1 cells after being treated with pure DOX Ti_3C_2, Ti_3C_2-CoNWs, and DOX@Ti_3C_2-CoNWs. Adapted with permission from Ref. [75]; Copyright (2020) Elsevier. **c** Schematic illustration of the synthesis of Ti_3C_2/CuO_2@BSA nanosheets. **d** Synergetic chemodynamic and sonodynamic therapeutic process of 2D-Ti_3C_2/CuO_2@BSA nanosheets under US irradiation. Adapted with permission from Ref. [77]; Copyright (2022) American Chemical Society. **e** Schematic diagram for photodynamic and photothermal mechanism between $Ti_2C_2T_X$ and Bi_2S_3. **f** Schematic illustration of Bi_2S_3/$Ti_3C_2T_X$ Schottky catalyst and antibacterial mechanism diagram of $Bi_2S_3/Ti_3C_2T_X$ under 808 nm irradiation. Adapted with permission from Ref. [80]; Copyright (2021) American Chemical Society

their magnetic properties. The heterostructured nanocarriers could efficiently deliver doxorubicin (DOX), which can be controlled by near-infrared (NIR) radiation or pH (4–6). Recently, in order to reduce the toxic side effects of drugs and promote the wound healing process, Yang et al. [76] developed 2D-MXene-based hydrogel systems for highly efficient photo and magnetic response delivery for deep chronic wound repair. An increase in the temperature under NIR and alternating magnetic field, triggered the drug release in a very controlled manner. In addition to drug delivery, MXenes are found to be very useful in anticancer applications. As conventional cancer therapy suffers from various disadvantages such as high doses resulting in unwanted side effects and poor drug availability, etc., the MXenes with excellent photothermal properties showed very good results for cancer treatment. Moreover, MXenes combined with chemo-photothermal therapy, a localized tumor can be effectively eradicated. Zhang et al. [77] reported a tumor microenvironment-specific in-situ generation of nanosonosensitizers on a Ti_3C_2Tx/CuO_2@BSA catalyst and achieved synergistic chemodynamic/sonodynamic tumor therapy (Fig. 13.6c, d). In-situ formed sonosensitizer TiO_2/Ti_3C_2 under tumor conditions enhanced the separation of electrons and holes and augmented the sonodynamic therapy (SDT) efficacy. Considering the lack of effectiveness, and off-target toxicities of current iron-based therapeutic strategies, Xu et al. [78] reported a dual-therapeutic prodrug,

DOXjade, which allows for iron chelation chemo-photothermal cancer therapy. Loading DOXjade onto ultrathin 2D-Ti_3C_2 MXene nanosheets produces a construct, Ti_3C_2-PVP@DOXjade, that allows the iron chelation and chemotherapeutic functions of DOXjade to be photo-activated at the tumor sites while potentiating a robust photothermal effect with photothermal conversion efficiencies up to 40%.

Investigations also showed that MXenes are promising candidates with antimicrobial activity higher than graphene oxide. Rasool et al. [79] investigated the antibacterial properties of single and few-layered Ti_3C_2 MXenes against *Escherichia coli* (*E. coli*) and *Bacillus subtilis* (*B. subtilis*). Direct contact with $Ti_3C_2T_x$ can disrupt cellular membranes leading to cell damage and eventual death. Li et al. [80] investigated the interfacial engineering and designed the eco-friendly photoresponsive Schottky junction of $Bi_2S_3/Ti_3C_2T_x$ to effectively eradicate bacterial infection (Fig. 13.6e–g). Furthermore, MXenes and their composites showed excellent antibacterial properties against *E. coli*, *B. subtilis*, *Staphylococcus aureus*, *Pseudomonas aeruginosa*, and *Shigella* [81]. MXenes functional groups are also reported to cause cell inactivation by preventing the intake of nutrients and thereby inhibiting the growth of bacteria [82].

13.3.5 Sensing Devices

In recent years, various types of sensors have been in increasing demand in the fields of our daily life including robotics, human healthcare monitoring, automation, and artificial intelligence [83–85]. So far extensive efforts have been made to find new materials and device designs that can be implemented into a high-performance sensor. The performance of a sensor device is predominantly determined by its active sensing materials. The tunable surface characteristics, tunable bandgap, and excellent mechanical strength make MXenes attractive for building various high-performance sensors. Although it emerged less than a decade ago, MXene and its derivatives are extensively explored to develop numerous sensors, including strain/stress sensors, gas sensors, biosensors, and electrochemical and humidity sensors [86–88]. The high electrical conductivity, excellent flexibility, and mechanical robustness make MXene a very suitable candidate for strain/stress and pressure sensors. Yang and co-workers developed a tensile strain sensor with the hybrid network of $Ti_3C_2T_x$ nanoparticles and nanosheets which exhibited a gauge factor of >178.4, a sensing range of 0–53%, a promising detection limit of 0.025%, and stable performance over 5000 cycles [89]. The developed strain sensor is used to demonstrate the real-time monitoring of the movement in various parts of the human body. In another study, Cheng et al. [90] fabricated a highly sensitive piezoresistive sensor with bioinspired microstructures using a simple approach of spray coating $Ti_3C_2T_x$ nanosheets on a PDMS surface. Significant contact resistance changes between microstructured PDMS and the interdigital electrodes provided high sensitivity (151.4 kPa^{-1}), short response time (<130 ms), detection limit of 4.4 Pa, and excellent stability. The fabricated strain sensor showed great potential in monitoring human physiological signals and remote monitoring of robotic motion (Fig. 13.7a). Along with the physical sensors, MXenes have also been well explored in chemical sensing applications such as gas sensors and humidity sensors [91–93]. The quantitative identification of several gaseous species, such as CO_2, NO, CO, NO_2, H_2S, and NH_3, is vital for human health, industry safety, and air pollution observation. Being a layered material with a high surface-to-volume ratio, MXenes provided an ideal platform for gas sensing [94, 95]. The suitability of Ti_2CO_2, M_2CO_2, and $Ti_3C_2T_x$ MXenes as sensing materials for various chemicals in gaseous forms such as methanol, acetone, and ammonia was demonstrated [96, 97]. Humidity sensing is another important and much-required prospect toward a number of potential applications such as agriculture, food storage, smart homes, and health care. $Ti_3C_2T_X$ MXene displayed significant hydration over a wide range of relative humidity in which water molecules get adsorbed on vacancies and hydrophilic groups. The insertion of cations such as Mg^{2+} and K^+ between MXene interlayers can further enhance water molecule diffusion [98].

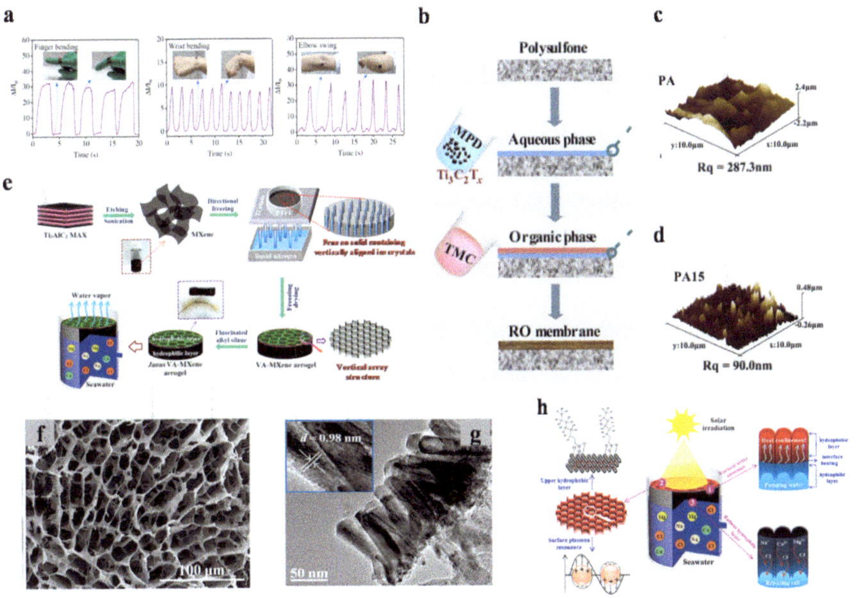

Fig. 13.7 **a** Applications of MXene-based piezoresistive sensor in real-time monitoring of human activity such as finger bending (left), wrist bending (middle), and elbow swing (right). Adapted with permission from Ref. [90]; Copyright (2022) American Chemical Society. **b** Schematic for $Ti_3C_2T_X$ embedded polyamide RO membrane preparation. **c–d** Surface AFM images of membranes. Adapted with permission from Ref. [99]; Copyright (2020) Elsevier. **e** Fabrication process of a janus VA-MXA with vertically aligned channels. **f** SEM image of upper layer of janus VA-MXA. **g** TEM image of the janus VA-MXA. **h** Schematic of high-efficiency solar steam generation and the salt resistance strategy. Adapted with permission from Ref. [100]; Copyright (2019) American Chemical Society

13.3.6 Water Treatment

The need for fresh water is indeed a requirement for global socioeconomic development. At present, reverse osmosis (RO) membranes are widely utilized for desalination, owing to their higher efficiency and operability. But a trade-off between water flux and salt rejection, as well as easy fouling and chlorine attack, suggests the development of new reverse osmosis membranes. Wang et al. [99] prepared polyamide-reverse osmosis (PA-RO) membranes containing two-dimensional $Ti_3C_2T_x$ nanosheets by an in-situ interface polymerization process (Fig. 13.7b–d). The fabricated membranes with improved surface hydrophilicity and decreased surface roughness showed remarkable stability to chlorine and superior fouling. The interaction between surface functional groups of layered $Ti_3C_2T_x$ nanosheets and active chlorine was proposed to contribute to improved chlorine resistance. Solar desalination is another effective way to convert solar energy to heat for seawater purification. Zhang et al. [100] utilized vertically aligned janus MXene-(VA-MXA)-based aerogels by directional freezing of ultrathin Ti_3C_2 nanosheets for solar desalination (Fig. 13.7e–h). VA-MXA showed a decreased heat loss and low transport resistance, enabling effective inhibition of salt crystallization due to rapid dissolution and fast pumping. Water contaminated with copper is very hazardous for living organisms, therefore, copper removal is an essential part of wastewater treatment. Shahzad et al. [101] investigated the delaminated (DL)-$Ti_3C_2T_x$, produced by ultrasonication of multilayer (ML)-MXenes, in the adsorption and copper removal from the aqueous media and compared the performance with that of (ML)-MXenes and activated carbon. A maximum adsorption capacity of 78.45 mg/g was observed for the DL-MXenes. The oxygenated functional group of the synthesized MXenes strongly reacts with copper resulting in the formation of Cu_2O and CuO, as confirmed with the XRD. Moreover, DL-$Ti_3C_2T_x$ was found to be a time-efficient adsorbent for heavy metals absorption along with a higher copper removal rate as compared to commercially available activated carbon. Heavy metal ions in industrial wastewater are not only harmful to the environment, but also to human lives. Therefore, the desalination of the industrial water is a very serious concern. Ma et al. [102] synthesized $Ti_3C_2T_x$ MXenes by LiF/HCl etching method and utilized for capacitive deionization, which is a promising desalination technology. The increased interlayer spacing of $Ti_3C_2T_x$ layers with a reduced number of –F termination groups favor the ion transport resulting in an ultrahigh desalination capacity. The synthesized MXenes showed an enhanced salt removal capacity of 67.7 mg/g. Moreover, a comparison of desalination performance obtained with the MXenes formed by HF etching method confirms the superiority of the process.

13.4 Future Prospects and Research Avenues

The MXenes were developed in 2011, and in the past few years, MXene-based materials have received much attention due to their potential for diverse applications. MXenes are considered to be excellent materials with high electrical conductivity, relatively large specific surface area, high redox capacity, and remarkable electrochemical and electrocatalytic properties [103]. Among the numerous advantages, the applicability of these materials is still at an early stage compared to other 2D nanomaterials. Materials have various challenges associated with them which require in-depth understanding to overcome those issues in order to utilize the material in an efficient way for commercial applications. In this section, we will discuss the various challenges associated with MXenes and possible strategies to overcome those challenges.

13.4.1 Synthesis

Various synthesis methods have been explored for the synthesis of MXenes, having their own advantages and disadvantages. Still, researchers are working to develop cost-effective and environment-friendly manufacturing processes. Several MXenes have been successfully synthesized at the laboratory scale, their large-scale production and commercialization should also be pursued. Two major factors such as high cost and use of hazardous chemicals like HF are the major drawbacks for industry scale production. Because of the high toxicity of HF, other methods using soft etchants like ammonium bifluoride (NH_4HF_2) and a combination of LiF/HCl should be explored further to make the process user-friendly and less hazardous to the environment [104]. Further research on fluoride-free etchants would be more environment friendly and prevent the –F terminations in the expansion of MXene production. The hydrothermal temperature and NaOH concentration are important variables that affect the quality (particularly purity) of the produced MXenes as well as the reaction kinetics. Even so, there are still serious dangers associated with such high-temperature hydrothermal treatment in very concentrated alkali solutions. Instead, a gentle, risk-free, yet effective method is electrochemical etching of the MAX phase. In one investigation, MXene nanosheets were produced electrochemically without the use of potentially harmful acid/alkali etchants [105]. The selective removal of A layers while preserving the 2D structure of MXenes depends on the electrolyte, voltage, temperature, and time. A variety of porous substrates can be used to build the MXene structures, they should be investigated further for their utilizations in high-performance devices. Currently, most of the preparation methods do not yield excellent MXene compounds. In general, it is difficult to manage the quantity of MXene nanosheet layers produced by etching, and the layers are quite nonuniform. Making high-quality MXenes continues to be difficult. The majority of top-down procedures, such as liquid and mechanical stripping, involve regulating the lateral dimensions of MXenes. These techniques can produce MXene materials that can be employed in biomedical applications, although bottom-up chemical vapor deposition is still the most used technique. The most efficient method is able to regulate MXen's lateral size, which can result in industrial production and is reasonably simple to use. It has a lot of potential uses in the fields of sensing, energy storage, and catalysis. The surface engineering of MXenes is also a very important and key factor for developing efficient MXenes. The presence of surface terminal groups can lead to different optical and electronic properties and also can change the metallic MXenes into semiconducting. Experimental conditions have a major impact on the surface chemistry of MXenes, therefore, new methods with variable processing conditions should be explored to tailor the surface termination of the MXenes, to develop MXenes with other functional groups besides –O, –OH, and –F. It will add the various other research avenues for the MXenes.

13.4.2 Simulations and Computational Studies

The simulations and computational studies can be very useful for the development and progress of MXenes, as these studies can predict various MXenes, estimate their properties, and suggest their application in various fields. Moreover, the studies can be helpful for developing the material with the required properties for specific applications. The simulation studies can provide favorable synthesis conditions to develop new MXenes materials with controlled properties for specific applications. Although simulation studies such as Monte Carlo and molecular dynamics have been utilized for the materials, these studies are very limited and performed only for a few fields. In past the simulation studies concerning the adsorption of biomolecules and various functionalized MXenes have been performed. Similar studies involving the interaction of biomolecules with MXenes can provide deeper insight and can be very helpful for the cytotoxic effects. Moreover, simulation studies can help in improving the thermal and oxidation stabilities of the MXenes which will expand the application of MXenes in the field of biomedicine. It is also interesting to point out that, although the simulation studies of more than 100 MXenes have been predicted, only a few are investigated experimentally. For the synthesis of nitride and carbide MXenes, the development of new MAX phases and layered 2D precursors is still a challenge. It suggests that only theoretical or experimental studies are not sufficient for exploring the full potential of the MXenes. Therefore, a parallel approach would be helpful for the development of suitable precursors as well as new MXenes phases.

13.4.3 Batteries and Supercapacitors

Considering the application of MXenes in supercapacitors and batteries, the specific surface area of MXene's rigid nanosheets is lower than that of graphene, which could be improved to potentially enhance the specific capacitance and energy density [106]. At the same time, it is also important to focus on how to modify morphology to increase the active sites. Numerous reports have been published on improving the properties of MXenes for electrochemical energy storage but still leaving a lot of scope to improve the performance of MXene-based electrochemical devices. Treatment with new combinations of concentrated acids and bases should be tested to induce pores in MXene-based electrodes. Pore-engineering can also be achieved by increasing the interlayer spacing between MXenes sheets. This can be achieved by entrapping small molecules such as DMF, DMSO, TBAB, and TBAOH between sheets. Conducting molecules such as tetracyanoquinodimethane (TCNQ) between the layers can potentially boost the energy density. Also, the use of inorganic columns/pillars between layers not only increase the distance between layers, but also the porosity of MXenes. This increased porosity provides space for the pseudocapacitive nanoparticles and ions of the high-voltage electrolyte to further increase the electrochemical behavior. Dissolved oxygen contained in an aqueous electrolyte leads to oxidation, another drawback related to MXenes stability, which also reduces the surface area and thus affects the performance of the product. So, a good way to stop oxidation is to saturate the dissolved oxygen in an aqueous electrolyte with nitrogen or argon. Oxidation usually begins at exposed areas such as the surface or edges. Surface passivation/coating with inorganic nanoparticles or polymers can be applied to protect against oxidation. Making composites with moisture-resistant polymers or 2D materials can effectively extend the life and recyclability of MXene-based materials. Solid electrolytes and non-aqueous electrolytes can be used to replace the aqueous electrolytes that can protect in-situ oxidation during the supercapacitance performance. According to recent reports, MXenes composed of transition metal elements from group 6 of the periodic table (Cr, Mo, W) are the most promising candidates for superconductivity, which need to be further advanced to achieve better response [107]. Recent experimental studies on Zn-air batteries with MXene-based electrodes show that MXenes are promising electrode materials [108]. However, there are only theoretical bases showing that MXenes are promising cathode materials for lithium-air batteries, but there is no experimental evidence using these theoretical findings [109]. Therefore, MXenes can be tested as an electrode material for Li-air batteries to achieve a remarkable breakthrough.

13.4.4 Sensors and Optoelectronics

The material synthesis is a prime concern with the MXenes-based biosensors. The successful implementation of a MXene-based sensor at the commercial level demands the development of novel and cost-effective synthesis strategies for producing materials in a controllable and scalable manner. As the selectivity and sensitivity of a sensor strongly depend on the surface characteristics, the production of MXene nanosheets with suitable surface terminations and size is highly desirable for distinct adsorption and catalytic activity selective molecules. The stability of

MXenes in oxygen-rich humid environments and high-temperature surroundings is another critical concern that needs to be addressed appropriately for the successful accomplishment of MXene-based sensors. The minimal sample-to-sample variation synthesis of MXenes is also not compatible with polymer-based flexible substrates due to high temperature. To tackle this problem, the 2D layers are grown separately and then transferred. It makes the process complicated and also adds to the cost. Therefore, strategies should be developed to produce low cost, scalable 2D layers with high uniformity as well as to grow these layers directly on the substrates. Similarly, for optoelectronic applications such as photo-sensing, photovoltaics, and imaging, more comprehension and investigation are required to utilize the full potential of MXenes. The most explored optoelectronic application of MXenes is as transparent electrodes, which require suitable work function. It has been theoretically predicted that the MXenes exhibit a wide range of work functions (1.6–8.0 eV), their exact experimental estimation is vital for the practical device. To be used as an efficient light absorbing material, particularly in photovoltaics, MXenes are supposed to have a semiconducting nature with optical absorption in a span of UV to NIR region. Most of the studies investigating the band gap nature of MXenes available in the literature are theoretical which are based on simulation and modeling. The realization of semiconducting MXenes in optoelectronic applications needs substantial experimental investigation with more precise theoretical results.

13.4.5 Biomedical Applications

The 2D MXene nanosheets display a wide range of physicochemical traits, including strong hydrophilicity, biocompatibility, a sizable surface area, and moderate biodegradability. They also have outstanding light-to-heat conversion efficiency and light absorption capacity. These distinct physicochemical features offer a fresh approach to the detection and management of significant clinical disorders since they are entirely distinct from those of bulk materials [110]. Because MXenes typically contain –OH, –O, or –F on their surfaces, some MXenes layers quickly accumulate and precipitate in the biological medium due to the surface's hydrophilicity and high surface charge. It frequently becomes unstable and isn't highly adaptable under physiological settings [111]. Therefore, the two tactics that are discussed in the previous section, Surface passivation/coating with polymers or inorganic nanoparticles, are the core emphasis of MXenes. For instance, the surface can be modified using PEG, PVP, SPs, folic acid, and hyaluronic acid, which not only improves the hydrophilicity and hydrophobicity of the MXene solution regulation, but also offers a therapeutic approach. The potential uses of MXenes can be increased by modifying them with multifunctional inorganic nanoparticles. Similarly, surface modification with inorganic nanoparticles, and nanoparticles like Fe_3O_4, MnO_x, Ag, Au, and others may find use in therapeutic and imaging. The interactions between MXene nanosheets and surroundings in organisms need to be thoroughly investigated. The assertion that MXene nanosheets are sufficiently biocompatible for clinical purposes is not sufficiently supported by the available research. Recent studies on 2D nanosheets have

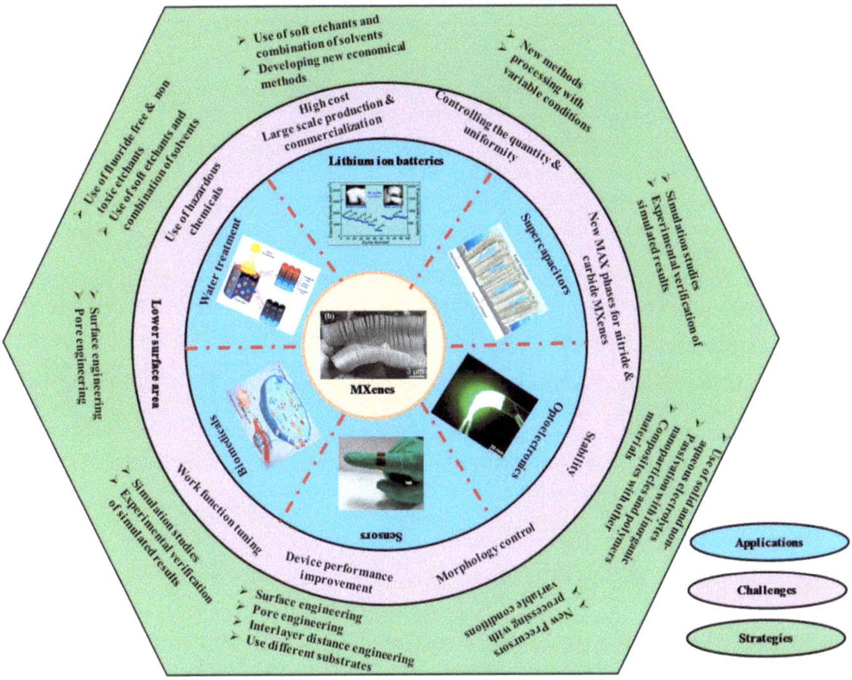

Fig. 13.8 Various research avenues of MXenes with associated challenges and possible strategies to address those challenges

mainly evaluated their biocompatibility using hematoxylin and eosin (H&E) stained pictures of significant organs [112]. However, additional research is required to determine their biodegradation, blood compatibility, histocompatibility, and excretion mechanisms in the body. It is necessary to conduct more research on issues including safe and effective preparation techniques, stability preservation, and quick in-vivo degradation. MXene materials have demonstrated excellent application potential in the crucial subject of medical diagnostics, even though it is clear that more studies and experimental verification on MXene nanosheets are still necessary.

In concluding remarks, it can be said that MXenes, a wonderful class of 2D materials, has emerged and extensively explored in a very short time for various avenues. The various research avenues of MXenes along with associated challenges and possible strategies to deal with those challenges are shown in Fig. 13.8. From the figure, it can be seen that various fields of application for the MXenes, indeed, make them future material but research is still in its early stage. Exploring their full potential for successful application in various fields including their commercialization demands more investigations. The computational studies associated with the experimental verification to develop new Mxenes, with different compositions and variable surface terminations, can be helpful to improve their properties and make them potential candidates for application in various fields.

References

1. S.J. Choi, I.D. Kim, Recent developments in 2D nanomaterials for chemiresistive-type gas sensors. Electron. Mater. Lett. **14**, 221–260 (2018)
2. H. Wang, X. Liu, P. Niu, S. Wang, J. Shi, L. Li, Porous two-dimensional materials for photocatalytic and electrocatalytic applications. Matter **2**, 1377–1413 (2020)
3. S. Panda, K. Deshmukh, S.K.K. Pasha, J. Theerthagiri, S. Manickam, M.Y. Choi, MXene based emerging materials for supercapacitor applications: recent advances, challenges, and future perspectives. Coord. Chem. Rev. **462**, 214518 (2022)
4. W. Tao, N. Kong, X. Ji, Y. Zhang, A. Sharma, J. Ouyang, B. Qi, J. Wang, N. Xie, C. Kang, H. Zhang, O.C. Farokhzad, J.S. Kim, Emerging two-dimensional monoelemental materials (Xenes) for biomedical applications. Chem. Soc. Rev. **48**, 2891 (2019)
5. G. Yang, F. Liu, J. Zhao, L. Fu, Y. Gu, L. Qu, C. Zhu, J.J. Zhu, Y. Lin, MXenes-based nanomaterials for biosensing and biomedicine. Coord. Chem. Rev. **479**, 215002 (2023)
6. M. Naguib, O. Mashtalir, J. Carle, V. Presser, J. Lu, L. Hultman, Y. Gogotsi, M.W. Barsoum, Two-dimensional transition metal carbides. ACS Nano **6**, 1322–1331 (2012)
7. M. Naguib, Y. Gogotsi, Synthesis of two-dimensional materials by selective extraction. Acc. Chem. Res. **48**, 128–135 (2015)
8. M.W. Barsoum, MAX phases are a large family of ternary layered carbide, nitride, or ternary layer carbonitride with general formula $M_{n+1}AX_n$. Prog. Solid State Chem. **28**, 201–281 (2000)
9. M. Naguib, V.N. Mochalin, M.W. Barsoum, Y. Gogotsi, 25th anniversary article: MXenes: a new family of two-dimensional aaterials. Adv. Mater. **26**, 992–1005 (2014)
10. V. Natu, M. Benchakar, C. Canaff, A. Habrioux, S. Celerier, M.W.A. Barsoum, Critical analysis of the X-ray photoelectron spectra of $Ti_3C_2T_z$ MXenes. Matter **4**, 1224–1251 (2021)
11. M. Naguib, V. Presser, D. Tallman, J. Lu, L. Hultman, Y. Gogotsi, M.W. Barsoum, On the topotactic transformation of Ti_2AlC into a Ti–C–O–F cubic phase by heating in molten lithium fluoride in air. J. Am. Ceram. Soc. **94**, 4556–4561 (2011)
12. S. Zada, W. Dai, Z. Kai, H. Lu, X. Meng, Y. Zhang, Y. Cheng, F. Yan, P. Fu, X. Zhang, Algae extraction controllable delamination of vanadium carbide nanosheets with enhanced near-infrared photothermal performance. Angew. Chem. Int. Ed. **59**, 6601–6606 (2020)
13. J. Mei, G.A. Ayoko, C. Hu, J.M. Bell, Z. Sun, Two-dimensional fluorine-free mesoporous Mo_2C MXene via UV-induced selective etching of Mo_2Ga_2C for energy storage. Sustain. Mater. Technol. **25**, e00156 (2020)
14. J. Wang, Y. Zhou, T. Liao, Z. Lin, Trend in crystal structure of layered ternary TAl-C carbides (T = Sc, Ti, V, Cr, Zr, Nb, Mo, Hf, W, and Ta). J. Mater. Res. **22**, 2685–2690 (2007)
15. R. Meshkian, L.A. Naslund, J. Halim, J. Lu, M.W. Barsoum, J. Rosen, Synthesis of two-dimensional molybdenum carbide, Mo_2C, from the gallium based atomic laminate Mo_2Ga_2C. Scr. Mater. **108**, 147–150 (2015)
16. X. Sang, Y. Xie, M.W. Lin, M. Alhabeb, K.L.V. Aken, Y. Gogotsi, P.R.C. Kent, K. Xiao, R.R. Unocic, Atomic defects in monolayer titanium carbide (Ti_3C_2Tx) MXene. ACS Nano **10**, 9193–9200 (2016)
17. M.A. Hope, A.C. Forse, K.J. Griffith, M.R. Lukatskaya, M. Ghidiu, Y. Gogotsi, C.P. Grey, NMR, reveals the surface functionalisation of Ti_3C_2 MXene. Phys. Chem. Chem. Phys. **18**, 5099–5102 (2016)
18. M. Ghidiu, M.R. Lukatskaya, M.Q. Zhao, Y. Gogotsi, M.W. Barsoum, Conductive two-dimensional titanium carbide 'clay' with high volumetric capacitance. Nature **516**, 78–81 (2014)
19. F. Liu, J. Zhou, S. Wang, B. Wang, C. Shen, L. Wang, Q. Hu, Q. Huang, A. Zhou, Preparation of high-purity V_2C MXene and electrochemical properties as Li-ion batteries. J. Electrochem. Soc. **164**, A709 (2017)
20. J. Zhu, E. Ha, G. Zhao, Y. Zhou, D. Huang, G. Yue, L. Hu, N. Sun, Y. Wang, L.Y.S. Lee, Recent advance in MXenes: a promising 2D material for catalysis, sensor and chemical adsorption. Coord. Chem. Rev. **352**, 306–327 (2017)

21. M. Alhabeb, K. Maleski, B. Anasori, P. Lelyukh, L. Clark, S. Sin, Y. Gogotsi, Guidelines for synthesis and processing of two-dimensional titanium carbide (Ti_3C_2Tx MXene). Chem. Mater. **29**, 7633–7644 (2017)
22. N. Li, J. Peng, W.-J. Ong, T. Ma, Arramel, P. Zhang, J. Jiang, X. Yuan, C. Zhang, MXenes: an emerging platform for wearable electronics and looking beyond. Matter **4**, 377–407 (2021)
23. J. Xuan, Z. Wang, Y. Chen, D. Liang, L. Cheng, X. Yang, Z. Liu, R. Ma, T. Sasaki, F. Geng, Organic-base-driven intercalation and delamination for the production of functionalized titanium carbide nanosheets with superior photothermal therapeutic performance. Angew. Chem. **128**, 14569–14574 (2016)
24. T. Li, L. Yao, Q. Liu, J. Gu, R. Luo, J. Li, X. Yan, W. Wang, P. Liu, B. Chen, Fluorinefree synthesis of high-purity Ti_3C_2Tx (T = OH, O) via alkali treatment. Angew. Chem. Int. Ed. **57**, 6115–6119 (2018)
25. P. Urbankowski, B. Anasori, T. Makaryan, E. Dequan, S. Kota, P.L. Walsh, M. Zhao, V.B. Shenoy, M.W. Barsoum, Y. Gogotsi, Synthesis of two-dimensional titanium nitride Ti_4N_3 (MXene). Nanoscale **8**, 11385 (2016)
26. S. Yang, P. Zhang, F. Wang, A.G. Ricciardulli, M.R. Lohe, P.W.M. Blom, X. Feng, Fluoride-free synthesis of two-dimensional titanium carbide (MXene) using a binary aqueous system. Angew. Chem. Int. Ed. **57**, 15491–15495 (2018)
27. W. Sun, S.A. Shah, Y. Chen, Z. Tan, H. Gao, T. Habib, M. Radovic, M.J. Green, Electrochemical etching of Ti_2AlC to Ti_2CT_x (MXene) in low-concentration hydrochloric acid solution. J. Mater. Chem. A **5**, 21663 (2017)
28. T.B. Limbu, B. Chitara, J.D. Orlando, Y. Martha, G. Cervantes, S. Kumari, Q. Li, Y. Tang, F. Yan, Green synthesis of reduced Ti_3C_2Tx MXene nanosheets with enhanced conductivity, oxidation stability, and SERS activity. J. Mater. Chem. C **8**, 4722–4731 (2020)
29. J. Mei, G.A. Ayoko, C. Hu, S. Ziqi, Thermal reduction of sulfur-containing MAX phase for MXene production. Chem. Eng. J. **395**, 125111 (2020)
30. A.E. Ghazaly, H. Ahmed, A.R. Rezk, J. Halim, P.O.A. Persson, L.Y. Yeo, J. Rosen, Ultrafast, one-step, salt-solution-based acoustic synthesis of Ti_3C_2 MXene. ACS Nano **15**, 4287–4293 (2021)
31. L. Jiang, D. Zhou, J. Yang, S. Zhou, H. Wang, X. Yuan, J. Liang, X. Li, Y. Chen, H. Li, 2D single- and few-layered MXenes: synthesis, applications and perspectives. J. Mater. Chem. A **10**, 13651 (2022)
32. C. Xu, L. Wang, Z. Liu, L. Chen, J. Guo, N. Kang, X.L. Ma, H.M. Cheng, W. Ren, Large-area high-quality 2D ultrathin Mo_2C superconducting crystals. Nat. Mater. **14**, 1135–1141 (2015)
33. D. Geng, X. Zhao, L. Li, P. Song, B. Tian, W. Liu, J. Chen, D. Shi, M. Lin, W. Zhou, K.P. Loh, Controlled growth of ultrathin Mo_2C superconducting crystals on liquid Cu surface. 2D Mater. **4**, 011012 (2017)
34. D. Geng, X. Zhao, Z. Chen, W. Sun, W. Fu, J. Chen, W. Liu, W. Zhou, K.P. Loh, Direct synthesis of large-area 2D Mo_2C on in situ grown graphene. Adv. Mater. **29**, 1700072 (2017)
35. J. Jia, T. Xiong, L. Zhao, F. Wang, H. Liu, R. Hu, J. Zhou, W. Zhou, S. Chen, Ultrathin N-doped Mo_2C nanosheets with exposed active sites as efficient electrocatalyst for hydrogen evolution reactions. ACS Nano **11**, 12509–12518 (2017)
36. Z. Zhang, F. Zhang, H. Wang, C.H. Chan, W. Lu, J. Dai, Substrate orientation induced epitaxial growth of face centered cubic Mo_2C superconductive thin film. J. Mater. Chem. C **5**, 10822–10827 (2017)
37. M. Naguib, J. Halim, J. Lu, K.M. Cook, L. Hultman, Y. Gogotsi, M.W. Barsoum, New two-dimensional niobium and vanadium carbides as promising materials for Li-ion batteries. J. Am. Chem. Soc. **135**, 15966–15969 (2013)
38. J. Nan, X. Guo, J. Xiao, X. Li, W. Chen, W. Wu, H. Liu, Y. Wang, M. Wu, G. Wang, Nanoengineering of 2D MXene-based materials for energy storage applications. Small **17**, 1902085 (2021)
39. R. Garg, A. Agarwal, M. Agarwal, A review on MXene for energy storage application: effect of interlayer distance. Mater. Res. Express **7**, 22001 (2020)

40. O. Mashtalir, M.R. Lukatskaya, M.Q. Zhao, M.W. Barsoum, Y. Gogotsi, Amineassisted delamination of Nb_2C MXene for Li-ion energy storage devices. Adv. Mater. **27**, 3501–3506 (2015)
41. J. Luo, X. Tao, J. Zhang, Y. Xia, H. Huang, L. Zhang, Y. Gan, C. Liang, W. Zhang, Sn^{4+} ion decorated highly conductive Ti_3C_2 MXene: promising lithium-ion anodes with enhanced volumetric capacity and cyclic performance. ACS Nano **10**, 2491–2499 (2016)
42. Z. Lin, D. Sun, Y.J. Huang, M.W. Barsoum, X. Yan, Carbon nanofiber bridged two-dimensional titanium carbide as a superior anode for lithium-ion batteries. J. Mater. Chem. A **3**, 14096–14100 (2015)
43. B. Ahmed, D.H. Anjum, Y. Gogotsi, H.N. Alshareef, Atomic layer deposition of SnO_2 on MXene for Li-ion battery anodes. Nano Energy **34**, 249–256 (2017)
44. G. Zou, Z. Zhang, J. Guo, B. Liu, Q. Zhang, C. Fernandez, Q. Peng, Synthesis of MXene/Ag composites for extraordinary long cycle lifetime lithium storage at high rates. ACS Appl. Mater. Interfaces **8**, 22280–22286 (2016)
45. X. Wang, S. Lin, H. Tong, Y. Huang, P. Tong, B. Zhao, J. Dai, C. Liang, H. Wang, X. Zhu, Two-dimensional V_4C_3 MXene as high performance electrode materials for supercapacitors. Electrochim. Acta **307**, 414–421 (2019)
46. Y.D. Agnese, M.R. Lukatskaya, K.M. Cook, P.L. Taberna, Y. Gogotsi, P. Simon, High capacitance of surface-modified 2D titanium carbide in acidic electrolyte. Electrochem. Commun. **48**, 118–122 (2014)
47. Q. Tao, M. Dahlqvist, J. Lu, S. Kota, R. Meshkian, J. Halim, J. Palisaitis, L. Hultman, M.W. Barsoum, P.O.A. Persson, Two-dimensional $Mo_{1.33}C$ MXene with divacancy ordering prepared from parent 3D laminate with in-plane chemical ordering. Nat. Commun. **8**, 1–7 (2017)
48. Q. Shan, X. Mu, M. Alhabeb, C.E. Shuck, D. Pang, X. Zhao, X.F. Chu, Y. Wei, F. Du, G. Chen, Two-dimensional vanadium carbide (V_2C) MXene as electrode for supercapacitors with aqueous electrolytes. Electrochem. Commun. **96**, 103–107 (2018)
49. S. Zhao, C. Chen, X. Zhao, X. Chu, F. Du, G. Chen, Y. Gogotsi, Y. Gao, Y.D. Agnese, Flexible Nb_4C_3Tx film with large interlayer spacing for highperformance supercapacitors. Adv. Funct. Mater. **30**, 2000815 (2020)
50. Y. Gogotsi, R.M. Penner, Energy storage in nanomaterials–capacitive, pseudocapacitive, or battery-like? ACS Nano **12**, 2081–2083 (2018)
51. M.R. Lukatskaya, S. Bak, X. Yu, X. Yang, M.W. Barsoum, Probing the mechanism of high capacitance in 2D titanium carbide using in situ X-ray absorption spectroscopy. Adv. Energy Mater. **5**, 2–5 (2015)
52. Y. Zhu, K. Rajoua, S.L. Vot, O. Fontaine, P. Simon, F. Favier, Modifications of MXene layers for supercapacitors. Nano Energy **73**, 104734 (2020)
53. X. Huang, J. Huang, D. Yang, P. Wu, A multi-scale structural engineering, strategy for high-performance MXene hydrogel supercapacitor electrode. Adv. Sci. **8**, 2101664 (2021)
54. K. Hantanasirisakul, Y. Gogotsi, Electronic and optical properties of 2D transition metal carbides and nitrides (MXenes). Adv. Mater. **30**, 1804779 (2018)
55. A.K. Jena, A. Kulkarni, T. Miyasaka, Halide perovskite photovoltaics: background, status, and future prospects. Chem. Rev. **119**, 3036–3103 (2019)
56. S.A. Olaleru, J.K. Kirui, D. Wamwangi, K.T. Roro, B. Mwakikunga, Perovskite solar cells: the new epoch in photovoltaics. Sol. Energy **196**, 295–309 (2020)
57. H. Su, L. Zhang, Y. Liu, Y. Hu, B. Zhang, J. You, Polarity regulation for stable 2D-perovskite-encapsulated high-efficiency 3D-perovskite solar cells. Nano Energy **95**, 106965 (2022)
58. H. Min, D.Y. Lee, J. Kim, G. Kim, K.S. Lee, J. Kim, M.J. Paik, Perovskite solar cells with atomically coherent interlayers on SnO_2 electrodes. Nature **598**, 444–450 (2021)
59. Z. Guo, L. Gao, Z. Xu, S. Teo, C. Zhang, Y. Kamata, S. Hayase, T. Ma, High electrical conductivity 2D MXene serves as additive of perovskite for efficient solar cells. Small **14**, 1802738 (2018)
60. D. Xu, T. Li, Y. Han, X. He, S. Yang, Y. Che, J. Xu, H. Zou, X. Guo, J. Wang, X. Lei, Fluorine functionalized MXene QDs for near–record–efficiency $CsPbI_3$ solar cell with high open–circuit voltage. Adv. Funct. Mater. **32**, 2203704 (2022)

61. Y. Wang, P. Xiang, A. Ren, H. Lai, Z. Zhang, Z. Xuan, Z. Wan, J. Zhang, X. Hao, L. Wu, M. Sugiyama, U. Schwingenschlogl, C. Liu, Z. Tang, J. Wu, Z. Wang, D. Zhao, MXene-modulated electrode/SnO_2 interface boosting charge transport in perovskite solar cells. ACS Appl. Mater. Interfaces **12**, 53973–53983 (2020)
62. L. Yang, D.C. Agnese, Y. Agnese, G. Chen, Y. Gao, Y. Sanehira, A.K. Jena, X.F. Wang, Y. Gogotsi, T. Miyasaka, Surface-modified metallic Ti_3C_2Tx MXene as electron transport layer for planar heterojunction perovskite solar cells. Adv. Funct. Mater. **29**, 1905694 (2019)
63. T. Chen, G. Tong, E. Xu, H. Li, P. Li, Z. Zhu, J. Tang, Y. Qi, Y. Jiang, Accelerating hole extraction by inserting 2D Ti_3C_2-MXene interlayer to all inorganic perovskite solar cells with long-term stability. J. Mater. Chem. A **7**, 20597–20603 (2019)
64. H.-C. Fu, V. Ramalingam, H. Kim, C.-H. Lin, X. Fang, H.N. Alshareef, J.-H. He, MXene-contacted silicon solar cells with 11.5% efficiency. Adv. Energy Mater. **9**, 1900180 (2019).
65. L. Yu, A.S.R. Bai, T.S.L. Grace, M. Batmunkh, J.G. Shapter, $Ti_3C_2T_x$ (MXene)-silicon heterojunction for efficient photovoltaic cells. Adv. Energy Mater. **9**, 1901063 (2019)
66. Z. Yu, W. Feng, W. Lu, B. Li, H. Yao, K. Zeng, J. Ouyang, MXenes with tunable work functions and their application as electron- and hole-transport materials in non-fullerene organic solar cells. J. Mater. Chem. A **7**, 11160–11169 (2019)
67. C. Hou, H. Yu, C. Huang, Solution-processable $Ti_3C_2T_x$ nanosheets as an efficient hole transport layer for high-performance and stable polymer solar cells. J. Mater. Chem. C **7**, 11549–11558 (2019)
68. H. Tang, H. Feng, H. Wang, X. Wan, J. Liang, Y. Chen, Highly conducting MXene−silver nanowire transparent electrodes for flexible organic solar cells. ACS Appl. Mater. Interfaces **11**, 25330–25337 (2019)
69. Z. Kang, Y. Ma, X. Tan, M. Zhu, Z. Zheng, N. Liu, L. Li, Z. Zou, X. Jiang, T. Zhai, Y. Gao, MXene–silicon Van Der Waals heterostructures for high-speed self-driven photodetectors. Adv. Electron. Mater. **3**, 1700165 (2017)
70. S. Guo, S. Kang, S. Feng, W. Lu, MXene-enhanced deep ultraviolet photovoltaic performances of crossed Zn_2GeO_4 nanowires. J. Phys. Chem. C **124**, 4764–4771 (2020)
71. K. Montazeri, M. Currie, L. Verger, P. Dianat, M.W. Barsoum, B. Nabet, Beyond gold: spin-coated Ti_3C_2-based MXene photodetectors. Adv. Mater. **31**, 1903271 (2019)
72. Y. Yang, J. Jeon, J.-H. Park, M.S. Jeong, B.H. Lee, E. Hwang, S. Lee, Plasmonic transition metal carbide electrodes for high-performance InSe photodetectors. ACS Nano **13**, 8804–8810 (2019)
73. S. Ahn, T.-H. Han, K. Maleski, J. Song, Y.-H. Kim, M.-H. Park, H. Zhou, S. Yoo, Y. Gogotsi, T.-W. Lee, A 2D titanium carbide MXene flexible electrode for high-efficiency light-emitting diodes. Adv. Mater. **32**, 2000919 (2020)
74. F. Cao, M. You, L. Kong, Y. Dou, Q. Wu, L. Wang, B. Wei, X. Zhang, W.-Y. Wong, X. Yang, Mixed-dimensional MXene-based composite electrodes enable mechanically stable and efficient flexible perovskite light-emitting diodes. Nano Lett. **22**, 4246–4252 (2022)
75. Y. Liu, Q. Han, W. Yang, X. Gan, Y. Yange, K. Xie, L. Xie, Y. Deng, Two-dimensional MXene/cobalt nanowire heterojunction for controlled drug delivery and chemo-photothermal therapy. Mater. Sci. Eng. C **116**, 111212 (2020)
76. X. Yang, C. Zhang, D. Deng, Y. Gu, H. Wang, Q. Zhong, Multiple stimuli-responsive mxene-based hydrogel as intelligent drug delivery carriers for deep chronic wound healing. Small **18**, 2104368 (2022)
77. M. Zhang, D. Yang, C. Dong, H. Huang, G. Feng, Q. Chen, Y. Zheng, H. Tang, Y. Chen, X. Jing, Two-dimensional MXene-originated in situ nanosonosensitizer generation for augmented and synergistic sonodynamic tumor nanotherapy. ACS Nano **16**, 9938–9952 (2022)
78. Y. Xu, Y. Wang, J. An, A.C. Sedgwick, M. Li, J. Xie, W. Hu, J. Kang, S. Sen, A. Steinbrueck, B. Zhang, L. Qiao, S. Wageh, J.F. Arambula, L. Liu, H. Zhang, J.L. Sessler, J.S. Kim, 2D-ultrathin MXene/DOXjade platform for iron chelation chemo-photothermal therapy. Bioact. Mater. **14**, 76–85 (2022)
79. K. Rasool, M. Helal, A. Ali, C.E. Ren, Y. Gogotsi, K.A. Mahmoud, Antibacterial activity of $Ti_3C_2T_x$ MXene. ACS Nano **10**, 3674–3684 (2016)

80. J. Li, Z. Li, X. Liu, C. Li, Y. Zheng, K. Wai, K. Yeung, Z. Cui, Y. Liang, S. Zhu, W. Hu, Y. Qi, T. Zhang, X. Wang, S. Wu, Interfacial engineering of $Bi_2S_3/Ti_3C_2T_x$ MXene based on work function for rapid photo-excited bacteria-killing. Nat. Commun. **12**, 1224 (2021)
81. A. Koyappayil, S.G. Chavan, Y.-G. Roh, M.-H. Lee, Advances of MXenes. Perspect. Biomed. Res. Biosens. **12**, 454 (2022)
82. M. Khatami, P. Iravani, S.G. Jamalipour, S. Iravani, MXenes for antimicrobial and antiviral applications: recent advances. Mater. Technol. **37**, 1–16 (2021)
83. Z. Hui, R. Chen, J. Chang, Y. Gong, X. Zhang, H. Xu, Y. Sun, Y. Zhao, L. Wang, R. Zhou, F. Ju, Q. Chen, J. Zhou, J. An, G. Sun, W. Huang, Solution-processed sensing textiles with adjustable sensitivity and linear detection range enabled by twisting structure. ACS Appl. Mater. Interfaces **12**, 12155–12164 (2020)
84. T.-S. DinhLe, J. An, Y. Huang, Q. Vo, J. Boonruangkan, T. Tran, S.-W. Kim, G. Sun, Y.-J. Kim, Ultrasensitive antiinterference voice recognition by bio-inspired skin-attachable self cleaning acoustic sensors. ACS Nano **13**, 13293–13303 (2019)
85. L. Li, P. Shi, L. Hua, J. An, Y. Gong, R. Chen, C. Yu, W. Hua, F. Xiu, J. Zhou, G. Gao, Z. Jin, G. Sun, W. Huang, Design of a wearable and shape-memory fibriform sensor for the detection of multimodal deformation. Nanoscale **10**, 118–123 (2018)
86. B.-M. Jun, S. Kim, J. Heo, C.M. Park, N. Her, M. Jang, Review of MXenes as new nanomaterials for energy storage/delivery, and selected environmental applications. Nano Res. **12**, 471–487 (2018)
87. A. Sinha, H. Dhanjai, Y. Zhao, X. Huang, J. Lu, J.R. Chen, MXene: an emerging material for sensing and biosensing. TrAC Trends Anal. Chem. **105**, 424–435 (2018)
88. S.J. Kim, H.J. Koh, C.E. Ren, O. Kwon, K. Maleski, S.Y. Cho, Metallic $Ti_3C_2T_x$ MXene gas sensors with ultrahigh signal-to-noise ratio. ACS Nano **12**, 986–993 (2018)
89. Y. Yang, L. Shi, Z. Cao, R. Wang, J. Sun, Strain sensors with a high sensitivity and a wide sensing range based on a $Ti_3C_2T_x$ (MXene) nanoparticle-nanosheet hybrid network. Adv. Funct. Mater. **29**, 1807882 (2019)
90. Y. Cheng, Y. Ma, L. Li, M. Zhu, Y. Yue, W. Liu, L. Wang, S. Jia, C. Li, T. Qi, J. Wang, Y. Gao, Bioinspired microspines for a high-performance spray $Ti_3C_2T_x$ MXene-based piezoresistive sensor. ACS Nano **14**, 2145–2155 (2020)
91. E. Lee, D.J. Kim, Recent exploration of two-dimensional MXenes for gas sensing: from a theoretical to an experimental view. J. Electrochem. Soc. **167**, 037515 (2020)
92. M. Jacoby, MXenes sense gas maximally. Chem. Eng. News **96**, 11 (2018)
93. K. Rui, X. Wang, M. Du, Y. Zhang, Q. Wang, Z. Ma, Q. Zhang, D. Li, X. Huang, G. Sun, J. Zhu, W. Huang, Dual-function metal-organic framework-based wearable fibers for gas probing and energy storage. ACS Appl. Mater. Interfaces **10**, 2837–2842 (2018)
94. M. Khazaei, A. Ranjbar, M. Ghorbani-Asl, M. Arai, T. Sasaki, Y. Liang, S. Yunoki, Nearly free electron states in MXenes. Phys. Rev. B: Condens. Matter. Mater. Phys. **93**, 205125 (2016)
95. R.A. Soomro, S. Jawaid, Q. Zhu, Z. Abbas, B. Xu, A mini review on MXenes as versatile substrate for advanced sensors. Chin. Chem. Lett. **31**, 922–930 (2020)
96. X. Yu, Y. Li, J. Cheng, Z. Liu, Q. Li, W. Li, X. Yang, B. Xiao, Monolayer Ti_2CO_2: a promising candidate for NH_3 sensor or capturer with high sensitivity and selectivity. ACS Appl. Mater. Interfaces **7**, 13707–13713 (2015)
97. B. Xiao, Y.-C. Li, X.-F. Yu, J.-B. Cheng, MXenes: reusable materials for NH_3 sensor or capturer by controlling the charge injection. Sens. Actuators B **235**, 103–109 (2016)
98. E.S. Muckley, M. Naguib, H.W. Wang, L. Vlcek, N.C. Osti, R.L. Sacci, X. Sang, R.R. Unocic, Y. Xie, M. Tyagi, E. Mamontov, K.L. Page, P.R.C. Kent, J. Nanda, I.N. Ivanov, Multimodality of structural, electrical, and gravimetric responses of intercalated MXenes to water. ACS Nano **11**, 11118–11126 (2017)
99. X. Wang, Q. Li, J. Zhang, H. Huang, S. Wu, Y. Yang, Novel thin-film reverse osmosis membrane with MXene $Ti_3C_2T_x$ embedded in polyamide to enhance the water flux, anti-fouling and chlorine resistance for water desalination. J. Membr. Sci. **603**, 118036 (2020)
100. Q. Zhang, G. Yi, Z. Fu, H. Yu, S. Chen, X. Quan, Vertically aligned janus MXene-based aerogels for solar desalination with high efficiency and salt resistance. ACS Nano **13**, 13196–13207 (2019)

101. A. Shahzad, K. Rasool, W. Miran, M. Nawaz, J. Jang, K.A. Mahmoud, D.S. Lee, Two-dimensional $Ti_3C_2T_x$ MXene nanosheets for efficient copper removal from water. ACS Sustain. Chem. Eng. **5**, 11481–11488 (2017)
102. J. Ma, Y. Cheng, L. Wang, X. Dai, F. Yu, Free-standing $Ti_3C_2T_x$ MXene film as binder-free electrode in capacitive deionization with an ultrahigh desalination capacity. Chem. Eng. J. **384**, 123329 (2020)
103. K.A. Papadopoulou, A. Chroneos, D. Parfitt, S.-R.G. Christopoulos, A perspective on MXenes: their synthesis, properties, and recent applications. J. Appl. Phys. **128**, 170902 (2020)
104. F. Ming, H. Liang, W. Zhang, J. Ming, Y. Lei, A.-H. Emwas, H.N. Alshareef, Porous MXenes enable high performance potassium ion capacitors. Nano Energy **62**, 853–860 (2019)
105. S.Y. Pang, Y.T. Wong, S. Yuan, Universal strategy for HF-free facile and rapid synthesis of two-dimensional MXenes as multifunctional energy materials. J. Am. Chem. Soc. **141**, 9610–9616 (2019)
106. R. Gautam, N. Marriwala, R. Devi, A review: study of Mxene and graphene together. Meas. Sens. **25**, 100592 (2023)
107. J. Bekaert, C. Sevik, M.V. Milošević, First-principles exploration of superconductivity in MXenes. Nanoscale **12**, 17354–17361 (2020)
108. Q. Xue, Z. Pei, Y. Huang, M. Zhu, Z. Tang, H. Li, C. Zhi, Mn_3O_4 nanoparticles on layer-structured Ti_3C_2 MXene towards the oxygen reduction reaction and zinc–air batteries. J. Mater. Chem. A **5**, 20818–20823 (2017)
109. Z. Wang, X. Chen, F. Shen, X. Hang, C. Niu, TiC MXene high energy density cathode for lithium-air battery. Adv. Theory Simul. **1**, 1800059 (2018)
110. H. Lin, Y. Chen, J. Shi, Insights into 2D MXenes for versatile biomedical applications: current advances and challenges ahead. Adv. Sci. **5**, 1800518 (2018)
111. B. Lu, Z. Zhu, B. Ma, W. Wang, R. Zhu, J. Zhang, 2D MXene nanomaterials for versatile biomedical applications: current trends and future prospects. Small **17**, 2100946 (2021)
112. S. Pan, J. Yin, L. Yu, C. Zhang, Y. Zhu, Y. Gao, Y. Chen, 2D MXene-integrated 3D-printing scaffolds for augmented osteosarcoma phototherapy and accelerated tissue reconstruction. Adv. Sci. **7**, 1901511 (2020)

MIX
Papier aus verantwortungsvollen Quellen
Paper from responsible sources
FSC® C105338

If you have any concerns about our products,
you can contact us on
ProductSafety@springernature.com

In case Publisher is established outside the EU,
the EU authorized representative is:
Springer Nature Customer Service Center GmbH
Europaplatz 3, 69115 Heidelberg, Germany

Printed by Libri Plureos GmbH
in Hamburg, Germany